This book is due for return on or before the last date shown below.

1 4 APR 2016

Nanotechnologies for the
Life Sciences
Volume 4

**Nanodevices for the
Life Sciences**

Edited by
Challa S. S. R. Kumar

Nanotechnologies for the Life Sciences
Volume 4

Nanodevices for the Life Sciences

Edited by
Challa S. S. R. Kumar

1st Edition

WILEY-VCH
VCH

WILEY-VCH Verlag GmbH & Co. KGaA

The Editor

Dr. Challa S. S. R. Kumar
The Center for Advanced Microstructures
and Devices
(CAMD)
Louisiana State University
6980 Jefferson Highway
Baton Rouge, LA 70806
USA

Cover

Cover design by G. Schulz based on
micrograph courtesy of M. W. Hällström
and C. Prinz, both Nanometer Structure
Consortium, Solid State Physics,
Lund University, Lund, Sweden.

Library of Congress Card No.: applied for
British Library Cataloguing-in-Publication Data:
A catalogue record for this book is available
from the British Library.

**Bibliographic information published by
Die Deutsche Bibliothek**
Die Deutsche Bibliothek lists this publication in
the Deutsche Nationalbibliografie; detailed
bibliographic data is available in the Internet at
⟨http://dnb.ddb.de⟩.

Printed in the Federal Republic of Germany.
Printed on acid-free paper.

Typesetting Asco Typesetter, Hong Kong
Printing Strauss GmbH, Mörlenbach
Binding Litges & Dopf GmbH, Heppenheim
Cover Design Grafik-Design Schulz,
Fußgönheim

ISBN-13: 978-3-527-31384-6
ISBN-10: 3-527-31384-2

Contents

Preface

Welcome to the world of nanoscale devices! The fourth volume of the series on Nanotechnologies for the Life Sciences is in front of you, providing glimpses of the exciting possibilities that exist in the world of tiny devices. Nanotechnology and nature are intimately intertwined. Such an intimate partnership is critical for reaping the benefits of nanotechnology by unraveling the mysteries of nature. Therefore, this volume, *Nanodevices for the Life Sciences*, is timely and provides a broader perspective to this partnership and enlightens us on the theory, physics, chemistry, biology and engineering of nanodevices that are being constructed in the laboratory as well those that are already being utilized by nature. See for example a recent article in Science (Vol 312, pp 860–861, 2006) that describes the possibilities to utilize biomolecular motors in nanometer-scale devices to perform mechanical work. Three chapters (chapters 7, 8 & 13) in the book are specifically dedicated to provide glimpses of the power of such natural nanoscale devices and I am certain that these chapters will catalyze development of new ideas and tools for non-biologists interested in utilizing the underlying principles.

Theory and experiments will have to go hand in hand as deeper understanding of the complexities associated with nanoscale devices bring us a step closer to designing devices that are as efficient as in nature. Therefore, the first two chapters have been dedicated to provide theoretical and computational understanding of nanoscale devices with potential applications in life sciences. The first chapter illuminates on *The Physics and Modeling of Biofunctionalized Nanolelectromechanical Systems*. Two leading theoreticicans, M. R. Paul from Virginia Polytechnic Institute and State University in Virginia, USA, and J. E. Solomon from the California Institute of Technology in Pasadena take readers to the realm of theoretical challenges associated with modeling of BioNEMS. I have no doubt that the information provided in this chapter will form a strong basis for deeper understanding of many other nanoscale systems that one encounters in our laboratories as well as in biological systems. The second chapter is a contribution from the laboratories of Vittorio Cristini from the University of California at Irvine delving on intricacies of modeling various components associated with nanoscale drug delivery for the treatment of cancer. The chapter, *Mathematical and Computational Modeling: Towards the Development and Application of Nanodevices for Drug Delivery*, is a great source of information on mathematical models and computer simulations of important steps in the journey of intravenously injected nanovectors into tumoral

Nanotechnologies for the Life Sciences Vol. 4
Nanodevices for the Life Sciences. Edited by Challa S. S. R. Kumar
Copyright © 2006 WILEY-VCH Verlag GmbH & Co. KGaA, Weinheim
ISBN: 3-527-31384-2

tissue in order to deliver drug in the most effective manner. This chapter is a must for all those interested in utilizing nanotechnologies for drug delivery.

Moving from utilizing theory, modeling and computational tools for fabrication of nanodevices, the rest of the book is a testimony to the rapid advances being made in the application of a variety of experimental techniques and tools for building nanoscale devices and for application of such devices in a number of fields ranging from biosensors to bioelectronics. In the third chapter, which is contributed by J. C. Garno and co-workers from Louisiana State University in Baton Rouge, USA, a detailed description is given of how scanning probe techniques are proving to be versatile tools for fabricating arrays of self-assembled monolayers (SAMs) and proteins. In this chapter, *Nanolithography: Towards Fabrication of Nanodevices for the Life Sciences*, the authors describe how nanolithography is revolutionizing the fabrication of nanoscale biomolecular devices in general and proteins in particular through precise control over chemical functionality, shape, dimensions and spacing on the nanometer scale. Continuing on a similar theme, H. D. Espinosa, K.-H. Kim and N. Moldovan from Northwestern University in Illinois, USA, have carried out a remarkable job in delineating the importance of microcantilevers for biopatterning and biosensing in scanning probe microscopy (SPM)-based techniques. Their contribution in the fourth chapter, entitled *Microcantilever-based Nanodevices in the Life Sciences*, is very unique in the sense that it covers not only various approaches for fabrication of microcantilevers but also their applications in the emerging field of bionanotechnology. The chapter clearly demonstrates the fact that microcantilevers are fundamental tools for biopatterning and biosensing in SPM-based techniques, and with the possibility of integrating micro/nano-fludics into microcantilevers, they are in the process of revolutionizing the field of bioanalytical nanodevices.

The fifth chapter in this volume, *Nanobioelectronics*, is a testimony to the fact that there have been several advances made in the field of molecular electronics over the last decade particularly in utilizing biomolecules for fabrication of molecular-scale devices and integrated computers. The authors from the University of Lecce in Italy, R. Rinaldi and G. Maruccio reviewed these advances in the field of nano-biomolecular electronics describing the fabrication of devices such as rectifiers, amplifiers, information storage devices based on biomoleucles in general and DNA and proteins in particular. Highlight of the chapter is the information on interconnecting biomolecules and exploitation of their self-assembly properties leading to nanobiodevices. It is particularly heartening to see that the progress made so far in the field of nanobiolelectronics is very promising and is likely to fill the void in the face of current limitations with CMOS devices and post-optical lithographies. While the fifth chapter focuses on electronic devices using DNA, the sixth chapter provides complete information on the most important properties of DNA and how these properties are being exploited in building functional devices such as DNA-based molecular motors and automata with possible applications in the life sciences. Friedrich Simmel from Ludwig-Maximilians-Universität München, Germany, has provided an up to date review on this subject and is very optimistic that the recent advances are likely to lead to the development of autonomous molecular-scale devices which can sense environmental information, perform computations and act independently as molecular motors, drug reservoirs, or as signal transducers. The chapter is aptly titled as *DNA Nanodevices: Prototypes and Applications*. While the fifth and sixth chapters provide a broader perspective to build nanoscale devices from DNA and proteins, the next two chapters contain very specific information on nanodevices made from G-protein coupled receptors and Kinesin-microtubule systems respectively. The seventh chapter, *Towards the Realization of Nanobiosensors Based on G-protein-coupled Receptors*, a contribution from the laboratories of Cecilia Pennetta, also from Lecce University, provides a thorough review on G-protein-coupled receptors (GPCRs) including different techniques to prepare and immobilize them on a substrate, followed by utilization of the electrochemical impedance spectroscopy (EIS) technique for the detection of biosensing events at the electrodes. A very unique aspect of the chapter is that it covers several theoretical aspects investigating the current response to an applied AC voltage of a nanodevice realized by a single GPCR embedded in its membrane and in contact with two functionalized metallic nanoelectrodes. The chapter is extremely valuable for nanotechnologists exploring applications in life sciences as GPCRs are one of the widest groups of receptor proteins known and they can be activated by a large variety of extracellular signals, such as light, odorant molecules, hormones, peptides, lipids, neurotransmitters and nucleotides. GPCRs mediate the sense of vision, smell, taste and pain, and are involved in an extraordinary number of physiological processes. Competing for prominence with GPCRs are the Kinesin-microtubule-driven systems as they hold significant potential as molecular motors due to their compactness, high efficiency in vitro in extracting energy from the aqueous environment. The eighth chapter, *Protein-based Nanotechnology: Kinesin–Microtubule-driven Systems for Bioanalytical Applications*, assumes enormous importance in this volume as it has valuable information on how kinesin molecular motors can be integrated with microtubule tracks into microdevices for bioanalytical applications. The chapter is an important contribution for the book as the author, William Hancock from Pennsylvania State University, USA, covers wide-ranging topics from cell biology and biophysics, in vitro assays, theoretical aspects, biofunctionalization of the kinesin–microtubule system in addition to experimental approaches to integrating into functional microscale devices for potential analytical applications.

Carbon Nanotubes (CNTs) are finding extraordinary applications in the field of life sciences especially in biosensing, drug delivery, diagnosis, imaging and so on. These applications are further complimented by recent efforts in fabricating CNT-based nanodevices especially field-effect transistors (FETs) having very interesting performance characteristics. The ninth chapter, *Self-assembly and Bio-directed Approaches of Carbon Nanotubes: Towards Device Fabrication*, begins with a review on important characteristic of CNTs followed by the synthesis methods reported in the literature. The central theme of the chapter, written by Arianna Filoramo from the Laboratory of Electronic Materials in Gif sur Yvette, France, is however the utilization of self-assembly approaches (bio as well as non-bio directed) for fabrication of CNT devices for application in the electronics industry. Alternative strategies to CMOS technologies such as bio-inspired technologies for nanoscale devices as described in this chapter as well as in chapter five are likely to revolutionize the electronics industry in the near future. Focusing primarily on biosensing, chapter ten explores the possibility of fabricating nanodevices based on nanophotonic/ optoelectronic platforms and on nanomechanical platforms. The chapter, *Nanodevices for Biosensing: Design, Fabrication and Applications*, contributed by L. M. Lechuga and co-workers from the Microelectronics National Center (CNM) in Spain, is a valuable source of information for design, fabrication and testing of nanosensors and their integration with microfluidics, optical and electronic functions on a single chip. Chapter eleven, *Fullerene-based Devices for Biological Applications*, written by Dirk M. Guldi from Friedrich Alexander University in Erlangen-Nürnberg, Germany, and his collaborators is complimentary to the ninth chapter, describing in detail the solubility, toxicity and major biological applications of fullerenes and their potential application in nanoscale devices. It provides a basic framework for fabricating fullerene-based nanoscale devices in near future. A more general approach for utilization of nanotechnological principles for the fabrication of biomedical devices is presented in the twelfth chapter, wherein Lars Montelius from Lund University in Sweden describes various types of nanotechnologies that are employed in the biomedical field in general and biomedical engineering in particular, together with suitable examples. The chapter, *Nanotechnology for Biomedical Devices*, provides a fundamentally strong backbone for the realization of nanoscale devices for biomedical applications.

Finally, the book ends with a chapter dedicated to providing an overview of nanoscale devices that nature utilizes. Chapter thirteen, *Nanodevices in Nature*, written by Alexander Volkov and Courtney Brown from Oakwood College in Huntsville, USA, complements more specialized information in chapters seven and eight. The chapter elegantly delineates the role of various nanodevices in a wide variety of biological processes focusing more specifically on cytochrome oxidase, photosynthesis and phototropism, membrane transport, molecular motors, and electroreceptors. In my view the last chapter is a grand finale to the excellent source of information that the authors of this book gathered and reminds me of a great statement from *Vedas*, one of the oldest Indian scriptures, which describes universal power as the smallest of the smallest and biggest of the biggest. Truly, the power that encompasses the universe comes from nanodevices!

As I conclude this preface, there is no doubt in my mind that a book of this magnitude and high quality would not have been possible without the timely contributions of all authors, and I am always grateful to them for sharing my vision for this book as well as for the rest of the series. I am glad to let you know that Volumes 1–3 and 5–6 of this exciting series have already been published and you might have seen them in your library or obtained a personal copy. The remainder of the series, volumes 7–10, is currently in press and will be available to you before the year ends. In addition to the authors, a project of this magnitude is not possible but for unwavering support from my employer, family, friends and Wiley-VCH publishers. This is yet another opportunity for me to convey my thanks to them. Before I take leave, I would like to request you, the reader, who is sharing the knowledge with me and rest of the authors, to let me know your comments, suggestions and constructive criticism to make further improvements to this exciting series.

May 2006, Baton Rouge *Challa S. S. R. Kumar*

List of Contributors

Vladimir Akimov
INRA
Neurobiologie de l'Olfaction et de la Prise
Alimentaire
Equipe Récepteurs et Communication
Chimique
78352 Jouy en Josas Cedex
France

Eleonora Alfinito
National Nanotechnology Laboratory
Istituto Nazionale di Fisica della Materia
Dipartimento di Ingegneria dell'Innovazione
Università di Lecce
73100 Lecce
Italy

Courtney L. Brown
Department of Chemistry
Oakwood College
Huntsville, AL 35896
USA

Laura G. Carrascosa
Instituto de Microelectrónica de Madrid IMM
CNM-CSIC
28760 Tres Cantos
Madrid
Spain

Ignacio Casuso
Laboratory of NanoBioEngineering
University of Barcelona
08028 Barcelona
Spain

Vittorio Cristini
Department of Biomedical Engineering and
Mathematics
University of California at Irvine
Irvine, CA 92697-2715
USA

Tatiana Da Ros
Dipartimento di Scienze Farmaceutiche
Università di Trieste
34127 Trieste
Italy

Abdelhamid Errachid
Laboratory of NanoBioEngineering
University of Barcelona
08028 Barcelona
Spain

Horacio D. Espinosa
Department of Mechanical Engineering
Northwestern University
Evanston, IL 60208-3111
USA

Giorgio Ferrari
Dipartimento di Elettronica ed Informazione
Politecnico di Milano
20133 Milano
Italy

Arianna Filoramo
Laboratoire d'Electronique Moleculaire
SPEC/DRECAM/DSM
CEA Saclay
91191 Gif Sur Yvette
France

Hermann B. Frieboes
Department of Biomedical Engineering
University of California at Irvine
Irvine, CA 92697-2715
USA

Laura Fumagalli
Dipartimento di Elettronica ed Informazione
Politecnico di Milano
20133 Milano
Italy

Jayne C. Garno
Chemistry Department
Louisiana State University
Baton Rouge, LA 70803
USA

Gabriel Gomila
Laboratory of NanoBioEngineering
University of Barcelona
08028 Barcelona
Spain

Tatiana Gorojankina
INRA
Neurobiologie de l'Olfaction et de la Prise
Alimentaire
Equipe Récepteurs et Communication
Chimique
78352 Jouy en Josas Cedex
France

Dirk M. Guldi
Institute for Physical Chemistry
Friedrich-Alexander-Universität
Erlangen-Nürnberg
91058 Erlangen
Germany

William O. Hancock
Department of Bioengineering
Penn State University
University Park, PA 16802
USA

Yanxia Hou
IFOS-Laboratoire d'Ingégnerie et
Fonctionnalisation
Ecole Centrale de Lyon
69134 Ecully Cedex
France

Nicole Jaffrezic-Renault
IFOS-Laboratoire d'Ingégnerie et
Fonctionnalisation
Ecole Centrale de Lyon
69134 Ecully Cedex
France

Keun-Ho Kim
Department of Mechanical Engineering
Northwestern University
Evanston, IL 60208-3111
USA

Laura M. Lechuga
Instituto de Microelectrónica de Madrid IMM
CNM-CSIC
28760 Tres Cantos
Madrid
Spain

Jie-Ren Li
Chemistry Department
Louisiana State University
Baton Rouge, LA 70803
USA

Giuseppe Maruccio
National Nanotechnology Laboratory
Istituto Nazionale di Fisica della Materia
Dipartimento di Ingegneria dell'Innovazione
Università di Lecce
73100 Lecce
Italy

Jasmina Minic
INRA
Neurobiologie de l'Olfaction et de la Prise
Alimentaire
Equipe Récepteurs et Communication
Chimique
78352 Jouy en Josas Cedex
France

Nicolaie Moldovan
Department of Mechanical Engineering
Northwestern University
Evanston, IL 60208-3111
USA

Lars Montelius
Department of Physics
Division of Solid State Physics & The
Nanometer Consortium
Lund University
221 00 Lund
Sweden

Miguel Moreno
Instituto de Microelectrónica de Madrid IMM
CNM-CSIC
28760 Tres Cantos
Madrid
Spain

Johnpeter Ndiangui Ngunjiri
Chemistry Department
Louisiana State University
Baton Rouge, LA 70803
USA

Edith Pajot-Augy
INRA
Neurobiologie de l'Olfaction et de la Prise
Alimentaire
Equipe Récepteurs et Communication
Chimique
78352 Jouy en Josas Cedex
France

Mark Paul
Department of Mechanical Engineering
Virginia Polytechnic and State University
Blacksburg, VA 24061
USA

Cecilia Pennetta
National Nanotechnology Laboratory
Istituto Nazionale di Fisica della Materia
Dipartimento di Ingegneria dell'Innovazione
Università di Lecce
73100 Lecce
Italy

Marie-Annick Persuy
INRA
Neurobiologie de l'Olfaction et de la Prise
Alimentaire
Equipe Récepteurs et Communication
Chimique
78352 Jouy en Josas Cedex
France

Lino Reggiani
National Nanotechnology Laboratory
Istituto Nazionale di Fisica della Materia
Dipartimento di Ingegneria dell'Innovazione
Università di Lecce
73100 Lecce
Italy

Ross Rinaldi
National Nanotechnology Laboratory
Istituto Nazionale di Fisica della Materia
Dipartimento di Ingegneria dell'Innovazione
Università di Lecce
73100 Lecce
Italy

Oscar Ruiz
Laboratory of NanoBioEngineering
University of Barcelona
08028 Barcelona
Spain

Roland Salesse
INRA
Neurobiologie de l'Olfaction et de la Prise
Alimentaire
Equipe Récepteurs et Communication
Chimique
78352 Jouy en Josas Cedex
France

Josep Samitier
Laboratory of NanoBioEngineering
University of Barcelona
08028 Barcelona
Spain

Marco Sampietro
Dipartimento di Elettronica ed Informazione
Politecnico di Milano
20133 Milano
Italy

Sandeep Sanga
Department of Biomedical Engineering
University of California at Irvine
Irvine, CA 92697-2715
USA

Ginka H. Sarova
Institute for Physical Chemistry
Friedrich-Alexander-Universität Erlangen-
Nürnberg
91058 Erlangen
Germany

Friedrich C. Simmel
CeNS & Sektion Physik der LMU
Gruppe Friedrich Simmel
80539 München
Germany

John P. Sinek
Department of Mathematics
University of California at Irvine
Irvine, CA 92697-2715
USA

Balakrishnan Sivaraman
Department of Biomedical Engineering and
Mathematics
University of California at Irvine
Irvine, CA 92697-2715
USA

Jerry E. Solomon
Condensed Matter Physics
California Institute of Technology
Pasadena, CA 91125
USA

Alexander G. Volkov
Department of Chemistry
Oakwood College
Huntsville, AL 35896
USA

Kirill Zinoviev
Instituto de Microelectrónica de Madrid IMM
CNM-CSIC
28760 Tres Cantos
Madrid
Spain

1
The Physics and Modeling of Biofunctionalized Nanoelectromechanical Systems

Mark R. Paul and Jerry E. Solomon

1.1
Introduction

Experimental fabrication and measurement are rapidly approaching the nanoscale (see, e.g. Refs. [1–4]). With this comes the potential for many important discoveries in both the physical and life sciences, with particularly intense attention in the fields of medicine and biology [5]. As Richard Feynman famously predicted in the early 1960s, there is indeed plenty of room at the bottom [6, 7]. A particularly promising avenue of research with the potential to make significant contributions is that involving what we will call biofunctionalized nanoelectromechanical systems [(BIO)NEMS]. This is a large and burgeoning field, and we do not attempt to present a survey, but rather we have picked an interesting example of a (BIO)NEMS device in order to highlight the dominant physics and types of modeling issues that arise.

This is not to imply that molecular-scale science is something new – scientists and engineers have been manipulating atoms and molecules for decades. (For an interesting discussion about where nanotechnology fits in with molecular science, see Ref. [8].) However, one new and exciting feature is the ability to fabricate micro- and nanoscale structures that can be used to manipulate, interact and sense biological systems at the single-molecule level.

For a better perspective of the length scales in question it is useful to place micro- and nanometers on biological scales; a human hair has a diameter of about 1 mm, a red blood cell has a diameter of about 10 µm ($1 \mu m = 1 \times 10^{-6}$ m), the diameter of the bacteria *Escherichia coli* is about 1 µm, the diameter of the protein lysozyme is about 5 nm ($1 nm = 1 \times 10^{-9}$ m) and the diameter of a single hydrogen atom is about 0.1 nm. When viewed in this context, a device with a characteristic length scale of 100 nm falls in the middle of these biological length scales, i.e. the device is quite large compared to single atoms, yet quite small when compared to single cells or large molecules. It is important to consider this when modeling these systems, as we illustrate below.

The force landscape descriptive of biological and chemical interactions occurs at the piconewton scale ($1 pN = 10^{-12}$ N). Biologically relevant force magnitudes are

Nanotechnologies for the Life Sciences Vol. 4
Nanodevices for the Life Sciences. Edited by Challa S. S. R. Kumar
Copyright © 2006 WILEY-VCH Verlag GmbH & Co. KGaA, Weinheim
ISBN: 3-527-31384-2

related to the breaking and manipulation of chemical bonds. For example it takes hundreds of piconewtons to break covalent bonds, and on the order of 10 pN to break a hydrogen bond or to describe the entropic elasticity of a polymer (see, e.g. Refs. [9, 10]).

The dominant biological time scales of small numbers of molecules are also dictated by their chemical interactions. The time scales of chemical reactions vary over many orders of magnitude, e.g. protein conformational changes can take of the order of milliseconds, binding reactions such as those that occur between transcription factors and genes or between enzymes and substrates are of the order of seconds, covalent bond modifications such that occur with phosphorylation is of the order of minutes, and new protein synthesis in a cell can take tens of minutes.

With these biological length, force and time scales in mind it becomes clear that a major challenge facing the successful development of a single-molecule biosensor is to measure on the order of tens of piconewtons on microsecond time scales. Despite the rapid advancement of new technologies such as surface plasmon resonance [11], optical tweezers [12, 13], microneedles [14, 15] and scanning force microscopy [16–19], detailed knowledge of the real-time dynamics of biomolecular interactions remains a current challenge.

An attractive device with the potential to measure the biophysical properties of a single molecule is based upon the dynamics of nanoscale cantilevers in fluid. In discussing the physics and modeling of (BIO)NEMS we will focus on this type of device. In some aspects this device can be thought of as the miniaturization of atomic force microscopy (AFM) which depends upon the response of micron-scale cantilevers. The invention of AFM [20] has revolutionized surface science, paving the way for direct measurements of intermolecular forces and topographical mapping with atomic precision for a wide array of materials, including semiconductors, polymers, carbon nanotubes and (CNTs) biological cells (see Refs. [21, 22] for current reviews). AFM is most commonly performed in one of three different driven modalities; contact mode, noncontact mode and tapping mode. In contact mode, the cantilever remains in contact with the surface and direct measurements are made based upon the cantilever response as it interacts with the sample. Despite its great success, contact-mode microscopy raises concerns about strong adhesive forces, friction and the damage of soft materials. In response to these issues emerged the noncontact- and tapping-mode modalities which are often referred to as dynamic AFM [23–28].

In noncontact mode, an oscillating cantilever never actually makes impact with the surface, yet its response alters due to an interaction between the cantilever tip and surface forces. The noncontact mode allows the measurement of electric, magnetic and atomic forces. In tapping mode, the cantilever oscillates near the sample surface making very short intermittent contact. Commonly the oscillation amplitude is held fixed through a feedback loop and as the cantilever moves over topographical features of the sample, the change in deflection is measured and related to the surface features. As a result of this minimal impact, and by greatly reducing the effects of adhesion and friction, the tapping mode has become the method of

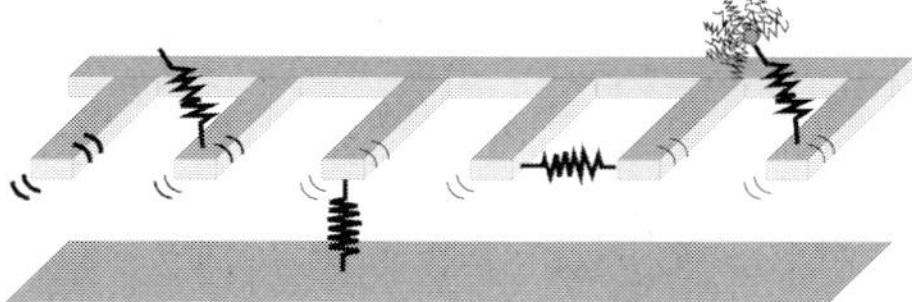

Figure 1.1. Schematic illustrating possible single-molecule detection modalities using small-scale cantilevers immersed in fluid.

choice for high-resolution topographical measurements of soft and fragile materials that are difficult to examine otherwise.

We would like to focus here upon something quite different – the stochastic dynamics of a passive cantilever in fluid. By passive we mean that the cantilever is not being dragged along a surface or forced to tap a surface, but that the cantilever is simply immersed in the fluid. However, recall that at microscopic length scales there is a sea of random thermal molecular noise. Experimental measurement is often limited by this inherent thermal noise; however, with current technology this noise can be exploited to make extremely sensitive experimental measurements [29] including the highly sensitive measurements to be made by gravitational wave detectors [30].

The basic idea is illustrated in Fig. 1.1. A small cantilever placed in fluid will exhibit stochastic dynamics due to the continual buffeting by water molecules that are in constant thermal motion (Brownian motion). In Fig. 1.1, all of the cantilevers will exhibit such oscillations; the four dark lines around each cantilever tip are meant to indicate these oscillations and their degree of shading represents the relative magnitude of these oscillations. One way to measure such oscillations in the laboratory would be through the use of optical methods. The cantilever on the far left is bare and is simply a reference cantilever placed in fluid. The adjacent cantilevers suggest various detection modalities that could also be considered. The fundamental idea is that in the presence of a biomolecule, either attached directly to a single cantilever or between the a cantilever and something else, the cantilever response will change. Measuring this change can then be used to detect the presence of a single biomolecule or, in more prescribed situations, details of the response will yield information about the dynamics of the molecule being probed.

An important advantage of this approach is that small cantilevers have large resonant frequencies, allowing the measurement of these dynamics on natural chemical time scales. In fact, nanoscale cantilevers immersed in water can have resonant frequencies in the megahertz range. The last cantilever on the right shows the case where the target biomolecule is bound between the cantilever and a very large molecule. The purpose of this would be to take advantage of its large surface area and, as a result, its increased fluid drag to enhance the change in response.

An additional complication is that the stochastic dynamics of cantilevers placed in an array, such as those shown in Fig. 1.1, will become coupled to one another through the resulting fluid motion. In other words, if one cantilever moves this will cause the fluid to move, which will cause the adjacent cantilevers to move and *vice versa*. At first this may appear as just another component of background noise to contend with. However, it is interesting to point out that this correlated noise can be exploited to significantly increase the sensitivity of these measurements. Consider measuring the cross-correlation of the fluctuations between two cantilevers in fluid supporting a tethered biomolecule. By examining only the correlated motion of the two cantilevers we have effectively eliminated the random uncorrelated component of the noise acting on each cantilever. In fact, this approach has been used to measure femtonewton forces ($1\ \text{fN} = 10^{-15}\ \text{N}$) on millisecond time scales between two micron-scale beads placed in water [12]. Additionally, this approach was used to quantify the Brownian fluctuations of an extended piece of DNA tethered between the beads leading to the resolution of some long-standing issues concerning the dynamics of single biomolecules in solution [13].

Whatever the manner in which the measurements will actually be made in the laboratory, it will be essential to have a firm understanding of the complex and sometimes counterintuitive physics at work on the these scales in order to interpret them (for an excellent introduction to the modeling of micro- and nanoscale systems, see Ref. [31]).

The purpose of this chapter is to shed some light upon this for the particularly illustrative case where the Brownian noise of small cantilevers in fluid is exploited for potential use as a single-molecule biosensor. Before these measurements can be made and understood, the following questions must be answered:

(a) What are the stochastic dynamics of an array of nanoscale cantilevers immersed in fluid in the absence of the target biomolecules?
(b) How much analyte will arrive at the sensor and what are the time scales for its capture?
(c) Successful measurements will require the discernment between the noise when the biomolecule is attached and the background noise. What signal processing schemes can be used to make these measurements?

We address these questions in the following sections.

1.2
The Stochastic Dynamics of Micro- and Nanoscale Oscillators in Fluid

1.2.1
Fluid Dynamics at Small Scales

The dynamics of fluid motion at small scales contains many surprises when compared with what we are accustomed to in the macroscopic world. In fact most of life involves the interactions of small objects in fluidic environments (see Ref. [32] for an introduction or Refs. [33, 34] for a detailed discussion).

At the molecular scale a fluid is clearly composed of individual molecules. However, most fluid analysis is done assuming that the fluid is a *continuum*. What this implies is that at any particular point in space (no matter how small) the properties of the fluid (velocity, pressure, etc.) are well defined and well behaved. Another way to think of this is that for any experimental measurement in question we assume that our probe is effectively sampling the average behavior of many molecules.

As our domain of interest becomes smaller it is clear that this assumption will eventually break down. This raises the difficult question – at which point does the continuum approximation become invalid? An approximate answer can be provided by physical reasoning. In the continuum limit one would like the mean free path of collisions of the fluid molecules to be much smaller that a characteristic fluid length scale. This idea is captured by the Knudsen number $Kn = \lambda/L$, where λ is the mean free path and L is a characteristic length scale. For the case of water, and of liquids in general, the molecules are always in very close contact with one another and the characteristic mean free path can be approximated by the diameter of a single molecule. For water this yields $\lambda \approx 0.3$ nm.

For the stochastic oscillations of small cantilevers we will use the cantilever half-width $w/2$ as the characteristic length. This is because while a cantilever oscillates most of the fluid flows around spanwise over the cantilever. Assuming the cantilever has a width of $w = 1$ μm and is immersed in water yields $Kn \approx 6 \times 10^{-4}$. Since $Kn \ll 1$ this indicates that the continuum approximation is good for the fluid dynamics even at these small scales. A quantitative understanding of when the continuum approximation breaks down and what the effects will be is currently an active and exciting area of research with many open questions (see, e.g. Ref. [35]).

The classical equations of fluid dynamics in the continuum limit are the well-known Navier–Stokes equations (see Ref. [36] for a thorough treatment):

$$R_\omega \frac{\partial \vec{u}}{\partial t} + R_u \vec{u} \cdot \vec{\nabla} \vec{u} = -\vec{\nabla} p + \nabla^2 \vec{u}, \tag{1}$$

$$\vec{\nabla} \cdot \vec{u} = 0 \tag{2}$$

Equation (1) is an expression of the conservation of momentum (we have neglected the body force due to gravity). Equation (2) expresses the conservation of mass for an incompressible fluid. We have written the equations in nondimensional form using L, U and T as characteristic length, velocity and time scales, respectively. Two nondimensional parameters R_ω and R_u emerge in Eq. (1) that multiply the two inertial terms on the left-hand side.

It is worthwhile discussing these two parameters in more detail, which will lend some insight into the dominant physics at small scales in fluids. The parameter:

$$R_u = \frac{UL}{v_f} \tag{3}$$

expresses the ratio between convective inertial forces and viscous forces (where v_f is the kinematic viscosity of the fluid, for water $v_f \approx 1 \times 10^{-6}$ m^2 s^{-1}). This is the

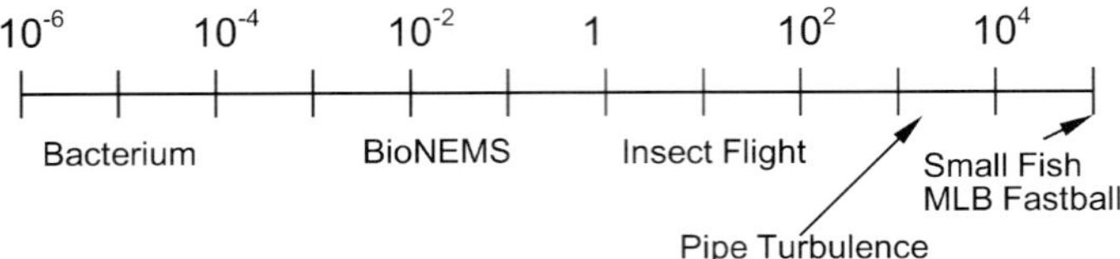

Figure 1.2. Examples of different phenomena occurring over a range of 10 orders of magnitude in the velocity-based Reynolds number, R_u.

velocity-based Reynolds number. It is clear that for micron or nanoscale devices both the characteristic velocity and length scales become quite small, resulting in what is commonly referred to as the low Reynolds number regime. A precise definition of what is meant by "low" is not clear. For perspective, Fig. 1.2 illustrates the Reynolds numbers for some particular cases of interest. Note the vast range of phenomena that occurs over 10 orders in magnitude of the Reynolds number. As the Reynolds number decreases, the effects of viscosity dominate inertial effects. For example, if a 1-μm microorganism swimming in water at 10 μm s^{-1} suddenly turns off its source of thrust, say by flagellar or cilia motion, it will come to rest in the fraction of an angstrom. This is nothing like what we are used to on the macroscale. An important consequence when $R_u \ll 1$ is that the nonlinear convective inertial term $\vec{u} \cdot \vec{\nabla}\vec{u}$ becomes negligible. As a result, the equations become linear, greatly simplifying the analysis. The parameter:

$$R_\omega = \frac{L^2}{v_f T} \tag{4}$$

expresses the ratio between inertial acceleration forces and viscous forces. Notice that if we take the characteristic velocity to be simply L/T, the frequency and the velocity-based Reynolds numbers become equivalent. However, it is useful not to make this assumption here because we want to consider further the case where the oscillations are imposed externally and the inverse frequency of these oscillations is taken as the time scale. The result is the frequency-based Reynolds number,

$$R_\omega = \frac{\omega w^2}{4 v_f} \tag{5}$$

where again we have used the cantilever half-width, $w/2$, as the characteristic length scale.

The frequency-based Reynolds is the appropriate Reynolds number to describe micro- or nanoscale cantilevers immersed in fluid. Let us consider further the type of cantilever currently under consideration for the next generation of biosen-

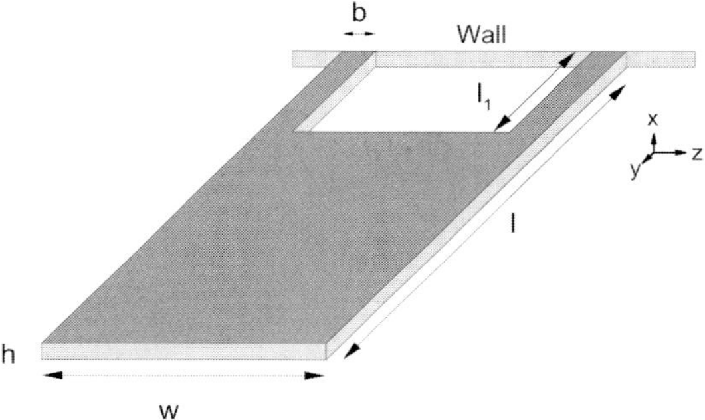

Figure 1.3. Schematic of a proposed cantilever geometry for use as a single-molecule biosensor (not drawn to scale): $l = 3$ µm, $w = 100$ nm, $l_1 = 0.6$ µm, $b = 33$ nm. The cantilever is silicon with a density $\rho_s = 2330$ kg m^{-3}, Young's modulus $E_s = 125$ GPa and spring constant, $k = 8.7$ mN m^{-1} [2, 41].

sors. Approximate values for the cantilever geometry are a width $w \approx 1$ µm, height $h \approx 100$ nm, resonant frequency $\omega \approx 2\pi \times 1$ MHz and we will assume water is the working fluid. As we show later, the maximum cantilever deflection due to Brownian motion will be of the order of $0.01h$ (and often much less depending upon the particular geometry in question). Using these numbers the characteristic velocity is $U = 0.01h\omega$, which yields a velocity-based Reynolds number of $R_u = 3 \times 10^{-3}$. Since $R_u \ll 1$, the nonlinear inertial term can be neglected. However, the frequency-based Reynolds number is $R_\omega = 1.6$. As a result, the first inertial term must be kept in Eq. (1), making the resulting linear analysis more difficult. The governing equations are now:

$$R_\omega \frac{\partial \vec{u}}{\partial t} = -\vec{\nabla}p + \nabla^2 \vec{u} \tag{6}$$

$$\vec{\nabla} \cdot \vec{u} = 0 \tag{7}$$

These equations are known as the time-dependent Stokes equations. In what follows we will drop the subscript ω on R_ω and assume that R represents the frequency-based Reynolds number. Although these equations are linear it is still a formidable challenge to derive an analytical solution for all but the simplest scenarios. One such example is when the cantilever is modeled as an oscillating two-dimensional (2-D) cylinder (discussed in more detail in Section 1.2.3). However, even in simple cases the fluid-coupled motion of arrays of oscillating objects still presents a challenge. This is in addition to the fact that most experimental geometries are not simple, which further complicates the analysis (e.g., see Fig. 1.3). This has led to the development of an experimentally accurate numerical approach to calculate the stochastic dynamics of small-scale cantilevers [37] (discussed below).

1.2.2
An Exact Approach to Determine the Stochastic Dynamics of Arrays of Cantilevers of Arbitrary Geometry in Fluid

At first sight the determination of the stochastic dynamics of an array of fluid-coupled nanoscale oscillators appears quite challenging. Considering the nature of the equations (a system of coupled partial differential equations) and the complex geometries under consideration for experiments, the appeal of a numerical solution is apparent. However, the important question then arises of how to carry out such a numerical investigation? One approach that may come to mind is to perform stochastic simulations of the precise geometries in question that resolves the Brownian motion of the fluid particles as well as the motion of the cantilevers. In principle, this could be done in the context of a molecular dynamics simulation. However, this would be extremely difficult, if possible at all. Two major problems with this approach are:

(a) There are simply too many molecules. A small box with side length of $L = 10$ µm will contain of the order of 10^{13} water molecules. For low Reynolds number flows fluid disturbances are long range and will be of the order of microns even if the oscillators are nanoscale. The length scale of the fluid disturbance scales as approximately $\sqrt{v_f/\omega}$. This length scale describes the distance from the oscillating cylinder over which the bulk of the fluid momentum is able to diffuse.

(b) There are vastly disparate time scales. For every oscillation of the cantilever, many water collisions will have had to occur. On average, a water molecule undergoes a collision every picosecond ($1 \text{ ps} = 1 \times 10^{-12}$ s). However, the cantilever oscillates about once per microsecond. In other words, a million water molecules collide with the cantilever for every single cantilever oscillation – imposing considerable overhead upon our numerical scheme. To make matters worse, in order to get good statistics the numerical solution will have to run for many cantilever oscillations or, equivalently, many numerical simulations will have to be run for different initial conditions and averaged.

However, there is a much better approach if one exploits the fact that the system is in *thermodynamic equilibrium*. This allows the use of powerful ideas from statistical mechanics and, in particular, the fluctuation–dissipation theorem, which relates equilibrium fluctuations with the way a system, that has been slightly perturbed out of equilibrium, returns to equilibrium. In other words, if one understands how a systems dissipates near equilibrium, one understands how that same system fluctuates at equilibrium. The fluctuation–dissipation theorem was originally discussed by Callen and Greene [38, 39]; also see Chandler [40] for an accessible introduction.

It has recently been shown that the fluctuation–dissipation theorem allows for the calculation of the stochastic equilibrium fluctuations of small-scale oscillators using only standard *deterministic* numerical methods [37]. For the case of small cantilevers in fluid, the dissipation is mostly due to the viscous fluid although internal elastic dissipation of the cantilever could be included if desired.

We will introduce the use of this approach for the case of two opposing cantilevers as shown in Fig. 1.7(a). Consider one dynamic variable to be the displacement of the cantilever on the left $x_1(t)$. This is a classical system, so $x_1(t)$ will be a function of the microscopic phase space variables consisting of $3N$ coordinates and conjugate momenta of the cantilever, where N is the number of particles in the cantilever. We now take the system to a prescribed excursion from equilibrium and observe how the system returns to equilibrium, which, in effect, quantifies the dissipation in the system. A particularly convenient way to accomplish this is to consider the situation where a force $f(t)$ has been applied to the cantilever on the left at some time in the distant past and is removed at time zero. The step force is represented by:

$$f(t) = \begin{cases} F_1 & \text{for } t < 0 \\ 0 & \text{for } t \geq 0 \end{cases} \tag{8}$$

This force couples to $x_1(t)$ causing a deflection in the cantilever. For this case the Hamiltonian of the system H is given by:

$$H = H_0 - f x_1 \tag{9}$$

We only consider the case of small f so the response of $x_1(t)$ remains in the linear regime. In the linear response regime, the change in the average value of a second dynamical quantity $X_2(t)$ (here we will use the displacement of the cantilever on the right, which is again a function of the $3N$ coordinates and conjugate momenta) from its equilibrium value in the absence of f is given by:

$$\Delta X_2(t) = \frac{F_1}{k_B T} \langle \delta x_1(0) \delta x_2(t) \rangle_0 \tag{10}$$

where k_B is Boltzmann's constant ($k_B = 1.38 \times 10^{-23}$ J K^{-1}) and T is the absolute temperature. The equilibrium fluctuations are given by:

$$\delta x_1 = x_1 - \langle x_1 \rangle_0 \tag{11}$$

$$\delta x_2 = x_2 - \langle x_2 \rangle_0 \tag{12}$$

where the average $\langle \ \rangle_0$ denotes the equilibrium average in the absence of the force f. However, for our case the cantilevers fluctuate about an equilibrium of zero deflection, $\langle x_1 \rangle_0 = \langle x_1 \rangle_0 = 0$, which then implies that $\delta x_1 = x_1$ and $\delta x_2 = x_2$. The average behavior of the cantilever deflection in the linear response regime is:

$$\Delta X_2(t) = X_2(t) - \langle x_2(t) \rangle_0 \tag{13}$$

However, as just mentioned, $\langle x_2(t)\rangle_0 = 0$, which also implies $\Delta X_2(t) = X_2(t)$ and yields:

$$X_2(t) = \frac{F_1}{k_B T}\langle x_1(0)x_2(t)\rangle_0 \tag{14}$$

Using this result we can calculate a general equilibrium cross-correlation function in terms of the linear response as:

$$\langle x_1(0)x_2(t)\rangle_0 = \frac{k_B T}{F_1}X_2(t) \tag{15}$$

Similarly, the autocorrelation of the fluctuations is given by:

$$\langle x_1(0)x_1(t)\rangle_0 = \frac{k_B T}{F_1}X_1(t) \tag{16}$$

where $X_1(t)$ is the average behavior of the deflection of the cantilever in which the force was applied. The spectral properties of the correlations can be found by taking the cosine Fourier transform of the auto- and cross-correlation functions. This yields the noise spectra, $G_{11}(v)$ and $G_{12}(v)$, given by:

$$G_{11}(v) = \int_0^\infty \langle x_1(0)x_1(t)\rangle \cos(\omega t)\, dt, \tag{17}$$

$$G_{12}(v) = \int_0^\infty \langle x_1(0)x_2(t)\rangle \cos(\omega t)\, dt \tag{18}$$

where v is the frequency defined by $\omega = 2\pi v$. The noise spectra are important because they are precisely what would be measured in an experiment.

This result is exact with the only assumptions being classical mechanics and linear behavior. Equations (15) and (16) are extremely useful in that they relate the *stochastic* cantilever dynamics on the left-hand side to its *deterministic* response to the removal of a step force on the right-hand side. In other words, Eq. (16) relates the equilibrium fluctuations of the cantilever to its average deflection as it returns to equilibrium from a prescribed excursion to a nonequilibrium state.

With this in mind, the remaining challenge is to calculate the deterministic quantities $X_1(t)$ and $X_2(t)$ for use in Eqs. (15) and (16). Since the dynamic variables of interest are macroscopic (after all they are the cantilever deflections X_1 and X_2), they can be calculated using the deterministic macroscopic equations which govern the fluid and solid dynamics. This can be from analytics, simplified models or large-scale numerical simulation.

To summarize, the scheme consists of the following steps in a deterministic calculation:

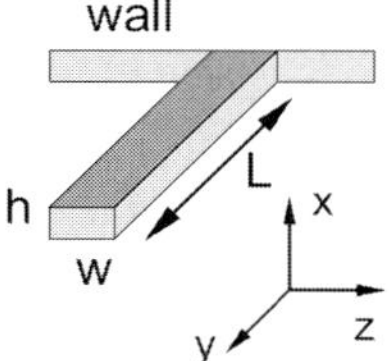

Figure 1.4. Schematic of a simple cantilevered beam of length L, width w and height h.

(a) Apply an appropriate force f that is constant in time and small enough so that the response remains linear. An appropriate force is one that couples to the variable of interest X_1. After applying the force, allow the system to come to steady state.
(b) Turn off the force at a time labeled $t = 0$.
(c) Measure some dynamical variable $X_2(t)$ (which might be the same as X_1 to yield an autocorrelation function) to yield the correlation function of the equilibrium fluctuations via Eqs. (15) and (16).

For the case of small cantilevers in fluid, the fluid motion can be calculated using the incompressible Navier–Stokes equations and the dynamics of the solid structures can be computed from the standard equations of elasticity. Using the sophisticated numerical tools developed for such calculations it is possible to find accurate results for realistic experimental geometries that may be quite complex, e.g. the triangular cantilever design often used in commercial AFM or the paddle geometries currently under investigation for use as detectors of single biomolecules as shown in Fig. 1.3.

1.2.3
An Approximate Model for Long and Slender Cantilevers in Fluid

Let us first consider a long and slender cantilever ($L \gg w, h$) that is fixed at its base and free at its tip with the simple beam geometry as shown in Fig. 1.4. This configuration is particularly useful because this geometry is commonly used for AFM. A simplified and effective model analysis is available for this case [42, 43]. In this model, the dynamics of the beam motion is described using classical elasticity theory:

$$\mu \frac{\partial^2 w(y, t)}{\partial t^2} + EI \frac{\partial^4 w(y, t)}{\partial y^4} = F_{\mathrm{f}}(y, t) \tag{19}$$

where $w(y, t)$ is the displacement of the beam as a function of distance y along the length of the beam and time t, μ is the mass per unit length of the cantilever, E is Young's modulus, I is the moment of inertia of the cantilever, and F_{f} is the force acting on the cantilever due to the fluid. In this expression we have neglected internal dissipation in the elastic body, tensile forces leading to a stressed or strained state when the cantilever is at equilibrium and gravity forces (it is straightforward to show that small cantilevers do not bend significantly in a gravitational field). It is important to note that Eq. (19) is coupled with the fluid equations Eqs. (6) and (7) through the force F_f, and that the coupled system of equations are linear. The equations governing beam dynamics are well studied and well understood (see Ref. [44] for an excellent reference on the theory of elasticity). The viscous dissipation in a low Reynolds number fluid is quite large and will dominate any other modes of dissipation such as internal elastic dissipation in the beam itself.

This leaves the important question of how to determine the flow field. Since the beam is long and slender, most of the fluid will interact with the beam by flowing around the sides as opposed to flowing over the beam tip. In this case one can assume that the cantilever is infinite in length and consider only the flow over a 2-D cross-section of the beam [42]. It can then be shown that it is a small correction to then assume that the usually rectangular cross-section of the beam is cylindrical. This is particularly convenient because an analytical solution for the flow field over an oscillating cylinder is available. In fact, the fluid problem was fist solved in 1851 by Stokes; however, for a modern treatment, see Ref. [45].

Since the fluidic damping dominates the cantilever motion we can further simplify the analysis by considering only the fundamental mode of the beam dynamics (the higher harmonics will be damped out by the fluid). This additional simplification aids in clarifying the approach without significantly affecting the results (for the analysis using the full beam equation, see Ref. [42]). The equation of motion describing the fundamental mode of a beam immersed in fluid then becomes:

$$m_e \ddot{x} + kx = F_f + F_B \tag{20}$$

where x represents the deflection of the cantilever tip, m_e is the effective mass of the beam in vacuum, k is the effective spring constant of the beam and F_B is the random force due to Brownian motion. Notice that F_f contains both the fluid damping as well as the fluid loading due to the additional fluid mass that the beam "carries" as it moves.

It is convenient to transform into frequency space by taking the Fourier transform of this equation to give:

$$(-m_e \omega^2 + k)\hat{x} = \hat{F}_f + \hat{F}_B \tag{21}$$

where:

$$\hat{F}_f = m_{\text{cyl,e}} \omega^2 \Gamma(\omega)\hat{x} \tag{22}$$

and:

$$m_{\text{cyl,e}} = 0.243 m_{\text{cyl}} = 0.243 \rho_1 \left(\frac{\pi}{4} w^2 L \right) \tag{23}$$

which is the effective mass of a fluid cylinder of radius $w/2$, where ρ_l is the fluid density. The prefactor of 0.243 ensures that mode-shape mass is equivalent for the mass of the cantilever, the fluid loaded mass and the fluid damping. The Fourier transform convention we are using is:

$$\hat{x}(\omega) = \int_{-\infty}^{\infty} x(t)e^{-i\omega t}\,dt \tag{24}$$

$$x(t) = \frac{1}{2\pi}\int_{-\infty}^{\infty} \hat{x}(\omega)e^{i\omega t} \tag{25}$$

Here, $\Gamma(\omega)$ is the hydrodynamic function and is defined to be:

$$\Gamma(\omega) = 1 + \frac{4iK_1(-i\sqrt{iR})}{\sqrt{iR}K_0(-i\sqrt{iR})} \tag{26}$$

where K_1 and K_0 are Bessel functions. Note that by this definition the arguments on the right-hand side are R and not the frequency ω. The cantilever is effectively loaded by the fluid which can be characterized by an effective mass, m_f, larger than m_e that takes into account the fluid mass that is also being moved. The fluid also damps the motion of the cantilever, which can be expressed as an effective damping γ_f. Relations for m_f and γ_f can be found by expanding $\Gamma(\omega)$ into its real and imaginary parts Γ_r and Γ_i in Eq. (21), and rearranging such that:

$$-m_f(\omega)\omega^2\hat{x} - i\omega\gamma_f(\omega)\hat{x} + k\hat{x} = \hat{F}_B \tag{27}$$

to give:

$$m_f = 0.243m_c(1 + T_0\Gamma_r) \sim \Gamma_r(\omega) \tag{28}$$

$$\gamma_f = 0.243m_{cyl}\omega\Gamma_i \sim \omega\Gamma_i(\omega) \tag{29}$$

Notice that both the fluid loaded mass of the cantilever and the fluidic damping are functions of frequency. The ratio of the mass of the fluid-loaded cantilever to the effective mass of the cantilever in vacuum, m_e, as a function of frequency is shown in Fig. 1.5. The cantilever has a mass of nearly 20 times the effective value at $R \approx 1$. Over 4 orders of magnitude in frequency the mass changes by a factor of about 200. The fluidic damping is shown in Fig. 1.5. There is a slight frequency dependence, over 4 orders of magnitude in frequency the damping changes by a factor of 7, which is much less than the frequency dependence of the mass loading.

From the fluctuation–dissipation theorem the spectral density of the fluctuating force, $G_{F_B}(\nu)$, can be related to the dissipation due to the fluid and is given by:

$$G_{F_B}(\nu) = 4k_B T m_e T_0\omega\Gamma_i(\omega) \tag{30}$$

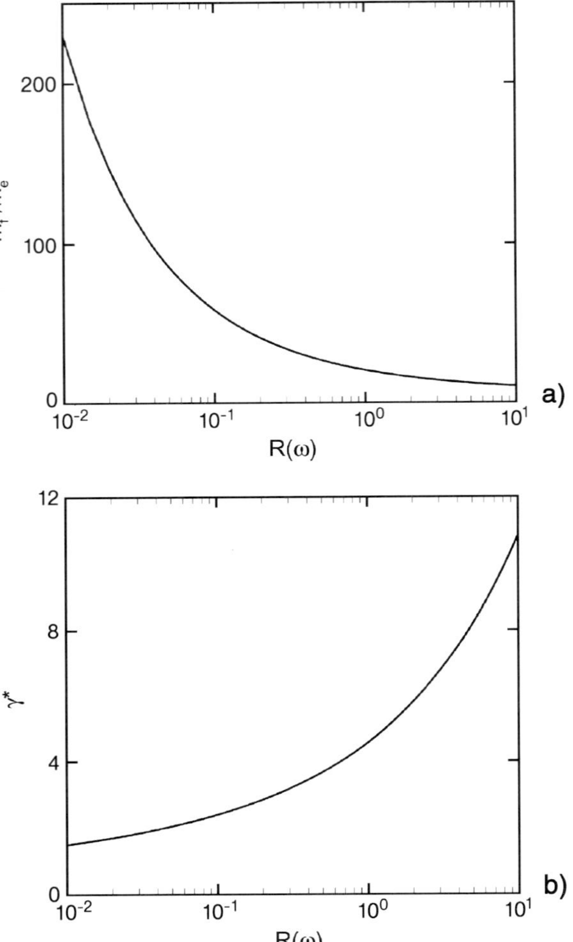

Figure 1.5. (a) The ratio of the mass of a fluid-loaded cantilever to the effective mass of the cantilever in vacuum as a function of the frequency-based Reynolds number. (b) The fluidic damping of a cantilever immersed in fluid as a function of the frequency-based Reynolds number. Shown is the nondimensional damping $\gamma^* = R\Gamma_i(R)$.

where T_0 is the ratio of the mass of fluid contained in a cylindrical volume of radius $w/2$ to the mass of the cantilever. The analysis of Ref. [42] does not take into account the frequency dependence of the damping and assumes that the numerator is constant. Although the frequency dependence of the damping is not large as shown in Fig. 1.5, it should be accounted for. Solving for the spectral density of the displacement fluctuations, $G_x(v)$, from Eqs. (21) and (30) yields:

$$G_x(\nu) = \frac{4k_B T}{k} \frac{1}{\omega_0} \frac{\tilde{\omega} T_0 \Gamma_i(R_0\tilde{\omega})}{[(1 - \tilde{\omega}^2(1 + T_0\Gamma_r(R_0\tilde{\omega})))^2 + (\tilde{\omega}^2 T_0\Gamma_i(R_0\tilde{\omega}))^2]}$$

where $\tilde{\omega} = \omega/\omega_0$ is the frequency relative to the vacuum resonance frequency $\omega_0 = \sqrt{k/m}$ and R_0 is the frequency-based Reynolds number using ω_0. Using the equipartition of energy theorem and applying it to the cantilever's potential energy, we arrive at:

$$\frac{1}{2} k \langle x^2 \rangle = \frac{1}{2} k_B T \tag{31}$$

Using this we scale $G_x(\nu)$ in Eq. (31) such that:

$$\int_0^\infty |\hat{x}(\omega)|^2 \, d\omega = \frac{k_B T}{k} \tag{32}$$

The value of ω at the maximum value of $|\hat{x}(\omega)|^2$ yields a theoretical prediction of the fundamental frequency in fluid ω_f. Once ω_f is known, an approximation for the quality factor of the oscillator, Q, is:

$$Q \approx \frac{\frac{1}{T_0} + \Gamma_r(\omega)}{\Gamma_i(\omega)} \tag{33}$$

Equation (33) is valid only for $Q \gtrsim 1/2$ because it neglects to account for the frequency dependence of the mass and fluid loading in Eq. (27) (by considering only the explicit frequency dependence) which become very important for highly overdamped cantilevers (i.e. $R \lesssim 1$).

Using what we have discussed so far let us quantify the stochastic dynamics of an AFM placed in water. We consider a cantilever with the simple beam geometry as shown in Fig. 1.4. The cantilever dimensions are length $L = 197$ μm, width $w = 29$ μm and height $h = 2$ μm. These are chosen so that we can compare with the analytical and experimental results of Ref. [43]. From beam theory, the effective spring constant of a cantilever is:

$$k = \frac{3EI}{L^3} \tag{34}$$

which, for the cantilever in question, yields $k = 1.3$ μN m^{-1}. Using the approach described in Section 1.2 we use a step force $F_1 = 26$ nN and calculate the deterministic response of the cantilever, $X_1(t)$, as it returns to equilibrium. For detailed information on the particular computation algorithm we used to solve the deterministic fluid–solid equations, see Refs. [46, 47]. The value of $\langle x_1(0)x_1(0)\rangle^{1/2}$ is interesting in that it yields the magnitude of the deflections that would be expected

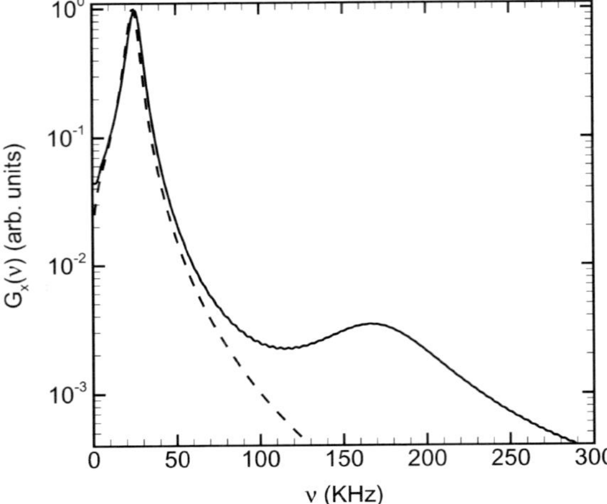

Figure 1.6. The noise spectrum as calculated from full finite element deterministic numerical simulation (solid line) and the noise spectrum from the approximate analytical theory (dashed line) for an AFM immersed in water (for experimental results see cantilever c2 in Ref. [43]). The full numerical simulations include all of the cantilever modes including two that are shown in the frequency range of the figure identified by the two peaks in the simulation results. The analytical model only considers the fundamental mode of the cantilever oscillation resulting in only one peak. Note that more modes could be included if desired; however, as shown in the figures, the higher-frequency modes are strongly damped and will not be significant in experiment. The micron-scale cantilever used for this calculation is of the geometry shown in Fig. 1.4, and has a length $l = 197$ μm, width $w = 29$ μm and height $h = 2$ μm. The applied step force is $F_1 = 26$ nN.

in an experiment. For this case we find that $\langle x_1(0)x_1(0)\rangle^{1/2} = 3.16 \times 10^{-21}$ m^2. This indicates that the deflection of the cantilever due to Brownian motion in an experiment is about 0.056 nm or about 0.003% of the thickness of the cantilever – an extremely small value even on an atomistic scale. Multiplying this quantity by the spring constant gives an estimate of the force sensitivity of 73.1 pN, which is clearly too large to be used as a biological force detector (recall biological force scales are around 10 pN). The noise spectrum is shown in Fig. 1.6, where there is good agreement with the approximate analytical theory available for this case.

1.2.4
The Stochastic Dynamics of a Fluid-coupled Array of (BIO)NEMS Cantilevers

We now use this approach to find the auto- and cross-correlation functions for the equilibrium fluctuations in the displacements of the tips of two nanoscale cantilevers with the experimentally realistic geometries depicted in Fig. 1.3. For this case we would like to emphasize that no analytical expressions or simplified models are currently available. However, we can again use full numerical simulations and exploit the fluctuation theorem, which remains exact.

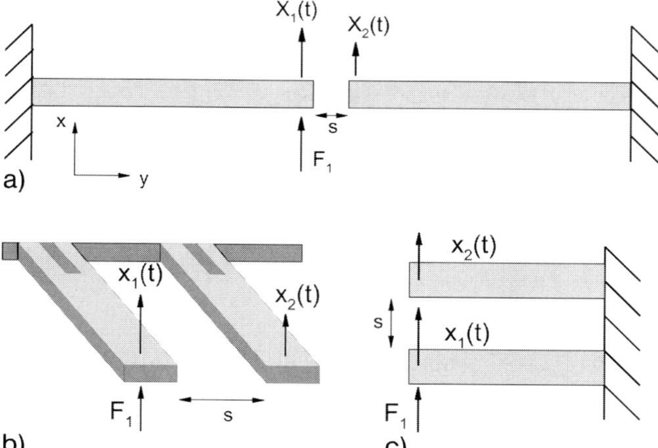

Figure 1.7. Schematic showing various cantilever configurations. In all configurations the step force F_1 is released at $t = 0$, resulting in the cantilever motion referred to by $X_1(t)$. The motion of the neighboring cantilever is $X_2(t)$ and is driven through the response of the fluid. (a) Two cantilevers with ends facing, (b) side-by-side cantilevers and (c) cantilevers separated along the direction of the oscillations.

To do this we again calculate the deterministic response of the displacement of each cantilever tip, which we call $X_1(t)$ and $X_2(t)$ after switching off at $t = 0$ a small force applied to the tip of the first cantilever, F_1, given by Eq. (8). Various possible cantilever configurations are shown in Fig. 1.7(a–c); however, we will only consider the case where two cantilevers face one another end-to-end as shown in Fig. 1.7(c).

Again, the equilibrium auto- and cross-correlation functions for the fluctuations x_1 and x_2 are given by Eqs. (15) and (16), and the noise spectra $G_{11}(v)$ and $G_{12}(v)$ are given by Eqs. (17) and (18). The cantilever autocorrelation function and the two cantilever cross-correlation function are shown in Fig. 1.8(b and c, respectively). The value of $\langle x_1(0)x_1(0)\rangle$ is 0.471 nm², indicating that the deflection of the cantilever due to Brownian motion in an experiment would be 0.686 nm or about 2.3% of the thickness of the cantilever. Multiplying this quantity by the spring constant gives an estimate of the force sensitivity of 6 pN; therefore, a (BIO)NEMS cantilever with this geometry is capable of detecting the breakage of a single hydrogen bond, indicating its potential as a single-molecule biosensor. The cross-correlation of the Brownian fluctuations of two facing cantilevers is small compared with the individual fluctuations. The largest magnitude of the of the cross-correlation is -0.012 nm² for $s = h$ and -0.0029 nm² for $s = 5h$. The noise spectra for both the one- and two-cantilever fluctuations are shown in Fig. 1.9(a and b).

The variation in the cross-correlation behavior with cantilever separation as shown in Fig. 1.8(c) can be understood as an inertial effect resulting from the non-zero Reynolds number of the fluid flow. The flow around the cantilever can be separated into a long-range potential component that propagates instantaneously in

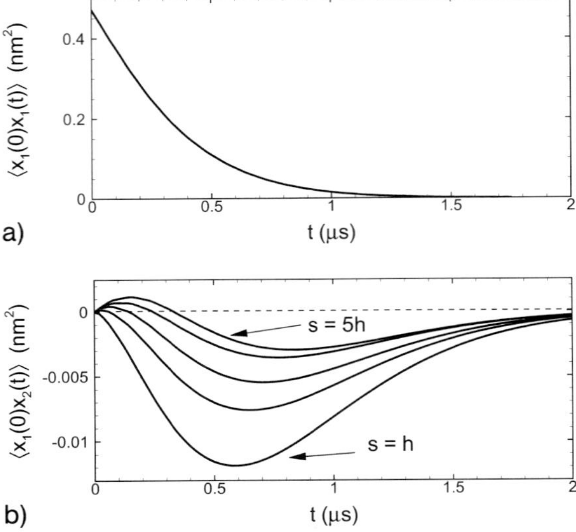

Figure 1.8. Predictions of the auto- and cross-correlation functions of the equilibrium fluctuations in displacement of the cantilevers shown in Figs. 1.3 and 1.7(a). The step force applied to the tip of the first cantilever is $F_1 = 75$ pN. (a) Autocorrelation and (b) cross-correlation of the fluctuations (5 separations are shown for $s = h, 2h, 3h, 4h$ and $5h$, where only $s = h$ and $5h$ are labeled, and the remaining curves lie between these values in sequential order).

the incompressible fluid approximation and a vorticity containing component that propagates diffusively with diffusion constant given by the kinematic viscosity v_f. For step forcing, it takes a time $\tau_v = s^2/v_f$ for the vorticity to reach distance s. For small cantilever separations the viscous component dominates, for nearly all times,

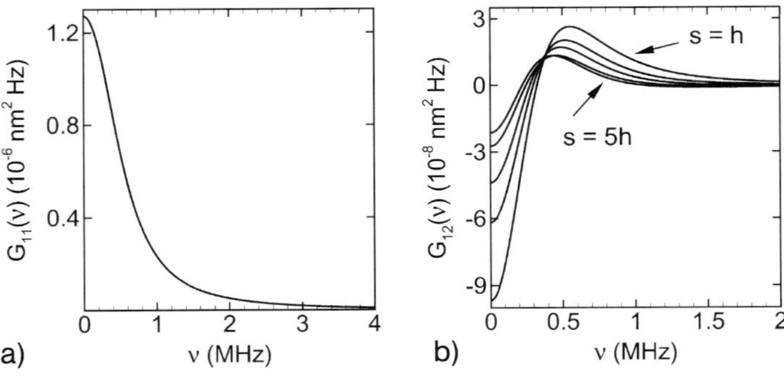

Figure 1.9. (a) The noise spectrum $G_{11}(v)$ and (b) the noise spectrum $G_{12}(v)$ as a function of cantilever separation s for two adjacent experimentally realistic cantilevers. Five separations are shown for $s = h, 2h, 3h, 4h$ and $5h$, where only $s = h$ and $5h$ are labeled, and the remaining curves lie between these values in sequential order.

and results in the anticorrelated response of the adjacent cantilever in agreement with [12]. However, as *s* increases, the amount of time where the adjacent cantilever is only subject to the potential flow field increases, resulting in the initial correlated behavior.

The complex fluid interactions between individual cantilevers in an array are still an area of active research. Nevertheless, using the thermodynamic approach described here it is now possible to describe quantitatively, with experimental accuracy, the stochastic dynamics of micro- and nanoscale oscillators in fluid. A compelling feature about these results is that the proposed experiments are just beyond the reach of current technologies, making the theoretical results that much more important, as the insight gained will be critical in guiding future efforts.

1.3
The Physics Describing the Kinetics of Target Analyte Capture on the Oscillator

Now that we have developed the methods necessary to understand the stochastic dynamics of small cantilevers in fluid we turn to the physics describing the capture of target analyte. In order to provide analyte specificity, cantilever surfaces are generally *functionalized* to contain an array of receptor molecules complementary to the target analyte (ligand). This functionalization is carried out by constructing a self-assembling monolayer (SAM), consisting of alkanethiol chains, to which specific receptor molecules are linked. Among other things, the overall performance of (BIO)NEMS cantilever-type sensors will depend on the analyte–receptor capture kinetics and we now discuss a number of issues related to this problem. The basic situation for analyte binding to the functionalized surface of a cantilever is shown in Fig. 1.10.

The binding of analyte from bulk solution to a fixed array of receptors located on a cantilever tip can be described by the kinetic equations relevant to the case of ligand binding to cell surface-bound receptors [48–50], i.e.:

$$\frac{dB}{dt} = [k_{on} L_o R_o - (k_{on} L_o + k_{off}) B] \left[1 + \frac{k_{on}(R_o - B)}{k_+} \right]^{-1} \tag{35}$$

where B is the number of analyte–receptor bound complexes, L_o is the analyte concentration and R_o is the total number of receptors in the functionalized array. This model equation describes the reversible biochemical reaction:

$$R + L \rightleftharpoons B \tag{36}$$

The parameters k_{on} and k_{off} are the usual forward and reverse rate constants for analyte–receptor binding, and k_+ is the so-called diffusion rate constant, which for the case at hand is just $k_+ = 4\pi \mathscr{D} a_c$. The quantity $\mathscr{D}$ is the analyte diffusion coefficient and a_c is a length which characterizes the size of the functionalized area,

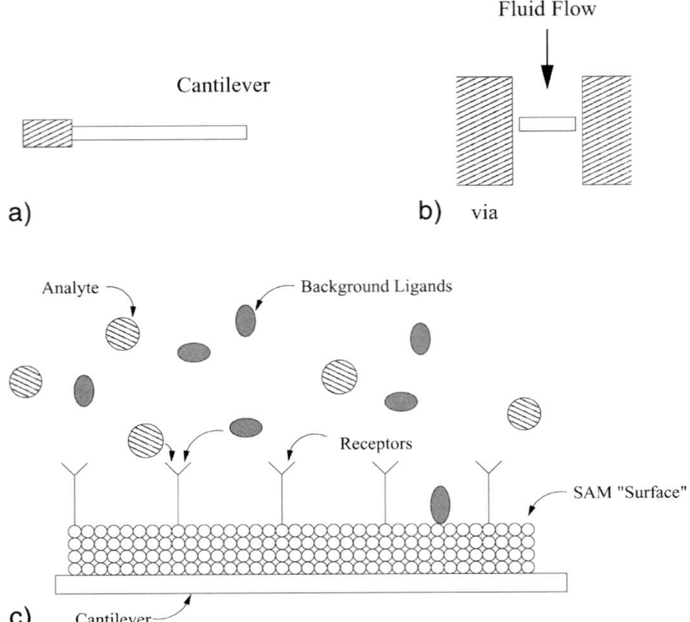

Figure 1.10. (a) The side view of a single cantilever. (b) A schematic placing the cantilever in a via. Fluid flows through the via and around the cantilever shown as a rectangular box in the center. (c) A closeup view of a cantilever tip that has been biofunctionalized.

e.g. its width. Defining new variables, $u = B/R_o$ and $\tau = k_{off} t$, this equation may be put into the more useful nondimensional form:

$$\frac{du}{d\tau} = \frac{K' - (1 + K')u}{[1 + \beta(1 - u)]} \tag{37}$$

with dimensionless parameters $K' = L_o k_{on}/k_{off}$ and $\beta = R_o k_{on}/k_+$. This rather simple kinetic equation describes the analyte–receptor binding under reaction-diffusion conditions, where the parameter β indicates the extent to which the binding is reaction limited ($\beta \ll 1$) or diffusion limited ($\beta \gg 1$).

To give the reader some quantitative insight into this problem, consider the case of biotin–streptavidin ligand–receptor binding. The functionalized region of the cantilever tip is taken to have an area of 1 μm^2, with a total of 10^4 receptors linked to the SAM surface. Note that receptor densities achievable using SAM construction are several orders of magnitude larger than those observed for specific receptors found on biological cell surfaces. The forward binding rate constant is approximately $k_{on} = 5 \times 10^6$ M^{-1} s^{-1}, with a reverse rate constant of $k_{off} \sim 10^{-3}$ s^{-1}. In addition, we find that $k_+ \approx 10^{12}$ M^{-1} s^{-1}; thus, $\beta = 0.05$ and the ligand–receptor binding process is essentially reaction limited. In fact, for cantilever-type devices designed for detection of biomolecules, we find that the capture process is almost invariably reaction limited. This means that the capture kinetics is dominated by the k_{on} and k_{off} rates of the analyte–receptor pair.

There are two issues of importance when it comes to evaluating the performance of these devices:

(a) The ultimate sensitivity, which will depend on the total number of analytes captured on the cantilever surface;
(b) The time required to achieve a specified sensitivity, which is determined by the capture kinetics.

If no other processes of significance are involved in analyte capture, then ultimate sensitivity can be estimated from a steady-state solution of the model equation given above. Thus, at steady-state, the fraction of total surface receptors that are bound by analyte is given by:

$$u_s = \frac{L_o K_a}{(1 + L_o K_a)} \tag{38}$$

where K_a is the analyte–receptor binding *affinity*. However, depending on the actual analyte–receptor rate constants and the analyte concentration, this may take a considerable time to achieve.

In addition to the basic model equation describing the capture of analyte from bulk solution to surface-bound receptors, there are at least two additional processes that should be considered in connection with sensor performance evaluation:

(a) The effects of background *contaminant* biomolecules;
(b) possible surface-enhanced analyte–receptor binding.

Interference by contaminant biomolecules may arise from two distinct mechanisms. The first of these is by *competitive* binding with the surface receptors, thus lowering the number of receptors available for analyte capture. Competitive binding effects may be analyzed by using a straightforward extension of the basic model equation discussed above. Results of such analyses show that these effects may generally be neglected even for background biomolecule concentrations approaching 10 times the analyte concentration. This is of course largely due to the fact that binding affinities for such biomolecules are 1–3 orders of magnitude smaller than the analyte–receptor binding affinities. The second mechanism involves non-specific binding of contaminant biomolecules to the SAM surface itself and would be important if analyte detection were accomplished by mass-loading effects. Even though achievable receptor surface densities for these devices approach 10^{12} cm^{-2}, a molecule in solution still "sees" mostly bare SAM surface. Thus, contaminant biomolecules may become attached to the cantilever through nonspecific surface binding. If one treats the alkanethiol end groups as discrete binding sites on the SAM surface, then this problem may be handled by a model equation analogous to the analyte–receptor capture kinetics equation.

Since the concept was first introduced by Adam and Delbruck [51], the possibility of so-called surface-enhanced ligand–receptor binding has been studied by a number of investigators [48, 52, 53]. This mechanism involves a two-step process:

(a) Nonspecific binding of ligand from bulk solution to a surface;
(b) Ligand–receptor binding following 2-D diffusion along the surface.

Although this process may easily be modeled by a pair of coupled kinetic equations, actual quantitative assessment is made difficult by the lack of reliable values for the relevant parameters, i.e. surface nonspecific binding rate and the so-called *collision-coupling* rate constant, k_c [48]. The parameter k_c is the rate constant for a surface-diffusing ligand to bind with a surface-bound receptor and is a difficult quantity to measure experimentally. Nevertheless, it may be useful to attempt to estimate the magnitude of this effect for a particular device implementation since it can result in significant enhancement of the analyte capture efficiency for certain combinations of parameters.

Without considering the parameters of a fully specified sensor it is difficult to give general estimates for cantilever capture performance. However, if we consider the k_{on}, k_{off} rates of the biotin–actin system described above, then the binding affinity will be $K_a = 5 \times 10^9$ M^{-1}. Thus, the steady-state receptor coverage is expected to be 33% for an analyte concentration of 0.1 nM. While this represents a very substantial capture efficiency, it should be noted that for these parameter values it will take many 10s of seconds to approach this coverage. This simple example points up an important issue that often arises when attempting to implement specific sensors of this type: One must usually make a trade-off between achievable sensitivity and the time required to make a measurement. A number of applications of these sensors require that detection of the presence of analyte be accomplished in times that are less than 1 s; not infrequently one wishes to achieve millisecond (or less) detection times.

So far the discussion has assumed that analyte transport to the cantilever is accomplished by diffusion only; however, most proposed cantilever sensor implementations involve the use of a microfluidic system to provide constant flow of analyte in a carrier fluid. Thus, in principle, one must consider analyte capture in the context of a reaction–diffusion–convection problem and examine the impact of convection on analyte capture efficiency. Given the previous discussion regarding the reaction rate-limited nature of the analyte capture process, one expects that convection will not have a significant impact on capture efficiency. We can also arrive at this conclusion based on two different fluid dynamics arguments. If one can show that diffusion effects dominate over convection effects in the system, then our previous argument regarding the reaction-limited character of the process still holds and convection cannot contribute significantly to analyte capture. A dimensionless parameter, the Peclet number:

$$\mathrm{Pe} = \frac{LU}{\mathscr{D}} \tag{39}$$

measures the relative importance of convective flow versus diffusive transport; here, L is a characteristic length of the system, U is the flow velocity and $\mathscr{D}$ is the diffusion coefficient. For example, if we take $L = 1\ \mu\mathrm{m}$, $U = 10\ \mu\mathrm{s}^{-1}$, and $\mathscr{D} = 100\ \mu\mathrm{m}^2\ \mathrm{s}^{-1}$, then we have $\mathrm{Pe} = 0.1$, and diffusion is the dominant transport process. We may also observe that for laminar flow perpendicular to the cantilever surface a diffusion boundary layer of thickness δ is formed. This boundary layer thickness is given approximately by:

$$\delta \approx L(1/\mathrm{Pe})^{1/3} \tag{40}$$

For the parameter values just used this yields a diffusion boundary layer thickness of about 2.2 μm; thus, at this flow rate essentially all analyte transport to the cantilever surface must be by diffusion. Of course one may also consider significantly increasing the fluid flow velocity; however, nanoscale cantilevers can easily be damaged by high flow rates. Even if the flow velocity is not high enough to actually damage a cantilever, it can result in a "bending bias" of the cantilever which can interfere with detection of binding events. We should point out, however, that these arguments should be re-examined when considering specific sensor implementations.

The use of mass-action-derived kinetic equations for the purpose of analyzing analyte capture performance is completely adequate for analyte concentrations down to about 0.1–1.0 nM. However, when we consider analyte concentrations in the picomolar (or smaller) range, concentration fluctuations may become important in describing the overall performance of the sensor system. Recall that an analyte concentration of 1 nM corresponds to a molecular density of slightly less that 1 molecule $\mu\mathrm{m}^{-3}$. In this event one must resort to stochastic methods for describing the reaction–diffusion process of analyte capture. For this case we mention an approach originally developed by Gillespie [54, 55] for "exact" stochastic simulation of coupled chemical reactions. Since the approach has been extended by Stundzia and Lumsden [56] to incorporate diffusion effects, the combined algorithm is suitable for providing a stochastic analysis of the analyte capture problem.

The Gillespie approach is based on the fact that at the microscopic level chemical reactions consist of discrete events that may be described by a joint probability density function (PDF). Thus, given a total of $\mu = 1, 2, \ldots, M$ coupled reactions, consisting of a total of $\nu = 1, 2, \ldots, N$ species, the appropriate joint PDF is $P(\mu, \tau)$, where τ is the time interval between reactions. This is simply the joint probability that the μth reaction occurs after a time interval of τ, which may be written as $P(\mu, \tau) = P(\mu)P(\tau)$. Expressions for the individual probabilities are readily derived; these expressions may then be used to implement a rather simple computer algorithm that simulates the evolution of the discrete species concentrations as a function of time, thus yielding the stochastic kinetics for the system. As Gillespie has shown [57], the resulting algorithm is an "exact" simulation of the stochastic master equation describing the coupled chemical system.

As mentioned above, as long as analyte concentrations are expected to be in a range where concentration fluctuations are not important, i.e. greater that about 0.1–1.0 nM, then use of the usual mass-action-derived kinetic equations is perfectly satisfactory in estimating the capture kinetics of the sensors considered here. However, since by its nature the mass-action-derived kinetics computes average values, this method cannot give one any insight into the stochastic behavior of the system. While there are no well-defined rules as to when one must consider fluctuations, it is generally true that when the total number of reactant molecules (ligands) in the reaction volume is only of the order of several hundred, then one should begin to suspect that fluctuations may play an important role in the system behavior. In such cases it is advisable to investigate this possibility through the use of a stochastic simulation algorithm such as the one described above.

1.4
Detecting Noise in Noise: Signal-processing Challenges

Although space does not permit a detailed analysis of the various signal-processing methods that may be used in conjunction with (BIO)NEMS cantilever-type sensors, we present a simple analysis of the most basic signal detection method that one might employ. For this analysis we assume a single passive cantilever that utilizes a piezoresistive transducer to sense the fluctuations in the cantilever tip. As discussed before, the term passive simply means that we do not actively drive the cantilever motion in order to provide for a lock-in detector-type processing system. Under these assumptions, and with no analyte bound to the cantilever, the mean-square displacement of the cantilever tip due to fluid fluctuations is given by:

$$\langle x^2(t) \rangle = \frac{4 k_B T \gamma_e}{k^2} \tag{41}$$

where k_B is Boltzmann's constant, T is the temperature, γ_e is the effective damping constant for the cantilever and k is the effective spring constant for the cantilever. The mean-square voltage signal into the front end of a signal-processing system is then just:

$$\langle v^2(t) \rangle = |\mathcal{G} \cdot I|^2 \langle x^2(t) \rangle \tag{42}$$

with $\mathcal{G}$ being the transducer conversion coefficient and I being the piezoresistive bias current.

We next assume that the presence of bound analyte on the cantilever tip appears as a change in the effective cantilever damping constant, i.e. $\gamma_e \rightarrow \gamma_e^b$. Note that in this situation our "signal" appears as a change in the mean-square fluctuations of the cantilever tip. From a signal detection theory standpoint we are attempting to discriminate against the presence of two random voltages, both being Gaussian distributed but having different variances. Our expressions for the mean-square voltage fluctuations yield a (power) signal-to-noise ratio, $(\text{SNR})_\text{p}$, of:

$$(\text{SNR})_\text{p} = \frac{\gamma_\text{e}^\text{b}}{\gamma_\text{e}} \tag{43}$$

Note that since our expression for the mean-square displacement fluctuations was essentially derived from a fluctuation–dissipation theorem, these expressions are for a system with infinite bandwidth. Our expression for SNR_p may thus be called an *inherent* signal-to-noise ratio for this detection modality. The simplest possible processing of this signal then amounts to sending it through a low-noise root mean square (r.m.s.) detector with threshold. The threshold is set to achieve the desired balance between probability of detection and false-alarm probability (*cf.* Ref. [58]).

Of course, since it is usually required that one achieve the highest possible system sensitivity, more sophisticated signal-processing techniques than the simple r.m.s. detector are usually required. We will mention only two such possibilities:

(a) Passive detection using a *reference* cantilever;
(b) Active detection using a reference cantilever and lock-in (phase) detection.

In the first case we incorporate an additional cantilever, which is *not* functionalized, into the system. One may then use a technique which is analogous to one developed in the early days of radio astronomy. In this implementation one periodically switches between the reference and sensing cantilevers to make what amounts to a phase-detection measurement of the "signal" power. The method allows one to eliminate the front-end electronics noise and to make a much better estimate of the no-signal power, thus allowing an improved signal-to-noise ratio. In the second approach we move to an active system where the reference and sensing cantilevers are subjected to periodic deflection forces that are $90°$ out of phase. This allows one to directly utilize lock-in amplifier (phase detector) technology to achieve significant enhancements to the achievable signal-to-noise ratio. For details on these and other more sophisticated signal-processing approaches to the detection of cantilever sensor signals, the reader is referred to Refs. [58–60].

1.5
Concluding Remarks

The physics and modeling of (BIO)NEMS devices poses many theoretical challenges that must be faced as experiment continues to push measurement to the nanoscale. In this chapter we have just scratched the surface of this exciting new field. In picking one particular example to focus upon it was our intent to leave the reader with an idea of some of the physics and modeling issues that one may encounter.

Acknowledgments

Our research in the modeling of MEMS and NEMS has benefited from many fruitful discussions with the Caltech BioNEMS effort (M. L. Roukes, PI) and we gratefully acknowledge extensive interactions with this team.

References

1 M. B. VIANI, T. E. SCHÄFFER, A. CHAND. Small cantilevers for force spectroscopy of single molecules. *J. Appl. Phys. 86*, 2258–2262, **1999**.

2 M. L. ROUKES. *Nanoelectromechanical Systems.* condmat/0008187, **2000**.

3 C. BUSTAMANTE, J. C. MACOSKO, G. J. L. WUITE. Grabbing the cat by the tail: manipulating molecules one by one. *Nature 1*, 130–136, **2000**.

4 C. WANG, M. MADOU. From MEMS to NEMS with carbon. *Biosens. Bioelectron. 20*, 2181–2187, **2005**.

5 C. ZANDONELLA. The tiny toolkit. *Nature 423*, 10–12, **2003**.

6 R. P. FEYNMAN. There is plenty of room at the bottom. *J. Microelectromech. Syst. 1*, 60–66, **1992**.

7 R. P. FEYNMAN. Infinitesimal machinery. *J. Microelectromech. Syst. 2*, 4–14, **1993**.

8 G. M. WHITESIDES. The 'right' size in nanbiotechnology. *Nat. Biotechnol. 21*, 1161–1165, **2003**.

9 H. CLAUSEN-SCHAUMANN, M. SEITZ, R. KRAUTBAUER, H. E. GAUB. Force spectroscopy with single bio-molecules. *Curr. Opin. Chem. Biol. 4*, 524–530, **2000**.

10 M. DOI, S. F. EDWARDS. *The Theory of Polymer Dynamics (International Series of Monographs on Physics 73).* Oxford Science Publications, Oxford, **1986**.

11 C. TASSIUS, C. MOSKALENKO, P. MINARD, M. DESMADRIL, J. ELEZGARAY, F. ARGOUL. Probing the dynamics of a confined enzyme by surface plasmon resonance. *Physica A 342*, 402–409, **2004**.

12 J.-C. MEINERS, S. R. QUAKE. Direct measurement of hydrodynamic cross correlations between two particles in an external potential. *Phys. Rev. Lett. 82*, 2211–2214, **1999**.

13 J.-C. MEINERS, S. R. QUAKE. Femtonewton force spectroscopy of single extended DNA molecules. *Phys. Rev. Lett. 84*, 5014–5017, **2000**.

14 A. KISHINO, T. YANAGIDA. Force measurements by micromanipulation of a single actin filament by glass needles. *Nature 334*, 74–76, **1988**.

15 A. ISHIJIMA, H. KOJIMA, H. HIGUCHI, Y. HARADA, T. FUNATSU, T. YANAGIDA. Multiple- and single-molecule analysis of the actomyosin motor by nanometer piconewton manipulation with a microneedle: unitary steps and forces. *Biophys. J. 70*, 383–400, **1995**.

16 M. RADMACHER, M. FRITZ, H. HANSMA, P. K. HANSMA. Direct observation of enzymatic activity with the atomic force microscope. *Science 265*, 1577–1579, **1994**.

17 N. H. THOMSON, M. FRITZ, M. RADMACHER, J. CLEVELAND, C. F. SCHMIDT, P. K. HANSMA. Protein tracking and detection of protein motion using atomic force microscopy. *Biophys. J. 70*, 2421–2431, **1996**.

18 M. B. VIANI, L. I. PIETRASANTA, J. B. THOMPSON, A. CHAND, I. C. GEBESHUBER, J. H. KINDT, M. RICHTER, H. G. HANSMA, P. K. HANSMA. Probing potein–protein interactions in real time. *Nat. Struct. Biol. 7*, 644–647, **2000**.

19 D. A. WALTERS, J. P. CLEVELAND, N. H. THOMSON, P. K. HANSMA, M. A. WENDMAN, G. GURLEY, V. ELINGS. Short cantilevers for atomic force microscopy. *Rev. Sci. Instrum. 67*, 3583–3590, **1996**.

20 G. BINNIG, C. F. QUATE, Ch. GERBER.

Atomic force microscope. *Phys. Rev. Lett. 56*, 930–933, **1986**.

21 F. J. GIESSIBL. Advances in atomic force microscopy. *Rev. Mod. Phys. 75*, 949–983, **2003**.

22 N. JALILI, K. LAXMINARAYANA. A review of atomic force microscopy imaging systems: applications to molecular metrology and biological sciences. *Mechatronics 14*, 907–945, **2004**.

23 Y. MARTIN, C. C. WILLIAMS, H. K. WICKRAMASINGHE. Atomic force microscope force mapping and profiling on a sub 100-A scale. *J. Appl. Phys. 61*, 4723–4729, **1987**.

24 T. R. ALBRECHT, P. GRUTTER, D. HORNE, D. RUGAR. Frequency-modulation detection using high-Q cantilevers for enhanced force microscope sensitivity. *J. Appl. Phys. 69*, 668–673, **1991**.

25 M. RADMACHER, R. W. TILLMAN, M. FRITZ, H. E. GAUB. From molecules to cells: imaging soft samples with the atomic force microscope. *Science 257*, 1900–1905, **1992**.

26 Q. ZHONG, D. INNISS, K. KJOLLER, V. B. ELINGS. Fractured polymer/silica fiber surface studied by tapping mode atomic force microscopy. *Surf. Sci. 290*, L688–L692, **1993**.

27 P. K. HANSMA, J. P. CLEVELAND, M. RADMACHER, D. A. WALTERS, P. E. HILLNER, M. BENZANILLA, M. FRITZ, D. VIE, H. G. HANSMA, C. B. PRATER, J. MASSIE, L. FUKUNAGE, J. GURLEY, V. ELINGS. Tapping mode atomic force microscopy in liquids. *Appl. Phys. Lett. 64*, 1738–1740, **1994**.

28 R. GARCIA, R. PEREZ. Dynamic atomic force microscopy methods. *Surf. Sci. Rep. 197–301*, **2002**.

29 S. KOS, P. LITTLEWOOD. Hear the noise. *Nature 431*, 29, **2004**.

30 Y. LEVIN. Internal thermal noise in the LIGO test masses: a direct approach. *Phys. Rev. D 57*, 659–663, **1998**.

31 J. A. PELESKO, D. H. BERNSTEIN. *Modeling MEMS and NEMS.* Chapman & Hall/CRC, London, **2003**.

32 E. M. PURCELL. Life at low Reynolds number. *Am. J. Phys. 45*, 3–11, **1977**.

33 J. HAPPEL, H. BRENNER. *Low Reynolds Number Hydrodynamics.* Springer, Berlin, **1983**.

34 C. POZRIKIDIS. *Boundary Integral and Singularity Methods for Linearized Viscous Flow.* Cambridge University Press, Cambridge, **1992**.

35 G. KARNIADAKIS, A. BESKOK, N. ALURU. *Micro Flows.* Springer, Berlin, **2001**.

36 R. L. PANTON. *Incompressible Fluid Flow.* Wiley, New York, **2005**.

37 M. R. PAUL, M. C. CROSS. Stochastic dynamics of nanoscale mechanical oscillators in a viscous fluid. *Phys. Rev. Lett. 92*, 235501, **2004**.

38 H. B. CALLEN, T. A. WELTON. Irreversibility and generalized noise. *Phys. Rev. 83*, 34–40, **1951**.

39 H. B. CALLEN, R. F. GREENE. On a theorem of irreversible thermo-dynamics. *Phys. Rev. 86*, 702–710, **1952**.

40 D. CHANDLER. *Introduction to Modern Statistical Mechanics.* Oxford University Press, Oxford, **1987**.

41 J. ARLETT et al. BioNEMS: biofunctionalized nanoelectromechanical systems. To be published.

42 J. E. SADER. Frequency response of cantilever beams immersed in viscous fluids with applications to the atomic force microscope. *J. Appl. Phys. 84*, 64–76, **1998**.

43 J. W. M. CHON, P. MULVANEY, J. SADER. Experimental validation of theoretical models for the frequency response of atomic force microscope cantilever beams immersed in fluids. *J. Appl. Phys. 87*, 3978–3988, **2000**.

44 L. D. LANDAU, E. M. LIFSHITZ. *Theory of Elasticity.* Butterworth-Heinemann, London, **1959**.

45 L. ROSENHEAD. *Laminar Boundary Layers.* Oxford University Press, Oxford, **1963**.

46 H. Q. YANG, V. B. MAKHIJANI. A strongly-coupled pressure-based CFD algorithm for fluid–structure interaction. In: *AIAA-94-0179*, pp. 1–10, **1994**.

47 CFD Research Corp., Huntsville, AL 35805.

48 D. A. LAUFFENBURGER, J. J. LINDER-MAN. *Receptors*. Oxford University Press, New York, **1993**.

49 H. C. BERG, E. M. PURCELL. Physics of chemoreception. *Biophys. J. 20*, 193–219, **1977**.

50 O. G. BERG, P. H. VON HIPPEL. Diffusion-controlled macromolecular interactions. *Annu. Rev. Biophys. Biophys. Chem. 14*, 131–160, **1985**.

51 G. ADAM, M. DELBRUCK. *Structural Chemistry and Molecular Biology*, A. RICH, N. DAVIDSON (Eds.). Freeman, San Francisco, CA, **1968**.

52 D. WANG, S.-Y. GOU, D. AXELROD. Reaction rate enhancement by surface diffusion of adsorbates. *Biophys. Chem. 43*, 117–137, **1992**.

53 D. AXELROD, M. D. WANG. Reduction-of-dimensionality kinetics at reaction-limited cell surfaces. *Biophys. J. 66*, 588–600, **1994**.

54 D. T. GILLESPIE. A general method for numerically simulating the stochastic time evolution of coupled chemical reactions. *J. Appl. Phys. 22*, 403–434, **1976**.

55 D. T. GILLESPIE. Exact stochastic simulation of coupled chemical reactions. *J. Phys. Chem. 81*, 2340–2361, **1977**.

56 A. B. STUNDZIA, C. J. LUMSDEN. Stochastic simulation of coupled reaction–diffusion processes. *J. Comp. Phys. 127*, 196–207, **1996**.

57 D. T. GILLESPIE. Concerning the validity of the stochastic approach to chemical kinetics. *J. Stat. Phys. 16*, 311–318, **1977**.

58 H. L. VAN TREES. *Detection, Estimation, and Modulation Theory: Part I*. Wiley, New York, **2001**.

59 A. D. WHALEN. *Detection of Signals in Noise*. Academic Press, New York, **1971**.

60 J. L. STENSBY. *Phase-Locked Loops: Theory and Applications*. CRC Press, Boca Raton, FL, **1997**.

2
Mathematical and Computational Modeling: Towards the Development and Application of Nanodevices for Drug Delivery

John P. Sinek, Hermann B. Frieboes, Balakrishnan Sivaraman, Sandeep Sanga, and Vittorio Cristini

2.1
Introduction

Within recent decades, quickening research and development of liposomal and nanoparticle delivery systems has made Paul Ehrlich's dream of *zauberkugeln* – therapeutic magic bullets – a reality. Although these bilipid and polymeric fabrications of the modern laboratory never received the scrutiny of his microscope, their potential to seek out and destroy specific pathogens while leaving the body's healthy tissues relatively unharmed promises to fulfill the paradigm of targeted drug delivery that he envisaged. A critical advantage afforded by the use of molecularly targeted nanovectors over conventional free-drug and antibody-based therapy is highly tunable selectivity, which greatly increases the therapeutic index of any given drug.

A plethora of excellent experimental work has been undertaken ranging from surface modification to prolong circulation, to ligand–particle conjugation to augment selectivity [1–17]. Mathematical and computational modeling can complement this experimental work by providing insight and guidance in both the fabrication and the performance of nanotechnology. A popular concern of such modeling is exemplified in work [18, 19] in which drug-release behavior of nanodevices is modeled according to the laws of mass balance and Fickian diffusion. However, the performance of micro- and nanodevices must be considered in the context of a dynamic biological environment, spanning several scales and modes, including the intravascular, the intratumoral and even the intracellular. Therefore, it is not merely what such devices do in isolation that requires investigation, but also what they do in the body, and what the body does, or attempts to do, to them. From this perspective, a principal consideration in the optimization of nanodevice performance is a thorough understanding of those bodily environments and systems with which the devices will interact. Thus, we do not merely use mathematics and computation to model the nanodevice, but rather to model the performance of the nanodevice/body system. The implications for improvement in not only the devices themselves, but also modes of delivery and possible adjuvant treatments to maximize performance (see, e.g. Ref. [20]), can be readily appreciated.

The treatment of cancer employing liposomes and nanoparticles provides fertile

Nanotechnologies for the Life Sciences Vol. 4
Nanodevices for the Life Sciences. Edited by Challa S. S. R. Kumar
Copyright © 2006 WILEY-VCH Verlag GmbH & Co. KGaA, Weinheim
ISBN: 3-527-31384-2

ground for demonstrating this approach. For example, the modeling and simulation of vasculogenesis and hemodynamics [21–23] point out difficulties in homogenously delivering nanovectors to tumoral lesions. This has consequences for their design specifications, such as circulation time, loading and release kinetics. Furthermore, fundamental performance limitations imposed by the biological environment must be defined in order for the direction of future development to be determined. Work on opsonization prevention by authors such as Torchilin and coworkers [24], and work on receptor–ligand binding by Bell [25] and Cozens-Roberts and coworkers [26, 27], has a direct influence on the design of liposomes and nanoparticles. Overprotection from protein adsorption may interfere with desirable receptor–ligand binding and therefore both potentials must be mutually optimized. Alternately, methods could be developed that circumvent interference. As yet another example, *in silico* modeling of nanoparticle chemotherapy, such as that performed by Sinek and coworkers [28], can demonstrate potential strengths as well as weaknesses in the particle-vectored delivery paradigm. Knowing the obstacles and options along one's path is at least half of what is required in planning one's journey.

The treatment of cancer motivates the lion's share of nanodevice drug delivery research and provides excellent modeling opportunities in the spirit of what has been discussed above. In this chapter we identify four critical scales or environments which intravenously injected nanovectors must navigate in order to extravasate in sufficient quantity into tumoral tissue and deliver drug in the most efficacious manner. The corresponding functions at each of the scales they must successfully perform are

- To avoid uptake by the reticuloendothelial system (RES) while in circulation.
- To navigate irregularities of tumoral vasculature and homogenously extravasate.
- To selectively bind to cancer cells and undergo endocytosis.
- To release drug at a level and on a timescale that optimizes cell kill without precipitating tumoral fragmentation.

Mathematical models and computer simulations regarding each of the four phases are discussed in the following sections. The mathematics used ranges from simple force-balance systems to stochastic processes and sophisticated reaction–diffusion solvers. To our knowledge, while facets of nanodevice drug delivery modeling have been expertly treated, no attempt has been made at integrated modeling encompassing the many scales of the problem. Defining those scales while providing examples of models that address each of them is a beginning to unification.

2.2
RES Avoidance

The RES is a system of macrophages and specialized cells lining the liver, spleen, bone marrow and lymphatic tissue. Unprotected colloidal moieties, including liposomes and nanoparticles, are sequestered and removed from circulation by the RES too rapidly for them to be effective [29]. A representative accumulation due to sequestration by the reticuloendothelial system is in the liver (60–90% of injected dose), spleen (2–10%), lungs (3–20%) and bone marrow (above 1%) [30, 31]. Not only does removal from circulation prevent particles from reaching their intended target, but also the accumulation in unintended sites could present a toxic threat.

A critical factor in avoiding this uptake is the prevention of opsonization, i.e. the accrual of proteins on the particle surface. The principal way in which this is achieved is by making suitable surface modifications to the particles [12]. Developments chiefly revolve around coating particles with hydrophilic polymers and surfactants. In an early study, Wilkins and Myers [1–3] treated polystyrene particles with polylysyl gelatin and gum Arabic, resulting in an altered distribution throughout the RES, but ultimately the same total sequestered fraction as with untreated particles. Later, Tröster and coworkers [4] and Tröster and Kreuter [5] performed an extensive study of 13 surfactants and polymers as coatings for nanoparticles, and were able to significantly reduce total RES uptake at 30 min post-injection, with uptake increasing to the same level as with uncoated nanoparticles after 7 days. Bazile and coworkers [6] developed nanoparticles based on methoxy poly(ethylene glycol) (PEG)–poly(lactic acid) (PLA) blends, and, by employing ^{14}C labeling, demonstrated a reduction in their capture by cultured THP-1 monocytes. Poloxamine- and poloxamer-coated nanoparticles have also been studied with respect to liver and spleen uptake and circulation longevity in rabbits and rodents with favorable results [7–11]. Today, popular coating materials are PEG, poly(ethylene oxide) (PEO), poloxamer, poloxamine, polysorbate (Tween-80) and lauryl ethers (Brij-35) [12].

Insight can be gained in modeling the mechanisms by which polymers like PEG and PEO reject protein adhesion to nanovectors. Lasic and coworkers [32] offer a qualitative model of particle rejection in which steric repulsion is generated by a surface of hydrated PEG chains that "brush" away incoming macromolecules. Indeed, the term "steric stabilization" has become standard in describing polymer-mediated protection. In what follows, we present models by Torchilin and coworkers [24] and Jeon and coworkers [33]. In the former, statistical simulations demonstrate the area of coverage of polymer chains as functions of their flexibility and length. The optimization of these parameters as well as the surface coating density is one goal. The latter model focuses on the balance of forces keeping proteins removed from the particle or liposome surface.

2.2.1
A Statistical Model of Nanovector Surface Coverage

Torchilin and coworkers [24] model nanoparticle protection as statistical "clouds" produced by surface-grafted PEG chains that rapidly transit within a "space" of conformations. The surface covered by these clouds is unavailable for blood protein binding, and therefore the density and area of coverage produced by each polymer chain is of interest.

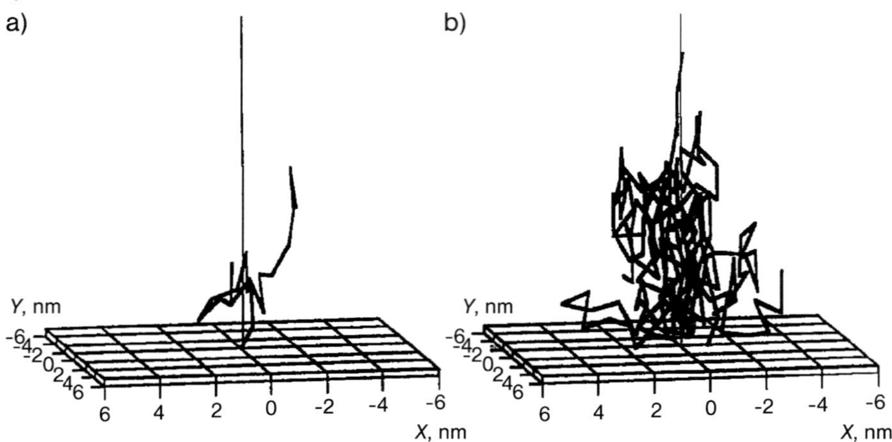

Figure 2.1. Model of a polymer chain attached to the particle or liposome surface. (a) Single conformation of one chain. (b) Superposition of 11 random conformations. (Reprinted from Ref. [24], p. 14, © 1994, with permission from Elsevier.)

A single polymer chain is modeled as being composed of a number of segments (e.g. 20), each a fixed unit in length (e.g. 1 nm). Each joint can be selectively articulated and, reckoning from joint 0 anchored into the particle, the mass of each segment is assumed to be concentrated at the distal end (Fig. 2.1). By simulating numerous conformations, an empirical probability distribution in the space directly above the liposome surface can be constructed. By coupling this distribution with the rate of conformation change, the apparent density of the cloud and, therefore, its ability to sterically hinder proteins can be known. Three parameters are critical – the degrees of freedom (or flexibility, as Torchilin calls it) of the polymer chain, its length and its rate of conformation change. To simulate a flexible polymer chain (one having many degrees of freedom), one would allow complete articulation at each joint. To simulate a less-flexible chain, some of the joints would be locked. The probability distributions in Fig. 2.2 show the effects of chain length and flexibility.

Torchilin's model can be used to optimize steric protection, which must be balanced against targeting affinity, if this is desired (see Section 2.4). In general, density and area of coverage increase as the length (weight) and flexibility of polymer chains increase. From the model simulations, the number of polymer chains to be attached to a given nanovector required for a specified degree of coverage, while still providing for accessibility to targeting ligands, can be calculated. Results are given in Fig. 2.3 and are in agreement with experimental data [24–35].

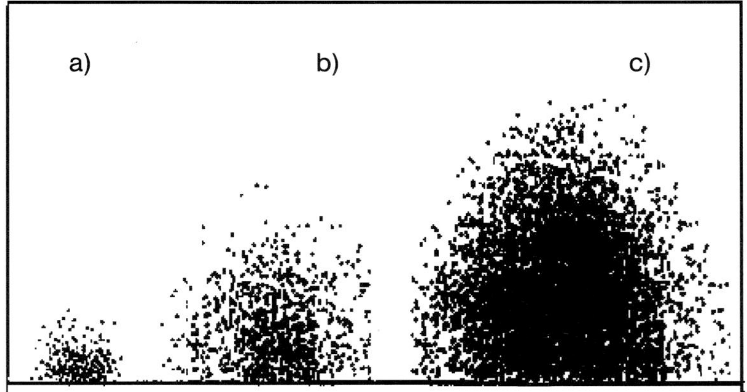

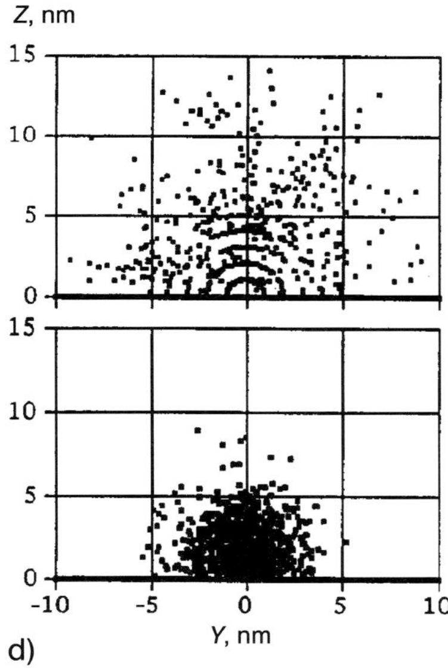

Figure 2.2. Simulated distributions of polymer conformations in space directly above the particle surface. (a–c) Effect of polymer length. (a) Short-chain polymer provides little protection. (b) Optimal chain length provides adequate protection while leaving enough surface exposed for receptor–ligand interaction. (c) Excessive length hinders the function of potential targeting ligands on the polymer surface. (d) Rigid polymer is simulated in the upper panel by assuming four segments of 5 nm each. Although the area of protection appears large, the density of coverage is compromised. Flexible polymer is simulated in the lower panel by assuming 20 segments of 1 nm each. All simulations assume the density of the segments is concentrated in their distal ends. (Reprinted from Ref. [24], pp. 16 and 15, © 1994, with permission from Elsevier.)

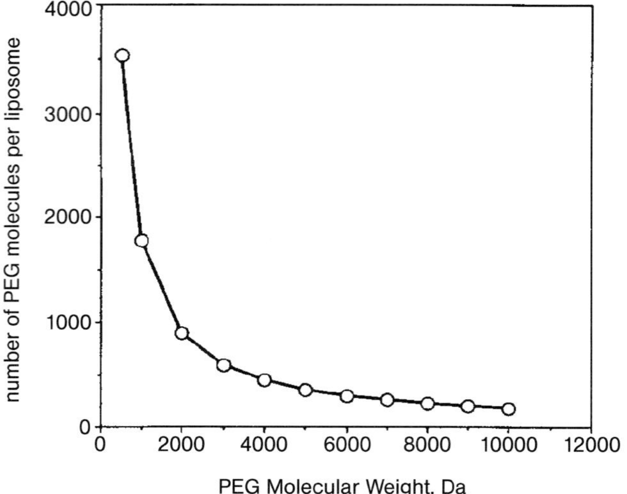

Figure 2.3. Model predictions for the minimum number of PEG molecules of a given weight needed for 100% coverage of a liposome of radius 100 nm. (Reprinted from Ref. [24], p. 17, © 1994, with permission from Elsevier.)

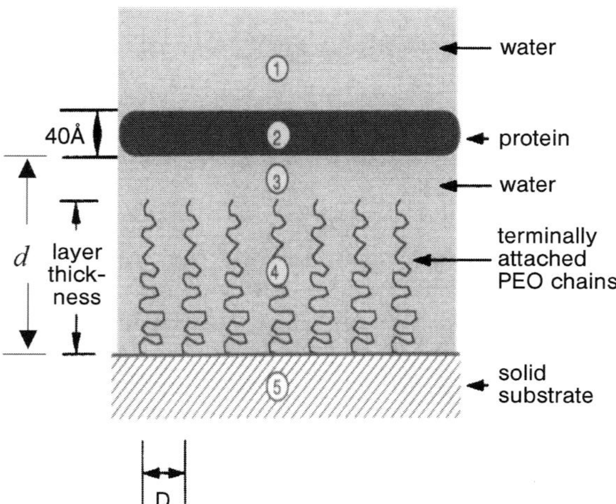

Figure 2.4. Pictorial description of the model used by Jeon and coworkers. The two principal forces that mediate binding are steric repulsion due to compression of the PEO chains, and hydrophobic attraction between the protein and solid substrate. Van der Waals attraction plays a minor role. (Reprinted from Ref. [33], p. 150, © 1991, with permission from Elsevier.)

2.2.2
Modeling the Forces Mediating Protein Approach and Binding

While surface availability is one determinant of opsonization, ultimately it is the forces between a particle and approaching protein that determine whether it binds. Jeon and coworkers [33] model the approach of a protein with a hydrophobic patch to a hydrophobic particle substrate surfaced with PEO chains as in Fig. 2.4.

The principal forces considered are steric repulsion due to compression of the PEO chains and hydrophobic attraction between the protein and the substrate, although van der Waals attraction plays a minor role. These forces are functions of PEO chain separation distance D (a measure of density) and polymerization N (a measure of length). Their corresponding free energies are given as:

$$F_{St} = K_1 ND^{-11/3} \left[\left(\left(\frac{K_2 ND^{-2/3}}{d} \right)^{5/4} - 1 \right) + \frac{5}{7} \left(\left(\frac{d}{K_2 ND^{-2/3}} \right)^{7/4} - 1 \right) \right]$$

$$F_{Hyd} = -K_3 e^{-d/14}$$

where the K's are positive constants, and d is the separation of protein and polymeric substrate.

Consulting Fig. 2.5, as the protein approaches the PEO-coated substrate, it experiences hydrophobic and, to a lesser extent, van der Waals attraction. The hydrophobic attraction free energy is negative and decreasing with approach, as shown in Fig. 2.5(a). However, strong repulsive steric forces generated by compression of the PEO chains are soon encountered and dominate (Fig. 2.5(b)). Depending on D and N, the sum of their free energies can produce an energy well as illustrated in Fig. 2.5(c). It is the depth of this well that determines how tightly bound the protein becomes. As might be expected, high surface density (low values of D) and long chain length of PEO (high values of N) are desirable for optimal protein resistance, with surface density having a greater effect. Furthermore, PEO retains an advantage among water-soluble synthetic polymers due to its low refractive index, resulting in low van der Waals interaction with the protein.

2.3
Tumoral Vasculature and Hemodynamics

All systemically administered drug therapy, whether free or nanovectored, relies upon the tumoral vasculature to gain access to malignant cells. As the quantity and uniformity of extravasated nanovectors is of pivotal importance to the success of therapy, models of tumoral vasculogenesis and hemodynamics are indispensable. However, the tumor vasculature is notorious for its irregularity [20, 36, 37]. The tumor vasculature does not follow the normal organizational pattern in which an artery connects to an arteriole to a capillary to a postcapillary venule to a venule to a vein. Instead, a tumor venule may connect to another venule via capillaries or

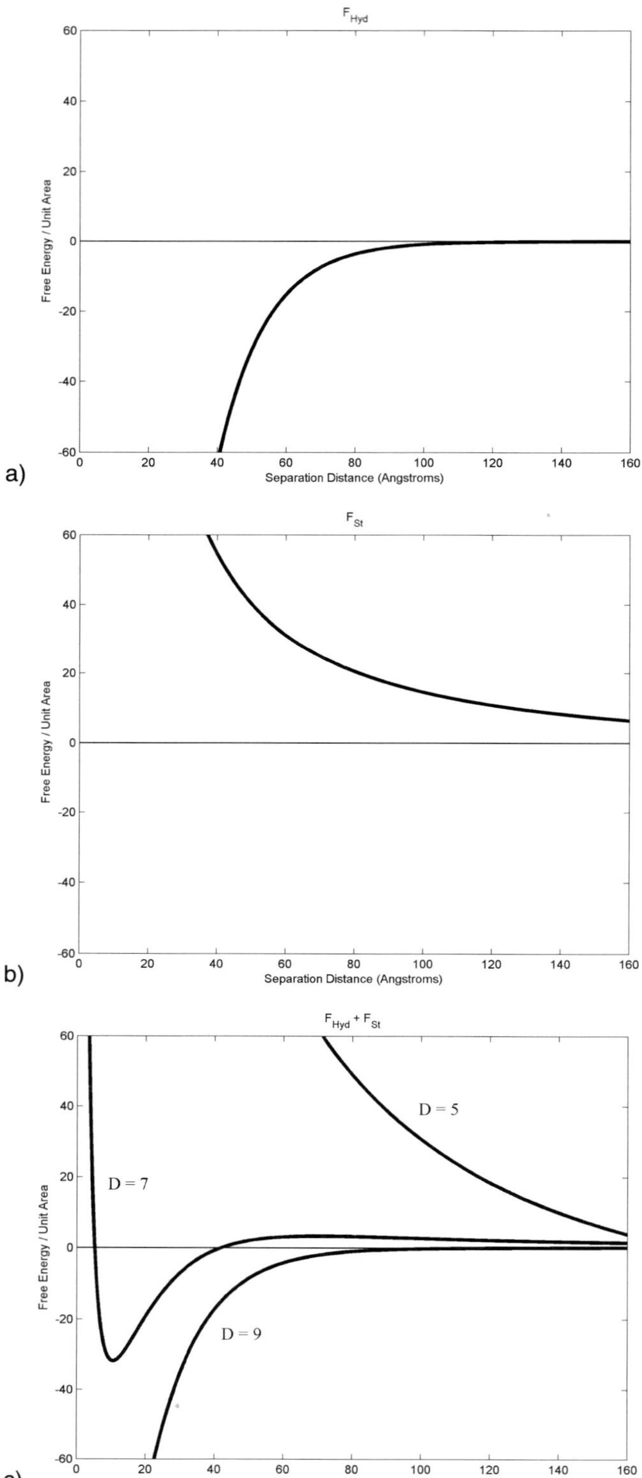

F_Hyd
Free Energy / Unit Area
Separation Distance (Angstroms)
a)
F_St
Free Energy / Unit Area
Separation Distance (Angstroms)
b)
F_Hyd + F_St
D = 5
D = 7
D = 9
Free Energy / Unit Area
Separation Distance (Angstroms)
c)

postcapillary venules. The organization may also be spatially and temporally heterogenous. The blood flow in tumors grown in transparent windows has been investigated and found to be intermittent, periodically abating and reversing [37, 38]. Tumor vessels are also dilated, saccular and hyper-fenestrated, often containing cancer cells within the vessel endothelial lining. Hobbs and coworkers [39] found that the pore sizes in one human and five murine tumors ranged from 380 to 780 nm, significantly higher than in normal tissue. While this pore size is used to advantage in the preferential extravasation of particles at lesion sites, it also leads to increased fluid extravasation and interstitial pressure. As extravasation from the vasculature depends in part on convection, this increased pressure may unfavorably influence transport. In addition to interstitial fluid pressure, a tumor has a separate mechanical pressure associated with cellular proliferation. Padera and coworkers [40] found that this mechanical stress plays a key role in the collapse of tumor vessels and further restriction of the blood supply in the tumor.

Mathematical models have revealed that the topology of the tumoral vasculature may have a significant impact on blood flow through the network. Secomb and Hsu [41] suggested that irregularities in the vascular geometry could lead to a 2-fold increase in the vascular resistance, relative to the resistance measured in a uniform tube with the same mean diameter. Baish and coworkers [22] have found similar characteristics. In a later work, Baish and coworkers [42] showed that the excessive compliance and leakiness of tumoral vasculature causes blood flow to be diverted from the center of the tumor to its periphery. Recently, Sinek and coworkers [28] have demonstrated that vasculature irregularities are as detrimental to particulate drug delivery systems as to free-drug administration. Results such as these suggest the need for therapies designed to "normalize" the vasculature [22, 20]. Pruning immature and inefficient blood vessels may lead to a more normal vasculature of vessels reduced in diameter, density and permeability, with the potential of restoring more normal hemodynamics.

We next consider models and simulations of both vasculogenesis as well as hemodynamics [21–23]. We furthermore review work performed [43] regarding erythrocyte and leukocyte dynamics within capillaries, which is highly nonlinear and cannot be inferred solely from the dynamics of a strictly Newtonian fluid.

2.3.1
An Invasion Percolation Model of Vasculogenesis and Hemodynamics

Baish and coworkers [22] used an invasion percolation model of vasculogenesis to investigate the heterogeneity of vessel perfusion and resistance to flow. The principle of invasion percolation is that vascular growth follows the gradient of a sub-

Figure 2.5. (a) Free energy due to hydrophobic attraction and (b) free energy of steric repulsion. Depending on the values of D and N, an energy well can be created as in panel (c). The depth of the well determines the force of binding. For low values of D (high density), steric repulsion dominates; for higher values, energy wells are produced. Strongest binding occurs for $D = 9$ or above. (Panel c reprinted from Ref. [33], p. 157, © 1991, with permission from Elsevier.)

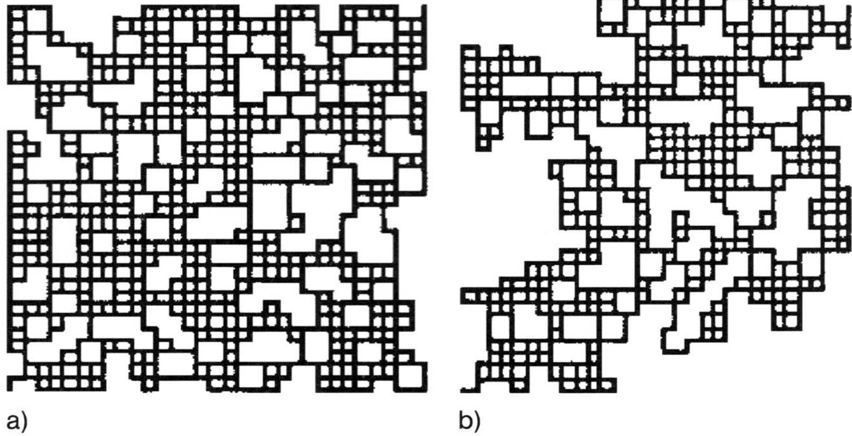

Figure 2.6. Examples of two networks produced by the invasion percolation model of Baish and coworkers, showing the network produced when 80 (a) or 60% (b) vessel occupancy is specified. (Reprinted from Ref. [22], p. 332, © 1996, with permission from Elsevier.)

strate's material weakness rather than responds to a physiological stimulus, such as the gradient of oxygen. To simulate the network, the model begins with a square array of lattice points to which material "strengths" are assigned. The network is then "seeded" at the lower left point. In subsequent iterations it extends to, or invades, the point adjacent to the network that has the lowest strength. This is repeated until a desired vessel "occupancy" (density) is attained (Fig. 2.6).

The structure of their simulated networks was characterized by two measurements propounded earlier by Gazit and coworkers [44] in their studies on mice – the fractal dimension $d_{\text{vasculature}}$ and the minimum path length d_{min}. The table reprinted here as Fig. 2.7 compares several known network growth processes to

	Known Growth Processes			Observed *in vivo*		
Dimension	Space-filling growth	Diffusion limited aggregation	Invasion percolation	Normal capillaries	Normal arteries and veins	Tumor vessels
$d_{\text{vasculature}}$	2.00	1.71	1.90	1.99 ± 0.01	1.70 ± 0.03	1.88 ± 0.04
d_{min}	1.00	1.00	1.13	1.00 ± 0.02	0.99 ± 0.02	1.10 ± 0.04

Figure 2.7. A comparison of the values of $d_{\text{vasculature}}$ and d_{min} for several known growth processes and processes observed *in vivo*. Note the agreement between the invasion percolation model and tumor vessels. (Reprinted from Ref. [22], p. 331, © 1996, with permission from Elsevier.)

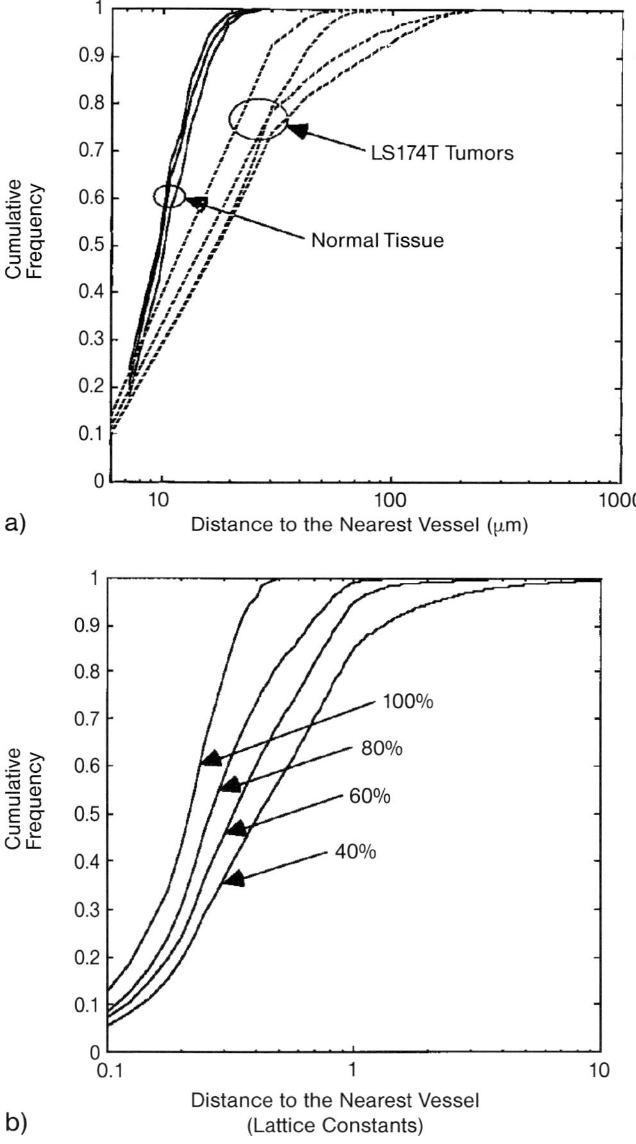

Figure 2.8. (a) Probability distributions of distance from the nearest vessel for both normal and tumoral tissue. (b) Probability distributions produced by the invasion percolation model set to various vessel occupancies. The long "tail" of the distributions is a hallmark of tumoral tissue and is well reproduced in the model simulations. (Reprinted from Ref. [22], pp. 334 and 335, © 1996, with permission from Elsevier.)

those observed *in vivo* by Gazit and coworkers [44], and demonstrates the close correspondence between the results of invasion percolation and tumoral vessels.

With respect to the heterogeneity of perfusion, Fig. 2.8 shows typical probability distributions of distance from the nearest vessel. The important characteristic here is the long tail seen in tumoral vasculature distributions, which is qualitatively reproduced by the percolation model. This means that a significant portion of tumoral tissue is distant from vessels, not only reducing oxygenation, but also the transport of chemotherapeutic agents, especially particles that are expected to deliver their death signal via endocytosis.

The percolation model also predicts inefficient and heterogeneous flow as well as the increased resistance found in tumoral vasculature. In contrast to the fairly uniform fluid flux found in a regular mesh, a few of the vessels of a percolation network carry a disproportionately large flux, while some of the vessels are almost stagnant. In line with this, tumors are typically associated with a higher flow resistance than normal tissue even though they contain a higher proportion of large-diameter vessels [38, 45–48]. For example, tumoral vessels are typically 50% greater in diameter than those of normal tissue. Since flow resistance varies inversely with the fourth power of diameter, one would expect the resistance to be 20% of that in normal tissue. Even decreasing the vascular density by half would only increase resistance 2-fold, still yielding a net 40% of resistance found in normal tissue. According to the percolation model, however, halving the vascular density increases the resistance *7-fold* – more than enough to offset the vessel diameter advantage.

2.3.2
Flow Simulations Using Anderson and Chaplain's Model

Unlike the previous angiogenesis model [22], Anderson and Chaplain's model [21] relies heavily upon physiological stimuli. In order to produce veridical networks, they note that tumor angiogenic factors (TAFs), such as vascular endothelial growth factor, and fibronectin, a large, non-diffusing molecular constituent of the extracellular matrix, play key roles. TAFs, produced by perinecrotic cells starved of nutrients, induce a chemotactic response in endothelial cells (the essential component of blood vessels), causing them to degrade their parent vessel's basement membranes and migrate towards the tumor. Fibronectin, existing naturally in most tissues and also produced and degraded by endothelial cells, forms an adhesive matrix upon which they can migrate. As the endothelial cells move chemotactically up the gradient of TAFs towards the tumor, consumption of fibronectin produces lateral gradients enabling them to spread via haptotaxis. The interaction of endothelial cells with the extracellular matrix is crucial to the model. In particular, without the interaction of endothelial cells and fibronectin, the lateral motion of the cells, necessary to form vessel loops, requires a much higher random diffusivity than is experimentally measured.

Anderson and Chaplain's is a hybrid continuum–discrete model. Endothelial cell density e, along with TAF c and fibronectin f, are modeled with a system of continuum reaction–diffusion equations incorporating chemotaxis and haptotaxis:

$$\frac{\partial e}{\partial t} = D_e \nabla^2 e - \nabla \cdot (\alpha e \nabla e) - \nabla \cdot (\beta e \nabla f)$$

$$\frac{\partial f}{\partial t} = v_f e - \eta_f ef$$

$$\frac{\partial c}{\partial t} = -\eta_c ec$$

The first term on the right-hand side in the first equation represents cell diffusion (relatively weak) with diffusivity D_e, while the second and third represent chemotaxis up the TAF gradient and haptotaxis up the fibronectin gradient. α and β can be constant; however, it is more realistic to have α be a decreasing function of TAF [21]. The terms in the second equation represent production and uptake of fibronectin by endothelial cells, with v_f and η_f being constant. The term in the last equation represents uptake of TAF, with constant rate η_c. Initial conditions for endothelial cell density are set by seeding several small regions of high-density "sprouts" along a parental vessel (see below in Fig. 2.10). Initial fibronectin and TAF are assumed to be produced from the parental vessel and the perinecrotic rim just within the tumor, respectively. This results in concentrations decaying with distance from their sources. (Evolution of endothelial cell density is shown in Fig. 2.10.)

The discrete portion of the model is a reinforced random walk of blood vessel tips that begin at the endothelial sprouts. These tips probabilistically follow endothelial cell density as shown in Fig. 2.9. Higher densities of cells bias the random walk of the tip in their direction. Further rules determine capillary branching and

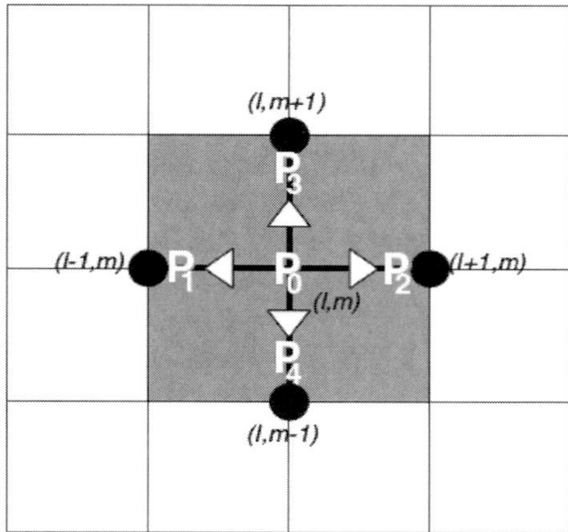

Figure 2.9. The vessel tip is initially in the middle. At the next time step it transits to one of the four neighboring grid points with probabilities P_1-P_4 or it stays at its present location with probability P_0. The probabilities are determined by the underlying endothelial cell density shown in Fig. 2.10. (Reprinted from Ref. [23], p. 679, © 2002, with permission from Elsevier.)

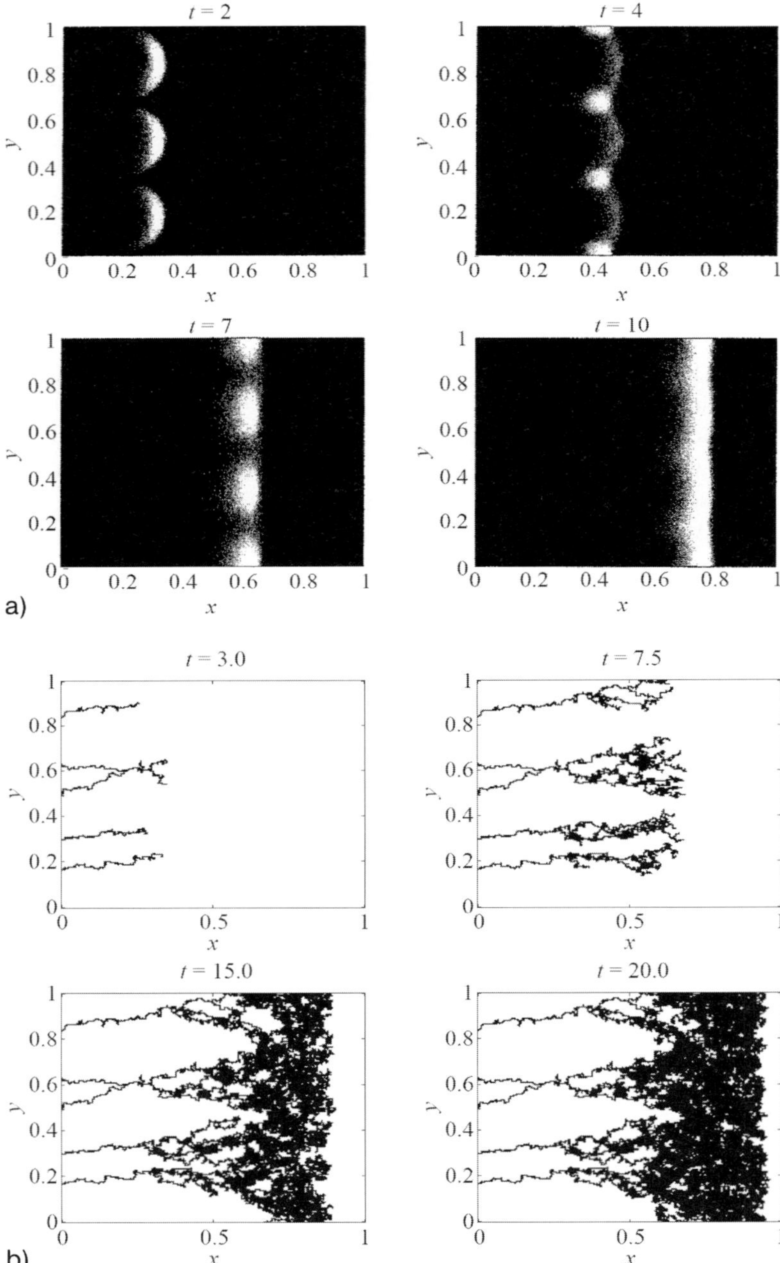

Figure 2.10. (a) The evolution of endothelial cell density of Anderson and Chaplain's model, beginning with three dense regions (sprouts) in the top left. Density proceeds across the region up the gradient of TAF towards the tumor, whose boundary is assumed on the right edge. Consumption of fibronectin produces gradients that allow for the lateral spreading of the cell density. (b) The evolution of the actual vessels, whose tips probabilistically follow endothelial cell density. (Reprinted from Ref. [21], pp. 870 and 883, © 1998, with permission from Elsevier.)

the formation of loops (anastamoses) enabling circulation. The result is a realistic capillary network with its essential dendritic structure as well as the reproduction of the experimentally observed "brush border", whereby extensive branching is observed just before the network penetrates the tumor [49, 50].

McDougall and coworkers [23] used Anderson and Chaplain's model to simulate blood and drug flow to tumors. They analyzed effects of blood viscosity, pressure drop across parental vessel, mean radius of capillaries and radius of parental vessel by tracking the total quantity of drug within the vasculature in time.

The specifics of the model are as follows. A network is generated using Anderson and Chaplain's algorithm as in Fig. 2.10, which is then mapped to a Cartesian grid. Then a radius R_{ij} is randomly assigned from a probability distribution to each element (vessel segment) joining nodes i and j on the grid. At each node i there exists a pressure P_i, and through each element joining nodes i and j there exists a flux Q_{ij}. This flux is assumed to obey Poiseuille's law:

$$Q_{ij} = \frac{\pi R_{ij}^4 \Delta P_{ij}}{8\mu L_{ij}}$$

where L_{ij} is the length of element ij and μ is fluid viscosity. Imposing mass conservation at each node i via:

$$\sum_{1 \leq j \leq 4} Q_{ij} = 0$$

where j varies over the four adjacent lattice nodes results in an exactly determined system of linear equations given the pressure drop across the parental vessel. The parental vessel feeds all capillaries going into the tumor.

To analyze drug flow, total drug mass M within the vasculature is tracked in time for values of four important parameters: blood viscosity, pressure drop across parental vessel, mean radius of capillaries and radius of parental vessel. A steady-state infusion as well as a bolus injection is shown in Fig. 2.11 for a base set of parameter values. The graph of M in time is shown for the continuous infusion and is compared to other graphs that result when, for example, mean capillary radius or viscosity are changed. When viscosity is raised, although the same limiting amount of drug mass is eventually reached within the vasculature M_∞, the time required to reach that level is greater. Lowering the mean radius of the capillary bed, however, decreases the carrying capacity of the network so that M_∞ is, itself, reduced.

It is apparent that understanding results such as these is crucial to optimizing nanovector delivery. With increasing precision of measurement of the tumoral environment so as to provide more accurate input parameters, such simulations may improve to the point that they can quantitatively model a given patient's lesion and contemplated therapy, bringing to bear the full weight of intervention on the disease.

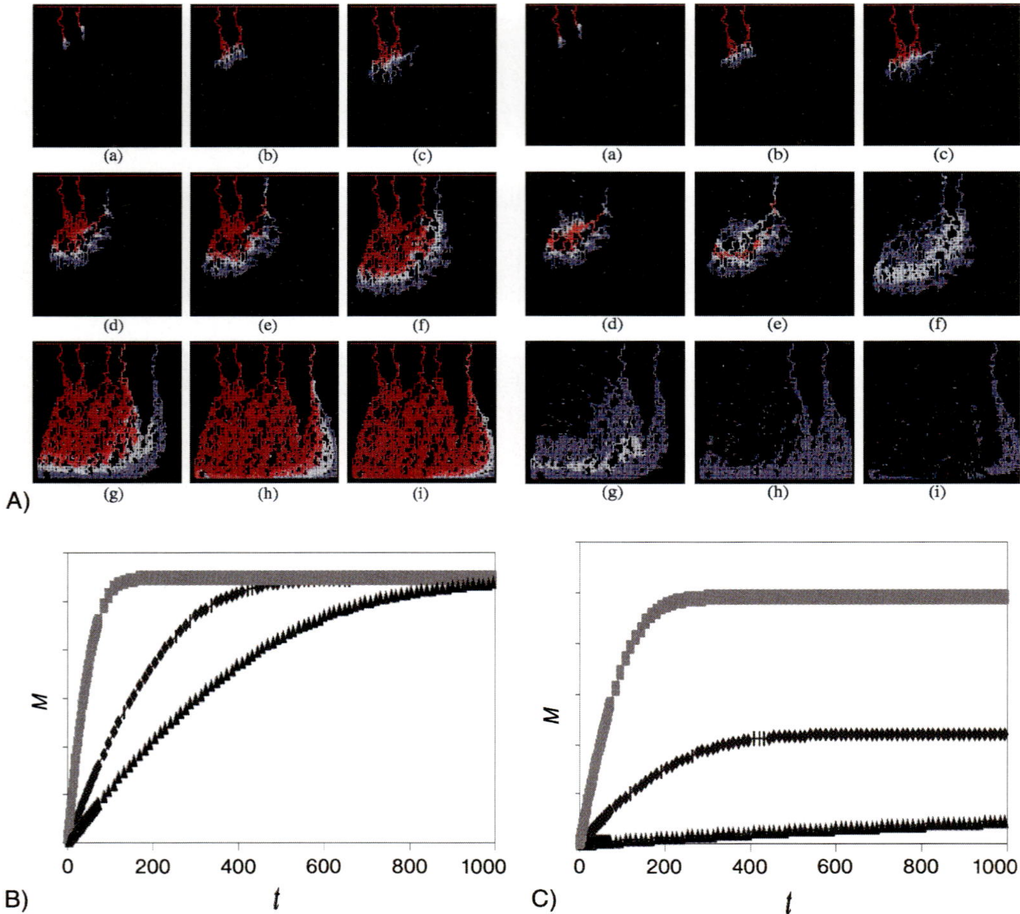

Figure 2.11. Simulations of drug delivery. The parental vessel runs across the top, with blood flowing from left to right. The tumor boundary runs along the bottom. (A) The left upper set of panels show a continuous infusion for a base set of parameters (blood viscosity, pressure drop across parental vessel, mean radius of capillaries and radius of parental vessel). Blood begins to flow into the vascular bed through the first two vessels in (b) and (c). It begins to flow out of the bed back towards the parental vessel in frames (d)–(f). Steady-state is reached by frame (i). The right upper set of panels show a bolus injection for the same base set of parameters. The lower set of graphs shows the effects of changing parameter values on the amount of drug in the vasculature M in scaled time t (continuous infusion). (B) From top to bottom shows the effect of increasing viscosity of blood: time to vascular saturation is lengthened although the saturation level remains constant. (C) From top to bottom shows the effect of decreasing the mean capillary radius: saturation capacity is severely reduced. (Reprinted from Ref. [23], pp. 689, 696, 690 and 691, © 2002, with permission from Elsevier.)

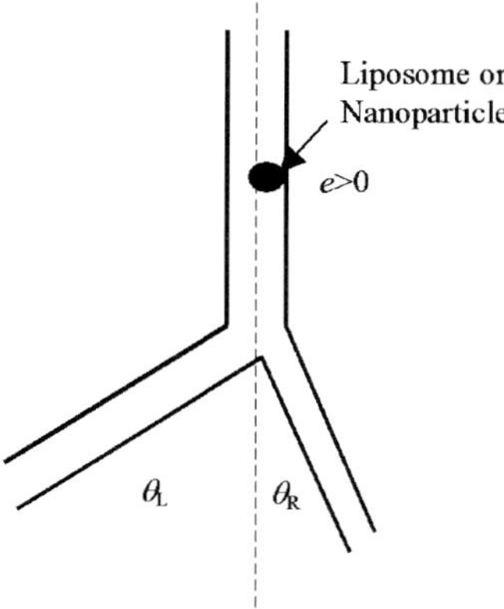

Figure 2.12. Diagram of a vessel bifurcation. Some important parameters are the branching angles θ_L and θ_R, the radii of the daughter vessels, and the eccentricity of the approaching liposome or particle. The latter is defined as axial displacement divided by vessel radius.

2.3.3
Particle Dynamics within the Tumoral Vasculature

The fluid flow dynamics discussed above become even more complex when particulate carriers are involved. Schmid-Schönbein and coworkers [43] model the flow of blood cells at vascular branches and find that the distribution of cells at a vessel bifurcation is nonlinearly related to bulk fluid flow through the branches. The analysis easily extends to liposomes and, to a lesser extent, nanoparticles. Figure 2.12 depicts the situation where a parental vessel splits into two daughter vessels. Here, important parameters are the left and right angles of separation θ_L and θ_R and the radii of the daughter vessels r_L and r_R. A particle that approaches the bifurcation may not be centered in the vessel, thus we measure its eccentricity e, defined as its distance from the centerline divided by the radius of the parental vessel. Letting the proportions of particles that enter either the left or right daughter vessel of a symmetric bifurcation ($\theta_L = \theta_R$ and $r_L = r_R$) be denoted by φ_L and φ_R ($\varphi_L + \varphi_R = 1$), and also letting the proportions of total flow (all particulate matter and plasma) that enter each daughter be ψ_L and ψ_R, we can write particulate distribution as a function of total flow distribution and eccentricity $\varphi = \varphi(\psi, e)$, "R" and "L" subscripts having been dropped. Note that in the symmetric case it is not necessary that $\psi_L = \psi_R = 0.5$ since blockages may exist downstream of one of the daughters. In the general case, this function will also depend upon the branching angles, daughter vessel radii and possibly other geometric information.

Several important results are obtained, perhaps the most notable being the nonlinear relationship shown in Fig. 2.13 in the case of symmetric branching. For particles whose diameter is small compared with the diameter of the vessel, $\varphi_R = \psi_R$.

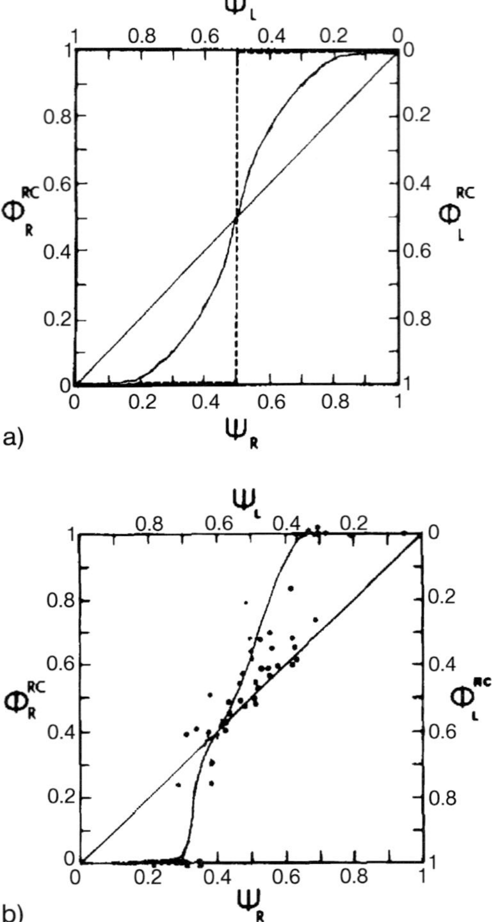

Figure 2.13. Relationship between bulk flow distribution into left and right daughter vessels (ψ_L and ψ_R) and particle flow distribution (φ_L and φ_R). (a) For particles that are small in comparison to vessel radius, bulk flow distribution and particle flow distribution are the same (diagonal line). As the relative particle size grows, a sigmoidal relationship becomes evident. At the limit, the step function results where $\varphi_R = 1$ for $\psi_R > 0.5$. (b) Data from an experiment performed by Schmid-Schönbein and coworkers on red blood cells (dots) together with a least-squares fit (solid curve). (Adapted from Ref. [43], pp. 24 and 27, © 1980, with permission from Elsevier.)

Otherwise, φ_R is very sensitive to deviations in ψ_R from 0.5, resulting in the steep curve. In the limit where the diameter of the particle is almost equal to the diameter of the vessel, this curve becomes vertical, so that for $\psi_R > 0.5$, all particles flow to the right, i.e. $\varphi_R = 1$. However, there are notable exceptions for particles that cling to the endothelium, such as leukocytes. In some experiments performed by

Schmid-Schönbein and coworkers [40], in spite of a high percentage of the bulk flow to the right ($\psi_R \geq 73\%$), white blood cells that were biased to the left, with some rolling along the endothelium, still entered the left branch. Although the eccentricity of the cells had an effect, this behavior was far less pronounced for red blood cells under similar conditions.

2.4
Receptor–Ligand-mediated Binding

Much effort has been devoted to conjugating cell-targeting ligands on nanoparticles and liposomes in an attempt to improve specificity. The objective is that particles whose surface is covered by a layer of "adhesive molecules" (ligands) will tether to target cells expressing receptors for the ligands. These targets may be cells of tumor endothelium or the tumor cells themselves, which typically overexpress certain receptors, e.g. folate receptors, in comparison to normal cells. Lee and Low [51] were among the first to suggest that liposomes could be targeted to cancerous KB cells by conjugating them with folic acid (Fig. 2.14). Folate conjugates have a high affinity for cell-surface folate receptors ($K_D \sim 10^{-10}$ M) [52]. Additionally, folate is inexpensive, stable during storage and *in vivo* circulation, and nontoxic. The steps in the folate-targeting of liposomes to cancer cells are shown in Fig. 2.15 and are as follows [17]:

- Liposomes pass through the tumor microvasculature.
- The increased vascular permeability of the tumor tissue enables the liposomes to extravasate into the tumor interstitial fluid.

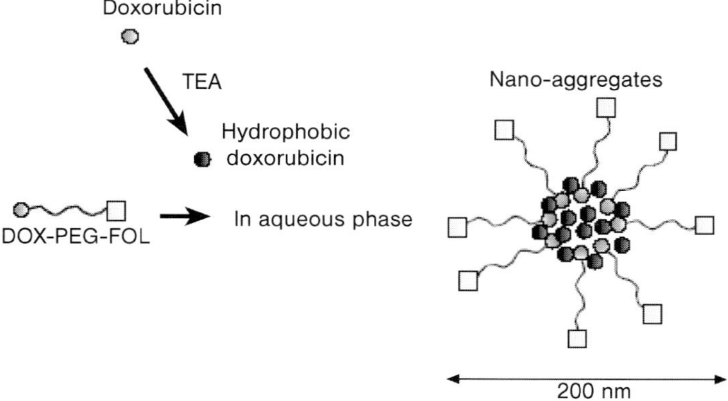

Figure 2.14. Scheme for creating folate-targeted doxorubicin(DOX–PEG–FOL) Nan conjugates. The folate-targeting moieties at the end of the PEG chains improve the targeting efficiency of the nano-aggregates. (Reprinted from Ref. [15], p. 248, © 2004, with permission from Elsevier.)

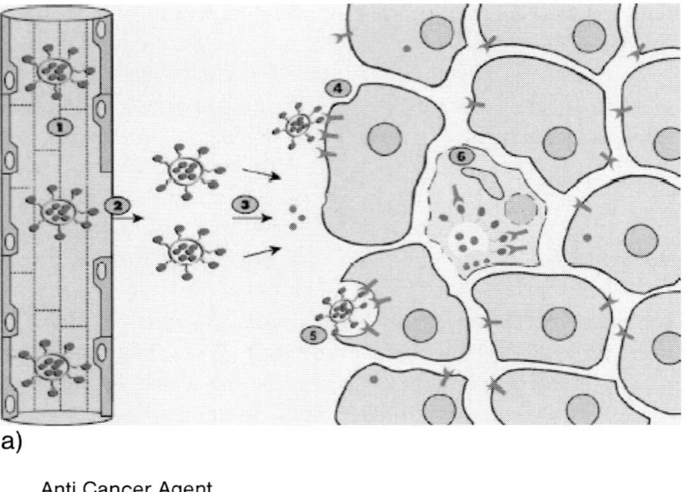

a)

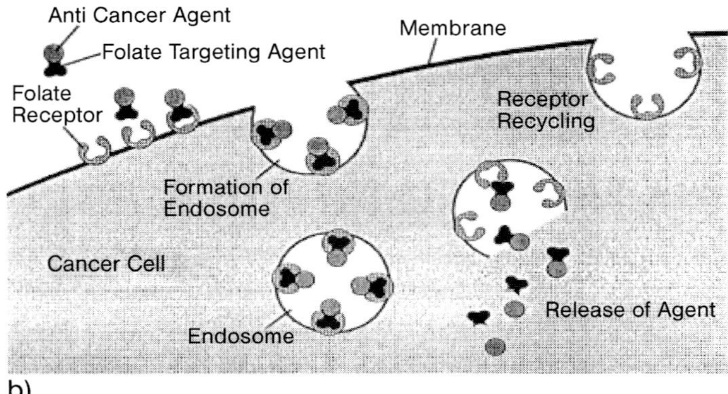

b)

Figure 2.15. (a) Schematic diagram illustrating the concept of folate-targeting of liposomes to cancer cells. After preferential extravasation from hyper fenestrated tumoral vasculature, conjugates bind to tumor cells and are endocytosed. (Reprinted from Ref. [17], p. 1179, © 2004, with permission from Elsevier.) (b) Folate receptor-mediated endocytosis of nanovectors containing anticancer drugs. The formation of endosomes that encapsulate the conjugates protects them from P-glycoprotein, MRP, GS-X and other efflux proteins. (Adapted from Ref. [53], p. 40, © 1998, with permission from Elsevier.)

- Drug is released gradually from the liposomes that remain in the interstitial fluid, enters the tumor cells as free drug, and exerts a cytotoxic effect.
- The other liposomes bind to the folate receptors on the surface of tumor cells via the folate ligand. It is important to note that due to the limited diffusion capacity of the liposomes, generally only the tumor cells closest to the blood vessels are associated with this binding.
- These liposomes are then internalized by the tumor cells via folate receptor-mediated endocytosis.

- The internalized liposomes then release the drug, which is free to exert its cytotoxic effect on the cells.

Research has shown that conjugates can overcome the multidrug resistance effect of the P-glycoprotein efflux pump due to their cellular uptake via folate receptor-mediated endocytosis [54]. According to Turek and coworkers [52], the process of endocytosis consists of receptor–ligand binding followed by internalization in clathrin-coated pits (clathrin is a protein that is the major constituent of the "coat" of the coated pits that are formed during the endocytosis of materials at the surface of cells) or "uncoated" caveolae (small invaginations in the plasma membrane that play a role in endocytosis as well as signal transduction and are observed especially in endothelial cells). As seen in Fig. 2.15, when the receptors bind to their target molecules, the pit deepens until a clathrin-coated vesicle is released into the cytosol. Due to the fact that the liposomes are encapsulated in these endosomal vesicles, they escape being effluxed by the P-glycoprotein efflux pumps.

Understanding the kinetics of receptor-mediated cell attachment would be of service in the design and optimization of ligand-conjugated nanovectors. An important question is, what are the strength and density of bonds needed to achieve adequate adhesion under various stresses? Bell [25] developed a deterministic model for cell attachment in the absence of fluid stress and cell detachment in the presence of fluid stress. We give an account of his work first, followed by a stochastic reanalysis. Another stochastic treatment of receptor–ligand binding can be found in Cozens-Roberts and coworkers [26].

2.4.1
Bell's Deterministic Model

Once a cell and nanoparticle or liposome have been brought into proximity via non-specific interactions such as Brownian motion and van der Waals forces, receptor–ligand bond formation may occur. The proximate cell and particle surface, each with its associated protein receptor or ligand available for contact, are shown in Fig. 2.16. The cell's receptor and liposome's ligand are free to diffuse in the plane of the phospholipid bilayer, while for a polymeric nanoparticle ligand is presumably limited to local gyrations. The most elementary formulations assume single receptor–ligand binding.

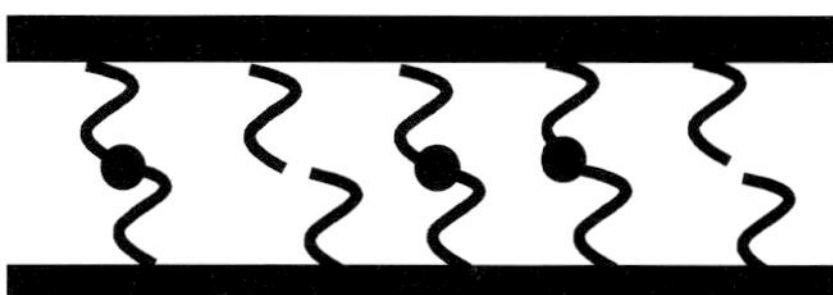

Figure 2.16. Proximate cell and particle or liposome surfaces with some free and bound receptors and ligands.

Letting N_1 and N_2 be the densities of total available receptor and ligand, respectively, and letting X_t be the density of bound receptor–ligand complex at time t with $N_1 \gg X_t$ for all t to simplify the exposition, bond formation is governed by:

$$\frac{dX_t}{dt} = k^+ N_1 (N_2 - X_t) - k^- X_t \tag{1}$$

where k^+ and k^- are the forwards and backwards kinetic rate constants. At equilibrium, we have:

$$X_\infty = \alpha N_2 (1 + \alpha)^{-1} \tag{2}$$

where $\alpha = KN_1$ and $K = k^+/k^-$ and is the association constant.

The forwards and backwards rate constants can in theory be derived from the planar diffusion constants D_1 and D_2 of membrane-bound receptor and ligand, although rotation and gyration of species also play a part. In order to do this, the binding is assumed to take place according to the two-step reaction:

$$R + L \underset{d^-}{\overset{d^+}{\rightleftharpoons}} R \circ L \underset{r^-}{\overset{r^+}{\rightleftharpoons}} RL \tag{3}$$

where $R \circ L$ represents an "encounter complex" in which receptor and ligand are within some critical radius C, and RL represents bound receptor–ligand. The d's and r's represent directional reaction rates for the steps. The former are related to the receptor and ligand planar diffusion constants according to:

$$d^+ = 2\pi(D_1 + D_2)$$
$$d^- = 2(D_1 + D_2)C^{-2} \tag{4}$$

In many cases the rate of change of the encounter complex is negligible. Setting $d[R \circ L]/dt = 0$ then allows for the net reaction rates in (1) to be approximated as $k^+ = d^+ r^+/(d^- + r^+)$ and $k^- = d^- r^-/(d^- + r^+)$. If, furthermore, $r^+ \gg d^-$, as is the case for receptors and ligands in a viscous membrane, then these rates reduce to $k^+ = d^+$ and $k^- = d^- r^-/r^+$.

This, then, provides the basis for determining the strength of bonding between, for example, a ligand-conjugated liposome and a receptor-rich cell. The question can be asked, what force is required to break a bound particle–cell complex? To answer this question, we assume that a force F is applied to separate the surfaces in Fig. 2.16. Under this condition kinetic theory provides that the reverse reaction rate k^- in Eq. (1) is modified to $k^- \exp(AF/X_t)$, where A is a constant, so that the equation becomes:

$$\frac{dX_t}{dt} = k^+ N_1 (N_2 - X_t) - k^- \exp(AF/X_t)X_t \tag{5}$$

From this expression it can be seen that as the force is increased, so does the reverse reaction rate and therefore the equilibrium number of bonds will be decreased. However, this rate also increases as the number of bonds X_t decreases because the force is being applied over fewer bonds. This is reflected by the presence of X_t in the denominator of the exponential. Therefore, the dynamics will be different than for the case of no applied force; for forces in excess of a critical minimum F_c, the net rate of bond formation dX_t/dt will always be negative.

This can be seen from Fig. 2.17, which shows dX_t/dt as a function of X_t for four different forces, all other parameters being held constant. In all cases (possibly unstable) equilibrium occurs where the curve crosses the horizontal axis. The upper curve represents the case where no force is applied. The three other curves in descending order represent the effects of a subcritical force F_{sub}, the critical force F_c and a supercritical force F_{sup}. For all curves, net bond formation is positive above the horizontal axis and we move to the right along the curve. Net bond formation is negative below the horizontal axis and we move to the left. It can be appreciated that for F_{sup} the curve is completely negative so that no new equilibrium can be reached, i.e. the particle–cell complex separates. The critical force F_c results in a curve that just touches the axis, but otherwise is completely negative, and any force greater than this results in separation. Assuming the association constant $K = 1$ and fixing the density of receptors at $N_1 = 10^3 \ \mu m^{-2}$, Bell calculates a representa-

Rate of Bond Formation vs. Number of Bonds

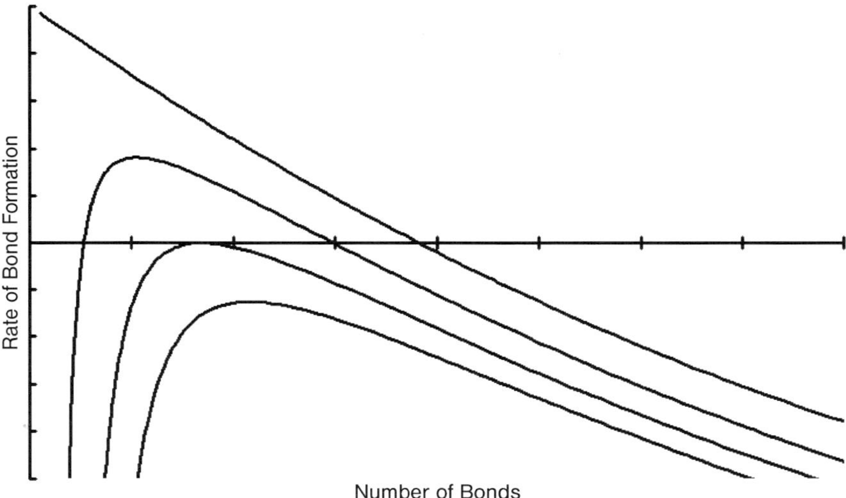

Figure 2.17. The *y*-axis is the right-hand side of Eq. (5) and represents the rate of bond formation under various separation forces. In all cases, when above the axis, the dynamics moves to the right along the curve; otherwise, it moves to the left. Forces represented from top to bottom are: none, a subcritical force F_{sub}, the critical force F_c and a supercritical force F_{sup}.

tive value of $F_c/N_2 \sim 4 \times 10^{-6}$ dynes per total available ligand. In the foregoing we have abused notation in that N_2 is here being used as the absolute number of available ligands in a specific 1-μm^2 patch.

This critical force can be compared to other biologically relevant forces, such as the stress created by a fluid flowing past a bound liposome or nanoparticle. Stokes' law says that for a spherical body of radius r the force generated by a fluid flowing past at velocity v is given by $F = 6\pi\eta r v$, where η is fluid viscosity. For a liposome of radius 4 μm this is calculated to be $5.3 \times 10^{-5}v$ dynes. Using the values of K and N_1 above, and again abusing notation, we set $4 \times 10^{-6} = F_c/N_2 = 5.3 \times 10^{-5}v/N_2$ to yield $N_2 \sim 13v$ potential bonding sites. Thus, the liposome would need only 13 available ligands to resist a fluid velocity of 1 cm s^{-1}.

2.4.2
A Stochastic Model

Stochastic formulations may be more appropriate for receptor–ligand binding, especially when the bonds are few. Again we base our model upon the deterministic equation (5) with the understanding that $N_1 \gg X_t$ at all times. This time, though, we must recognize that X_t is a stochastic process. Most importantly, this means that rather than assuming any particular value at a given time, it is a probability *distribution* of bonds. Even after having attained equilibrium as $t \to \infty$, it becomes a limiting distribution X_∞. In what follows we limit our attention to one square unit of area, since otherwise the values of X_t are not discrete. Nonetheless, the results are completely general. Following Bailey [55] and letting $X_{\Delta t} \equiv X_{t+\Delta t} - X_t$ we have, for $1 \leq X_t \leq N_2 - 1$:

$$P[X_{\Delta t} = j | X_t] = \begin{cases} k^+ N_1 (N_2 - X_t)\Delta t + o(\Delta t) & \text{if } j = 1 \\ k^- \exp(AF/X_t)X_t\Delta t + o(\Delta t) & \text{if } j = -1 \end{cases} \tag{6}$$

Making appropriate adjustments for $X_t = 0$ and $X_t = N_2$, this results in the system of partial differential equations for the discrete probabilities $p_i(t) \equiv P[X_t = i]$:

$$\frac{\partial p_0(t)}{\partial t} = k^- \exp(AF)p_1(t) - k^+ N_1 N_2 p_0(t)$$

$$\frac{\partial p_i(t)}{\partial t} = k^+ N_1 (N_2 - (i-1))p_{i-1}(t) + (i+1)k^- \exp\left(\frac{AF}{i+1}\right)p_{i+1}(t)$$

$$- \left(k^+ N_1(N_2 - i) + ik^- \exp\left(\frac{AF}{i}\right)\right)p_i(t) \quad \text{for } 0 < i < N_2 \tag{7}$$

$$\frac{\partial p_{N_2}(t)}{\partial t} = k^+ N_1 p_{N_2-1}(t) - N_2 k^- \exp\left(\frac{AF}{N_2}\right)p_{N_2}(t)$$

With the further restriction that $\sum p_i = 1$ we can find the distribution of X_∞ by solving for the steady-state solution of the above, yielding:

$$p_i(\infty) = \frac{N_2!}{(N_2 - i)!} \frac{\alpha^i p_0(\infty)}{\prod_{1 \le j \le i} i \exp\left(\frac{AF}{i}\right)}$$

$$p_0(\infty) = \left(\sum_{0 \le i \le N_2} \frac{N_2!}{(N_2 - i)!} \frac{\alpha^i}{\prod_{1 \le j \le i} i \exp\left(\frac{AF}{i}\right)} \right)^{-1} \tag{8}$$

where $\alpha = KN_1$ and $K = k^+/k^-$ is the association constant. In the case of no separating force, the probability generating function $P(\theta) = \sum_{i \ge 0} p_i \theta^i$ is discovered from this to be the Taylor expansion of $(1 + \alpha)^{-N_2}(1 + \alpha\theta)^{N_2}$. A quick check on the mean yields $\overline{X}_\infty = P'(1) = \alpha N_2(1 + \alpha)^{-1}$, in agreement with (2).

Although the variance as well as any chosen moment could also be calculated, it is perhaps of more interest to calculate the probability of complete disengagement of all bonds under various values of F. This means we are interested in p_0 (henceforth understanding this to mean $p_0(\infty)$), which, in the case of no applied force, is simply $(1 + \alpha)^{-N_2}$. For the values of $\alpha = 10^3$ and $N_2 = 13$ used for the example in Section 2.4.1 this yields 9.9×10^{-40}, hardly consequential. However, in contrast to the deterministic analysis, the stochastic analysis demonstrates that even a subcritical force that is little more than *half* of the critical force is sufficient to dislodge the liposome. Consulting Fig. 2.18, we see three curves that result from using the aforementioned values of α and N_2, and three different forces: none, F_{sub} ($\sim 0.5 F_c$) and F_c. (Strictly speaking, these are *scaled* curves of the rate of bond formation (Eq. (5)) because we do not know k^+ or k^- independently. Moreover, we

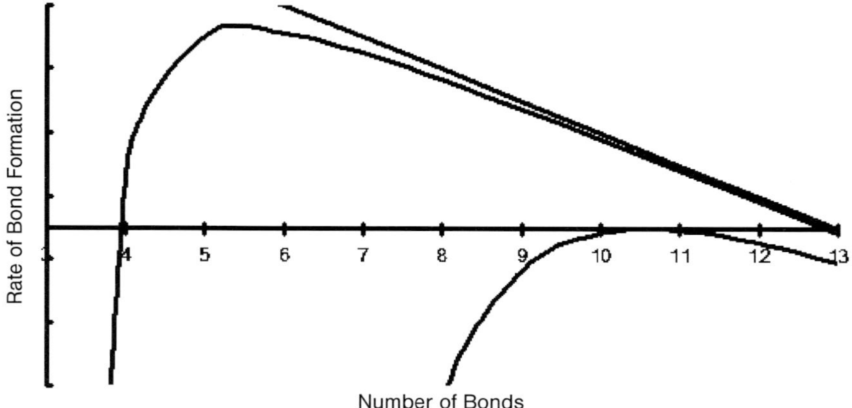

Figure 2.18. The example from Section 2.4.1 reanalyzed ($\alpha = 10^3$ and $N_2 = 13$). The top curve has no separation force applied. The bottom curve results from applying F_c. The middle curve results from applying $F_{\text{sub}} \sim 0.5 F_c$. The deterministic model predicts no separation until F_c is reached; however, the stochastic model predicts the probability of separation for F_{sub} to be greater than 0.9958.

do not quote units of force. Nonetheless, for the relevant computations to find probabilities at equilibrium, k^+ and k^- only enter as a ratio, and AF, as a product. Only one value of this product will result in a critical force curve. The two product values used in the calculations were $AF_c = 57.5$ and $AF_{sub} = 30$.) While it is no surprise to learn that F_c results in $p_0 = 1$ to more than 20 decimals, it may yet be surprising that F_{sub} results in $p_0 > 0.9958$.

The application of both the deterministic and stochastic models to the optimization of receptor–ligand binding is apparent. Subtle differences in the models may become significant in light of the necessity of balancing binding affinity with protein rejection needed to avoid RES uptake (Section 2.2).

2.5
Intratumoral and Cellular Drug Kinetics and Pharmacodynamics

Once nanovectors have successfully evaded the RES, navigated the tortuous topology of tumoral vasculature and extravasated at their intended site, cellular-level drug kinetics and pharmacodynamics determine modeling concerns. Liposomes are too large to penetrate much more than two or three cell layers into tumoral interstitium, while most nanoparticles fair no better. In many cases, lesion tissue is within 100 µm of the nearest vessel and a typical cell diameter is about 10 µm. Thus, for ligand-conjugated vectors, expected to enter into individual cells, penetration into and destruction of three cell layers means that 30% of the tumor can hypothetically be eradicated with one treatment. If, furthermore, as tissue is destroyed, inner layers are next exposed to unspent particles, more cell kill is potentially possible. For "plain" nanovectors, it is not penetration that is the issue, but rather the sustained release of sufficient concentration of drug. This may also be an issue for ligand-conjugated vectors, since some of their charge will be released into tumoral interstitium.

The physics of liposomal and nanoparticle drug release is well researched, with the Higuchi, power law and Weibull models sometimes used as phenomenological approximations. Although these models are not considered to be mechanistic, Kosmidis and coworkers [19] make the case for such an interpretation of the Weibull model, which is a single exponential asymptotically approaching 100% release in time. In fact, nanoparticle release profiles frequently evince a simple bi-exponential release pattern described by $C_t = C_\infty - (Ae^{-\alpha t} + Be^{-\beta t})$, where C_∞ is the total drug, C_t is the amount of drug released by time t, A is the rapidly released portion of drug with rate constant α, and $B = C_\infty - A$ is the slowly released portion of drug with rate constant β [28, 29, 56]. If the release can be sustained long enough, then the bi-exponential becomes approximately linear with release rate $B\beta$.

Even with the above simplification, cellular-level drug kinetics and transport is highly nonuniform not only because of the inhomogeneous transport of vectors through and extravasation from tumoral vasculature discussed in Section 2.3, but also because of drug gradients due to cellular uptake and metabolism. Below, we present models and simulations of chemotherapy [28, 57] that highlight intratumoral and cellular-level drug release and kinetics.

2.5.1
A Two-Dimensional Model of Chemotherapy

In 2004, Sinek and coworkers simulated nanoparticle-mediated chemotherapy
and tumor response using a two-dimensional multiscale tumor simulator due to
Zheng and coworkers [58]. The simulator is built upon continuum-scale reaction–
diffusion equations for its growth component following the previous work of Byrne
and Chaplain [59, 60] and Cristini and coworkers [61], together with a combined
continuum–discrete model of angiogenesis based upon the work of Anderson and
Chaplain [21]. It is capable of tracking cancer progression from its avascular stage,
through the transition from avascular to vascular growth and into the later stages
of invasion of normal tissue. Sinek's group focused on the case of glioblastoma
multiforme by using microphysical parameters characterizing malignant glioma
cells obtained from *in vitro* experiments by Frieboes and Cristini [62] and clinical
data. Glioblastoma is an aggressive brain tumor that may present as the last stage
of astrocytoma progression or *de novo*. It is extremely recalcitrant to all forms of
therapy, whether surgical, genetic, chemical or radiological. This is in part due to
the high motility of glial cells, rendering the tumor highly diffuse on the periphery
[63].

They first simulated the growth of a highly perfused lesion of glioma. This was
then exposed to simulated chemotherapy in which nanoparticles were assumed
to remain at their point of extravasation from the vasculature and function as a
constant source of drug along the vessels. Extravasation, diffusion and cellular uptake of both drug and nutrient was simulated according to the quasi-steady-state
reaction–diffusion equations:

$$
\begin{aligned}
0 &= v_s \delta_{\mathrm{V}} + D_s \nabla^2 s - \eta_s s \\
0 &= v_n \delta_{\mathrm{V}} + D_n \nabla^2 n - \eta_n s
\end{aligned}
\tag{9}
$$

where s and n are the local concentrations of drug and nutrient, respectively, the v's
are (spatially and temporally variable) production rates related to release of drug
and supply of nutrient, the η's are uptake rates by cancer cells, and the D's are diffusion coefficients. δ_{V} is the Dirac delta function indicating the location of the vasculature. Drug action was then modeled as cell kill being proportional to normal-
ized drug concentration $\bar{s}$ acting on the fraction of cycling cells, given by the
normalized nutrient $\bar{n}$. When combined with the growth of cells, modeled as the
product of a mitosis constant and normalized nutrient $\lambda_{\mathrm{M}} \bar{n}$, the net local growth
or regression of tumor cells (the velocity field divergence) becomes:

$$
\nabla \cdot \mathbf{u} = \lambda_{\mathrm{M}} \bar{n} - \lambda_{\mathrm{D}} \overline{sn}
\tag{10}
$$

where λ_{D} is the killing power of the drug.

Sinek's group performed two classes of simulations. The first recognized ex-
travasational difficulties due not only to irregular vascular topology generated
using Anderson and Chaplain's model (Section 2.3.2), but also pressure variations within tumor interstitium. The latter was accomplished by using $v_n =$

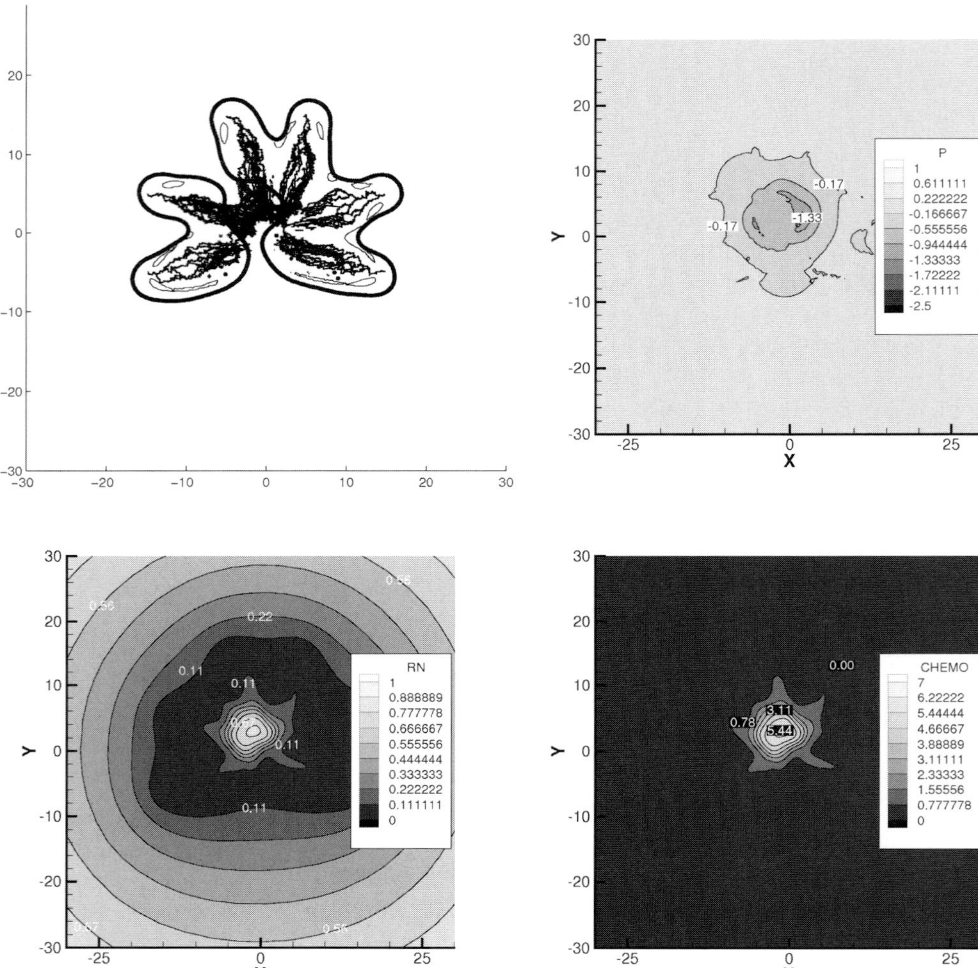

Figure 2.19. Simulations of nanoparticle chemotherapy. Clockwise, from upper left corner: tumor at equilibrium after many months of simulated continuous therapy, pressure contours, drug concentration contours and nutrient contours. Inhomogeneities in nutrient delivery and initial nanoparticle extravasation and subsequent diffusion and uptake of drug result in stable equilibrium at significant tumor mass. (Adapted from Ref. [28], Fig. 4(b), © 2004 Kluwer Academic Publishers. With kind permission from Springer Science and Business Media.)

$v'_n(p_V - p)(n_V - n)$ in Eq. (9), where v'_n is constant, p_V and p are the pressures in the vasculature and tumor, respectively, and n_V and n are the nutrient concentration in the vasculature and tumor, respectively. A similar function for v_s was used for the initial extravasation of particles. This model qualitatively demonstrated that inhomogeneities in drug delivery and action, even using nanoparticles releasing drug at a constant rate, had the potential to diminish chemotherapeutic efficacy, still leaving substantial regions of tumor unharmed after months of simulated therapy (Fig. 2.19).

The second class of simulations addressed the possible benefits of improving drug delivery via the use of adjuvant antiangiogenic drugs to "normalize" tumor vasculature [20, 40, 64]. To do this, Sinek's group let v_s in Eq. (9) be constant so that the source of drug was uniform along the vasculature. Again, although cell kill was significantly greater, inhomogeneities due to drug diffusion and cellular uptake resulted in nonuniform tumor regression, and eventually fragmentation at significant equilibrium masses (Fig. 2.20). These results compare favorably with experimental data. Antiangiogenic and chemotherapeutic treatments have been observed to induce tumor mass fragmentation, cancer cell migration and tissue invasion [65–67]. Cristini and coworkers [68] have termed this behavior "diffusional instability". It is a phenomenon known to materials scientists and is analogous to the processes whereby dendritic ice crystals form or water droplets in ice migrate via a melt-and-freeze process over an imposed temperature gradient. The existence of this phenomenon with regard to tumor response would have significant influence on the design of all modes of chemotherapy.

2.5.2
Refinements of the Model

In Eq. (9), the two parameters D_s and η_s are critical to the penetration of drug into the tumor. One criticism of the previously described simulations is that, although drug gradients formed for the values chosen, these values were not experimentally measured. If the ratio D_s/η_s should actually be relatively high, then drug would be uniformly presented to tumor cells and it is possible that the simulations would have revealed complete tumor regression with no fragmentation. A subsequent work of Sinek and coworkers [57] obtains values of the drug diffusion coefficient D_s, the uptake rate η_s and the killing power λ_D for the drugs, cisplatin and doxorubicin.

In order to more accurately model the cellular drug kinetics of cisplatin and doxorubicin, compartmental models were constructed by Sinek and coworkers [57] following earlier work of Dordal and coworkers [69, 70], Jackson [71] and El-Kareh and Secomb [72]. Three compartments are assumed (Fig. 2.21): Compartment 1 is extracellular interstitium, Compartment 2 is intracellular free drug and Compartment 3 is intracellular DNA-bound drug. For cisplatin, the system of equations determining drug kinetics is:

$$\frac{\partial s_1}{\partial t} = v_s \delta_V + D_s \nabla^2 s_1 - k_{12} s_1 + k_{21} s_2$$

$$\frac{\partial s_2}{\partial t} = k_{12} s_1 - k_{21} s_2 - k_{23} s_2 \tag{11}$$

$$\frac{\partial s_3}{\partial t} = k_{23} s_2 - k_3 s_3$$

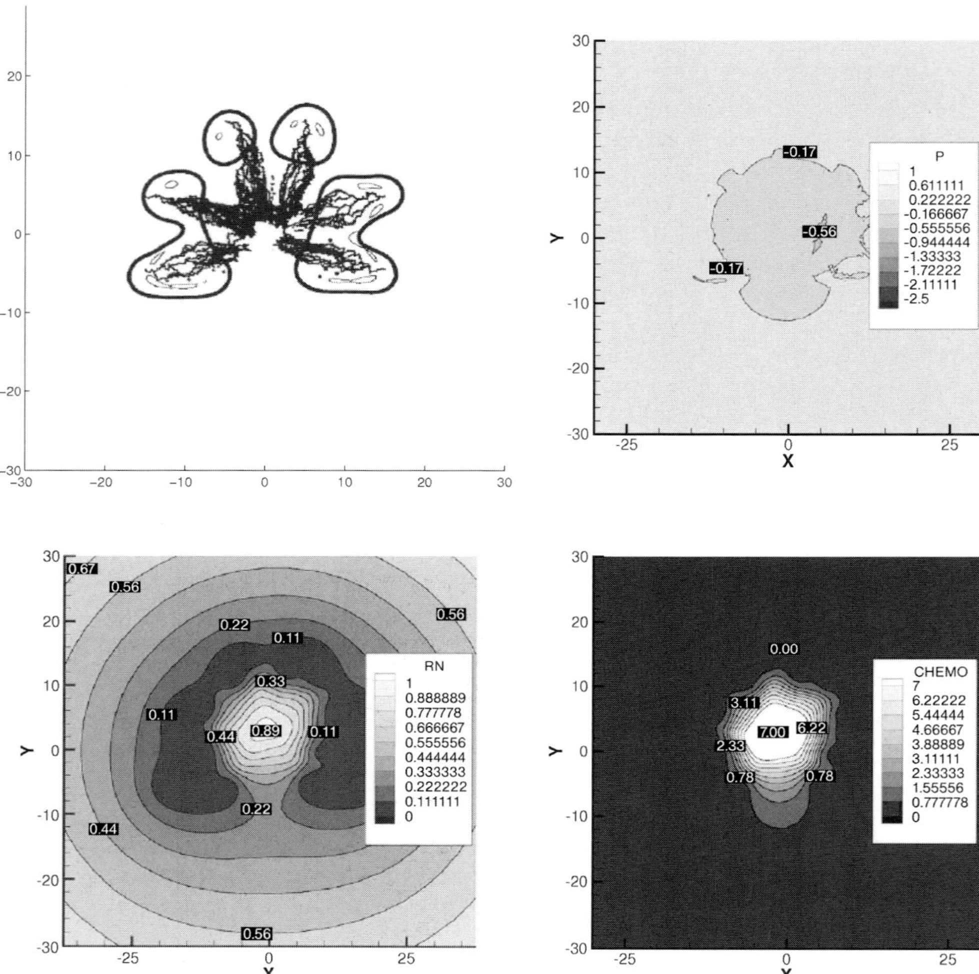

Figure 2.20. Nanoparticle chemotherapy after antiangiogenic "normalization", simulated by making drug release constant along the vasculature. Clockwise, from upper left: tumor at equilibrium after many months of simulated continuous therapy, pressure contours, drug concentration contours and nutrient contours. Although tumor kill is greater than that in Fig. 2.19, diffusional instability (see text) results in fragmentation and aggressive phenotype. (Adapted from Ref. [28], Fig. 5, © 2004 Kluwer Academic Publishers. With kind permission from Springer Science and Business Media.)

where s_i is drug concentration in Compartment i, and the k_{ij}'s are transfer rates from Compartment i to j. In particular, k_{12} is determined by the "inward" cell membrane permeability, k_{21} by the "outward" permeability and k_{23} by the drug–DNA binding affinity. k_3 is a repair rate of DNA platinum adducts. The system for

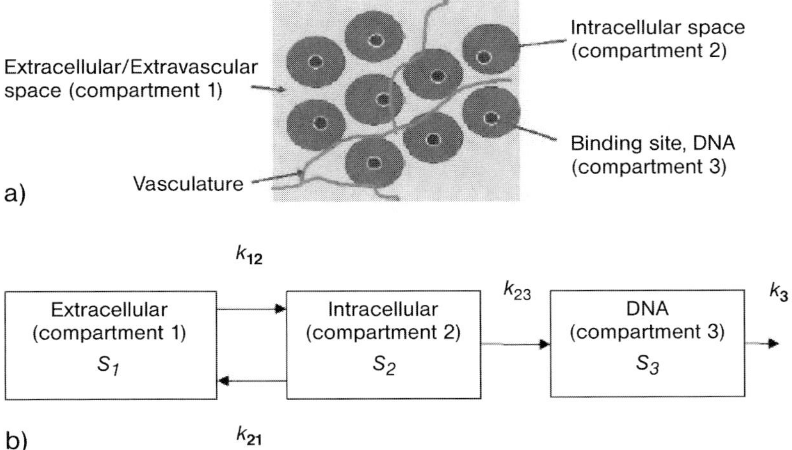

Figure 2.21. The compartment model used by Sinek and coworkers [28]. k's represent transfer rates, while s's represent concentrations. Cell death is ultimately due to s_3.

doxorubicin is similar; for purposes of illustration we shall refer to the cisplatin model.

It is well established that cisplatin induces apoptosis by forming DNA–platinum adducts and that doxorubicin kills cells via DNA intercalation [73, 74]. Thus, of critical importance is the value of s_3. If we reduce the situation to one dimension with a boundary condition at $x = 0$ representing the vessel source, then the steady-state form of Eq. (11) can be written in terms of s_3 alone:

$$0 = D_s(k_{21} + k_{23})\frac{\mathrm{d}^2 s_3}{\mathrm{d}x^2} - k_{12}k_{23}s_3 \tag{12}$$

The solution of this equation is $s_3 = C_1 e^{-\alpha x} + C_{12} e^{\alpha x}$, where:

$$a = \sqrt{\frac{k_{12}}{D_s(k_{21}/k_{23} + 1)}}$$

The constants are determined according to the boundary value of drug at the vessel source along with a no-flux condition at approximately 100 μm from the vessel (since another vessel is assumed to exist approximately 200 μm from the one being modeled). It is the negative exponential in which we are most interested, with high values of a resulting in gradients. Thus, high cell permeabilities (k_{12}) as well as poor interstitial diffusion (D_s) result in gradients and, thus, presumably poor lateral penetration into tumor (Fig. 2.22). This is in line with studies that demonstrate doxorubicin penetration is particularly poor (severe gradients) in comparison with cisplatin penetration. Doxorubicin enters cells much more rapidly than cispla-

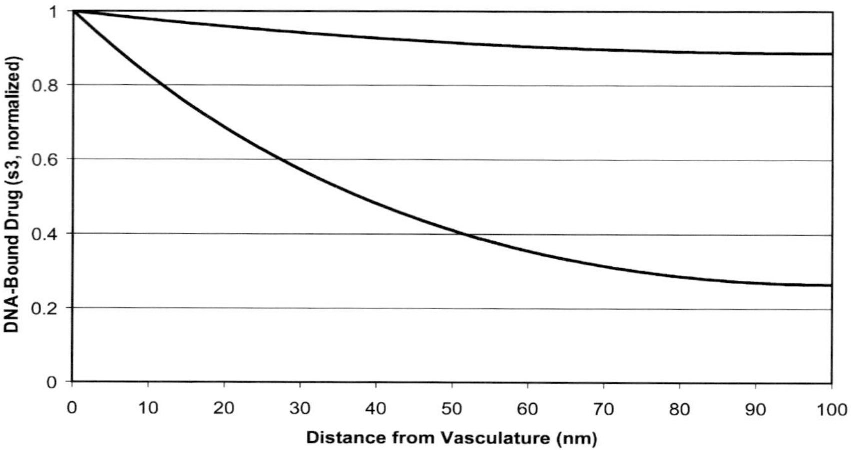

Figure 2.22. The normalized steady-state DNA-bound drug (s_3) as a function of distance from nearest vessel (located at distance $= 0$). Gradients are sensitive to drug–cell permeability. A drug like doxorubicin, which penetrates cells quickly and possesses high affinity for cell endosomes and DNA, would give rise to the lower graph.

tin, and it has particularly high affinities for cell endosomes and DNA [69, 70, 75–79].

The apparent paradox that high cell permeability should result in poor tissue penetration underscores the power of mathematical modeling in overcoming the prejudice of intuition. Upon reflection it can be appreciated that if drug is removed from interstitium quickly due to high cell permeability as it diffuses from its originating vessel, there will be correspondingly less to continue further along, hence low penetration. It should be noted, however, that significant *lateral* penetration at the expense of insignificant *cellular* penetration is also undesirable, but rather an optimal balance of the two must be sought.

Values of the transfer and repair rates in Eq. (11) can be obtained via several techniques including measuring the time uptake of intracellular and bound cisplatin concentrations for cells in monolayer exposed to constant serum drug concentrations. The repair rate k_3 is estimated from two different experiments performed by Sadowitz and coworkers [80], wherein the adducts per million nucleotides fell in 2 h from 75 to 5 in one case and from 185 to 40 in the other. Using the exponential repair model $s'_3 = -k_3 s_3$, k_3 is estimated to be between 0.013 and 0.023 min^{-1}. An initial estimate of k_{23} is made as follows. Sadowitz shows that cells incubated in 7 μM, in 2 h cells accumulate about 25 DNA adducts per million nucleotides. This converts to 1.03×10^{-19} moles of platinum docked on the DNA. Neglecting the cell membrane and supposing the DNA to be exposed directly to the *cis* solution, we solve the ODE $s'_3 = 7k_{23} - k_3 s_3$, substituting 0.015 min^{-1} for k_3. k_{23} is thereby estimated to be 2.6×10^{-16} min^{-1}. Finally, k_{12} and k_{21} are estimated by

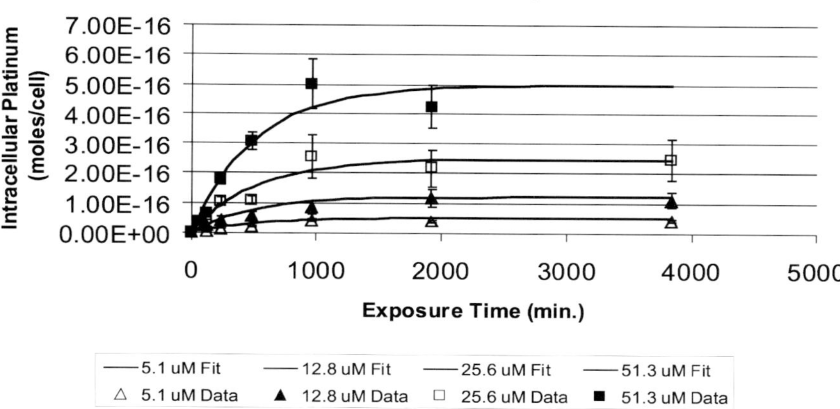

Figure 2.23. Troger and coworkers [81] plotted total intracellular $(s_2 + s_3)$ platinum versus time for cells exposed as monolayers to four different concentrations of cisplatin. Curves represent best fits obtained by adjusting k_{12} and k_{21} in the second two equations of (11). In the equations, s_1 is set to the concentrations that Troger used. (Adapted from Sinek and coworkers [57].)

fitting the second two equations in (11) with intracellular cisplatin time-uptake data of Troger and coworkers [81]. They exposed cells in monolayer to four different concentrations of cisplatin and then measured the total intracellular platinum content at selected times. The graph in Fig. 2.23 shows Troger and coworker's data (symbols) along with the fits (lines) for four different extracellular concentrations. For this purpose, in the equations s_1 was set to be constant and equal to the concentrations that Troger and coworkers used. Then the sum $s_2 + s_3$ was plotted, with values for k_{12} and k_{21} chosen to provide the best fits. Although these parameters are obtained from different experiments using different cell lines, they provide a starting point for producing quantitatively accurate simulations.

2.6
Conclusion

We have identified four critical phases of nanovector performance, each requiring a thorough understanding in order to optimize delivery and anticancer action. In the service of this understanding, mathematics and computation provides its own laboratory of pencil, paper and silicon chip, lending insight, providing guidance and offering testable prediction. The models of protein rejection in Section 2.2, in conjunction with the models of ligand binding in Section 2.4, can facilitate optimized surface preparation, yielding particles and liposomes that reject opsonization and therefore circulate long, yet bind effectively when finally presented to target cells.

The models of tumoral vasculature and hemodynamics presented in Section 2.3 reveal the "nature of the beast", at once starkly exposing the tremendous difficulties in homogenous nanovector delivery and extravasation, while offering hope in quantifying the improvement that proposed therapies, such as vascular normalization, might bring. Finally, multiscale models of nanoparticle chemotherapy in Section 2.5, incorporating vasculogenesis, particle extravasation, nutrient and drug diffusion and uptake, and cellular-level phenomena such as DNA repair, suggest the existence of powerful mechanisms, such as diffusional instability, that can possibly be tempered, or even used to advantage, with the right approach.

Biological systems are "murky", rarely providing simple, solid boundaries. Rather, their behavior is the sum of myriad processes spanning the scales of the whole body, tissues, cells and molecules. The nascent field of mathematical biology in conjunction with nanoscale therapeutics development is therefore presented with great opportunities. New technological designs and paradigms provide exciting grist for the mathematical mill. In turn, advanced mathematical methods along with computational techniques shine a bright and promising light on the future potential of technology. We hope that the foregoing has offered illuminating ideas that will guide this potential to fruition.

References

1 WILKINS, D. J., MYERS, P. A. Studies on the relationship between the electrophoretic properties of colloids and their blood clearance and organ distribution in the rat. *Br. J. Exp. Pathol.* **1966**, *47*, 568–576.

2 WILKINS, D. J. The biological recognition of foreign from native particle as a problem in surface chemistry. *J Colloid Interface Sci.* **1967**, *25*, 84.

3 WILKINS, D. J. Interaction of charged colloids with the RES. In: *The Reticuloendothelial System and Arteriosclerosis*, DI LUZIO, N. R., PAOLETTI, R. (Eds.). New York: Plenum Press, **1967**, p. 25.

4 TRÖSTER, S. D., MULLER, U., KREUTER, J. Modification of the body distribution of poly(methyl methacrylate) nanoparticles in rats by coating with surfactants. *Int. J. Pharm.* **1990**, *61*, 85–100.

5 TRÖSTER, S. D., KREUTER, J. Influence of the surface properties of low contact angle surfactants on the body distribution of 14-C-poly(methyl methacrylate) nanoparticles. *J. Microencapsul.* **1992**, *9*, 19–28.

6 BAZILE, D., PRUD HOMME, C., BASSOULLET, M., MARLARD, M., SPENLEHAUER, G., VEILLARD, M. Stealth Me-PEG–PLA nanoparticles avoid uptake by the mononuclear phagocyte system. *J. Pharm. Sci.* **1995**, *84*, 493–498.

7 ILLUM, L., DAVIS, S. S. Effect of the nonionic surfactant poloxamer 338 on the fate and deposition of polystyrene microspheres following intravenous administration. *J. Pharm. Sci.* **1983**, *72*, 1086–1089.

8 ILLUM, L., DAVIS, S. S. The organ uptake of intravenously administered colloidal particles can be altered using a non-ionic surfactant (poloxamer 338). *FEBS Lett.* **1984**, *167*, 79–82.

9 RUDT, S., MULLER, R. H. *In vitro* phagocytosis assay of nano- and microparticles by chemiluminescence. III. Uptake of differently sized surface-modified particles, and its correlation to particle properties and *in vivo* distribution. *Eur. J. Pharm. Sci.* **1993**, *1*, 31–39.

10 MOGHIMI, S. M., GRAY, T. A. Single dose of IV-injected poloxamine coated

long-circulating particle triggers macrophage clearance of subsequent doses in rats. *Clin. Sci.* **1997**, *93*, 371–379.

11 DEMOY, M., ANDREUX, J. P., WEINGARTEN, C., GOURITIN, B., GUILLOUX, V., COUVREUR, P. Spleen capture of nanoparticles: influence of animal species and surface characteristics. *Pharm. Res.* **1999**, *16*, 37–41.

12 SOPPIMATH, K. S., AMINABHAVI, T. M., KULKARNI, A. R., RUDZINSKI, W. E. Biodegradable polymeric nanoparticles as drug delivery devices. *J. Controlled Rel.* **2001**, *70*, 1–20.

13 SAPRA, P., ALLEN, T. M. Ligand-targeted liposomal anticancer drugs. *Prog. Lipid Res.* **2003**, *42*, 439–462.

14 WANG, S., LEE, R. J., CAUCHON, G., GORENSTEIN, D. G., LOW, P. S. Delivery of antisense oligodeoxyribonucleotides against the human epidermal growth factor receptor into cultured KB cells with liposomes conjugated to folate via polyethylene glycol. *Proc. Natl Acad. Sci. USA* **1995**, *92*, 3318–3322.

15 YOO, H. S., PARK, T. G. Folate-receptor-targeted delivery of doxorubicin nano-aggregates stabilized by doxorubicin–PEG–folate conjugate. *J. Controlled Rel.* **2004**, *100*, 247–256.

16 MATSUMURA, Y., GOTOH, M., MURO, K., YAMADA, Y., SHIRAO, K., SHIMADA, Y., OKUWA, M., MATSUMOTO, S., MIYATA, Y., OHKURA, H., CHIN, K., BABA, S., YAMAO, T., KANNAMI, A., TAKAMATSU, Y., ITO, K., TAKAHASHI, K. Phase I and pharmacokinetic study of MCC-465, a doxorubicin (DXR) encapsulated in PEG immuno-liposome, in patients with metastatic stomach cancer. *Ann. Oncol.* **2004**, *15*, 517.

17 GABIZON, A., SHMEEDA, H., HOROWITZ, A. T., ZALIPSKY, S. Tumor cell targeting of liposome-entrapped drugs with phospholipid-anchored folic acid–PEG conjugates. *Adv. Drug Deliv. Rev.* **2004**, *56*, 1177–1192.

18 JO, Y. S., KIM, M.-C., KIM, D. K., KIM, C.-J., JEONG, Y.-K., KIM, K.-J., MUHAMMED, M. Mathematical modelling on the controlled-release of indomethacin-encapsulated poly(lactic acid-*co*-ethylene oxide) nanospheres. *Nanotechnology* **2004**, *15*, 1186–1194.

19 KOSMIDIS, K., ARGYRAKIS, P., MACHERAS, P. A reappraisal of drug release laws using Monte Carlo simulations: the prevalence of the Weibull function. *Pharmac. Res.* **2003**, *20*, 988–995.

20 JAIN, R. K. Normalizing tumor vasculature with anti-angiogenic therapy: a new paradigm for combination therapy. *Nat. Med.* **2001**, *7*, 987–989.

21 ANDERSON, A., CHAPLAIN, M. Continuous and discrete mathematical models of tumor-induced angiogenesis. *Bull. Math Biol*, **1998**, *60*, 857–900.

22 BAISH, J. W., GAZIT, Y., BERK, D. A., NOZUE, M., BAXTER, L. T., JAIN, R. K. Role of tumor vascular architecture in nutrient and drug delivery: an invasion percolation-based network model. *Microvasc. Res.* **1996**, *51*, 327–346.

23 MCDOUGALL, S., ANDERSON, A., CHAPLAIN, M., SHERRATT, J. Mathematical modeling of flow through vascular networks: implications for tumour-induced angiogenesis and chemotherapy strategies. *Bull. Math. Biol.* **2002**, *64*, 673–702.

24 TORCHILIN, V. P., OMELYANENKO, V. G., PAPISOV, M. I., BOGDANOV, JR., A. A., TRUBETSKOY, V. S., HERRON, J. N., GENTRY, C. A. Poly(ethylene glycol) on the liposome surface: on the mechanism of polymer-coated liposome longevity. *Biochim. Biophys. Acta* **1994**, *1195*, 11–20.

25 BELL, G. I. Models for the specific adhesion of cells to cells. *Science* **1978**, *200*, 618–627.

26 COZENS-ROBERTS, C., QUINN, J. A., LAUFFENBURGER, D. A. Receptor-mediated cell attachment and detachment kinetics. I. Probabilistic model and analysis. *Biophys. J.* **1990**, *58*, 841–856.

27 COZENS-ROBERTS, C., LAUFFENBURGER, D. A., QUINN, J. A. Receptor-mediated cell attachment and detachment kinetics. II. Experimental model studies with the radial-flow detachment assay. *Biophys. J.* **1990**, *58*, 857–872.

28 SINEK, J. P., FRIEBOES, H. B., ZHENG, X., CRISTINI, V. Two-dimensional chemotherapy simulations demonstrate fundamental transport limitations involving nanoparticles. *Biomed. Microdevices* **2004**, *6*, 297–309.

29 KREUTER, J. (Ed.), *Colloidal Drug Delivery Systems*. Dekker, New York, **1994**.

30 KREUTER, J. Evaluation of nanoparticles as drug-delivery systems II: comparison of the body distribution of nanopoarticles with the body distribution of microspheres, liposomes, and emulsions. *Pharm. Acta Helv.* **1983**, *58*, 217.

31 KREUTER, J. Factors influencing the body distribution of polyacrylic nanoparticles. In: *Drug Targeting*, BURI, P., GUMMA, A. (Eds.). Elsevier, Amsterdam, **1985**, p. 51.

32 LASIC, D. D., MARTIN, F. J., GABIZON, A., HUANG, S. K., PAPAHADJOPOULOS, D. Sterically stabilized liposomes: a hypothesis on the molecular origin of the extended circulation times. *Biochim. Biophys. Acta* **1991**, *1070*, 187–192.

33 JEON, S. I., LEE, J. H., ANDRADE, J. D., DE GENNES., P. G. Protein–surface interactions in the presence of polyethylene oxide, I: simplified theory. *J. Colloid Interface Sci.* **1991**, *142*, 149–158.

34 TORCHILIN, V. P., KLIBANOV, A. L., HUANG, L., O'DONNELL, S., NOSSIFT, N. D., KHAW, B. A. Targeted accumulation of polyethylene glycol-coated immunoliposomes in infarcted rabbit myocardium. *FASEB J.* **1992**, *6*, 2716–2719.

35 ALLEN, T. M., HANSEN, C. Pharmacokinetics of stealth versus conventional liposomes: effect of dose. *Biochim. Biophys. Acta* **1991**, *1068*, 133–141.

36 HAROON, Z., PETERS, K. G., GREENBERG, C. S., DEWHIRST, M. W. Angiogenesis and blood flow in the solid tumors. In: *Antiangiogenic Agents in Cancer Therapy*, TEICHER, B. (Ed.).

Humana Press, Totowa, NJ, **1999**, pp. 3–21.

37 JAIN, R. K. Physiological barriers to delivery of monoclonal antibodies and other macromolecules in tumors. *Cancer Res. (Suppl.)* **1990**, *50*, 814s–819s.

38 JAIN, R. K. Determinants of tumor blood flow: a review. *Cancer Res.* **1988**, *49*, 2641–2658.

39 HOBBS, S. K., MONSKY, W. L., YUAN, F., ROBERTS, W. G., GRIFFITH, L., TORCHILIN, V. P., JAIN, R. K. Regulation of transport pathways in tumor vessels: Role of tumor type and microenvironment. *Proc. Natl Acad. Sci. USA* **1998**, *95*, 4607–4612.

40 PADERA, T. P., STOLL, B. R., TOOREDMAN, J. B., CAPEN, D., DI TOMASO, E., JAIN, R. K. Cancer cells compress intratumour vessels. *Nature* **2004**, *427*, 695.

41 SECOMB, T. W., HSU, R. Motion of red blood cells in capillaries with variable cross-sections. *J. Biomech. Eng.* **1996**, *118*, 538–544.

42 BAISH, J. W., NETTI, P. A., JAIN, R. K. Transmural coupling of fluid flow in microcirculatory network and interstitium in tumors. *Microvasc. Res.* **1997**, *53*, 128–141.

43 SCHMID-SCHÖNBEIN, G. W., SKALAK, R., USAMI, S., CHIEN, S. Cell distribution in capillary networks. *Microvasc. Res.* **1980**, *19*, 18–44.

44 GAZIT, Y., BERK, D., LEUNIG, M., BAXTER, L. T., JAIN, R. K. Scale-invariant behavior and vascular network formation in normal and tumor tissue. *Phys. Rev. Lett.* **1995**, *75*, 2428–2431.

45 DEWHIRST, M. W., TSO, C. Y., OLIVER, R., GUSTAFSON, C. S., SECOMB, T. W., GROSS, J. F. Morphologic and hemodynamic comparison of tumor and healing normal tissue microvasculature. *Int. J. Radiat. Oncol. Biol. Phys.* **1989**, *17*, 91–99.

46 LESS, J. R., POSNER, M. C., SKALAK, T. C., WOLMARK, N., JAIN, R. K. Geometric resistance and microvascular network architecture of human colorectal carcinoma. *Microcirculation* **1997**, *4*, 25–33.

47 Leunig, M., Yuan, F., Menger, M. D., Boucher, Y., Goetz, A. E., Messmer, K., Jain, R. K. Angiogenesis, microvascular architecture, microhemodynamics, and interstitial fluid pressure during early growth of human adenocarcinoma LS174T in SCID mice. *Cancer Res.* **1992**, *52*, 6553–6560.

48 Skinner, S. A., Tutton, P. J. M., O'Brien, P. E. Microvascular architecture of experimental colon tumors in the rat. *Cancer Res.* **1990**, *50*, 2411–2417.

49 Gimbrone, M. A., Cotran, R. S., Leapman, S. B., Folkman, J. Tumor growth and neovascularization: an experimental model using the rabbit cornea. *J. Natl Cancer Inst.* **1974**, *52*, 413–427.

50 Ausprunk, D. H., Folkman, J. Migration and proliferation of endothelial cells in preformed and newly formed blood vessels during tumour angiogenesis. *Microvasc. Res.* **1977**, *14*, 53–65.

51 Lee, R. J., Low, P. S. Delivery of liposomes into cultured KB cells via folate receptor-mediated endocytosis. *J. Biol. Chem.* **1994**, *269*, 3198–3204.

52 Turek, J. J., Leamon, C. P., Low, P. S. Endocytosis of folate-protein conjugates: ultrastructural localization in KB cells. *J. Cell Sci.* **1993**, *106*, 423–30.

53 Wang, S., Low, P. S. Folate-mediated targeting of antineoplastic drugs, imaging agents, and nucleic acids to cancer cells. *J. Controlled Rel.* **1998**, *53*, 39–48.

54 Yoo, H. S., Lee, K. H., Oh, J. E., Park, T. G. *In vitro* and *in vivo* antitumor activities of nanoparticles based on doxorubicin-PLGA conjugates. *J. Controlled Rel.* **2000**, *68*, 419–431.

55 Bailey, N. T. J. *The Elements of Stochastic Processes*. Wiley, New York, **1964**.

56 Feng, S. S., Chien, S. Chemotherapeutic engineering: application and further development of chemical engineering principles for chemotherapy of cancer and other diseases. *Chem. Eng. Sci.* **2003**, *58*, 4087–4114.

57 Sinek, J. P. Cristini, V. Modeling and simulation of multi-drug resistance mechanisms. In preparation.

58 Zheng, X., Wise, S. M., Cristini, V. Nonlinear simulation of tumor necrosis, neo-vascularization and tissue invasion via an adaptive finite-element/level-set method. *Bull. Math. Biol.* **2005**, *67*, 211–259.

59 Byrne, H., Chaplain, M. Growth of nonnecrotic tumors in the presence and absence of inhibitors. *Math. Biosci.* **1995**, *130*, 151–181.

60 Byrne, H., Chaplain, M. Growth of necrotic tumors in the presence and absence of inhibitors. *Math. Biosci.* **1996**, *135*, 187–216.

61 Cristini, V., Lowengrub, J., Nie, Q. Nonlinear simulation of tumor growth. *J. Math. Biol.* **2003**, *46*, 191–224.

62 Frieboes, H. B., Zheng, X., Sun, C.-H., Tromberg, B., Gatenby, R., Cristini, V. An integrated computational experimental model of tumor invasion. *Cancer Res.* **2006**, *66*, 1597–1604.

63 Maher, E., Furnari, F., Bachoo, R., Rowitch, D., Louis, D., Cavenee, W., De-Pinho, R. Malignant glioma: genetics and biology of a grave matter. *Genes Dev.* **2001**, *15*, 1311–1333.

64 Jain, R. K. Delivery of molecular medicine to solid tumors: lessons from *in vivo* imaging of gene expression and function. *J. Controlled Rel.* **2001**, *74*, 7–25.

65 Pennacchietti, S., Michieli, P., Galluzzo, M., Mazzone, M., Giordano, S., Comoglio, P. Hypoxia promotes invasive growth by transcriptional activation of the met protooncogene. *Cancer Cell* **2003**, *3*, 347–361.

66 Bello, L., Lucini, V., Costa, F., Pluderi, M., Giussani, C., Acerbi, F., Carrabba, G., Pannacci, M., Caronzolo, D., Grosso, S., Shinkaruk, S., Colleoni, F., Canron, X., Tomei, G., Deleris, G., Bikfalvi, A. Combinatorial administration of molecules that simultaneously inhibit angiogenesis and invasion leads to increased therapeutic efficacy in mouse models of malignant glioma. *Clin. Cancer Res.* **2004**, *10*, 4527–4537.

67 KUNKEL, P., ULBRICHT, U., BOHLEN, P., BROCKMANN, M. A., FILLBRANDT, R., STAVROU, D., WESTPHAL, M., LAMSZUS, K. Inhibition of glioma angiogenesis and growth *in vivo* by systemic treatment with a monoclonal antibody against vascular endothelial growth factor receptor-2. *Cancer Res.* **2001**, *61*, 6624–6628.

68 CRISTINI, V., FRIEBOES, H. B., GATENBY, R., CASERTA, S., FERRARI, M., SINEK, J. P. Morphological instability and cancer invasion. *Clin Cancer Res.* **2005**, *11*, 6772–6779.

69 DORDAL, M. S., WINTER, J. N., ATKINSON JR., A. J. Kinetic analysis of P-glycoprotein-mediated doxorubicin efflux. *J. Pharmacol. Exp. Ther.* **1992**, *263*, 762–766.

70 DORDAL, M. S., HO, A. C., JACKSON-STONE, M., FU, Y. F., GOOLSBY, C. L., WINTER, J. N. Flow cytometric assessment of the cellular pharmacokinetics of fluorescent drugs. *Cytometry* **1995**, *20*, 307–314.

71 JACKSON, T. L. Intracellular accumulation and mechanism of action of doxorubicin in a spatio-temporal tumor model. *J. Theor. Biol.* **2003**, *220*, 201–213.

72 EL-KAREH, A. W., SECOMB, T. W. A mathematical model for cisplatin cellular pharmacodynamics. *Neoplasia* **2003**, *5*, 161–169.

73 JOHNSON, N. P., LAPETOULE, P., RAZAKA, H., VILLANI, G. Biological and biochemical effects of DNA damage caused by platinum compounds. In: *Biochemical Mechanisms of Platinum Antitumour Compounds*, MCBRIEN, D. C. H., SLATER, T. F. (Eds.). IRL Press, Oxford, **1986**, pp. 1–28.

74 ROBERTS, J. J., KNOX, R. J., FRIEDLOS, F., LYDALL, D. A. DNA as the target for the cytotoxic and antitumour action of platinum co-ordination complexes: comparative *in vitro* and *in vivo* studies of cisplatin and carboplatin.

In: *Biochemical Mechanisms of Platinum Antitumour Compounds*, MCBRIEN, D. C. H., SLATER, T. F. (Eds.). IRL Press, Oxford, **1986**, pp. 28–64.

75 PAUL, C., PETERSON, C., GAHRTON, G., LOCKNER, D. Uptake of free and DNA-bound daunorubicin and doxorubicin into human leukemic cells. *Cancer Chemother. Pharmacol.* **1979**, *2*, 49–52.

76 ERLANSON, M., DANIEL-SZOLGAY, E., CARLSSON, J. Relations between the penetration, binding and average concentration of cytostatic drugs in human tumour spheroids. *Cancer Chemother. Pharmacol.* **1992**, *29*, 343–353.

77 TANNOCK, I. F., LEE, C. M., TUNGGAL, J. K., COWAN, D. S. M., EGORIN, M. J. Limited penetration of anticancer drugs through tumor tissue: a potential cause of resistance of solid tumors to chemotherapy. *Clin. Cancer Res.* **2002**, *8*, 878–884.

78 JEKUNEN, A. P., SHALINSKY, D. R., HOM, D. K., ALBRIGHT, K. D., HEATH, D., HOWELL, S. B. Modulation of cis-platin cytotoxicity by permeabilization of the plasma membrane by digitonin *in vitro*. *Biochem. Pharmacol.* **1993**, *45*, 2079–2085.

79 DEMANT, E. J. F., FRICHE, E. Kinetics of anthracycline accumulation in multidrug-resistant tumor cells: relationship to drug lipophilicity and serum albumin binding. *Biochem. Pharmacol.* **1998**, *56*, 1209–1217.

80 SADOWITZ. P. D., HUBBARD, B. A., DABROWIAK, J. C., GOODISMAN, J., TACKA, K. A., AKTAS, M. K., CUNNINGHAM, M. J., DUBOWY, R. L., SOUID, A.-K. Kinetics of cisplatin binding to cellular DNA and modulations by thiol-blocking agents and thiol drugs. *Drug Metab. Dispos.* **2002**, *30*, 183–190.

81 TROGER, V., FISCHEL, J. L., FOMENTO, P., GIOANNI, J., MILANO, G. Effects of prolonged exposure to cisplatin on cytotoxicity and intracellular drug concentration. *Eur. J. Cancer* **1992**, *28*, 82–86.

3

Nanolithography: Towards Fabrication of Nanodevices for Life Sciences

Johnpeter Ndiangui Ngunjiri, Jie-Ren Li, and Jayne Carol Garno

3.1
Introduction: Engineering Surfaces at the Nanoscale

Tools for nanofabrication have begun to provide important contributions for life sciences investigations, for developing biochip and biosensing technologies, as well as supplying basic research in protein–protein interactions and protein function. Scanning probe microscopy (SPM) supplies tools for visualization, physical measurements and precise manipulation of atoms and molecules at the nanometer scale. Nanoscale studies can facilitate the development of new and better approaches for immobilization and bioconjugation chemistries, which are key technologies in manufacturing biochip and biosensing surfaces.

Protein patterning is essential for the integration of biological molecules into miniature bioelectronic and sensing devices. To fabricate nanodevices for the life sciences it is often necessary to attach biomolecules to surfaces with retention of structure and function. For example, controlling the interaction of proteins, biomolecules and cells with surfaces is important for the development of new biocompatible materials. Precisely engineered surfaces can be used for the exploration of biochemical reactions in controlled environments. Spatially well-defined regions of surfaces can be constructed with reactive or adhesive terminal groups for the attachment of biomolecules. Micropatterning of proteins has been applied for biosensors and biochips [1–4]. Direct applications of protein patterning include biosensing, medical implants, control of cell adhesion and growth, and fundamental studies of cell biology [5–7]. Protein patterning has been accomplished at the micrometer level using microcontact printing [8–13], photolithography [14–16] and microfluidic channels [17, 18]. Thus, the capabilities for micrometer-scale methods for controlling the spatial arrangements of biomolecules have been well established and offer valuable new research methodologies for life sciences investigations. Collectively, these techniques provide a means for assembling proteins at a size scale of hundreds of nanometers or larger.

To progress to even smaller sizes, atomic force microscopy (AFM)-based lithography can be applied to pattern surfaces at nanometer dimensions. Scanning probe lithography (SPL) provides versatile approaches for designing the chemistry of sur-

Nanotechnologies for the Life Sciences Vol. 4
Nanodevices for the Life Sciences. Edited by Challa S. S. R. Kumar
Copyright © 2006 WILEY-VCH Verlag GmbH & Co. KGaA, Weinheim
ISBN: 3-527-31384-2

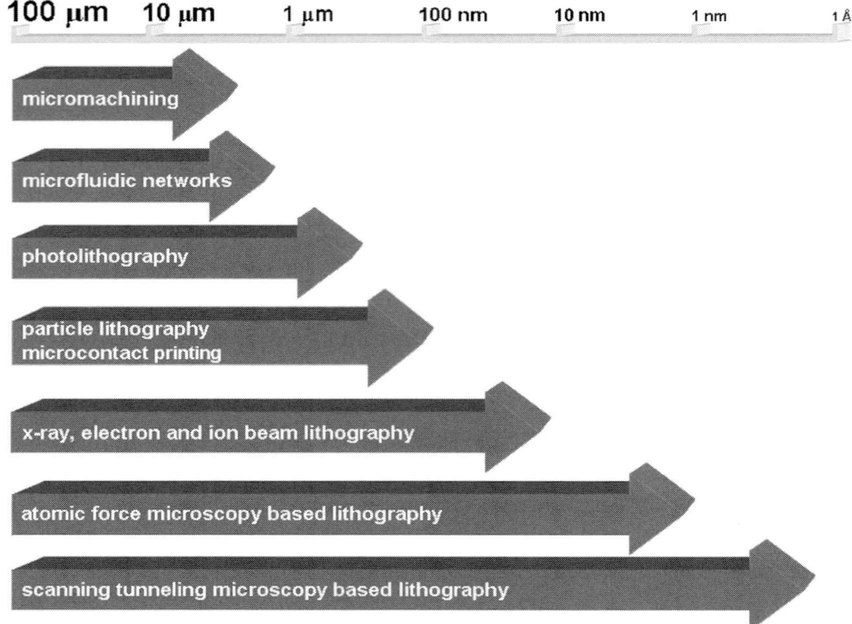

Figure 3.1. Overview of the hierarchy of dimensions which can be achieved using various micro- and nanopatterning methods.

faces at the nanoscale. Figure 3.1 shows the dimensions which can be achieved using various micro- and nanopatterning methods. Arrays of self-assembled monolayers (SAMs) and proteins can be fabricated via SPL, with precise control over chemical functionality, shape, dimensions and spacing on the nanometer scale. Combined with the capabilities for high-resolution imaging and characterization, SPM enables a molecular-level approach for directly investigating changes that occur on surfaces during biochemical reactions. The tools of SPL are accessible to investigators across a broad range of disciplines and do not require costly instrument modifications.

Cutting-edge research has begun to apply nanolithography for studying proteins on surfaces, possibly at the level of single-molecule detection. At present, nanodevices constructed by SPM-based lithography are being conceptualized and, to the best of our knowledge, SPL has not yet been applied for making nanodevices. Readers are referred to recent reviews which discuss potential nanoscale devices [19, 20]. Although there are also many studies which investigate peptides, DNA and cells, we limit the focus of this chapter to studies which apply nanoscale lithography to proteins and to applications using SPL for nanoscale protein assays. This chapter provides an overview of advances in the application of nanolithography using SPM and latex particle lithography for protein patterning. Beginning with a general introduction of the chemistry for immobilization of proteins on surfaces, the application of SAMs for coupling proteins to surfaces is presented. Nano-lithography methods including bias-induced lithography, AFM-based force-induced nanolithography, "dip-pen" nanolithography (DPN) and latex particle lithography are described, including examples of protein nanopatterning. The chapter concludes with a discussion of nanoscale detection of protein binding and future directions in cantilever array technology.

3.2
Immobilization of Biomolecules for Surface Assays

A number of factors need to be considered for choosing a successful protein-immobilization strategy, such as the efficiency and rate of binding, potential side-reactions, and the strength and resilience of the attachment. For protein assays, the binding site recognized by immunoglobulin G (IgG) on an antigen is relatively small; consisting of only 5–6 amino acids or several sugar residues. The recognition element is referred to as an antigenic determinant or epitope. Proteins must be attached in such an orientation that their active sites or binding domains are accessible for binding and not buried or blocked by the surface. The binding site is only a small part of the total surface area of the protein. Adsorption on a surface may impair or prevent the protein's activity. The eventual orientation of proteins on surfaces is determined by multiple factors such as the type of binding, the positions and composition of external residues on the protein surface, the isoelectric point of the protein, and the pH of buffers used during application.

Proteins have a three-dimensional (3-D) structure which is critical to their function and activity. Most proteins have both positively and negatively charged regions that interact with surfaces. Upon encountering a surface, intramolecular forces within proteins can be disrupted, causing the proteins to unfold and become denatured. Some proteins are known to lose activity when bound to a solid surface, due to a loss of tertiary structure. For example, the strong polarization forces at metal surfaces along with ionic or covalent interactions on many inorganic metal oxides and semiconductor surfaces may cause denaturation of biomolecules [21]. For retention of activity, chemistries for protein arrays should permit the immobilization of proteins on surfaces such that perturbation to the native 3-D structure is minimized. Using a spacer or linker molecule on the sensor surface often enables biomolecules to retain their functionality and 3-D structure. The tools of organic chemistry provide a wealth of chemical strategies and binding motifs for conjugating biomolecules such as proteins to solid surfaces [22, 23].

3.2.1
Strategies for Linking Proteins to Surfaces

Increasingly, researchers have begun to use the self-assembly of functionalized alkanethiol and alkylsilane molecules as model surfaces for protein binding. The terminal moieties of SAM surfaces mediate the type of binding, such as through

Figure 3.2. Strategies for linking proteins to surfaces include electrostatic interactions, covalent bonding, antigen–antibody recognition and biotin–streptavidin specific interactions.

electrostatic interactions, covalent binding, molecular recognition or via specific interactions (Fig. 3.2). The following sections introduce representative examples of chemical immobilization strategies which have been applied for protein patterning.

3.2.1.1 Electrostatic Immobilization

The strategy of functionalizing a surface through electrostatic assembly is often used to immobilize biomolecules on surfaces. Electrically charged amino acids are found mostly on the exterior of proteins and can mediate assembly on charged surfaces. Proteins contain both positively and negatively charged domains that interact with surfaces via long-range electrostatic forces. The electrostatic attraction between oppositely charged molecules is nonspecific and surfaces are negatively or positively charged, depending on the solution pH.

Electrostatic binding is physically mediated and proteins often retain their activity after immobilization. It is a direct, simple method for attaching proteins to surfaces without requiring multiple steps for chemical activation. Binding is reversible, since certain buffers and detergents can remove proteins from surfaces. However, a potential disadvantage of electrostatic immobilization is that the resulting orientation of proteins on surfaces is random; electrostatic-mediated binding does not provide a means for directing the protein assembly in a designed conformation. Representative examples of chemistries for the electrostatic immobilization of proteins which have been applied for nanopatterning proteins are summarized in Tab. 3.1. For example, alkanethiols or alkylsilanes terminated with functional groups, such as NH_2 or $COOH$, have been used to immobilize biomolecules through electrostatic interactions.

3.2.1.2 Covalent Immobilization

Covalent immobilization is important for applications in which displacement or desorption of proteins can be a problem. Covalent bonds occur when two molecules share atoms and form the strongest chemical bonds for surface immobilization. The methods of covalent attachment are boundless – thousands of proteins have been immobilized on hundreds of different solid supports for affinity-capture assays [22]. The best choice for covalent immobilization will depend on the functionalities of both the protein and the surface. Several amino acids provide suitable functional groups for covalent modification. Common functional groups of amino acids used for covalent immobilization include: amino groups from the side-chains of lysine and the N-terminus; carboxyl groups from the C-terminus, aspartic and glutamic acids; sulfhydryl groups of cysteine; hydroxyl groups of serine and threonine; and the phenyl groups of phenylalanine and tyrosine. Since proteins typically present a number of these groups, the chemical nature of the solid surface becomes a primary consideration. A specific chemical reaction is chosen to activate the surface and then proteins are immobilized upon exposure to the active surface groups. Examples of chemistries for covalent immobilization of proteins include activation of surface hydroxyl groups, carboxyl groups and amines. Also, bifunctional crosslinking reagents such as glutaraldehyde have been used to covalently couple proteins to various surfaces. Further examples of covalent immobilization chemistries are listed in Tab. 3.1.

An important factor to be considered in covalent attachment of proteins is the possibility of chemically altering the protein in such a way that its reactivity is reduced. For example, covalent approaches may be hindered by competing side-reactions. It is possible that groups associated with the active site or binding site of a protein could be involved in the reaction. In addition, chemical crosslinking within protein domains could occur, causing damage to the protein's tertiary structure.

3.2.1.3 Molecular Recognition and Specific Interactions

Highly specific interactions between binding pairs can be used effectively for protein immobilization. Examples include affinity capture ligands such as biotin–streptavidin binding and molecular recognition through antigen–antibody recognition. Such affinity ligands require either physical or covalent immobilization of one moiety of the affinity pair onto the surface. Small-molecule receptors such as biotin offer viable strategies for the immobilization of proteins. Further examples are listed in Tab. 3.1. A strong advantage of specific immobilization is to provide a means for directing the protein assembly in a designed conformation. The orientation of proteins on surfaces can be designated by selectively targeting certain amino acid residues of the protein for specific coupling.

3.2.1.4 Nonspecific Physical Adsorption to Surfaces

By far the most widely used method of protein immobilization for protein arrays uses nonspecific adsorption of proteins dried on solid supports. Forces which nonspecifically influence the binding of proteins to almost any substrate include ion bridging, hydration forces, hydrophobic forces and short-range attractive or repulsive forces. This approach produces randomly oriented proteins, some of which may be denatured. Surface assays typically include a blocking step, such as with the adsorption of bovine serum albumin (BSA) to prevent nonspecific binding of

Tab. 3.1 Strategies used for biomolecule immobilization applied for nanopatterning proteins.

Type of interaction	Surface derivatization	Proteins studied	Surface	SPL method	Dimensions	Reference
Chemisorption	S-Au attachment of C-terminal thiol groups	bundle metalloproteins	Au(111)	nanografting	100 nm	80
Chemisorption	gold surface	thiolated collagen	Au(111)	DPN	30–50 nm	92
Covalent	3-mercapto-1-propanal patterns in a decanethiol resist	lysozyme, IgG	Au(111)	nanografting	40–350 nm	74
Covalent	MHA passivated with EG$_3$-SH, then activated to form aldehyde groups	elastin-like polypeptide	Au(111)	DPN	200 nm dots	91
Covalent	1,2-diols cleaved to produce aldehydes	acetylcholine esterase–insulin	Au(111)	nanografting	50–200 nm	79
Covalent and specific	EDC activation of mixed hydroxyl and carboxyl SAMs, then biotin–streptavidin binding	anti-IgG, protein G	Au(111)	uCP	10 µm	10
Electrostatic	MHA SAM decanethiol resist	lysozyme	Au(111)	nanografting	100–400 nm	74
Electrostatic	MHA and dodecanethiol SAMs	lysozyme	Au(111)	natural assembly	<1 µm	137
Electrostatic	MHA passivated with ethylene glycol SAM	rabbit IgG, lysozyme	Au(111)	DPN	100–350 nm	86
Electrostatic	MHA	mouse anti-p24 IgG HIV-1 p24 antigen	Au(111)	DPN	60 nm	108
Electrostatic	gold surface	cytochrome *c*	Au(111)	DPN	200 nm dots	87

Electrostatic	mercaptoundecanoic acid passivated with octanethiol and glycol SAMs	rabbit IgG anti-rabbit IgG	Au(111)	nanografting	500 nm–1 μm	78
Electrostatic	mercaptohexanol, mercaptopropionic acid, N-(mercapto)hexylpyridinimum bromide thiols in matrix terminated with hexa(ethylene glycol) resist	lysozyme, bovine carbonic anhydrase, rabbit IgG	Au(111)	nanografting	200–400 nm	77
Electrostatic and specific	biotin–streptavidin on MHA, with oligoethylene glycol SAM passivation	biotinylated BSA	Au(111)	DPN	100–230 nm	89
Physical adsorption	PDMS stamping onto glass surfaces	rabbit IgGs, BSA, avidin	glass	uCP	40–100 nm	138
Physical adsorption	nickel oxide surface	ubiquitin and thioredoxin	nickel oxide	DPN	80 nm	95
Physical adsorption	direct writing on bare gold; passivation with PEG	lysozyme, rabbit IgG	Au(111)	DPN	45–200 nm	94
Physical adsorption	direct writing on modified SiO_2	IgG, anti-rabbit IgG	SiO_2	DPN	55–550 nm	93
Specific avidin–biotin	oligo-(ethylene glycol) SAM	avidin–biotin–BSA	Si(111)	bias-induced SPL	90 nm	66
Specific avidin–biotin	oxidized regions of PMMA layer spin-coated onto p-doped silicon wafer	biotin IgG–avidin	Si wafer	bias-induced SPL	0.5–1.5 μm	61
Specific maleimide–cysteine	maleimide substituted SAM as ink for specific immobilization of cysteine-labeled biomolecules	virus capsid particles	Au(111)	DPN	150 nm	90

proteins. BSA is a globular serum protein which is often used in bioassays to back-fill uncovered areas of surfaces where proteins did not attach.

3.2.2
SAM Chemistry

SAMs provide a chemical method for creating well-defined surfaces with controllable surface functionality [24]. Due to their stability, ease of preparation and well-ordered surface structures, SAMs of alkanethiols and alkylsilanes provide excellent models for studying protein binding, since layers of defined thickness and designed properties can be generated [25, 26]. Thiol endgroups of *n*-alkanethiols bond via chemisorption to metal surfaces. SAM surface properties can be flexibly controlled by changing the functional (head) groups of the alkyl chain (Fig. 3.3) and these end groups can also be used for further chemical reactions. The acidity, adhesion, wetting and structural properties of surfaces can be modified by choosing specific chemical headgroups (such as NH_2, OH, COOH, CH_3, glycol, etc.) [27, 28]. For example, surfaces can be made hydrophilic by introducing SAMs with polar moieties such as hydroxyl or carboxyl groups. Nonpolar functionalities such as methyl-terminated groups yield hydrophobic surfaces. The preparation, characterization and properties of SAMs have been described and reviewed previously [28–32]. SAMs have promising applications in biosensing, corrosion inhibition, lubrication, surface modification and molecular device fabrication. This section introduces the chemistry and structure of SAMs of alkanethiols and alkylsilanes (Fig. 3.3), which are often applied for nanolithography with proteins.

Close-packed *n*-alkanethiol SAMs can be readily prepared with high reproducibility to present functional groups such as alkyls, amides, esters, alcohols, etc., on

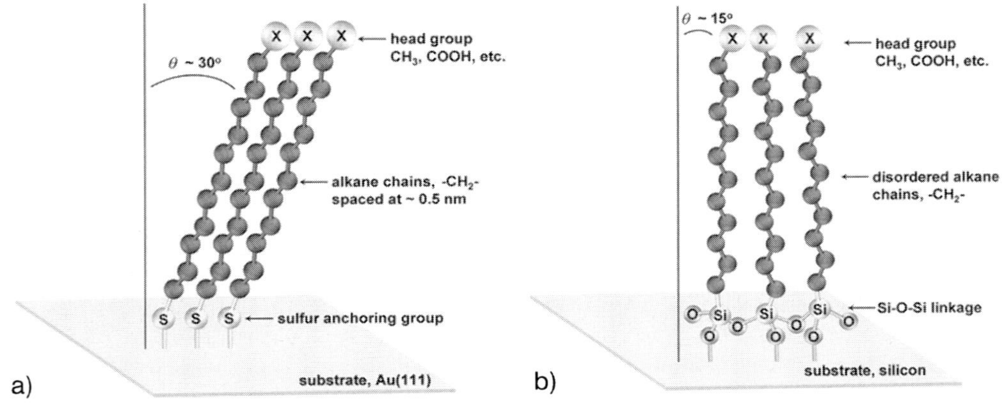

Figure 3.3. General structural features of self-assembled monolayers of (A) *n*-alkanethiols/Au(111) and (B) *n*-alkylsilanes/Si(111).

surfaces of gold or coinage metals. Typically, alkanethiol SAMs are formed by soaking gold thin films in dilute (0.1–1.0 mM) solutions of thiols dissolved in solvents such as 2-butanol, hexane or ethanol. Typically, substrates can be stored in a thiol solution for 1–7 days at room temperature to ensure the formation of mature monolayers. Alkanethiols on Au(111) form a close packed, commensurate ($\sqrt{3} \times \sqrt{3}$)R30° lattice on Au(111) surfaces [33–36]. In surface assemblies of alkanethiol SAMs, according to studies by IR, near-edge X-ray absorption fine structure (NEXAFS) spectroscopy and grazing incidence X-ray diffraction (GIXD), the alkyl chains of thiol molecules are tilted approximately 30° from surface normal (Fig. 3.3A) [36–38]. The sulfur atoms of alkanethiol molecules are considered to bind at the triple hollow sites of Au(111) lattices [29].

STM and AFM studies have confirmed the long-range order and periodicity of alkanethiol monolayers, and have provided a direct view of defects such as domain boundaries, etch pits, steps and dislocations within SAM films [29, 39]. SPM images visualize the intricate details of the surface topography of SAMs. Figure 3.4 displays a typical topographic view of an octadecanethiol SAM/Au(111) acquired in ethanol by AFM. These molecular landscapes may appear somewhat rough, because at the atomic scale most surfaces are not truly smooth and flat, and contain defects. Considering that the height of gold steps is 0.25 nm, the overall surface roughness of the underlying gold substrates for these images is less than 1 nm. The monolayer surfaces consist of domains of closely packed thiol molecules decorated with etch pits. Readers are referred to several works using STM for a more detailed discussion of the morphology and packing of *n*-alkanethiol SAMs [29, 39–41].

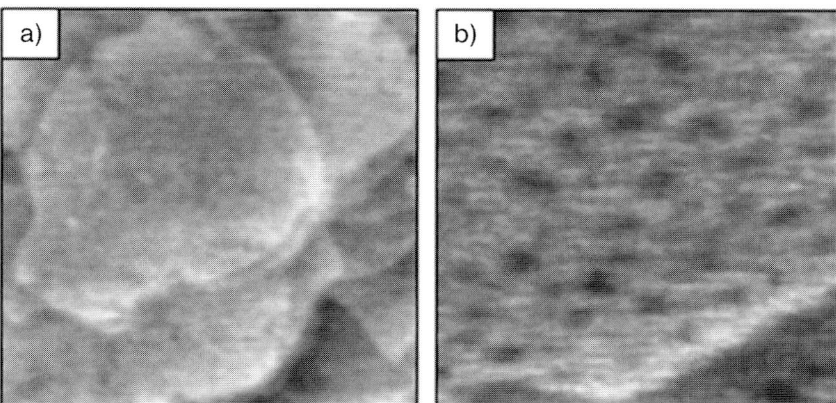

Figure 3.4. Contact-mode AFM topographs of an octadecanethiol self-assembled monolayer on Au(111). (A) Terrace arrangement of flat gold steps coated with octadecanethiol SAM (400 × 400 nm²). (B) Zoom-in view displays etch pits (80 × 80 nm²).

Similar to alkanethiols, the chain length and terminal moieties of alkylsilane SAMs can be tailored to meet experimental requirements; however, the properties of alkylsilane assemblies are quite different from SAMs of alkanethiols. SAMs of alkylchlorosilanes, alkylalkoxysilanes and alkylaminosilanes require hydroxylated surfaces to form polysiloxane, which is connected to surface silanol groups (–SiOH) through a network of Si–O–Si bonds. Substrates on which silane SAMs have been prepared include silicon oxide, aluminum oxide, quartz, glass, mica, zinc selenide and germanium oxide [32]. High-quality alkylsilane SAMs are not as simple to produce as thiol SAMs, because of the need to carefully control the presence of water in solutions. Reproducibility can be a problem, since the quality of the monolayers formed is very sensitive to reaction conditions. Silane monolayers on mica typically consist of domains separated by boundaries. Within domains, silane molecules form structures without long-range order or periodicity [42–44]. The headgroups of silane SAMs crosslink into a Si–O network and the chains are estimated to tilt 15° from surface normal (Fig. 3.3B) [42, 43].

Mixed SAM surfaces can be engineered to avoid nonspecific protein adsorption, yet make specific interactions with targeted proteins to be assayed, by choosing the appropriate buffered conditions as well as an effective matrix layer, resistive to protein adsorption (such as glycol-terminated SAMs). Very few surfaces resist protein adsorption and it remains a challenge to understand the mechanisms that contribute to protein resistance or adhesion to surfaces. To prepare monolayers that resist protein adsorption, the groups of Whitesides [45–48], Mrksich [49] and Grunze [50] have conducted systematic studies of functionalized SAMs to determine the molecular characteristics that impart resistance to protein adsorption. The factors that determine the resistance to protein adsorption were found to include characteristics such as the hydrophilicity of the terminal group, lateral packing density, the presence of hydrogen bond accepting groups and the absence of hydrogen bond donor groups, and terminal groups with overall electrical neutrality.

Approaches which use chemical methods for the activation of SAM surfaces are beginning to gain importance for the surface coupling of biomolecules. Thus far, most reactions for the surface activation of SAMs for protein adsorption have been accomplished after the SAM has been formed with monolayers terminated with carboxyl, amino or hydroxyl groups. Hundreds of synthetic pathways can be applied for *in situ* activation chemistry, including reagents such as *N*-hydroxysuccinimide (NHS), 1-ethyl-3-(3-dimethylaminopropyl)carbodiimide hydrochloride (EDC) and dithiobis(succinimidyl undecanoate) (DSU) [10, 51].

3.3
Methods for Nanolithography with Proteins

With the invention and continuing development of scanning probe techniques, such as scanning tunneling microscopy (STM) [52] and AFM [53], surface changes became evident when too much force was applied by an AFM tip, or if the applied bias voltages exceeded certain thresholds using STM or conductive AFM imaging.

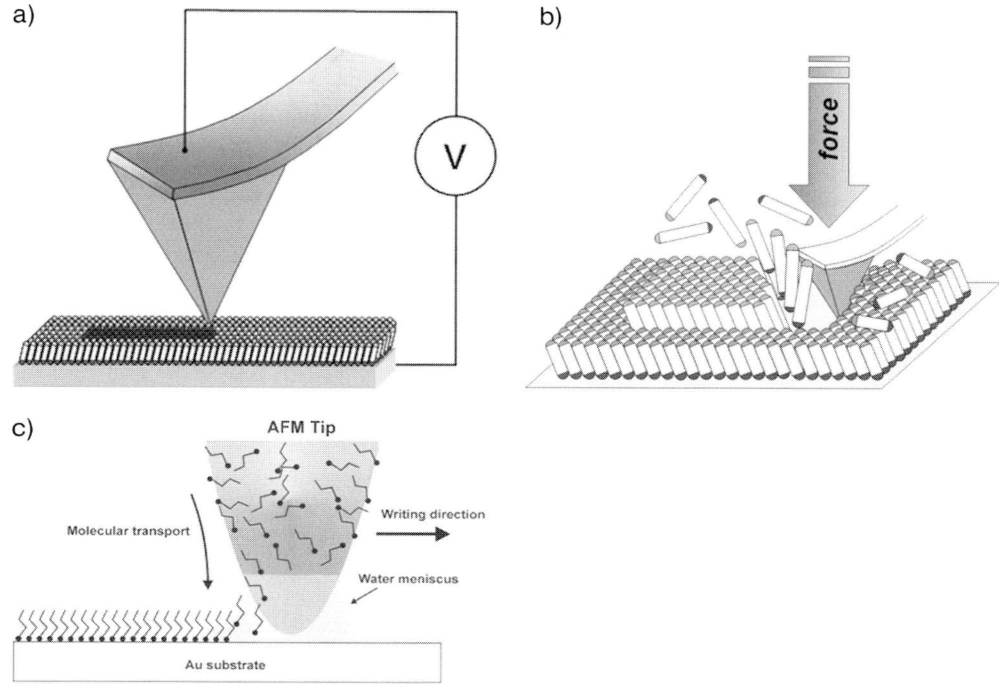

Figure 3.5. Schematic representations of three AFM-based nanofabrication techniques: (A) bias-induced lithography, (B) force-induced nanolithography (nanografting) and (C) DPN. (Panel C reproduced with permission from Ref. [56].)

Researchers began experimenting to deliberately and selectively control these alterations. Molecules of SAMs can be written precisely on surfaces using a variety of different SPL methods. Figure 3.5 illustrates the fabrication principles of the three most predominant SPL methods applied for patterning proteins. SPL provides flexible and convenient approaches to construct SAM nanopatterns with designated functionalities in selected nanosized areas. These nanoengineered surfaces can then be used to selectively immobilize desired proteins through covalent, electrostatic or specific recognition approaches.

A common feature of all SPL methods is that an SPM tip is used as a tool for both nanofabrication and characterization of surfaces. A helpful analogy for describing SPL methods with SAMs is an SPM tip (pen) which writes with molecules (ink) on various surfaces (paper). SPL provides exquisite control of surface chemistry including parameters such as the spatial arrangement, chemical composition and the written density of molecular ligands. The shape and dimensions of the tip dictate the detailed resolution of written nanostructures – SAM patterns as small as 5 nm have been reported and it has become routine to achieve patterns of 20–50 nm (or larger). Since the dimensions of proteins range from tens to hundreds of nanometers, SPL methods are ideally suited for surface studies of protein binding. Particle lithography is another promising method for protein nanopatterning, which can produce arrays of protein nanostructures. Table 3.2 provides a comparison of approaches which have been successfully applied for protein patterning. The next sections of this chapter present further details of these nanolithography methods, including nanografting [54], bias-induced lithographies [55], DPN [56] and latex particle lithography [57].

As a tool for high-resolution characterization, the same SPM tips used to write nanopatterns on surfaces are also used to explore the morphology of nanopatterns after protein adsorption. Both AFM and STM are highly suitable, well-established methods for visualizing surfaces with high resolution. The new tools of AFM and STM have emerged as significant and powerful techniques for imaging surfaces at the molecular scale. Unlike electron microscopy methods which require high vacuum environments and conductive coating of specimens, *in situ* AFM/STM experiments can be accomplished under physiological conditions in aqueous buffered environments. Experiments using SPM provides exquisite resolution for the detailed characterization of molecular structures, has versatility in imaging modes and can be used for local modification of surfaces by lithography. Topographic images provide direct visualization of changes on surfaces after proteins bind to nanopatterns. Commercial advances continue to improve SPM resolution by providing consistently higher-quality probes at lower cost and by the ongoing development of imaging modes for viewing chemical contrast differences for surfaces. Various AFM imaging modes can be applied for probing friction, softness, surface charge, polarizability, magnetic domains and viscoelasticity at the atomic scale.

3.3.1
Bias-induced Nanolithography of SAMs

When an electric field is applied at elevated bias voltages between a conductive SPM tip and sample, local chemical or physical changes occur in the area under the tip. Depending on the nature of the surface and environmental conditions (ambient versus UHV), the "bias-induced" changes may result from electrochemistry (oxidation) at either the tip or sample which occurs from electric field effects [58], or the changes may result from ohmic heating, which induces evaporation or desorption of organic layers [59, 60]. This section describes the method of bias-induced lithography and then presents an example of bias-induced nanofabrication applied for protein nanopatterning.

Figure 3.5(A) displays the general principle of bias-induced lithography. For bias-induced SPL, short (microsecond to millisecond) pulses of bias voltage are applied between a conductive SPM tip placed very near, but not in contact with, the surface. The size of the surface features are determined by the duration and magnitude of the electric field, and also by the dimensions of the area probed by the SPM tip. Often, with bias-induced oxidation, the chemical changes produced by an electric field do not manifest height changes and thus are not detectable by top-

Tab. 3.2 Comparison of methods applied for nanopatterning proteins.

Nanopatterning method	Pen	Paper	Mechanism	Surface chemistry	Reference
Bias-induced oxidation	biased tip STM/AFM	conductive or semiconductive substrate	surface oxidation	oxidization of surfaces or SAM terminal groups	66
Bias-induced replacement lithography	biased tip STM/AFM in solution	conductive or semiconductive substrate with SAM	displacement of SAMs under elevated bias	replace matrix SAM with new molecules	55, 62
Nanografting	bare AFM tip in a SAM solution	thiol SAMs	force and solution replacement	replace matrix with diverse functional groups of new SAMs	75, 80
DPN	ink-coated AFM tip in air	clean, uncoated surface	meniscus liquid transfer	write diverse functional groups of SAMs and other nanomaterials	86
Latex particle lithography	(not an SPL-based method)	mica(0001)/Au(111)	physical adsorption	self-ordering of monodisperse spheres as a structural template	57

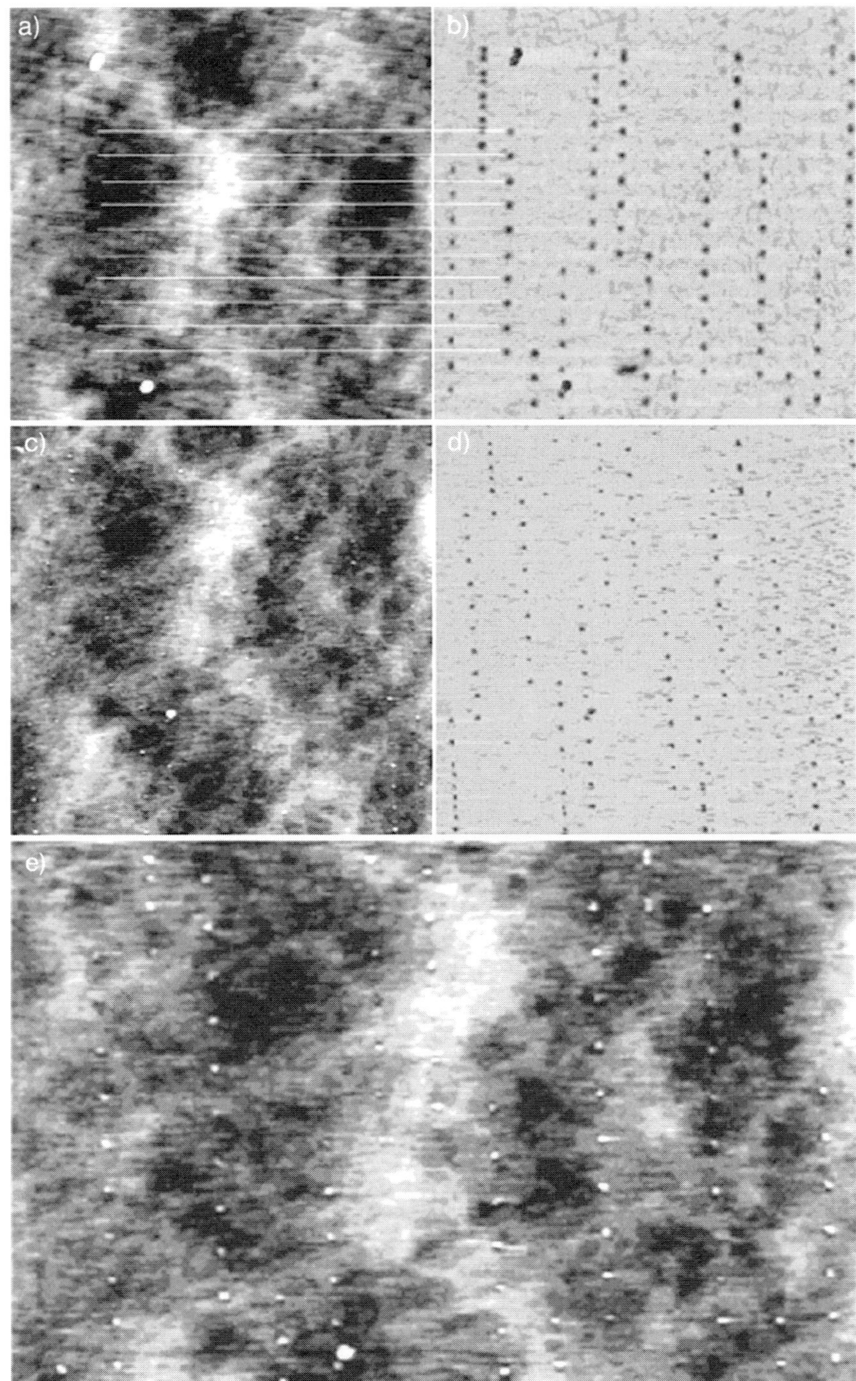

ographic imaging. However, SPM imaging modes which display contrast between different terminal groups, such as force modulation, lateral force imaging and current/electrical force images, can clearly differentiate areas that are modified based on chemical changes.

Bias-induced lithography is emerging as a flexible and convenient means for nanofabrication of designed surface components, using either silane or thiol SAMs. Requirements for bias-induced nanofabrication include a conductive or semiconductive substrate and a conductive SPM probe. To prepare conductive AFM tips, a thin film of metal (usually gold) is sputter-coated onto the surface of probes precoated with a precursor binding layer of chromium or titanium. Conductive tips and cantilevers comprised of doped silicon exhibit sufficient electrical conductivity for bias-induced modification of surfaces without requiring metal coatings. Bias-induced SPL methods are now accessible techniques for most SPM users, as a result of improvements in instruments, and in the quality, cost and availability of commercial AFM probes, which now include coatings of cobalt, diamond-like carbon, doped diamond, platinum, platinum/iridium, tungsten carbide, titanium nitride and nickel.

Researchers have begun to apply biased-induced SPL to pattern proteins. Bias-induced lithography was used directly for protein patterning by attachment of IgG–biotin to an array of 30-dot patterns generated using 1-ms voltage pulses (40–80 V) [61]. The substrate was a poly(methyl methacrylate) (PMMA) layer spin-coated on polished p-doped silicon wafer. Differences in hydrophobicity between the patterned areas and the substrate provided a driving force for the selective electrostatic deposition of proteins on nanopatterns. After reaction with avidin–fluorescein isothiocyanate (FITC), micron-sized patterned regions could then be imaged by fluorescence microscopy.

To further extend the capabilities of bias-induced lithography for patterning SAMs with designated surface chemistries, methods which include replacement or addition of new molecules from solution have recently been developed [55]. After voltage pulsing, small areas of the surface were exposed for adsorption of new molecules using bias-induced replacement lithography [62–64]. In another approach, bias conditions which selectively oxidize SAM terminal groups (tip-induced electro-oxidation) were used to generate surface oxides of SAMs. Oxidized areas then were used to chemically attach new molecules with desired functional groups [58, 65]. Nanopatterned protein arrays were fabricated by Cai and coworkers using bias-induced oxidation followed by protein adsorption (Fig. 3.6) [66]. Bias-induced SPL was applied to oxidize the headgroups of monolayers of

---

Figure 3.6. More than 100 protein dots produced by bias-induced nanofabrication. (A) AFM height and (B) friction image of nanoholes produced by bias-induced nanofabrication after treatment with EDAC/avidin $(4 \times 4\ \mu m^2)$. The lines provide a reference for corresponding features. The dots are approximately 90 nm in diameter. (C) Topography and (D) friction images of the same area after incubation with biotinylated BSA. (E) After the nanopatterns of biotinylated BSA were reacted with avidin a positive height was observed for the nanodot arrays. (Reproduced with permission from Ref. [66].)

α-hepta(ethylene glycol) methyl ω-undecenyl ether on Si(111) substrates. Bursts of 1-ms pulses of +17 V were applied to the sample, to generate rows of spots (90 nm diameter) separated by about 270 nm. The nanopatterned templates were then used to attach avidin followed by biotinylated BSA.

Although bias-induced lithography has not yet been widely applied for nanopatterning proteins, the newly introduced capabilities of customizing surface chemistry by tethering coupling agents to oxidized surfaces holds promise for new applications of bias-induced lithography in future investigations.

3.3.2
Force-induced Nanolithography of SAMs

Nanofabrication of SAMs can be accomplished by applying mechanical force to an AFM tip during scans. An intrinsic advantage of AFM instruments is the superb control of forces applied between the tip and sample, ranging from pico- to nanonewtons. For high-resolution and faithful imaging of surface topography it is critical to apply minimal, nondestructive forces. When too much force is applied by an AFM tip, areas of the surface can be swept clean or "nanoshaved" [59]. Nanografting was first invented in 1997 and combines nanoshaving with the simultaneous replacement of matrix SAM molecules by the self-assembly of new molecules [54, 67]. A broad range of thiolated molecules have been nanografted to provide tremendous flexibility in choosing the desired molecular lengths and terminal groups for experimental designs [68–72]. This section describes the procedure for nanografting SAMs, presents an example using automated nanografting with SAMs and then reviews examples which apply force-induced lithography (nanografting) for protein patterning.

Nanografting (Fig. 3.5B) is accomplished in dilute SAM solutions containing the selected molecule to be patterned by exerting a high local force on an AFM tip, pushing through the matrix SAM to contact the underlying gold surface. During scanning, pressure between the tip and surface displaces the SAM matrix molecules underneath the tip. As matrix molecules are removed, new thiol molecules from solution immediately adsorb onto the uncovered areas of the substrate to form nanopatterns, following the scanning track of the tip. SPM controllers can be programmed for automated lithography, to rapidly and consistently generate desired surface arrangements of arrays of SAM nanopatterns [73]. Commercial instruments typically include software with capabilities to control the length, direction, speed, bias pulse duration, residence time and the applied force of the scanning motion of the SPM tip, analogous to a pen-plotter. Automated SPL offers tremendous advantages for the speed and reproducibility of nanopatterning, and can produce highly sophisticated pattern arrangements and geometries, with superb precision and reproducibility for the alignment, spacing and shapes of nanopatterns. Examples of SAM nanopatterns generated by force-induced AFM-based lithography (nanografting) are shown in Fig. 3.7.

The AFM contact-mode topograph (Fig. 3.7A) displays 16 nanopatterns of 11-mercaptoundecanoic acid (11-MUA) written within a resist of octadecanethiol.

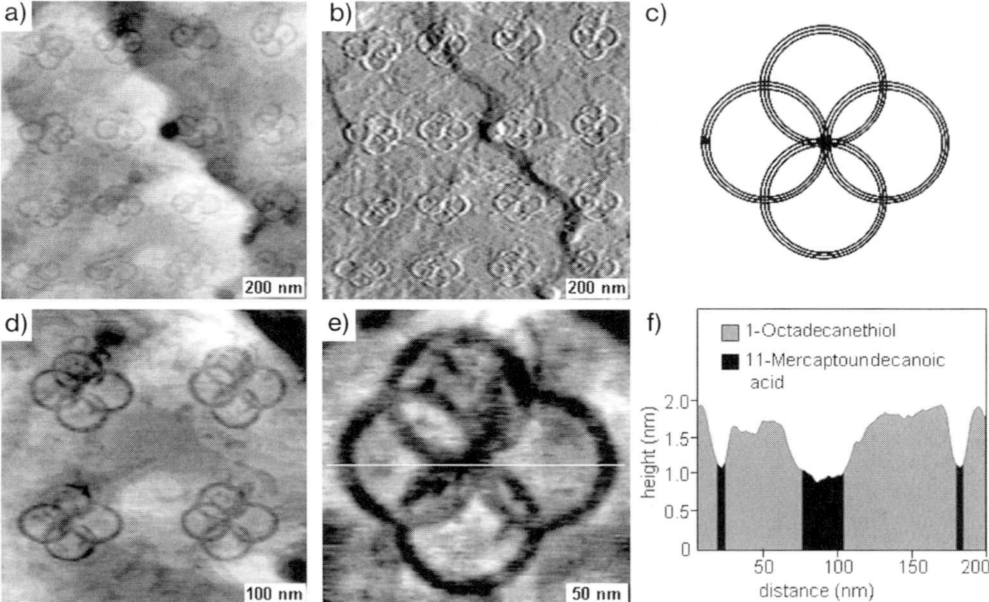

Figure 3.7. Nanopatterned array generated by automated nanografting. (A) Topography and (B) corresponding friction images (1.0 × 1.0 μm²) of nanopatterns of 11-mercaptoundecanoic acid grafted in octadecanethiol/ Au(111). (C) Design of nanopatterned (100 nm) ring elements. (D) Close-up view of four patterns. (E) Zoom-in view of a single pattern (250 × 250 nm²). (F) Cross-section taken along the line in (E).

The corresponding frictional force image (Fig. 3.7B) more clearly displays the arrangement and shapes for the nanopatterned array of circular designs. Nanografting was executed using a programmable computer module interfaced to the AFM controller for translating the tip rapidly and uniformly across the surface to create designed arrangements of nanopatterns (controllers from RHK Technology, Troy, MI). A computer script (written in-house) was used to apply a higher load on the tip to inscribe the pretzel-shaped designs. Each design was generated by writing four 100-nm diameter rings as in Fig. 3.7(C). The rings were inscribed by outlining each circle 3 times with the AFM tip, beginning with the bottom ring and moving in a clockwise direction around the center intersection. It required approximately three minutes to complete the entire 4 × 4 array, (around 12 s to write each pattern). After nanografting, AFM images were acquired under normal imaging conditions using minimal force. Figure 3.7(D) displays a close-up view of four nanopatterns, exhibiting nearly perfect alignment and symmetry. The high-resolution topograph of a single nanopattern in Fig. 3.7(E) reveals the exquisite capabilities of AFM-based nanolithography to write and visualize surface details. The height difference between the (ODT) nanopattern and the 11-MUA matrix SAM is indicated by the cursor profile (Fig. 3.7F) to be approximately 0.7 ± 0.1 nm, in close agreement with the theoretical height difference (0.7 nm). The line width of the rings is approximately 10 nm. Commercial silicon nitride cantilevers with an average spring constant of 0.58 N m^{-1} were used for both imaging and lithography (Veeco Probes, Santa Barbara, CA). The patterning and imaging experiments for Fig. 3.7 were conducted *in situ*, in a liquid environment.

Typically, it requires less than 1 min to fabricate individual nanopatterns using force-induced SPL such as nanoshaving and nanografting. Hundreds of nanopatterns can be written during an experiment without evidence of tip damage, provided that a minimal threshold force is applied. The fabrication forces used for force-induced SPL typically range from 2 to 30 nN, depending on the system under investigation and the geometries and spring constants of the cantilevers. Of course, if far too much force is applied the tip or substrate can be damaged, so it is critical to determine the minimum force for nanofabrication with each experiment. Commercially available soft Si_3N_4 cantilevers have mostly been used for nanofabrication by mechanical force, with force constants ranging from 0.03 to 2.0 N m^{-1}. When imaging in liquid, the total force applied typically is less than 1 nN, to prevent damage to substrate layers.

The first studies using nanografting to immobilize proteins were conducted in 1999 by Gang-Yu Liu and coworkers using either electrostatic or covalent interactions to immobilize lysozyme, rabbit IgG and BSA on SAM nanopatterns [74]. Since then, a growing number of investigators have taken advantage of the flexibility of nanografting in liquids for surface studies with biomolecules. The typical general steps of an *in situ* protein binding experiment are (a) to fabricate nanopatterns of adhesive tethering molecules, (b) bind proteins to these nanopatterns and (c) test the activity of the immobilized proteins by introducing a second antibody or protein that will bind specifically to the surface-bound protein.

An important advantage of nanografting is the capability to conduct experiments *in situ*, viewing the successive changes in surface topography after the steps of nanopatterning SAMs, rinsing, and introducing buffers and proteins. With *in situ* nanografting, the protein patterns are not subjected to air exposure, and remain in a carefully controlled environment by rinsing and exchanging solutions within the liquid cell. As molecules bind to nanopatterns, *sequential real-time AFM images expose reaction details at a molecular level*, uncovering critical details of the adsorption of proteins to nanostructured surfaces. Figure 3.8 illustrates the basic steps of an *in situ* protein adsorption experiment using nanografting.

In the initial investigations of protein immobilization on nanografted SAMs, Liu et al., used functionalized alkanethiol SAMs to mediate electrostatic and covalent binding of IgG and lysozyme [74]. The reactivity and stability of protein nanopatterns was studied in further reports, and included investigation of the retention of specific activity of the immobilized proteins for binding antibodies [75, 76]. Protein patterns sustained washing with buffer and surfactant solutions and were stable for at least 40 h of AFM imaging. The smallest protein feature yet produced by nanografting is a 10×150 nm^2 line containing three proteins [74, 76].

The first *in situ* antigen–antibody binding AFM experiment with nanofabricated SAMs was conducted using nanografting to direct protein immobilization [75].

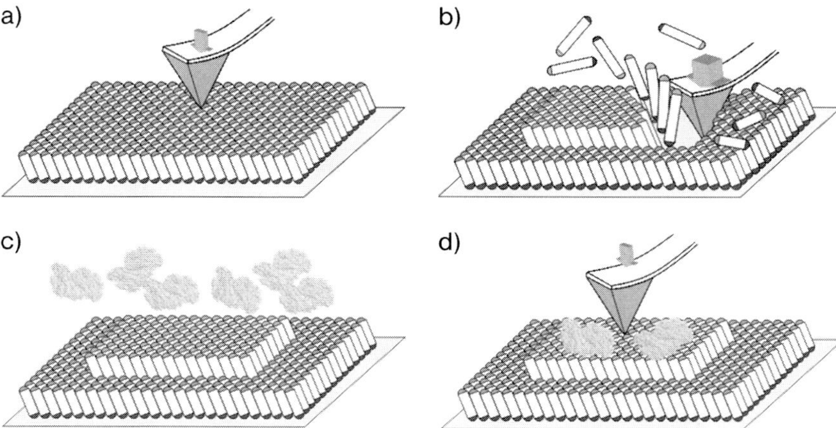

Figure 3.8. Steps for nanopatterning proteins using force-induced lithography. (A) A flat area is chosen for writing nanopatterns by imaging at low, nondestructive forces. (B) Nanofabrication is accomplished by applying higher force to write new SAM molecules. (C) Proteins are introduced by exchanging liquids. (D) Protein nanostructures can be characterized under low force.

The activity of covalently immobilized rabbit IgG was tested by reactivity toward mouse anti-rabbit IgG. AFM topographs of protein binding on nanografted patterns of an aldehyde-terminated SAM are shown in Fig. 3.9. Several aldehyde-terminated nanopatterns, a1–a5, were first grafted into a dodecanethiol SAM matrix. The depth of these patterns measured 0.8 ± 0.2 nm and images display dark contrast where the rectangular patterns were inscribed (Fig. 3.9A). After injecting rabbit IgG and rinsing with a surfactant solution, selective adsorption was observed on all six nanopatterns (Fig. 3.9B) in which the bright contrast indicates heights taller than the matrix SAM. In the next step, mouse anti-rabbit IgG was introduced (Fig. 3.9C), showing further height increases. By comparing the height of nanopatterns before and after secondary IgG binding, it was observed that immobilized IgG may adopt various configurations (Fig. 3.9G).

Several investigators have applied nanografting to write nanopatterns for protein immobilization. Abell and coworkers conducted a side-by-side comparison of protein adsorption on multifunctionalized surfaces at the nanoscale using nanografting. Protein adsorption on three differently charged linkers nanografted within a hexa(ethylene glycol) terminated alkanethiol resist SAM was monitored *in situ* by AFM at various pH values [77]. The adsorption of proteins onto nanografted patterns (400×400 nm^2) of 6-mercaptohexan-1-ol (MCH), *n*-(6-mercapto hexyl) pyridinium bromide (MHP) and 3-mercaptopropionic acid (MPA) was studied with lysozyme, IgG and carbonic anhydrase II. They conclude that in addition to the overall charge of protein molecules, the charge of local domains of the proteins plays a role in immobilization. In the same paper, Abell and coworkers used nanografting to assemble multilayered protein G/IgG/anti-IgG nanostructures through

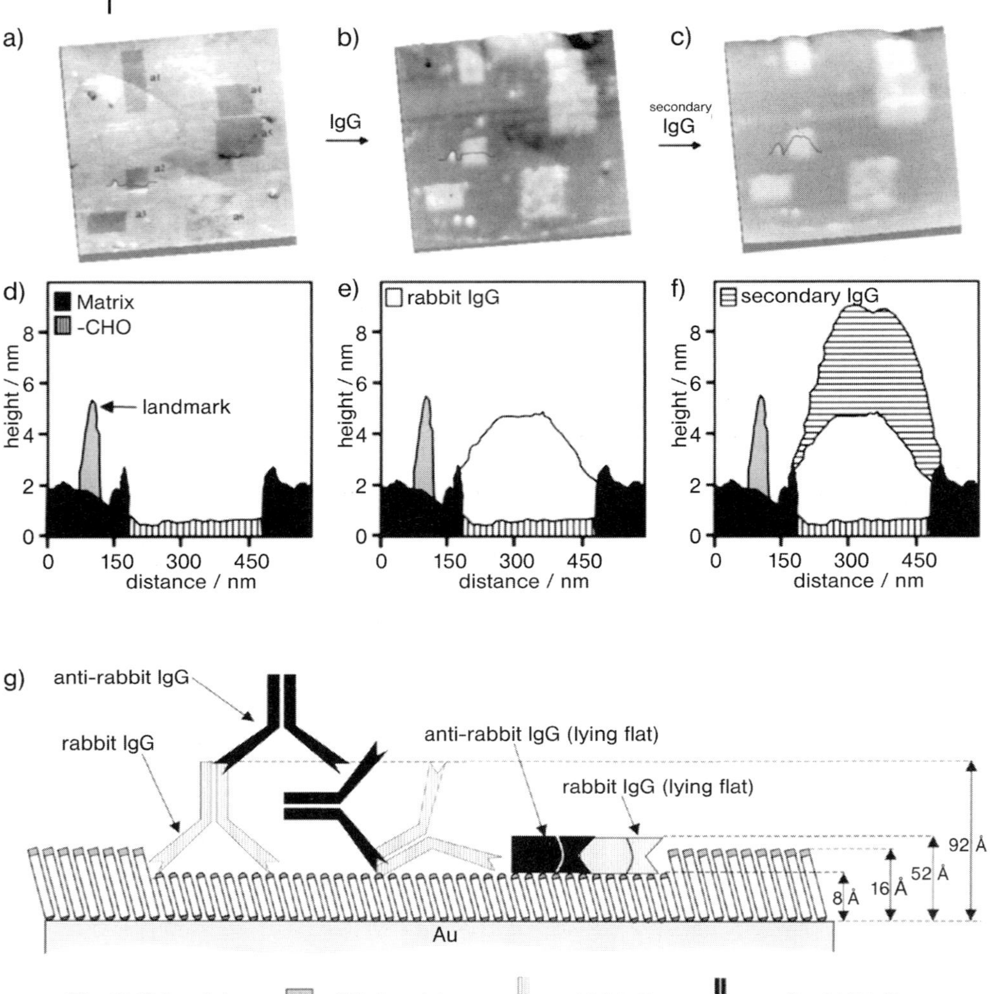

Figure 3.9. The steps of protein binding and molecular recognition with nanografted patterns captured by AFM topographic images. (A) Five nanopatterns of 3-mercapto-1-propanal were written in a dodecanethiol SAM. (B) The image contrast changed after rabbit IgG bound covalently to the aldehyde-terminated nanopatterns. (C) After introducing mouse anti-rabbit IgG, the patterns display further height changes, indicating the antibody binds specifically to the protein nanopatterns. Cursor traces across pattern a2 indicate the height changes (D) after nanografting, (E) after injecting IgG and (F) after introducing anti-rabbit IgG. (G) Map for understanding the evolution of molecular height changes during the steps of this *in situ* experiment. (Reproduced with permission from Ref. [75].)

electrostatic interactions as a potential means to orient IgG molecules for antibody-based biosensor surfaces.

Using force-induced SPL methods of nanografting and nanoshaving, Porter and coworkers compared three approaches for protein patterning [78]. They successfully combined force-induced SPL with immobilization of IgG via EDC activation of 11-mercaptoundecanoic acid; through direct adsorption of Fab'-SH fragments to nanoshaved regions of an EG$_3$-OMe matrix, and through chemisorption of a disulfide coupling agent, DSU. Ducker and coworkers applied nanografting to immobilize insulin and acetylcholinase esterase on nanografted 1,2-diols which were activated by sodium periodate to produce aldehyde groups [79]. Retention of catalytic activity was demonstrated for nanopatterned enzymes. Nanografting was applied to directly pattern designed metalloproteins by Au-S chemisorption by Scoles and coworkers [80]. The bundle protein structure was designed to present the C-termini of three helices, terminated with D-cysteine residues for assembly in a vertical orientation, normal to the Au(111) substrate.

A potential disadvantage for nanografting is that exchange takes place between solution molecules and the surface matrix SAM for some systems of alkanethiols. Natural self-exchange is an issue particularly when nanografting longer chain thiols into a shorter chain matrix layer, thus it is important to use very dilute (below 0.1 mM) solutions for nanografting. Exchange can be detected within 2–4 h (depending on the age of the matrix SAM) when molecules from solution adsorb onto defect sites and at step edges.

Although not yet practical for high-throughput applications and manufacturing, combining SPL with protein immobilization enables new approaches for directly investigating changes that occur on surfaces during biochemical reactions. Nano-engineered surfaces are useful for viewing antigen–antibody binding at the nanometer scale, to assess the specificity of selective binding, and to evaluate protein orientation and the accessibility of ligands for binding. Advantages of force-induced SPL include the ability to precisely produce nanometer-sized patterns of bioreceptors, and to successively image and conduct fabrication *in situ*, within well-controlled environments. For protein nanopatterning, force-induced SPL can be applied to either directly write proteins on surfaces via nanoshaving, or can be applied to write molecules for attaching proteins to surfaces through electrostatic, covalent or specific binding chemistries.

3.3.3
DPN of SAMs and Proteins

DPN, developed by Chad Mirkin and coworkers in 1999, has emerged as an important and versatile method for producing multicomponent arrays of SAM nanopatterns, as well as other molecules and nanomaterials [81]. This section describes the DPN nanofabrication method and then presents examples of DPN applied for protein nanopatterning. Protein nanopatterning has been accomplished by several different approaches via DPN. For SAM molecules written directly by DPN, proteins may be attached to nanopatterns through electrostatic, covalent or specific interactions on nanopatterns after surface passivation. Another strategy is to use surface activation of nanopatterned amine, carboxylate or hydroxyl groups for protein immobilization. Direct writing of proteins has also been accomplished by DPN using modified AFM tips.

In DPN, an AFM tip (pen) is coated with a molecular "ink" to write on clean gold substrates or "paper" under ambient conditions in air [56, 82]. Ink molecules migrate from the coated AFM tip through a capillary meniscus to the substrate by diffusion (Fig. 3.5C). Capillary transport of molecules from the AFM tip to the substrate can be used to directly write arrays of SAM nanopatterns. Additional mechanical force is not applied to the AFM tip or "pen." When an AFM tip is used in air to image a surface, the narrow gap between the tip and surface forms a tiny capillary meniscus from the condensation of water. Nanopatterns such as individual lines, dots, grids and arrays of alkanethiols have been written on bare gold surfaces [56, 83, 84]. The size of the water meniscus that bridges the tip and substrate depends on the tip shape and the relative humidity [85]. In DPN, the meniscus is used to transport molecules from the tip to the surface. The resolution of DPN depends on several parameters, such as the geometry of the AFM tip, the humidity of the ambient environment, as well as the duration over which the inked tip is placed in contact with the surface – typically of the order of approximately 1–10 s. With commercial cantilevers, DPN routinely generates feature sizes down to 15 nm.

Protein arrays were produced with DPN by patterning 16-mercaptohexadecanoic acid (MHA) for immobilizing lysozyme and IgG through electrostatic interactions by Mrksich and coworkers (Fig. 3.10) [86]. After writing MHA dots (diameter 100–350 nm) the gold surface was passivated with 11-mercaptoundecyl-tri(ethylene glycol). The MHA patterns were exposed to proteins by immersing substrates in a solution containing the desired protein. The arrays were then rinsed with buffer and imaged in air by tapping-mode AFM. The reaction of IgG patterns with rabbit antibody and with mixtures of proteins was also studied using AFM to investigate and detect nonspecific binding to nanopatterns and to passivated areas of the substrate. In another study, Choi and coworkers immobilized cytochrome c through electrostatic adsorption on nanopatterns of MHA written by DPN [87]. The areas surrounding the MHA nanopatterns were passivated with octadecanethiol.

Covalent attachment of synthetic peptides was accomplished by Ivanisevic and Cho using DPN [88]. First, nanopatterns of amine-terminated silane molecules (3-aminopropyl-triethoxysilane) (APTES) were written on SiO_x surfaces by DPN. Next, the heterobifunctional cross-linker N-hydroxysuccinimide ester (SMPB) was conjugated to the APTES nanopatterns. In the final step, a TAT peptide was covalently linked to SMPB via a cysteine residue in the peptide sequence.

The molecular recognition-mediated, stepwise fabrication of patterned protein nanostructures was accomplished by Zauscher and colleagues using DPN [89]. First, a SAM of MHA was patterned on gold by DPN, shown in Fig. 3.11. Next, the surrounding regions were passivated with a protein-resistant oligoethylene glycol-terminated alkanethiol SAM [11-mercaptoundecyl-tri(ethylene glycol) (EG_3-SH)]. Nonspecific adsorption of proteins on the background was prevented by pas-

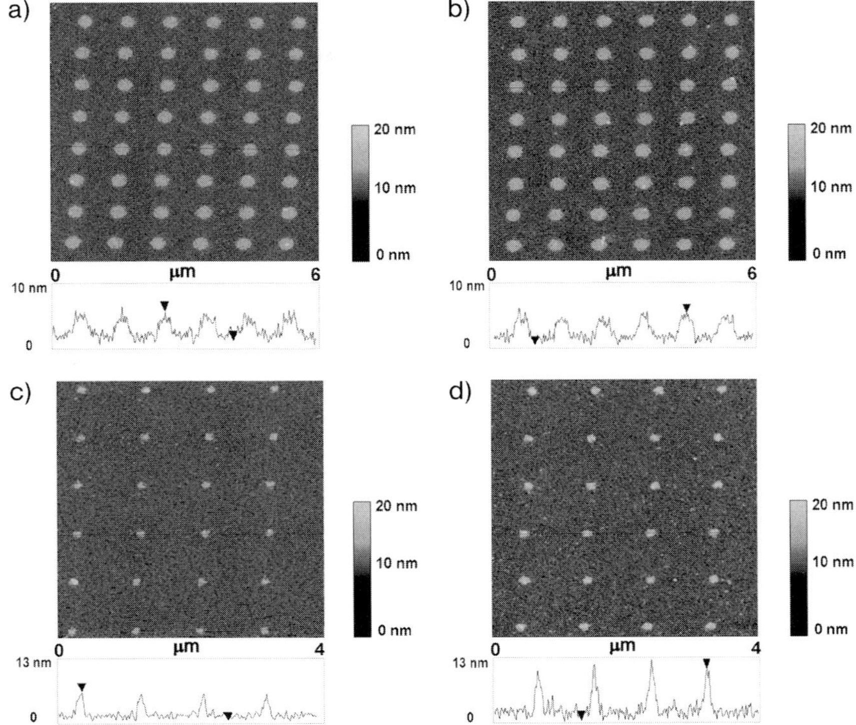

Figure 3.10. Snapshots of protein patterns generated by DPN captured *ex situ* by tapping mode AFM images in air and corresponding AFM height profiles. (A) MHA dot array written by DPN after rabbit IgG adsorption. (B) Same IgG array after treatment with a mixture of lysozyme, retronectin, goat/sheep anti-IgG and human IgG. No height change or nonspecific adsorption is detected. (C) A rabbit IgG array before and (D) after treatment with a mixture containing lysozyme, goat/sheep anti-IgG, human anti-IgG and rabbit anti-IgG. The height of the nanopatterns increased from 6.5 ± 0.9 to 12.1 ± 1.3 nm, indicating the binding of anti-IgG onto IgG nanopatterns. (Reproduced with permission from Ref. [86].)

sivation with EG_3-SH. In the third step, an amine-terminated biotin derivative was covalently conjugated with the MHA nanopatterns through a reaction with NHS and 1-ethyl-3-(dimethylamino)propyl carbodiimide (EDAC). The surface was then incubated with streptavidin in a fourth step, mediated by molecular recognition between biotin and streptavidin. In the final step, protein nanopatterns were fabricated by molecular recognition-mediated immobilization of biotinylated protein (BSA) in solution. This method provides a generic platform for immobilization of biotin-tagged molecules mediated by biospecific interactions of biotin–streptavidin ligands.

For the specific immobilization of cysteine-labeled cowpea mosaic virus (CPMV) capsid particles, Mirkin and coworkers applied DPN to write a mixture of two dia-

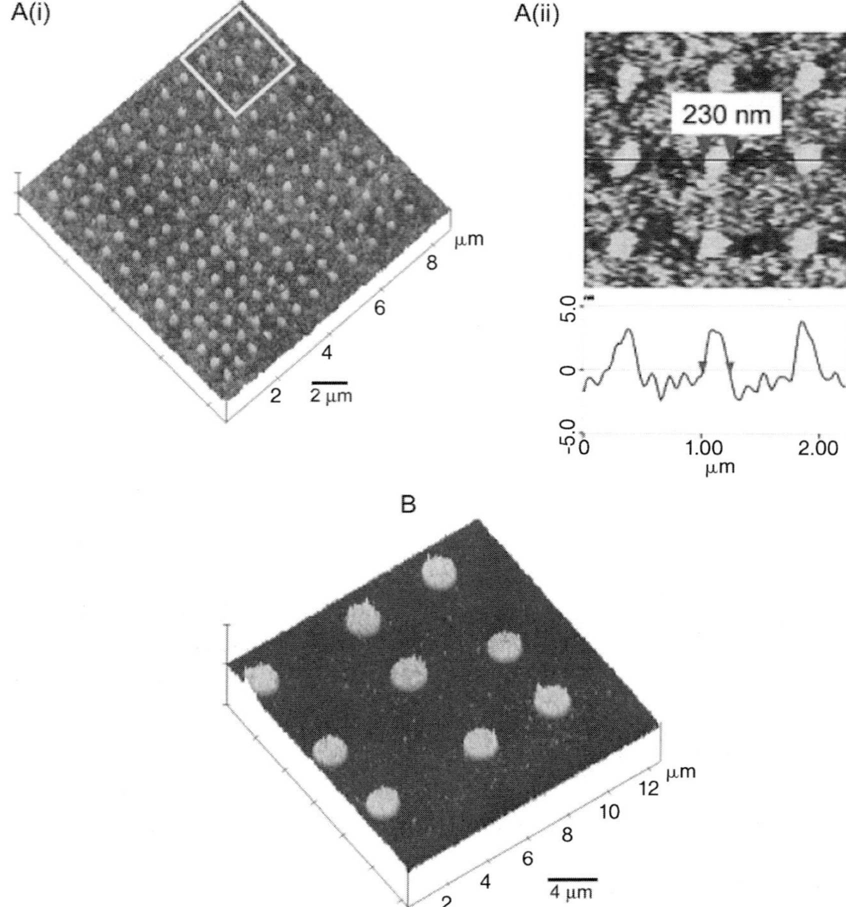

Figure 3.11. Nanopatterns of biotin–BSA imaged by tapping mode AFM. [A(i)] An array of 144 dots written by DPN; [A(ii)] zoom-in view of the area within the frame, showing the size of the dots. (B) The 3-D dot array with 1-μm features. (Reproduced with permission from Ref. [89].)

lkyl disulfides as ink [90]. The areas surrounding the nanopatterns were passivated with penta(ethylene glycol) groups. The density of maleimide groups provided efficient thiol capture through Michael addition of the thiol from cysteine residues of engineered CPMV particles to the nanopatterned maleimide groups.

Surface activation of nanopatterns of MHA written by DPN was used by Zauscher and coworkers to fabricate nanostructures of stimulus-responsive elastin-like polypeptide (ELP) [91]. First, patterns of MHA were written directly using DPN. The surface was then passivated with 11-mercaptoundecyl-tri(ethylene glycol). Next, the COOH groups of the nanopatterned MHA were reacted with

NHS and EDAC to covalently conjugate ELP to the surface. ELP was end-grafted to the surface through an amine group to dictate the surface orientation.

Mirkin and colleagues have presented several studies with direct writing of proteins using DPN. Thiolated collagen was used as ink for direct writing of collagen nanopatterns on gold substrates [92]. Using a modified AFM tip, IgG was written directly on either negatively charged SiO_2 or on aldehyde-modified SiO_2 surfaces by DPN [93]. Tips were coated with 2-[methoxypoly(ethyleneoxy)propyl]trimethoxysilane (Si-PEG), which forms a biocompatible and hydrophilic surface layer on AFM tips for protein inking. Protein arrays of IgG and lysozyme were nanopatterned by direct writing using metallized AFM tips (gold) with a thioctic acid coating, for protein adsorption [94]. Humidity is a critical variable in transporting proteins from tips to a surface, for direct writing of proteins; optimum results were obtained in a glove box at 80–90% humidity to achieve consistent transport properties. Direct writing of proteins on nickel oxide surfaces also was accomplished using tips modified with nickel. For direct write DPN, AFM tips were coated with 5 nm of nickel to facilitate transfer of histidine-tagged proteins (ubiquitin and thioredoxin) as inks [95]. High humidity enabled diffusion from the tip to the surface and also served to minimize the denaturation of protein structures on nickel substrates. Oxidized nickel has a high affinity for polyhistidine residues and patterns could not be generated for protein inks without histidine tags. The binding activity of the nanopatterns was investigated by reaction with fluorescently labeled antibodies, indicating that surface structures remained active for fluorescent labeling.

Another strategy for DPN nanopatterning is accomplished by combining bias-induced electrochemistry and DPN [96]. Electrochemical DPN (E-DPN) was used to immobilize histidine-tagged proteins on nickel substrates [97]. Silicon AFM probes were coated with polyhistidine-tagged peptides, proteins and free-base porphyrins for nanopatterning. By applying a negative bias (−2 to −3 V) to the coated AFM tips, nanopatterns could be written on nickel surfaces using tapping-mode AFM. Without an applied potential, protein deposition was not observed.

DPN provides methods for directly writing chemical inks on surfaces for complex, multistep fabrication of nanostructures of proteins. Chemical reagents can be delivered directly to nanosized areas of a surface and then the surrounding uncovered areas can be passivated with resistive SAMs. The DPN method is amenable to conducting experiments at ambient temperatures in air and ink transfer is facilitated by controlling humidity. After each reaction step, samples can be characterized *ex situ* by AFM imaging. Different molecules can be deposited by exchanging tips to produce multicomponent arrays of nanopatterns. For protein nanopatterning, DPN can be applied to either directly write proteins on surfaces using modified tips or to nanopattern SAM molecules for attaching proteins to surfaces through electrostatic, covalent or specific binding.

3.3.4
Latex Particle Lithography with Proteins

Particle or nanosphere lithography is an approach for nanopatterning which uses physical adsorption of materials to surfaces. Monodisperse latex particles self-assemble into periodic structures on flat surfaces, which have then been used as structural templates or photomasks for defining the deposition of proteins or other materials. The latex particles are removed by various approaches, such as calcination, solvent dissolution or simple rinsing with water. Particle lithography has been successfully applied for patterning metals, sols, polymers and inorganic materials [98–102]. Researchers have also applied colloidal lithography with latex beads as photomasks to construct functional surfaces for selective protein adsorption on lithographically defined regions [102, 103]. This section describes a method of particle lithography which can be applied directly for controlling the organization of proteins on surfaces through physical adsorption [57].

Particle lithography can be used to construct arrays of protein nanostructures on surfaces, with superb control of the distribution of proteins within a single layer over micron-sized areas [57]. An outline of the steps for patterning proteins using latex particle lithography is shown in Fig. 3.12. First, the protein and latex are mixed together in an aqueous solution. For best results, the solution containing protein and latex should be allowed to remain at room temperature for time intervals not longer than 4 h. (To maintain protein activity the solutions should be freshly prepared and to minimize contaminants the latex particles should be prewashed to ensure that the particles are free of surfactants.) In the second step, a small volume (10 μL cm^{-2}) of the colloidal suspension is deposited at the center of the substrate surface, using a pipette. The liquid spreads out into a thin layer across the surface as it dries. The protein and latex mixture forms ordered assem-

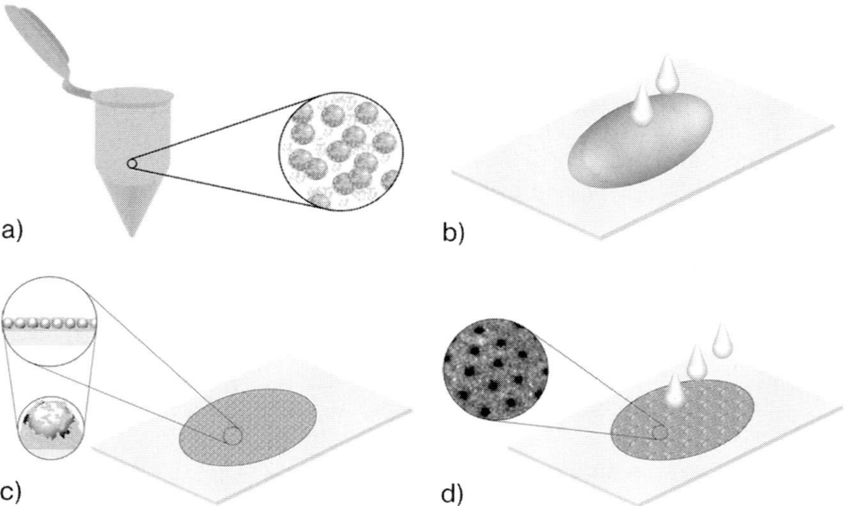

Figure 3.12. Fabrication steps for creating arrays of protein nanostructures via particle lithography. (A) Monodisperse latex particles are mixed with protein. (B) The mixture is deposited on a flat surface such as mica(0001). (C) After the droplet dries, an ordered crystalline layer is formed. (D) Latex are removed by rinsing with deionized water, leaving a layer of protein nanostructures on the surface.

blies supported by mica or gold. During drying, the convective motion of water as it evaporates pulls the latex together into close-packed assemblies. After the deposits have dried, the latex is rinsed away with deionized water to leave a single layer of protein nanostructures on the surface. The assembly of latex particles and the protein nanostructures can be characterized using AFM throughout the fabrication process.

Example images of BSA arrays formed from 500- and 200-nm diameter spheres are shown in Fig. 13(A and B, respectively). The 2-D AFM topographs reveal an organized arrangement of circular dark holes (uncovered areas of mica) surrounded by clusters of BSA. The periodicity of the resulting nanopatterns depends on the separation of latex spheres, which is observed to be 10–15% smaller than the original latex diameters. This is likely attributable to the shrinking and deformation of latex particles during drying. Using 500 nm particles, the ratio of BSA:latex was 55 000:1 which roughly corresponds to a single layer of proteins encapsulating a sphere. For the 200 nm particles, the ratio of BSA:latex was 9000:1, also corresponding to a monolayer shell. The ratios for successful lithography have

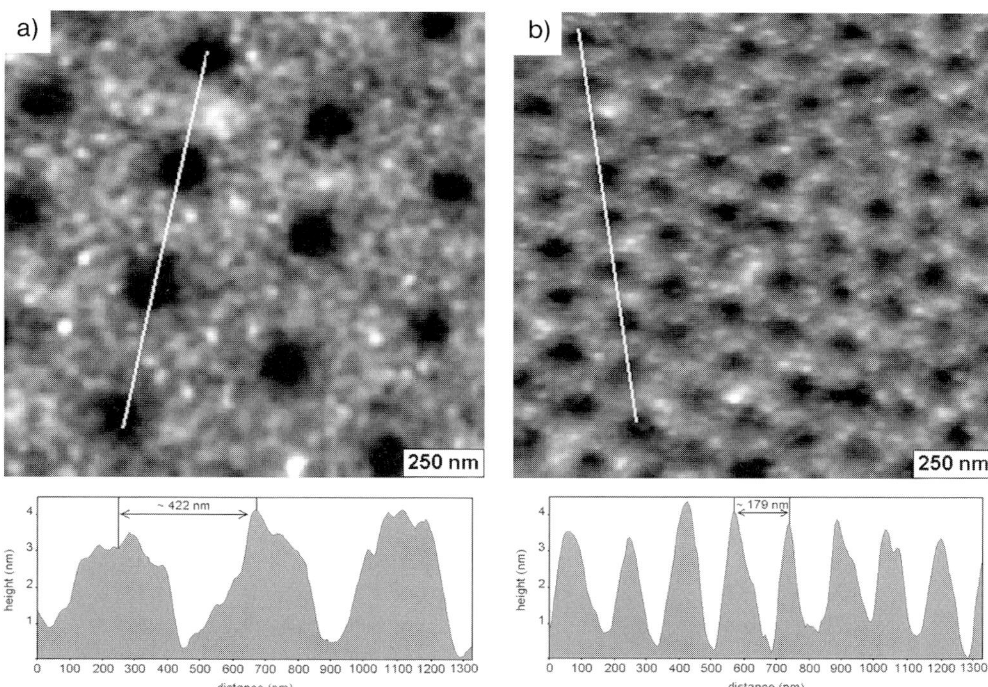

Figure 3.13. Periodic arrays of BSA nanostructures generated with latex nanoparticle lithography. (A) BSA nanopatterns from 500-nm particles, using a BSA:latex ratio of 55 000:1. The corresponding cursor indicates the periodicity of the BSA nanostructures is 422 ± 33 nm. (B) BSA nanostructures generated using 200-nm particles and a BSA:latex ratio of 9000:1. The periodicity of the BSA nanostructures measured 179 ± 21 nm.

ranged from approximately half of monolayer coverage of spheres to that of two layers, yielding different distributions and surface morphologies. Particle lithography uses mild conditions (ambient temperatures, buffers), yet provides nanometer-level control of the spatial distribution of proteins organized within a single surface layer. Particle lithography was also successfully applied for nanopatterning rabbit IgG [57]. Both IgG and BSA nanopatterns retained the ability to bind corresponding specific antibodies.

Particle lithography is a highly reproducible and robust method for patterning proteins, and serves as an excellent starting point for continuing to develop more complex bioassays with different surfaces and proteins. Using latex particles to control the arrangement of proteins on surfaces is a practical technology which is amenable to microspotting or immersion methods used for protein microarrays and biochips. Latex bead immobilization has been applied in spotting solutions to create microarrays for detection of antibodies [104]. Particle lithography offers the advantages of nanometer precision and high throughput, since a small vial of solution can produce hundreds of replicate samples. Future investigations will address the suitability of particle lithography to other surfaces and to other proteins, for application in surface-bound immunoassays.

3.4
Detection of Protein Binding at the Nanoscale

Nanoengineering approaches for the development of nanoscale bioassays capitalize on the unique *in situ* and high-resolution capabilities of SPM. Designed surfaces can be created with precisely placed proteins, which are subsequently monitored during the binding of secondary antibodies. SPM can be applied directly for detection, measuring changes in the heights of nanopatterns with protein binding. The height changes indicate the side-on or end-on orientation of immobilized proteins (Fig. 3.14) [74]. In some investigations, fluorescent labels were conjugated to antibodies for microscopic examination of samples. Optical microscopy can detect changes in fluorescence after antigen–antibody binding. An important question to address is whether or not the tagging entity hinders the affinity and efficiency of antigen–antibody binding. Many fluorescent dyes currently used are hydrophobic, which substantially decreases the solubility of protein–dye conjugates. This could adversely influence signal intensity for fluorescent detection [105]. Nanoscale studies can be applied to refine critical parameters used to link and organize proteins on surfaces of biochips and biosensors. SPM images of protein binding can be beneficial for evaluating the effectiveness of different bioconjugation chemistries for biomarkers.

At the core of biosensing is detection of biomolecular binding events with high selectivity and sensitivity. Typically, bioassays for surface-bound proteins are not as sensitive as approaches which use solution chemistry, due in part to the accessibility of molecules for binding. Pressing the limits of protein patterning to the nanometer scale will furnish direct views of the differences in immobilization chemis-

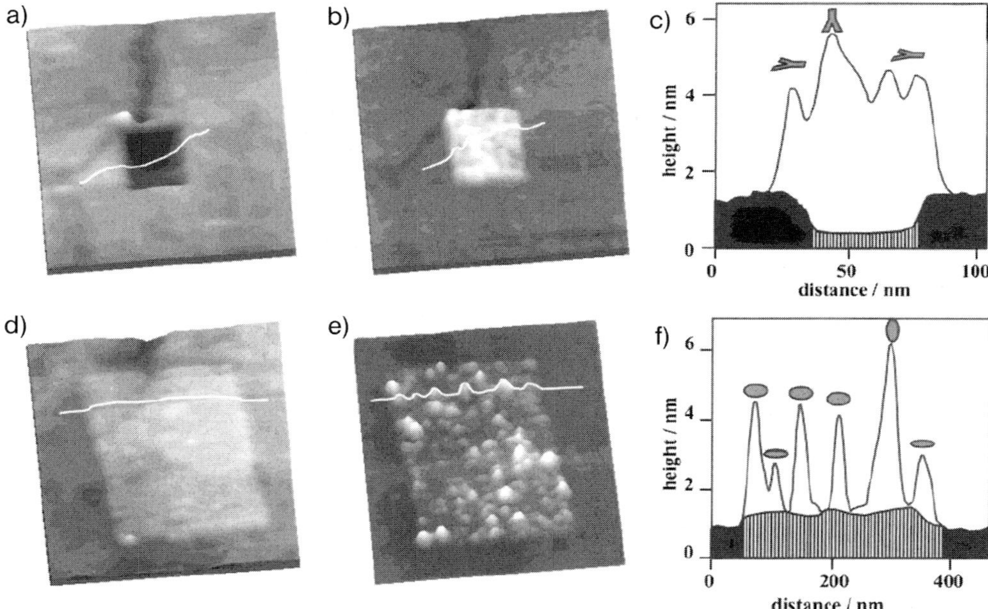

Figure 3.14. Nanografted patterns of aldehyde-terminated SAMs were used to co-valently immobilize proteins via imine bonds. (A) Nanopattern of 3-mercapto-1-propanal (150×150 nm^2) written in decanethiol; (B) after *in situ* immersion in buffer containing rabbit IgG; (C) combined cursor profiles for lines in (A) and (B). (D) Nanografted rectangle of mercaptodecanal (340×300 nm^2) written in hexanethiol; (E) after immersion in lysozyme solution. (F) Corresponding cursor profiles across (D) and (E). The ellipsoidal lysozyme and Y-shaped IgG molecules may adopt various orientations, as shown in the schematics above the cursor plots. (Reproduced with permission from Ref. [74].)

tries. Conventionally, biosensors and biochips rely on microspotting or solution deposition to place proteins on various surfaces, without control of the placement and arrangement of target proteins. There is a requirement for efficient yet mild immobilization chemistries which preserve tertiary structure and maximize the activity of fragile biomolecules.

Biomolecules immobilized on a surface serve as the receptor and in some cases as the signal transducer in biosensors. Therefore, the placement of biological ligands in precisely defined locations can increase the density of sensor elements and lead to improved detection limits with molecular-level control of the surface reactivity [106, 107]. As a proof-of-concept, Wolinsky and Mirkin have reported a nanometer-scale antibody array prepared by DPN to test for the presence of the human immunodeficiency virus type 1 (HIV-1) in blood samples [108]. The HIV-1 antibodies were immobilized on the MHA nanopatterns for hybridizing (HIV-1 p24) antigen and bound proteins. With a nanoarray of 100-nm features written by DPN, the three-component sandwich assay exceeded the limit of detection of conventional enzyme-linked immunosorbent assay (ELISA) based immunoassays by 1000-fold.

3.5
Future Directions

It can be anticipated that array-based technologies in proteomics including protein-based biochip and biosensing devices will significantly advance biotechnology, clinical diagnostics, tissue engineering and targeted drug delivery [109–111]. Ultra-small protein patterns can be used in biosensing, control of cell adhesion and growth and in biochip fabrication [5, 6, 112]. Methods of high-throughput protein analysis offer immense potential for fast, direct and quantitative detection, including the possibility of screening thousands of proteins within a single sample to test for protein, ligand and drug interactions. Improved binding to surfaces onto which capture proteins are arrayed and improved sensitivity of detection are technical challenges advancing protein microarray technology. The next section first discusses the motivation for advancing beyond microtechnology to the nanoscale frontier and then describes new technologies which are being developed for multiplexing AFM systems for parallel processes.

3.5.1
Advantages of Nanoscale Detection

Tools for nano- and microfabrication will provide important contributions in developing biochip and biosensing technologies, as well as supply basic research in protein–protein interactions. With the rapid progress in development of large sets of characterized antibodies, protein and antibody arrays will provide tremendous advantages for diagnostics and medical science. Miniaturization provides rewards of reduced quantities of analytes and reagents, increased density of sensor and chip elements, and more rapid reaction response [107, 113–115]. Multiplex screening of many interactions in a parallel fashion reduces analysis time and gives insight into the multiplicity of factors involved in diseases. Protein microarrays used in experiments based on AFM detection may soon reach capabilities for routinely achieving single-molecule detection.

Nanotechnology offers advantages not only for array production, but also for sample detection for bioarrays. In the nanoscale size regime, material properties are different than at macroscopic scales, exhibiting phenomena such as electromagnetic field enhancement, narrow emission band fluorescence, surface plasmon resonance, and conductivity and signal amplification. These properties enable new signaling and recognition capabilities for use in sensor systems. For example, bioconjugates of nanoparticles and quantum dots may provide improved stability and sensitivity for quantitative fluorescence detection with advantages over the conventional approach of fluorescent staining which suffers from the disadvantage of photobleaching [116]. With regard to biomolecule detection, new strategies based on functionalized metal nanoparticles, in combination with magnetic detection have been reported, using superparamagnetic beads coated with antibodies for detection [117].

3.5.2
Development of Cantilever Arrays

Using SPM-based nanofabrication, the single pen-and-ink approach is far too slow for cost-effective manufacturing of protein arrays. AFM-based lithography offers the ultimate capabilities for nanometer-scale control of surfaces with extremely high spatial precision; however, it has the limitation of relatively low throughput by fabricating each pattern individually. If higher throughput can be accomplished for nanoscale biochips, such arrays would offer immense capabilities. Miniaturization of protein sensing to nanodimensions will require techniques for rapid, efficient and high-throughput writing of biomolecules and SAMs. New designs for AFM probe arrays are being developed which will provide parallel and multiplexing capabilities for surface characterization and fabrication. Readers are referred to two recent reviews which detail the developments in multiple probe systems [81, 118].

Successful lithography with probe arrays has been demonstrated by several researchers, and representative examples are summarized in Tab. 3.3. Initial designs of tip arrays used feedback from a single cantilever for operation. This can sometimes be problematic for lithography because of variations in tip geometries and difficulties for precise alignment [119]. Micromachined arrays of cantilevers operated in parallel ($2 \times 1, 5 \times 1, 10 \times 1, 32 \times 1$ and 50×1) were used for bias-induced oxidation of silicon as reported by Quate and coworkers [119–121]. Proof-of-concept experiments were presented for operation of parallel AFM probes for bias-induced lithography using as many as 50 probes. Arrays with integrated z-axis control were found to improve the quality and reproducibility of writing. There are several approaches for control and actuation of individual probes within cantilever arrays, including piezoelectric sensing [121–123], optical interferometric detection [124], thermal actuation of bimetallic tips [125–129] and conductivity sensing of tip–surface contact [130].

DPN has been advanced to parallel processes through the use of 1-D cantilever arrays [131, 132]. Control of the horizontal and vertical movements of AFM probes can be achieved using the laser signal feedback of a closed-loop AFM scanner. Using an eight-cantilever array, a miniaturized combinatorial chemical sensor was produced with DPN by writing multiple inks as sensor elements [133]. Also, examples have been reported using tip arrays for simultaneously writing multiple patterns of octadecanethiol using DPN [128–130].

Two-dimensional cantilever arrays have been developed and tested, including designs which combine passive (feedback from deflection of a single tip) and active control through actuation of multiple tips [134]. A group at IBM has implemented an addressable 32×32 probe array designed for high-density data storage [135, 136]. The "Millipede" array format with 1024 cantilevers measures 3×3 mm^2

Tab. 3.3 Examples of AFM probe arrays which have been successfully demonstrated for AFM imaging/lithography applications.

Year	Array size	Tip actuation method	Application	Reference
1995	2×1	integrated piezoelectric sensor and piezoresisitive sensor for both tips	parallel constant-force AFM imaging	139
1995	2×1, 5×1	single tip feedback (piezoresistive levers)	bias-induced oxidation of silicon (100–200 nm line patterns)	119
1996	2×1	integrated piezoelectric actuators for each tip	bias-induced oxidation of silicon (micron patterns)	121
1998	10×1, 32×1, 50×1	thermal actuation of individual probes, piezoresisitive sensors and integrated ZnO actuators	bias-induced oxidation of silicon (1.1-μm line patterns) and parallel AFM imaging	120
1999	5×5	integrated piezoresistive sensing using five actuators	multiplexed AFM imaging	134
2000	2×1	thermal bimorph actuator, metal-oxide-semiconductor (MOS) electronics	AFM imaging with constant height mode, tapping mode, constant force mode	126
2000	2×4	piezoresistive deflection sensors	parallel AFM imaging	123
2000	32×32	thermal actuation, corner sensors for z-feedback control of entire chip	data storage read/write operations	135, 136
2001	5×1	optical interferometric detection	parallel AFM imaging, constant height mode	124
2002	2×2, 2×7	piezoresisitive sensing with integrated electrical interconnects	parallel contact-mode imaging of large sample areas	122

2002	8 × 1, 32 × 1	single tip actuation and optical deflection feedback	DPN with octadecanethiol/gold	131
2003	8 × 1	single tip actuation and optical deflection feedback	DPN with sol inks for chemical sensing	133
2003	10 × 1	conductivity-based sensing of tip–surface contact	simultaneous DPN writing with eight probes	130
2004	10 × 1	integrated thermal actuation of probes with piezoresistive stress sensors	parallel constant-force AFM imaging	127
2004	12 × 1	thermal actuation of bimetallic probes, piezoresisitive Wheatstone bridge detection	AFM imaging and force–distance measurements	125
2004	10 × 1	individual tip control by thermal actuation of bimorph probes	simultaneous DPN writing with 10 probes	128, 129

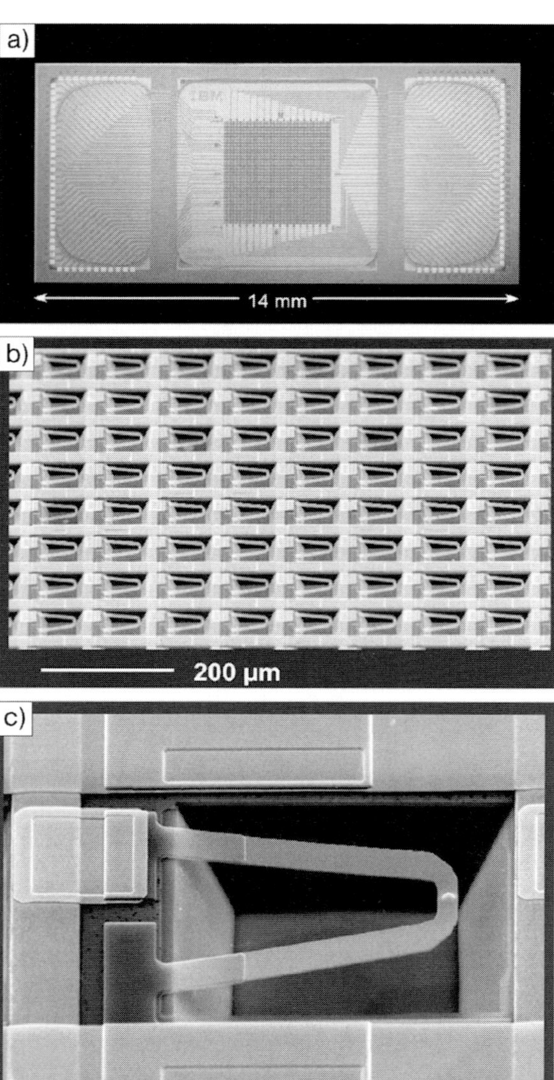

Figure 3.15. The "Millipede" array of 1024 cantilevers.
(A) Photograph of the entire chip. (B) SEM images of
cantilevers of the 32 × 32 array. (C) A single cantilever with
integrated silicon tip. (Reproduced with permission from
Ref. [135].)

and each cantilever is assigned to read and write areas of 100×100 μm^2. The Millipede approach is not based on individual z-feedback for each cantilever; feedback control is applied for the whole chip. This design requires stringent control of tip fabrication parameters via micromachining to generate uniform probe dimensions (Fig. 3.15). Although the Millipede design is intended for read/write data storage, other applications can be envisaged using SPL.

3.5.3
Concluding Remarks

Fundamental understanding of the interactions of protein binding to substrates or antibodies is essential for developing workable technologies for life sciences. The new capabilities to study and control processes on the nanometer scale are emerging as valuable assets in both fundamental and applied research. At present, SPM and SPL are primarily used as research tools in laboratories rather than as tools for manufacturing. However, in the future, nanoscale technology in manufacturing is predicted to bring an even greater impact and benefit to society than present-day microfabrication technologies. Potential applications include the development of a new generation of chemical and biosensors, biochips, and molecular electronic devices. We anticipate that nanoscale research will define new directions in areas such as biosensing, biomimetic surfaces for drug delivery and biomolecule-based electronics. This chapter provides insight on the tremendous versatility of several new SPL methods applied for protein nanopatterning. In addition, there are many new nanofabrication methods being developed which may be suitable in the future for engineering surfaces for nanoscale protein assays.

Applying the *in situ* tools of lithography with proteins will enable systematic evaluation of the differences in bioaffinity for various chemical immobilization strategies, with direct views of how the morphology and geometry of nanoengineered surfaces direct and influence the binding of antibodies and proteins. Conceptually, by arranging and orienting proteins on well-defined surfaces, the selectivity and sensitivity of surface-based protein assays can be substantially improved. These studies will facilitate the development of new and better approaches for immobilization and bioconjugation chemistries – key technologies used in manufacturing biochip and biosensing surfaces.

References

1 O'BRIEN, J. C., JONES, V. W., PORTER, M. D. Immunosensing platforms using spontaneously adsorbed antibody fragments on gold. *Anal. Chem.* **2000**, *72*, 703–710.

2 DELAMARCHE, E., SUNDARABABU, G., BIEBUYCK, H., MICHEL, B., GERBER, C., SIGRIST, H., WOLF, H., RINGSDORF, H., XANTHOPOULOS, N., MATHIEU, H. J. Immobilization of antibodies on a photoactive self-assembled monolayer on gold. *Langmuir* **1996**, *12*, 1997–2006.

3 ROWE, C. A., TENDER, L. M., FELDSTEIN, M. J., GOLDEN, J. P., SCRUGGS, S. B., MacCRAITH, B. D.,

CRAS, J. J., LIGLER, F. S. Array biosensor for simultaneous identification of bacterial, viral, and protein analytes. *Anal. Chem.* **1999**, *71*, 3846–3852.

4 LYNCH, M., MOSHER, C., HUFF, J., NETTIKADAN, S., JOHNSON, J., HENDERSON, E. Functional protein nanoarrays for biomarker profiling. *Proteomics* **2004**, *4*, 1695–1702.

5 SCOUTEN, W. H., LUONG, J. H. T., BROWN, R. S. Enzyme or protein immobilization techniques for applications in biosensor design. *Trends Biotechnol.* **1995**, *13*, 178–185.

6 ZHANG, S. G., YAN, L., ALTMAN, M., LASSLE, M., NUGENT, H., FRANKEL, F., LAUFFENBURGER, D. A., WHITESIDES, G. M., RICH, A. Biological surface engineering: a simple system for cell pattern formation. *Biomaterials* **1999**, *20*, 1213–1220.

7 DILLMORE, W. S., YOUSAF, M. N., MRKSICH, M. A photochemical method for patterning the immobilization of ligands and cells to self-assembled monolayers. *Langmuir* **2004**, *20*, 7223–7231.

8 KANE, R. S., TAKAYAMA, S., OSTUNI, E., INGBER, D. E., WHITESIDES, G. M. Patterning proteins and cells using soft lithography. *Biomaterials* **1999**, *20*, 2363–2376.

9 JAMES, C. D., DAVIS, R. C., KAM, L., CRAIGHEAD, H. G., ISAACSON, M., TURNER, J. N., SHAIN, W. Patterned protein layers on solid substrates by thin stamp microcontact printing. *Langmuir* **1998**, *14*, 741–744.

10 LAHIRI, J., OSTUNI, E., WHITESIDES, G. M. Patterning ligands on reactive SAMs by microcontact printing. *Langmuir* **1999**, *15*, 2055–2060.

11 BERNARD, A., RENAULT, J. P., MICHEL, B., BOSSHARD, H. R., DELAMARCHE, E. Microcontact printing of proteins. *Adv. Mater.* **2000**, *12*, 1067–1070.

12 BERNARD, A., DELAMARCHE, E., SCHMID, H., MICHEL, B., BOSSHARD, H. R., BIEBUYCK, H. Printing patterns of proteins. *Langmuir* **1998**, *14*, 2225–2229.

13 WHITESIDES, G. M., OSTUNI, E., TAKAYAMA, S., JIANG, X., INGBER, D. E. Soft lithography in biology and biochemistry. *Annu. Rev. Biomed. Eng.* **2001**, *3*, 335–373.

14 BLAWAS, A. S., OLIVER, T. F., PIRRUNG, M. C., REICHERT, W. M. Step-and-repeat photopatterning of protein features using caged-biotin–BSA: characterization and resolution. *Langmuir* **1998**, *14*, 4243.

15 NICOLAU, D. V., TAGUCHI, T., TANIGUCHI, H., YOSHIKAWA, S. Micron-sized protein patterning on diazonaphthoquinone/novolak thin polymeric films. *Langmuir* **1998**, *14*, 1927–1936.

16 DONTHA, N., NOWALL, W. B., KUHR, W. G. Generation of biotin/avidin/enzyme nanostructures with maskless photolithography. *Anal. Chem.* **1997**, *69*, 2619–2625.

17 DELAMARCHE, E., BERNARD, A., SCHMID, H., BIETSCH, A., MICHEL, B., BIEBUYCK, H. Microfluidic networks for chemical patterning of substrate: design and application to bioassays. *J. Am. Chem. Soc.* **1998**, *120*, 500–508.

18 PATEL, N., SANDERS, G. H. W., SHAKESHEFF, K. M., CANNIZZARO, S. M., DAVIES, M. C., LANGER, R., ROBERTS, C. J., TENDLER, S. J. B., WILLIAMS, P. M. Atomic force microscopic analysis of highly defined protein patterns formed by microfluidic networks. *Langmuir* **1999**, *15*, 7252–7257.

19 KINBARA, K., AIDA, T. Toward intelligent molecular machines: directed motions of biological and artificial molecules and assemblies. *Chem. Rev.* **2005**, *105*, 1377–1400.

20 FORTINA, P., KRICKA, L. J., SURREY, S., GRODZINSKI, P. Nanobiotechnology: the promise and reality of new approaches to molecular recognition. *Trends Biotechnol.* **2005**, *23*, 168–173.

21 KASEMO, B. Biological surface science. *Surf. Sci.* **2002**, *500*, 656–677.

22 WONG, S. S. *Chemistry of Protein Conjugation and Cross-linking.* CRC Press, Boca Raton, FL, **1991**.

23 HERMANSON, G. T. *Bioconjugate Techniques.* Academic Press, San Diego, CA, **1996**.

24 NUZZO, R. G., ALLARA, D. L. Adsorption of bifunctional organic

disulfides on gold surfaces. *J. Am. Chem. Soc.* **1983**, *105*, 4481–4483.

25 SCHREIBER, S. L. Self-assembled monolayers: from "simple" model systems to biofunctionalized interfaces. *J. Phys.: Condens. Matter* **2004**, *16*, R881–R900.

26 SAGIV, J. Organized monolayers by adsorption – formation and structure of oleophobic mixed monolayers on solid surfaces. *J. Am. Chem. Soc.* **1980**, *102*, 92–98.

27 ULMAN, A. *An Introduction to Ultrathin Organic Films: From Langmuir–Blodgett to Self-Assembly.* Academic Press, Boston, MA, **1991**.

28 WITT, D., KLAJN, R., BARSKI, P., GRZYBOWSKI, B. A. Applications, properties and synthesis of *w*-functionalized *n*-alkanethiols and disulfides – the building blocks of self-assembled monolayers. *Curr. Org. Chem.* **2004**, *8*, 1–35.

29 POIRIER, G. E. Characterization of organosulfur molecular monolayers on Au(111) using scanning tunneling microscopy. *Chem. Rev.* **1997**, *97*, 1117–1127.

30 SCHREIBER, F. Structure and growth of self-assembling monolayers. *Prog. Surf. Sci.* **2000**, *65*, 151–256.

31 DUBOIS, L. H., NUZZO, R. G. Synthesis, structure and properties of model organic surfaces. *Annu. Rev. Phys. Chem.* **1992**, *43*, 437–463.

32 ULMAN, A. Formation and structure of self-assembled monolayers. *Chem. Rev.* **1996**, *96*, 1533–1554.

33 PORTER, M. D., BRIGHT, T. B., ALLARA, D. L., CHIDSEY, C. E. D. Spontaneously organized molecular assemblies. 4. Structural characterization of normal-alkyl thiol monolayers on gold by optical ellipsometry, infrared-spectroscopy, and electrochemistry. *J. Am. Chem. Soc.* **1987**, *109*, 3559–3568.

34 FENTER, P., EBERHARDT, A., EISENBERGER, P. Self-assembly of *n*-alkyl thiols as disulfides on Au(111). *Science* **1994**, *266*, 1216–1218.

35 SCHWARTZ, P. K. Mechanisms and kinetics of self-assembled monolayer formation. *Ann. Rev. Phys. Chem.* **2001**, *52*, 107–137.

36 FENTER, P., EBERHARDT, A., LIANG, K. S., EISENBERGER, P. Epitaxy and chainlength dependent strain in self-assembled monolayers. *J. Chem. Phys.* **1997**, *106*, 1600–1608.

37 DANNENBERGER, O., WEISS, K., HIMMEL, H. J., JAGER, B., BUCK, M., WOLL, C. An orientation analysis of differently endgroup-functionalised alkanethiols adsorbed on Au substrates. *Thin Solid Films* **1997**, *307*, 183–191.

38 NUZZO, R. G., DUBOIS, L. H., ALLARA, D. L. Fundamental studies of microscopic wetting on organic surfaces. 1. Formation and structural characterization of a self-consistent series of polyfunctional organic monolayers. *J. Am. Chem. Soc.* **1990**, *112*, 558–569.

39 QIAN, Y., YANG, G., YU, J., JUNG, T. A., LIU, G.-Y. Structures of annealed decanethiol self-assembled monolayers on Au(111): an ultrahigh vacuum scanning tunneling microscopy study. *Langmuir* **2003**, *19*, 6056–6065.

40 CAMILLONE III, N., EISENBERGER, P., LEUNG, T. Y. B., SCHWARTZ, P., SCOLES, G., POIRIER, G. E., TARLOV, M. J. New monolayer phases of *n*-alkane thiols self-assembled on Au(111): preparation, surface characterization, and imaging. *J. Chem. Phys.* **1994**, *101*, 11031–11036.

41 YANG, G., LIU, G.-Y. New insights for self-assembled monolayers of organothiols on Au(111) revealed by scanning tunneling microscopy. *J. Phys. Chem. B* **2003**, *107*, 8746–8759.

42 XIAO, X.-D., LIU, G. Y., CHARYCH, D. H., SALMERON, M. Preparation, structure, and mechanical stability of alkylsilane monolayers on mica. *Langmuir* **1995**, *11*, 1600–1604.

43 PEANASKY, J., SCHNEIDER, H. M., GRANICK, S. Self-assembled monolayers on mica for experiments utilizing the surface forces apparatus. *Langmuir* **1995**, *11*, 953–962.

44 SCHWARTZ, D. K., STEINBERG, S., ISRAELACHVILI, J., ZASADZINSKI, J. A. N. Growth of a self-assembled monolayer by fractal aggregation. *Phys. Rev. Lett.* **1992**, *63*, 3354–3357.

45 CHAPMAN, R. G., OSTUNI, E., TAKAYAMA, S., HOLMLIN, R. E., YAN, L., WHITESIDES, G. M. Surveying for surfaces that resist the adsorption of proteins. *J. Am. Chem. Soc.* **2000**, *122*, 8303–8304.

46 OSTUNI, E., CHAPMAN, R. G., LIANG, M. N., MELULENI, G., PIER, G., INGBER, D. E., WHITESIDES, G. M. Self-assembled monolayers that resist the adsorption of proteins and the adhesion of bacterial and mammalian cells. *Langmuir* **2001**, *17*, 6336–6343.

47 HOLMLIN, R. E., CHEN, X., CHAPMAN, R. G., TAKAYAMA, S., WHITESIDES, G. M. Zwitterionic SAMs that resist nonspecific adsorption of protein from aqueous buffer. *Langmuir* **2001**, *17*, 2841–2850.

48 OSTUNI, E., CHAPMAN, R. G., HOLMLIN, R. E., TAKAYAMA, S., WHITESIDES, G. M. A Survey of structure–property relationships of surfaces that resist the adsorption of protein. *Langmuir* **2001**, *17*, 5605–5620.

49 LUK, Y.-Y., KATO, M., MRKSICH, M. Self-assembled monolayers of alkanethiolates presenting mannitol groups are inert to protein adsorption and cell attachment. *Langmuir* **2000**, *16*, 9604–9608.

50 HERRWERTH, S., ECK, W., REINHARDT, S., GRUNZE, M. Factors that determine the protein resistance of oligoether self-assembled monolayers – internal hydrophilicity, terminal hydrophilicity, and lateral packing density. *J. Am. Chem. Soc.* **2003**, *125*, 9359–9366.

51 TENGVALL, P., JANSSON, E., ASKENDAL, A., THOMSEN, P., GRETZER, C. Preparation of multilayer plasma protein films on silicon by EDC/NHS coupling chemistry. *Colloids Surf. B* **2003**, *28*, 261–272.

52 BINNIG, G., ROHRER, H., GERBER, C. Surface studies by scanning tunneling microscopy. *Phys. Rev. Lett.* **1982**, *49*, 57.

53 BINNIG, G., QUATE, C. F., GERBER, C. Atomic force microscope. *Phys. Rev. Lett.* **1986**, *56*, 930.

54 XU, S., LIU, G. Y. Nanometer-scale fabrication by simultaneous nano-shaving and molecular self-assembly. *Langmuir* **1997**, *13*, 127–129.

55 GORMAN, C. B., FUIERER, R. R., KRAMER, S. Scanning probe lithography using self-assembled monolayers. *Chem. Rev.* **2003**, *103*, 4367–4418.

56 PINER, R. D., ZHU, J., XU, F., HONG, S., MIRKIN, C. A. "Dip-pen" nanolithography. *Science* **1999**, *283*, 661–663.

57 GARNO, J. C., AMRO, N., WADU-MESTHRIGE, K., LIU, G.-Y. Production of periodic arrays of protein nanostructures using particle lithography. *Langmuir* **2002**, *18*, 8186–8192.

58 LIU, S., MAOZ, R., SAGIV, J. Planned nanostructures of colloidal gold via self-assembly on hierarchically assembled organic bilayer template patterns with in-situ generated terminal amino functionality. *Nano Lett.* **2004**, *4*, 845–851.

59 LIU, G.-Y., XU, S., QIAN, Y. Nanofabrication of self-assembled monolayers using scanning probe lithography. *Acc. Chem. Res.* **2000**, *33*, 457–466.

60 LEWIS, M. S., GORMAN, C. B. Scanning tunneling microscope-based replacement lithography on self-assembled monolayers. Investigation of the relationship between monolayer structure and replacement bias. *J. Phys. Chem. B* **2004**, *108*, 8581–8583.

61 NAUJOKS, N., STEMMER, A. Using local surface charges for the fabrication of protein patterns. *Colloids Surf. A* **2004**, *249*, 69–72.

62 CHEN, J., REED, M. A., ASPLUND, C. L., CASSELL, A. M., MYRICK, M. L., RAWLETT, A. M., TOUR, J. M., VANPATTEN, P. G. Placement of conjugated oligomers in an alkanethiol matrix by scanned probe microscope lithography. *Appl. Phys. Lett.* **1999**, *75*, 624–626.

63 GORMAN, C. B., CARROLL, R. L., HE, Y., TIAN, F., FUIERER, R. Chemically well-defined lithography using self-assembled monolayers and scanning tunneling microscopy in nonpolar organothiol solutions. *Langmuir* **2000**, *16*, 6312–6316.

64 ZHAO, J., UOSAKI, K. Formation of nanopatterns of a self-assembled monolayer (SAM) within a SAM of different molecules using a current sensing atomic force microscope. *Nano Lett.* **2002**, *2*, 137–140.

65 MAOZ, R., COHEN, S. R., SAGIV, J. Nanoelectrochemical patterning of monolayer surfaces: toward spatially defined self-assembly of nanostructures. *Adv. Mater.* **1999**, *11*, 55–61.

66 GU, J., YAM, C. M., LI, S., CAI, C. Nanometric protein arrays on protein-resistant monolayers on silicon surfaces. *J. Am. Chem. Soc.* **2004**, *126*, 8098–8099.

67 XU, S., LAIBINIS, P. E., LIU, G.-Y. Accelerating the kinetics of thiol self-assembly on gold – a spatial confinement effect. *J. Am. Chem. Soc.* **1998**, *120*, 9356–9361.

68 XU, S., MILLER, S., LAIBINIS, P. E., LIU, G.-Y. fabrication of nanometer scale patterns within self-assembled monolayers by nanografting. *Langmuir* **1999**, *15*, 7244–7251.

69 HACKER, C. A., BATTEAS, J. D. B., GARNO, J. C., MARQUEZ, M., RICHTER, C. A., RICHTER, L. J., VAN ZEE, R. D., ZANGMEISTER, C. D. Structural and chemical characterization of monofluoro-substituted oligo(phenylene-ethynylene) thiolate self-assembled monolayers on gold. *Langmuir* **2004**, *20*, 6195–6205.

70 LIU, J.-F., CRUCHON-DUPEYRAT, S., GARNO, J. C., FROMMER, J., LIU, G.-Y. Three-dimensional nanostructure construction via nanografting: positive and negative pattern transfer. *Nano Lett.* **2002**, *2*, 937–940.

71 BROWER, T. L., GARNO, J. C., ULMAN, A., LIU, G.-Y., YAN, C., GOLZHAUSER, A., GRUNZE, M. Self-assembled multilayers of 4,4-dimercaptobiphenyl formed by Cu(II) oxidation. *Langmuir* **2002**, *18*, 6207–6216.

72 LIU, M., AMRO, N. A., CHOW, C. S., LIU, G. Y. Production of nanostructures of DNA on surfaces. *Nano Lett.* **2002**, *2*, 863–867.

73 CRUCHON-DUPEYRAT, S., PORTHUN, S., LIU, G.-Y. Nanofabrication using computer-assisted design and automated vector-scanning probe lithography. *Appl. Surf. Sci.* **2001**, *175–176*, 636–642.

74 WADU-MESTHRIGE, K., XU, S., AMRO, N. A., LIU, G.-Y. Fabrication and imaging of nanometer-sized protein patterns. *Langmuir* **1999**, *15*, 8580–8583.

75 WADU-MESTHRIGE, K., AMRO, N. A., GARNO, J. C., XU, S., LIU, G.-Y. Fabrication of nanometer-sized protein patterns using atomic force microscopy and selective immobilization. *Biophys. J.* **2001**, *80*, 1891–1899.

76 LIU, G.-Y., AMRO, N. A. Positioning protein molecules on surfaces: a nanoengineering approach to supramolecular chemistry. *Proc. Natl Acad. Sci. USA* **2002**, *99*, 5165–5170.

77 ZHOU, D., WANG, X., BIRCH, L., RAYMENT, T., ABELL, C., AFM study on protein immobilization on charged surfaces at the nanoscale: toward the fabrication of three-dimensional protein nanostructures. *Langmuir* **2003**, *19*, 10557–10562.

78 KENSETH, J. R., HARNISCH, J. A., JONES, V. W., PORTER, M. D. Investigation of approaches for the fabrication of protein patterns by scanning probe lithography. *Langmuir* **2001**, *17*, 4105–4112.

79 JANG, C.-H., STEVENS, B. D., PHILLIPS, R., CALTER, M. A., DUCKER, W. A. A Strategy for the sequential patterning of proteins: catalytically active multiprotein nanofabrication. *Nano Lett.* **2003**, *3*, 691–694.

80 CASE, M. A., MCLENDON, G. L., HU, Y., VANDERLICK, T. K., SCOLES, G. Using Nanografting to achieve directed assembly of *de novo* designed metalloproteins on gold. *Nano Lett.* **2003**, *3*, 425–429.

81 GINGER, D. S., ZHANG, H., MIRKIN, C. A. The evolution of dip-pen nanolithography. *Angew. Chem. Int. Ed. Engl.* **2004**, *43*, 30–45.

82 HONG, S., ZHU, J., MIRKIN, C. A. A new tool for studying the *in situ* growth processes for self-assembled monolayers under ambient conditions. *Langmuir* **1999**, *15*, 7897–7900.

83 HONG, S., MIRKIN, C. A. A nanoplotter with both parallel and serial writing capabilities. *Science* **2000**, *288*, 1808–1811.

84 MIRKIN, C. A., HONG, S., DEMERS, L. Dip-pen nanolithography: controlling surface architecture on the sub-100 nanometer length scale. *ChemPhysChem* **2001**, *2*, 37–39.

85 SHEEHAN, P. E., WHITMAN, L. J. Thiol diffusion and the role of humidity in "dip pen nanolithography". *Phys. Rev. Lett.* **2002**, *88*, 1561041–1561044.

86 LEE, K.-B., PARK, S.-J., MIRKIN, C. A., SMITH, J. C., MRKSICH, M. Protein nanoarrays generated by dip-pen nanolithography. *Science* **2002**, *295*, 1702–1705.

87 KWAK, S. K., LEE, G. S., AHN, D. J., CHOI, J. W. Pattern formation of cytochrome *c* by microcontact printing and dip-pen nanolithography. *Mater. Sci. Eng. C* **2004**, *24*, 151–155.

88 CHO, Y., IVANISEVIC, A. SiO_x surfaces with lithographic features composed of a TAT peptide. *J. Phys. Chem. B* **2004**, *108*, 15223–15228.

89 HYUN, J., AHN, S. J., LEE, W. K., CHILKOTI, A., ZAUSCHER, S. Molecular recognition-mediated fabrication of protein nanostructures by dip-pen lithography. *Nano Lett.* **2002**, *2*, 1203–1207.

90 SMITH, J. C., LEE, K.-B., WANG, Q., FINN, M. G., JOHNSON, J. E., MRKSICH, M., MIRKIN, C. A. Nanopatterning the chemospecific immobilization of cowpea mosaic virus capsid. *Nano Lett.* **2003**, *3*, 883–886.

91 HYUN, J., LEE, W.-K., NATH, N., CHILKOTI, A., ZAUSCHER, S. Capture and release of proteins on the nanoscale by stimuli-responsive elastin-like polypeptide "switches". *J. Am. Chem. Soc.* **2004**, *126*, 7330–7335.

92 WILSON, D. L., MARTIN, R., HONG, S., CRONIN-GOLOMB, M., MIRKIN, C. A., KAPLAN, D. L. Surface organization and nanopatterning of collagen by dip-pen nanolithography. *Proc. Natl Acad. Sci. USA* **2001**, *98*, 13660–13664.

93 LIM, J.-H., GINGER, D. S., LEE, K.-B., HEO, J., NAM, J.-M., MIRKIN, C. A. Direct-write dip-pen nanolithography of proteins on modified silicon oxide surfaces. *Angew. Chem. Int. Ed. Engl.* **2003**, *42*, 2309–2312.

94 LEE, K.-B., LIM, J.-H., MIRKIN, C. A. Protein nanostructures formed via direct-write dip-pen nanolithography. *J. Am. Chem. Soc.* **2003**, *125*, 5588–5589.

95 NAM, J.-M., HAN, S. W., LEE, K.-B., LIU, X., RATNER, M. A., MIRKIN, C. A. Bioactive protein nanoarrays on nickel oxide surfaces formed by dip-pen nanolithography. *Angew. Chem. Int. Ed. Engl.* **2004**, *43*, 1246–1248.

96 LI, Y., MAYNOR, B. W., LIU, J. Electrochemical AFM "dip-pen" nanolithography. *J. Am. Chem. Soc.* **2001**, *123*, 2105–2106.

97 AGARWAL, G., NAIK, R. R., STONE, M. O. Immobilization of histidine-tagged proteins on nickel by electrochemical dip pen nanolithography. *J. Am. Chem. Soc.* **2003**, *125*, 7408–7412.

98 HAYNES, C. L., VAN DUYNE, R. P. Nanosphere lithography: a versatile nanofabrication tool for studies of size-dependent nanoparticle optics. *J. Phys. Chem. B* **2001**, *105*, 5599–5611.

99 FREY, W., WOODS, C. K., CHILKOTI, A. Ultraflat nanosphere lithography: a new method to fabricate flat nanostructures. *Adv. Mater.* **2000**, *12*, 1515–1519.

100 XIA, Y. N., GATES, B., YIN, Y. D., LU, Y. Monodispersed colloidal spheres: old materials with new applications. *Adv. Mater.* **2000**, *12*, 693–713.

101 JIANG, P., HWANG, K. S., MITTLEMAN, D. M., BERTONE, J. F., COLVIN, V. L. Template-directed preparation of macroporous polymers with oriented and crystalline arrays of voids. *J. Am. Chem. Soc.* **1999**, *121*, 11630–11637.

102 DENIS, F. A., HANARP, P., SURTHERLAND, D. S., DUFRENE, Y. F. Nanoscale chemical patterns fabricated by using colloidal lithography and self-assembled monolayers. *Langmuir* **2004**, *20*, 9335–9339.

103 MICHEL, R., REVIAKINE, I., SUTHERLAND, D., FOKAS, C., CSUCS, G., DANUSER, G., SPENCER, N. D., TEXTOR, M. A novel approach to produce biologically relevant chemical patterns at the nanometer scale: selective molecular assembly patterning combined with colloidal lithography. *Langmuir* **2002**, *18*, 8580–8586.

104 MARQUETTE, C. A., DEGIULI, A., IMBERT-LAURENCEAU, E., MALLET, F., CHAIX, C., MANDRAND, B., BLUM, L. J. Latex bead immobilisation in PDMS matrix for the detection of p53 gene point mutation and anit-HIV-1 capsid protein antibodies. *Anal. Bioanal. Chem.* **2005**, *381*, 1019–1024.

105 MACBEATH, G. Protein microarrays and proteomics. *Nat. Genet. (Suppl.)* **2002**, *32*, 526–532.

106 EGGERS, M., HOGAN, M., REICH, R. K., LAMTURE, J., EHRLICH, D., HOLLIS, M., KOSICKI, B., POWDRILL, T., BEATTIE, K., SMITH, S., VARMA, R., GANGADHARAN, R., MALLIK, A., BURKE, B., WALLACE, D. Microchip for quantitative detection of molecules utilizing luminescent and radioisotope reporter groups. *Biotechniques* **1994**, *17*, 516.

107 KUNZ, R. E. Miniature integrated optical modules for chemical and biochemical sensing. *Sens. Actuators B* **1997**, *38*, 13–28.

108 LEE, K.-B., KIM, E.-Y., MIRKIN, C. A., WOLINSKY, S. M. The use of nanoarrays for highly sensitive and selective detection of human immunodeficiency virus type 1 in plasma. *Nano Lett.* **2004**, *4*, 1869–1872.

109 WILSON, D. S., NOCK, S. Functional protein microarrays. *Curr. Opin. Chem. Biol.* **2002**, *6*, 81–85.

110 TEMPLIN, M. F., STOLL, D., SCHRENK, M., TRAUB, P. C., VOHRINGER, C. F., JOOS, T. O. Protein microarray technology. *Trends Biotechnol.* **2002**, *20*, 160–166.

111 SCHWEITZER, B., KINGSMORE, S. F. Measuring proteins on microarrays. *Curr. Opin. Chem. Biol.* **2002**, *13*, 14–19.

112 BLAWAS, A. S., REICHERT, W. M. Protein patterning. *Biomaterials* **1998**, *19*, 595–609.

113 BERGVELD, P. The future of biosensors. *Sens. Actuators A* **1996**, *56*, 65–73.

114 TEMPLIN, M. F., STOLL, D., SCHWENK, J. M., PÖTZ, O., KRAMER, S., JOOS, T. O. Protein microarrays: promising tools for proteomic research. *Proteomics* **2003**, *3*, 2155–2166.

115 WALT, D. R. Miniature analytical methods for medical diagnostics. *Science* **2005**, *308*, 217–219.

116 CHAN, W., C. W., NIE, S. Quantum dot bioconjugates for ultrasensitive nonisotopic detection. *Science* **1998**, *281*, 2016–2018.

117 BASELT, D. R., LEE, G. U., COLTON, R. J. Biosensor based on force microscope technology. *J. Vac. Sci. Technol. B* **1996**, *14*, 789–793.

118 WOUTERS, D., SCHUBERT, U. S. Nanolithography and nanochemistry: probe-related patterning techniques and chemical modification for nanometer-sized devices. *Angew. Chem. Int. Ed. Engl.* **2004**, *43*, 2480–2495.

119 MINNE, S. C., FLUECKIGER, P., SOH, H. T., QUATE, C. F. Atomic force microscope lithography using amorphous silicon as a resist and advances in parallel operation. *J. Vac. Sci. Technol. B* **1995**, *13*, 1380–1385.

120 MINNE, S. C., ADAMS, J. D., YARALIOGLU, G., MANALIS, S. R., ATALAR, A., QUATE, C. F. Centimeter scale atomic force microscope imaging and lithography. *Appl. Phys. Lett.* **1998**, *73*, 1742–1744.

121 MINNE, S. C., MANALIS, S. R., ATALAR, A., QUATE, C. F. Independent parallel lithography using the atomic force microscope. *J. Vac. Sci. Technol. B* **1996**, *14*, 2456–2461.

122 CHOW, E. M., YARALIOGLU, G., QUATE, C. F., KENNY, T. W. Characterization of a two-dimensional cantilever array with through-wafer electrical interconnects. *Appl. Phys. Lett.* **2002**, *80*, 664–666.

123 CHOW, E. M., SOH, H. T., LEE, H. C., ADAMS, J. D., MINNE, S. C., YARALIOGLU, G., ATALAR, A., QUATE, C. F., KENNY, T. W. Integration of through-wafer interconnects with a two-dimensional cantilever array. *Sens. Actuators A* **2000**, *83*, 118–123.

124 SULCHEK, T., GROW, R. J., YARALIOGLU, G., MINNE, S. C., QUATE, C. F. Parallel atomic force microscopy with optical interferometric detection. *Appl. Phys. Lett.* **2001**, *78*, 1787–1789.

125 HAFIZOVIC, S., BARRETTINO, D., VOLDEN, T., SEDIVY, J., KIRSTEIN, K.-U., BRAND, O., HIERLEMANN, A. Single-chip mechatronic microsystem for surface imaging and force response studies. *Proc. Natl Acad. Sci. USA* **2004**, *101*, 17011–17015.

126 AKIYAMA, T., STAUFER, U., DEROOIJ, N. F., LANGE, D., HAGLEITNER, C., BRAND, O., BALTES, H., TONIN, A., HIDBER, H. R. Integrated atomic force microscopy array probe with metal-oxide-semiconductor field effect transistor stress sensor, thermal bimorph actuator, and on-chip complementary metal-oxide-semiconductor electronics. *J. Vac. Sci. Technol. B* **2000**, *18*, 2669–2675.

127 VOLDEN, T., ZIMMERMAN, M., LANGE, D., BRAND, O., BALTES, H. Dynamics of CMOS-based thermally actuated cantilever arrays for force microscopy. *Sens. Actuators A* **2004**, *115*, 516–522.

128 BULLEN, D., CHUNG, S.-W., WANG, X., ZOU, J., MIRKIN, C. A., LIU, C. Parallel dip-pen nanolithography with arrays of individually addressable cantilevers. *Appl. Phys. Lett.* **2004**, *84*, 789–791.

129 WANG, X., BULLEN, D., ZOU, J., LIU, C., MIRKIN, C. A. Thermally actuated probe array for parallel dip-pen nano-lithography. *J. Vac. Sci. Technol. B* **2004**, *22*, 2563–2567.

130 ZOU, J., BULLEN, D., WANG, X., LIU, C., MIRKIN, C. A. Conductivity-based contact sensing for probe arrays in dip-pen nanolithography. *Appl. Phys. Lett.* **2003**, *83*, 581–583.

131 ZHANG, M., BULLEN, D., CHUNG, S.-W., HONG, S., RYU, K. S., FAN, Z., MIRKIN, C. A., LIU, C. A MEMS nanoplotter with high-density parallel dip-pen nanolithography probe arrays. *Nanotechnology* **2002**, *13*, 212–217.

132 HONG, S., MIRKIN, C. A. Multiple ink nanolithography: toward a multiple-pen nano-plotter. *Science* **1999**, *286*, 523–525.

133 SU, M., LI, S., DRAVID, V. P. Miniaturized chemical multiplexed sensor array. *J. Am. Chem. Soc.* **2003**, *125*, 9930–9931.

134 LUTWYCHE, M., ANDREOLI, C., BINNIG, G., BRUGGER, J., DRECHSLER, U., HABERLE, W., ROHRER, H., ROTHUIZEN, H., VETTIGER, P., YARALIOGLU, G., QUATE, C. F. 5 × 5 2D AFM cantilever arrays a first step towards Terabit storage device. *Sens. Actuators A* **1999**, *73*, 89–94.

135 KING, W. P., KENNY, T. W., GOODSON, K. E., CROSS, G. L. W., DESPONT, M., DURIG, U. T., ROTHUIZEN, H., BINNIG, G., VETTIGER, P. Design of atomic force microscope cantilevers for combined thermomechanical writing and thermal reading in array operation. *J. Microelectricalmech. Syst.* **2002**, *11*, 765–774.

136 VETTIGER, P., DESPONT, M., DRECHSLER, U., DURIG, U. T., HABERLE, W., LUTWYCHE, M., ROTHUIZEN, H., STUTZ, R., WIDMER, R., BINNIG, G. K. The "Millipede" – one thousand tips for future AFM data storage. *IBM J. Res. Dev.* **2000**, *44*, 323–340.

137 LI, L., CHEN, S., JIANG, S. Protein adsorption on alkanethiolate self-assembled monolayers: nanoscale surface structural and chemical effects. *Langmuir* **2003**, *19*, 2974–2982.

138 RENAULT, J. P., BERNARD, A., BIETSCH, A., MICHEL, B., BOSSHARD, H. R., DELAMARCHE, E., KREITER, M., HECHT, B., WILD, U. P. Fabricating arrays of single protein molecules on glass using microcontact printing. *J. Phys. Chem.* **2003**, *107*, 703–711.

139 MINNE, S. C., MANALIS, S. R., QUATE, C. F. Parallel atomic force microscopy using cantilevers with integrated piezoresistive sensors and integrated piezoelectric actuators. *Appl. Phys. Lett.* **1995**, *67*, 3918–3920.

4

Microcantilever-based Nanodevices in the Life Sciences

Horacio D. Espinosa, Keun-Ho Kim, and Nicolaie Moldovan

4.1
Introduction

Microcantilevers were initially developed for atomic force microscopy (AFM), where precise force sensing is critical for atomic-resolution imaging [1, 2]. Microcantilevers were required to have a low force constant and a high mechanical resonance frequency for a superior signal-to-noise ratio. Such requirements were met by cantilevers made by microfabrication techniques, which had been originally developed for integrated circuit (IC) process technology, but were later on applied to micrometer-scale silicon sensors and actuators [3, 4]. Microfabrication also permitted the integration of sharp tips at the free ends of microcantilevers for high lateral resolution in AFM scanning and manipulation. Microfabrication techniques further tailored microcantilevers to broaden the AFM techniques for probing, material delivery, manipulation, biomaterial sensing and lithography. Most of such applications take advantage of the high-precision positioning and subnanometer deflection-detection capabilities of scanning probe microscopy (SPM), which includes scanning tunneling microscopy (STM), AFM, near-field scanning optical microscopy (NSOM) and a plethora of conductive, capacitive, magnetic or thermal probing techniques. Among the extended use of microcantilever applications, patterning of biological materials at the submicron scale is of great importance to fabricating ultra-miniaturized bioanalytical tests and devices. Patterning using biological materials at the miniaturized scale has been pursued by many methods such as ink-jet printing [5, 6], photolithography [7], microcontact printing [8–10], microfluidic devices [11, 12] and "dip-pen" nanolithography (DPN) [13–15]. To date, the DPN technique, an AFM-based direct-write lithographic method, provides the best resolution. Features smaller than 100 nm in size containing patterned biomolecules have been obtained. In such direct-write patterning techniques, the tip sharpness is critical. Through micromachining, tip radii as small as a few nanometers have been obtained [16].

In the initial application of the DPN technique, microcantilevers were used to pattern self-assembled monolayers (SAMs) formed by the adsorption of alkanethiols onto gold surfaces [13]. The SAMs have extraordinary utility to control inter-

Nanotechnologies for the Life Sciences Vol. 4
Nanodevices for the Life Sciences. Edited by Challa S. S. R. Kumar
Copyright © 2006 WILEY-VCH Verlag GmbH & Co. KGaA, Weinheim
ISBN: 3-527-31384-2

facial characteristics, such as the adsorption of proteins and the attachment of cells, and for use in biological analysis including array-based high-throughput screening and diagnostic applications [14, 15]. Also, SAM-coated microcantilevers can detect specific chemical interactions via an optical beam-deflection technique [17]. Direct delivery of biomaterials without the use of a SAM as a binding agent has also been pursued. In biopatterning applications, it is usually required to maintain and transport biomaterials in a liquid environment. Hence, the incorporation of microfluidic systems has been attempted. For example, a commercial AFM tip with a small opening at its apex was reported [18]. These tips allow the on-demand deposition of small single droplets at a predetermined location on a sample. The demand to use liquid as a transport medium has created versatile micro- and nanofluidic tools, specifically engineered for biopatterning, based on the popular microcantilever architecture. Micromachining techniques made the development of such useful structures possible.

Practical implementation of biopatterning and sensing by SPM techniques requires high throughput. This is being pursued by means of microcantilever parallelization [19, 20]. A two-dimensional (2-D) array of independently addressable microcantilevers was developed for high-density data storage [21]. Although this design was conceived for thermal patterning, it is possible to imagine similar tools for biopatterning. With such a massively parallel biopatterning tool, the feature sizes of DNA [22] or protein [23] chips can be further miniaturized, leading to nanoscale assays. For instance, specialized microfluidic probes arranged in a 2-D array would be able to produce ultra-miniaturized, high-density assays, requiring only extremely small analyte volumes, and allowing the simultaneous processing and massively parallel integration of proteins. The massive parallelization at the nanoscale benefits the technique by decreasing the reaction time and by drastically increasing the statistical significance of the experiments. This has the potential to have a huge impact on applications in the life sciences, such as drug discovery and diagnosis devices.

In this chapter we describe how microfabrication techniques were utilized to build micro- and nanoscale devices which facilitated research and development in the life sciences. Microfabrication techniques in general are now available from many sources; books dedicated to topics of microfabrication for versatile SPM cantilever probes have also been published [16, 24, 25]. Furthermore, versatile principles and applications of the SPM techniques have been reviewed [26–30]. Without claiming to exhaust the field, we include a list of currently available books and review articles related to the construction and working principles of general microcantilever probes in Tab. 4.1 for easy reference. However, none of the existing books and reviews bring together the microfabrication aspects and life sciences applications, to discuss them in their close, technologically-targeted relationship. For example, the microfabrication of apertured tips was introduced [16], but their applications in life sciences were not considered for delivering biosamples. Similarly, articles on principles/applications lack information on how the microcantilevers were constructed. In addition, since the development of microfluidic probes for bio-nano applications is an emerging field, to the best of our knowledge, there are

Tab. 4.1. Books and review articles on microcantilever-based applications in the life sciences.

Subjects covered	Reference
Fabrication of cantilever probes for near-field optical applications	16
Principles of cantilever sensors for vibrating and force sensing applications; parallelization of recording; chemical surface sensing with nanochemical functionalized probes	24
Arraying probes and integration of sensing and actuation	25
Microfabrication	4
Artificial nose	32
AFM force spectroscopy	26–28
DPN applications	30
SPM for surface patterning	27–29
SPM imaging techniques	26, 28

no current review articles or books that provide a systematic treatment. We hope that this book chapter helps researchers to broaden their viewpoints in this hot topic, organize their thoughts to rise from microfabrication to principles and stimulate new ideas.

Our focus is on devices and methods for surface patterning that typically involve SPMs, although some other interesting devices and concepts will be briefly introduced, such as integration of microchannels. Section 4.2 gives a description of how conventional types of microcantilevers have been microfabricated and used in biotechnology. Beyond the use of conventional microcantilevers, many efforts have been made to integrate microfluidic systems in micro- and nanocantilever devices. Microfluidic systems are of interest because most applications in life sciences require liquid samples to be delivered or probed. Such advanced microcantilevers involving microfluidics are in Section 4.3 and their applications in biopatterning are presented in Section 4.4. Conclusions and outlook are given in Section 4.5.

4.2
Microcantilevers

The most commonly used cantilevers in SPM-based biopatterning and sensing are rectangular or V-shaped. The preferred materials in microcantilever fabrication are thin films, such as silicon oxide, silicon nitride, single crystal silicon, diamond or metals. Sharp protruding tips are necessary at the free end of the cantilever for high-resolution imaging or manipulation, allowing the interaction area between the tip and the sample surface to be minimized. The force constant of microcantilevers used in AFM has been accustomed by practical experience to be in the range from 10^{-2} to 10^2 N m^{-1} in order to achieve atomic resolution in the contact mode. The force constant k of a rectangular cantilever with fixed-free boundary condition is defined as follows:

$$k = \frac{Ewt^3}{4L^3} \tag{1}$$

where E is the modulus of elasticity, w is the width, t is the thickness and L is the length of the cantilever. Surface micromachining techniques are typically used to fabricate micro- and nanoscale cantilevers with or without integrated tips. The fundamental fabrication methods for AFM microcantilevers were established in the late 1980s and their basic concepts are still being used, but with improved features. However, since AFM started to be used for purposes other than for surface profiling, many modified and specific designs have been proposed, leading to a large diversity of micro- and nanocantilever probes. For example, microcantilevers with an apertured pyramidal tip were used to pattern surfaces with liquid contained in a small reservoir just above the tip.

In this section we describe microfabrication techniques of versatile micro/nanocantilevers and their applications in life sciences.

4.2.1
Microfabrication of Miniaturized Probes

The fabrication methods of cantilevers can be basically divided into two classes [31]. The first uses thin films deposited or diffused into a silicon substrate; in the second fabrication method, all parts of the cantilevered are micromachined out of bulk materials. In the first method, cantilevers beams were commonly formed from thin films such as silicon nitride and silicon oxide. Thin films were deposited on a (100) silicon substrate and then patterned through lithographic processes to define the cantilever shape. The cantilevers were released by etching the substrate, e.g. in aqueous potassium hydroxide (KOH) solutions or ethylenediamine/pyrocatechol/water mixtures (EDP) [4]. Depending on the requirements, sharp tips can be integrated on the cantilever, as discussed later.

Rectangular microcantilevers were employed as biological or biochemical detectors utilizing the bending by adsorption-induced surface stress. The advantage of cantilever-based biosensors is label-free detection and subnanometer deflection sensitivity. In biological sensing applications, the cantilever generates either a static deflection or a change in the resonance frequency when target molecules are adsorbed on the surface of the cantilever [32]. Measurements of such changes can be achieved by either electrical or optical means. Electrical methods include capacitance and piezoresistive sensing, whereas optical techniques include optical lever and interferometric methods.

Figure 4.1. SEM image of a microcantilever sensor array [33].

Arrays of silicon cantilevers were produced by a combination of dry and wet etching techniques [33], to form an "artificial nose", working on the principle of bending stress induced by surface adsorbed molecular species. Individual cantilevers were made (Fig. 4.1) with rectangular dimensions of 500 µm in length, 100 µm in width and 0.8 µm in thickness, resulting in a typical spring constant of 0.02 N m^{-1}.

A thiol-gold immobilization system [34] was used to form stressed monolayers to induce bending of the cantilevers in detection schemes for vapor thiols. For this application, silicon nitride cantilevers with a 20-nm gold layer evaporated on one side were used as sensors for gas-phase adsorption of alkanethiols [35]. The optical detection scheme of an AFM was used to measure deflections down to the picometer scale. Alkanethiol vapors were generated by placing a few microliters of alkanethiol in a closed glass beaker. After thermal equilibrium was reached, the cantilever was exposed to the alkanethiol vapor. The deflection was measured as a function of time to show how the chemisorbed alkanethiols caused compressive surface stress during the progressive self-assembly. With the same methodology, single-strand (ss) DNA [36, 37] and proteins [38, 39] were detected utilizing DNA hybridization and antigen–antibody interaction as a mechanism to cause surface stress change. The biomaterials were delivered onto the cantilever in liquid environment.

In addition to deflection detection, the change in resonance frequency was used for the detection of target materials adsorbed on a cantilever [40–42]. The cantilever was driven close to the resonance frequency by a piezoelectric actuator. The oscillation was detected by the AFM optical detection. The deposited material increased the mass of the cantilever system, thus decreased the resonance frequency. Assuming there is negligible stiffness change of the cantilever due to the layer of adsorbed materials, the mass change Δm can be calculated from the following equation [40]:

$$\Delta m = \frac{k}{0.72\pi^2} \left(\frac{1}{f_1^2} - \frac{1}{f_2^2} \right) \tag{2}$$

where k is the spring constant of the cantilever, and f_0 and f_1 are the resonance frequencies before and after adsorption, respectively. Simultaneous use of deflection and resonance frequency detection was also reported [43].

Furthermore, the force exerted by conformational changes of a surface-tethered DNA system was investigated on microfabricated cantilevers [44]. The ability to manipulate the direction and amplitude of bending of cantilevers envisioned potential for micromechanical machinery, including valves, switches and actuators triggered by molecular shape.

In order to integrate tips onto thin film cantilevers, molds for tips needed to be fabricated on the silicon substrate prior to the deposition of the thin film [45]. Either pits or convexities produced on a substrate were exploited as molds. Since molds were employed for tip integration, those types of techniques for thin-film cantilevers were frequently referred to as micromolding or microcasting techniques.

The fabrication of molding-pit cantilevers started with the formation of pits on a (100) silicon surface by an anisotropic etchant such as KOH solutions with a circular or square opening on a mask layer. Such an anisotropic etching produced a pyramidal pit delineated by (111) planes. Subsequently, a thin film was deposited to conformally coat the pit and patterned to delineate the cantilever shape. By removing the whole silicon substrate underneath the cantilever, a protruding tip integrated on the cantilever was obtained [45, 46]. Hybrid tip/cantilever structures were reported by replacing the tip materials with metals, such as tungsten [45] and gold [47], while keeping the cantilever body made of silicon nitride, with better mechanical properties. In a different design, Rasmussen and coworkers [48] reported having all parts of the probe, including the tip, cantilever and support, made out of electroformed metal. The general disadvantage of this type of molding technique is that the fabricated tips point into the substrate, which makes the substrate impractical to use as a chip handling body. Thus, the addition of an extra handling body is needed on the opposite side of the tip, prior to the releasing step. Wafer bonding [45] or electroplating techniques [48] were used to form a handling body. The sharpness of the released tip is determined by that of the molding pit on the silicon substrate, whose final tip angle is around $70°$ due to the characteristic of anisotropic etching of silicon [3]. However, the sharpness, as well as the aspect ratio, of the tips can be improved by sharpening the molding pit using thermal oxidation at around $950\,°C$ [49]. Oxidation at such a low temperature caused the growth of a nonuniform silicon dioxide film in the pit, resulting in a narrower angle between the walls of the pit, giving a sharper molded tip. With modified pits, tips with curvature radii as low as $110\,Å$ were obtained [49].

Convex tip molds were also used to integrate tips onto microcantilevers [45]. In this fabrication scheme, a precursor cap and isotropic etching were used to create a molding tip out of the silicon substrate. For this purpose, a masking layer (SiO_2 or Si_3N_4) was deposited on a (100) silicon wafer, followed by lithographically patterning the layer to produce a tip mask of square or circular shape. During a subsequent etching process, the tip mask was undercut to form a silicon post underneath. The silicon etching was a combination of isotropic and anisotropic plasma etching. Extended etching formed a sharp tip with the cap detached. Subsequent thermal oxidation at $1100\,°C$ forms a silicon oxide layer incorporating the tip. A cantilever was lithographically formed on the silicon oxide layer and finally released by removing the silicon substrate in a KOH solution. When a protruding tip is used as a mold, the molded tip faces in the opposite direction of the silicon substrate, which allows the substrate to be used to form a handling chip. An oxidation sharpening technique can also improve the geometry of convex-molded tips [50, 51]. This technique is based on an experimental study showing that the oxide growth is about 30% slower at silicon step edges than on flat surfaces at temperatures ranging between 900 and 950 °C [50, 51]. The effect is related to the stress-induced diffusion anisotropy of oxygen in silicon oxide, which allows oxidation sharpening to be used both for sharpening concave and convex silicon molds. The low temperature during the oxidation process is important, since at above 950 °C silicon oxide suffers a softening and flow, thus eliminating the stress and the diffusion anisotropy.

As for the materials for a cantilever/tip structure other than silicon oxide, single-crystal silicon doped with boron was used [52]. Also, a silicon nitride layer deposited onto a convex mold was reported [53].

Instead of thin film cantilevers, whole parts of the structure (i.e. tips, cantilevers and handling chips) can be micromachined entirely out of bulk silicon. In the scheme reported by Wolter and coworkers [31], the fabrication was done by etching a (100) silicon wafer through a rectangular etch widow in a mask layer on the reverse side until a thickness of twice the desired thickness of the cantilever remained. The cantilever was subsequently defined on a mask layer on the front side of the wafer, followed by resumed KOH etching until both sides met. The tip integration was achieved with a precursor and undercut. The thickness of the cantilever was controlled by the etching time in KOH. Accurate geometry control by timing the etching is generally impeded by the following factors: (a) the etching is not uniform on a wafer due to stirring-induced flow, liquid flow induced by bubble generation and strong proximity effects, and (b) the etch rate cannot be precisely controlled on every batch. Hence, etch stop techniques, such as boron doping [54, 55] and buried oxide [56, 57], were pursued to precisely define the cantilever thickness.

The molding techniques are particularly important for fabricating tips and cantilevers out of materials hard to pattern directly, such as diamond, silicon carbide, platinum, silicon elastomers, resins, etc. Some of these cases will be treated in more depth in the following sections.

Sharp integrated tips are essential in many SPM-based applications including lithography and manipulation of nano-objects in order to produce smaller nanopatterned or assembled structures. For example, such tips integrated on cantilevers allowed the manipulation of single molecules, and measurements of intermolecular [58–60] and intramolecular [61–63] forces. Force spectroscopy with SPM was typically achieved by functionalizing a tip with a partner suitable for specific interaction with a target molecule bound onto a substrate. When the tip and substrate were brought into contact, the molecular partners interacted. Subsequently, the cantilever was retracted to stretch the target molecule. Mechanical properties of the intermolecular bond established between partners were identified for biomolecular combinations including biotin–streptavidin, antibody–antigen, complemen-

Tab. 4.2. Types of fabrication methods for tip integration on cantilevers.

Fabrication type	Notes	Reference
Molding on a pyramidal pit	monolithic tip/cantilever	48, 83
	hybrid tip/cantilever	45, 47, 77
Molding on a convexity	pyramidal convexity	20, 53
	conical convexity	91
Monolithic etching	doping for etch stop	20, 54, 55
	buried oxide for etch stop	56, 57

tary strands of DNA and interactions between proteins. By stretching molecules bound between a tip and a substrate, unfolding events were measured to determine rupture forces associated with various numbers of base parings [61, 62]. Many applications in SPM force microscopy may be found in the literature [26].

The fabrication methods of the aforementioned types of cantilevers with integrated tips are summarized in Tab. 4.2. In the following sections we discuss how microcantilevers with integrated tips have been used for nanofabrication and manipulation.

4.2.2
Cantilever Probes for Nanopatterning

Tips integrated on microcantilevers may function as pens or quills, to locally deliver molecules previously present on the tip surface. The tip sharpness allows sub-100 nm patterning with various molecular species. The domain becomes relevant for life sciences if these molecules achieve a local functionalization of the surfaces, such that specific biochemical experiments can be conducted at these length scales. DPN is a direct-write lithographic technique which uses AFM tips to locally deposit materials on a variety of surfaces [13, 30]. The deposited material and the substrate usually have to be paired, such that a chemical reaction occurs upon delivery, or a SAM is formed, so that a reading is possible using the same AFM technique. In some cases, this reaction or SAM formation is not needed, such as the case of simple patterning with fluorescent dyes, when the reading can be achieved optically. In the DPN technique, the species form a thin molecular layer on the tip surface, such that the tip geometry is not significantly altered. When the tip is placed in contact with a target surface, the molecules migrate from the tip onto the surface and the time of surface contact directly correlates with the amount of materials transferred. DPN was originally reported to pattern gold surfaces with a solution of alkanethiols [13], and its applications have been subsequently extended to patterning surfaces with versatile types of materials such as biomolecules [14, 15, 64], polymers [65], small organic materials [66], sol precursors [67] and metal salts [68]. One of the advantages of the DPN technique is that biomolecules, such as DNAs or proteins, can be patterned both by direct-write and indirect assembly [14, 15, 69–72], which can be utilized to build nanoscale biomolecular sensor arrays with higher sensitivity and selectivity due to much smaller sample volumes. For example, modified ssDNAs with a thiol group on one end were patterned on a gold surface and used to capture complementary DNA sequences tagged with nanoparticles [14]. In this way a pattern of gold particles could be assembled (bottom up) on the surface, in a completely different way than by thin film deposition and etching or lift-off (traditional top-down techniques). Feature sizes ranging from a few micrometers to less than 100 nm were achieved.

The water meniscus formed by capillary condensation was suggested as the mechanism of molecular transfer from a tip to a substrate in DPN and the formation of such a meniscus was recently confirmed using an environmental scanning electron microscopy (SEM) technique [73]. Once transferred to the substrate, the molecules spread across the surface depending on the humidity, temperature, reactivity of the ink with the substrate and contact radius of the tip. Assuming the environmental conditions and types of ink and substrate are optimized, the patterning of small feature sizes is critically dependent on the radius of curvature of the tips integrated on AFM cantilevers. The most commonly used probes in the DPN technique are made of silicon nitride with integrated pyramidal tips, which are typically fabricated using the pyramidal-pit-molding technique [45]. For higher resolution of DPN writing, sharper tips improved by the oxidation-sharpening technique [49] are usually employed. The dimensions of the cantilevers meet the reported range of desired force constants of the cantilever 0.03–0.3 N m^{-1} [74].

Once a tip is coated with molecules of interest, patterning is typically controlled by commercially available AFM instruments to precisely deposit desired amounts of molecules at controlled locations. DPN writing has been typically performed with a single probe; however, patterning large areas with a single tip, due to its serial nature and limited AFM scan size, is very inefficient. In an effort to improve the throughput of the DPN technique, the feasibility of parallel DPN patterning with a commercially available tip array was demonstrated in a commercially available AFM [74].

Furthermore, linear arrays of high-density probes with integrated tips for DPN were microfabricated [19, 20]. Two types of DPN probe arrays were developed using surface micromachining techniques. The first type, or type-1 probe array, was made out of thin-film silicon nitride using the molding technique with a protruding tip, whereas the second type, or type-2 probe array, was fabricated from heavily boron-doped silicon (Fig. 4.2). The type-1 probe array consists of 32 straight probes in a 1-D arrangement with the space between consecutive probes being 100 μm. The dimensions of an individual cantilever were 400 μm long, 50 μm wide and 0.6 μm thick. The type-2 array had eight probes separated from each other by a

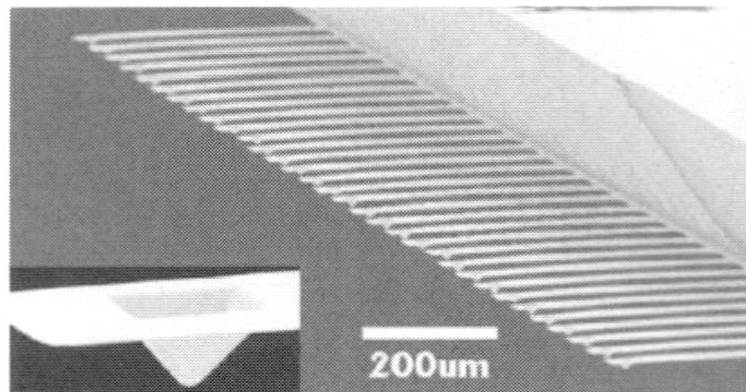
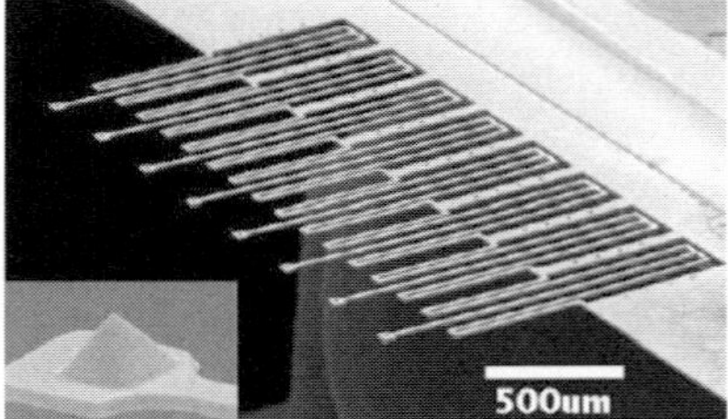

Figure 4.2. SEM micrographs of DPN probe arrays with the type-1 (left) and type-2 (right) arrays [20].

spacing of 310 µm, whereas each cantilever was in a multifold configuration. The dimensions are 1400 µm long, 15 µm wide and 10 µm thick.

The fabrication of the type-1 probe started with growing a thin oxide film on a (100) silicon wafer. The oxide film on the front side was patterned to serve as precursor caps for the subsequent etching (Fig. 4.3a). Anisotropic etching in KOH solution undercut the precursors to create tips protruding out of the silicon substrate until the tips had small flat top surfaces (Fig. 4.3b). Subsequently, a thermal oxidation process sharpened the tip and removed the precursor cap (Fig. 4.3c). As a result, uniform sharpness was obtained for the produced tip array. A layer of silicon nitride film was deposited through a low-pressure chemical vapor deposition (LPCVD) process and then patterned to form the shanks of the individual probes (Fig. 4.3d). The cantilevers were released in the ensuing anisotropic wet etching in ethylene-diamine pyrocatechol (EDP) solution at 95 °C (Fig. 4.3e).

The fabrication of the type-2 probes started from a silicon wafer having two extra layers on the top, consisting of a 10-µm layer heavily doped with boron and a 10-µm silicon layer grown epitaxially on top of the doped layer. The wafer was thermally oxidized to form a 500-nm oxide layer, from which tip precursors were lithographically patterned (Fig. 4.4a). Tips were created from the epitaxial silicon layer by underetching the silicon oxide precursor caps (Fig. 4.4b). Thermal oxidation followed to sharpen the tips (Fig. 4.4c). Another layer of oxide film was formed to protect the front side during the subsequent etching (Fig. 4.4d). The bulk silicon substrate was etched away using EDP until the boron-doped etch-stop layer was reached (Fig. 4.4e). A final reactive ion etching (RIE) process was performed to define the cantilever shape (Fig. 4.4f).

This fabricated array of probes was used to pattern gold surfaces with 16-mercaptohexadecanoic acid (MHA) and octadecanethiol (ODT), and to demonstrate parallel DPN capabilities in a conventional AFM. Only one cantilever was monitored by the optical beam-deflection scheme of the AFM, while the others in the array followed in a passive manner. Eight duplicate copies of patterns with a 60-nm feature size were generated by parallel-writing, while only one cantilever was used for the deflection sensing.

In addition, a linear array of 26 tips was operated in a parallel fashion to deposit patterns of MHA on a gold surface to form a mask for a subsequent gold-etching

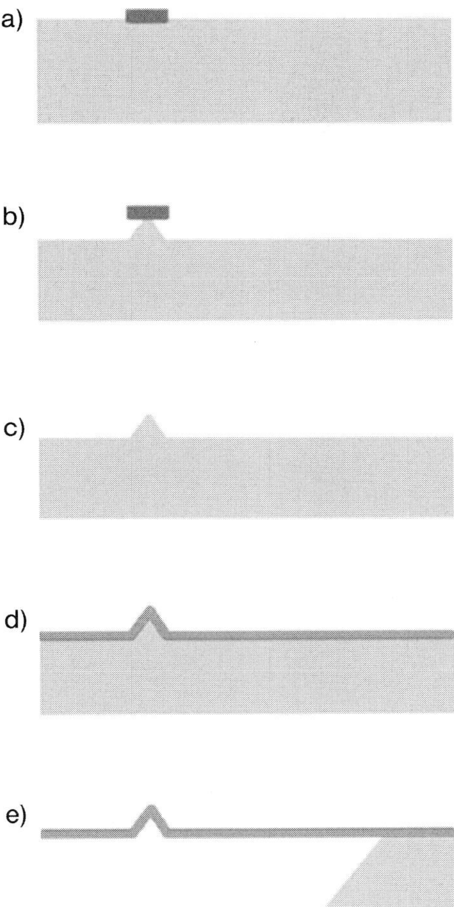

Figure 4.3. Fabrication process of a type-1 probe array [20].
The sequence is described in detail in the text.

step [75] (Fig. 4.5). Results obtained demonstrated that the line widths were nearly identical without any writing failure at individual tips (Fig. 4.6). Patterning with an array of three 26-tip arrays produced more than 34 000 dot features in 7.3 min. Furthermore, arrays containing up to 250 probes performed parallel DPN writing in a high-throughput fashion over the centimeter length scale. Patterning of multilayer organic thin films was also demonstrated in parallel fashion [76].

By addressing the throughput issue of DPN with microcantilever arrays, it was envisaged that parallel-probe lithography would lead to many applications in high-resolution patterning over large areas with the aforementioned versatile inks such as biomolecules, organic materials, polymers, sol precursors, metal salts, etc. [30].

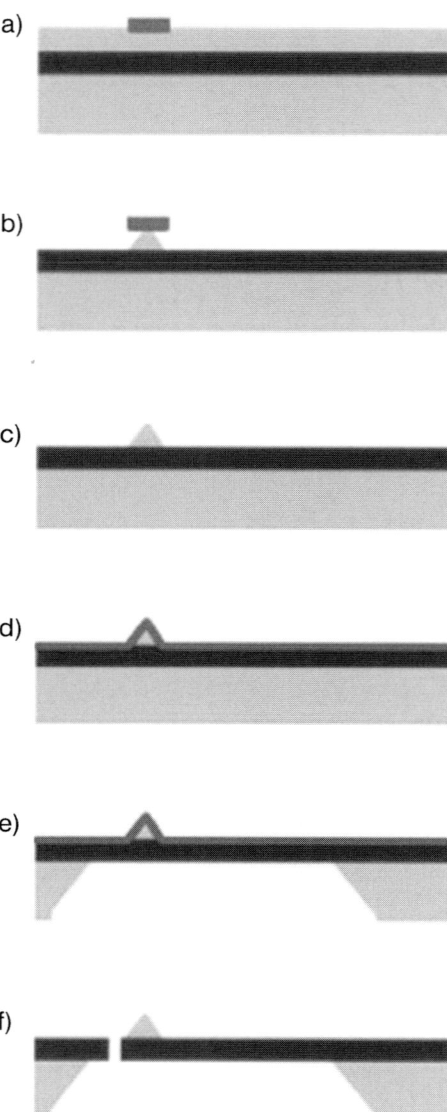

Figure 4.4. Fabrication process of a type-2 probe array [20]. The sequence is described in detail in the text.

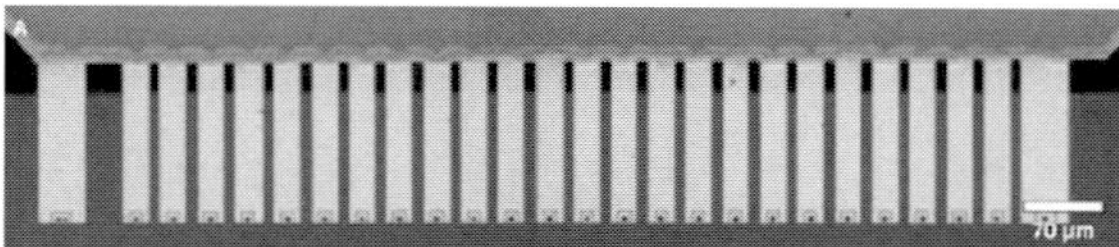

Figure 4.5. Optical micrograph of a 26-pen array [75].

Figure 4.6. SEM image of gold patterns generated through the parallel DPN technique using a linear 26-pen array [75].

4.2.3
Elastomeric AFM Probes

Hard materials, especially silicon nitride, have been the preferred tip materials for DPN patterning probes. However, tips made of soft materials such as silicon elastomers were also reported [77]. Poly(dimethylsiloxane) (PDMS) has been widely used in microcontact printing, and is known to be compatible with a wide range of chemicals and biological media [78]. A molded AFM tip was made of PDMS and integrated at the end of a polymide cantilever (Fig. 4.7). The microfabricated tip demonstrated DPN patterning capabilities with ODT as ink and achieved dot sizes as small as 330 nm in diameter.

The fabrication sequence of the PDMS probe is presented in Fig. 4.8. A tip mold was made on a (100) silicon wafer using an oxide mask layer via wet etching in EDP (Fig. 4.8a). After the oxide layer was etched, an aluminum film was deposited as a sacrificial layer (Fig. 4.8b). A 10:1 mixture of Sylgard 184 silicone elastomer base and curing agent was coated on the wafer (Fig. 4.8c). Excessive liquid-state PDMS was plowed out using a rubber blade leaving the PDMS only in the pit and curing was performed at 90 °C for 30 min (Fig. 4.8d). A thin polymide layer was spin-coated over the processed surface and patterned to define a cantilever shape

Figure 4.7. Optical micrograph of a probe chip with two cantilevers with elastomeric tips. Inset shows a SEM image of the integrated PDMS tip [77].

(Fig. 4.8e). A thick PDMS piece was bonded on the patterned polymide layer as a handling body (Fig. 4.8f). An aluminum etchant removed the sacrificial aluminum layer to release the cantilever/tip structure without damaging either the PMDS tip (Fig. 4.8g) or the silicon mold, which could be reused after cleaning.

The polyimide cantilever in this example was 400 μm long, 100 μm wide and 4 μm thick. The minimal printed ODT dot diameter was consistent with the PDMS probe radius of curvature of 300 nm.

Another use of PDMS for DPN printing was reported by Zhang and coworkers, who coated commercially available silicon nitride tips with PDMS and used them to pattern surfaces with sub-100-nm resolution [79]. The advantage of using PDMS as the tip material is that it provides higher coating efficiency, especially when the inks are macromolecules such as biomolecules and polymers, for which rich experience is available from the domain of microcontact printing with PDMS stamps [78, 80]. The advantage comes at the expense of a serious decrease in writing resolution, due to the soft nature of the contact between the tip and substrate. In contrast, DPN with silicon nitride tips needed functionalization of the tip surface to improve ink-coating efficiency when used for DNA patterning [14, 81], but achieved 100-nm resolution.

4.2.4
Monolithically Fabricated Conductive Diamond Probes

Since DPN is performed in contact-mode operation, wear of tips is expected during extended use. Tips made of the aforementioned materials such as silicon nitride, silicon oxide, silicon and PDMS would be good enough for laboratory-level experiments; however, the use of wear-resistant tips is of great importance to probe arrays for high-throughput manufacture of bioassays. Hard materials are typically employed to enhance the wear characteristics and diamond is the hardest known material. Furthermore, tailoring its surface properties would allow diamond to become a platform material to construct bioinorganic interfaces [82].

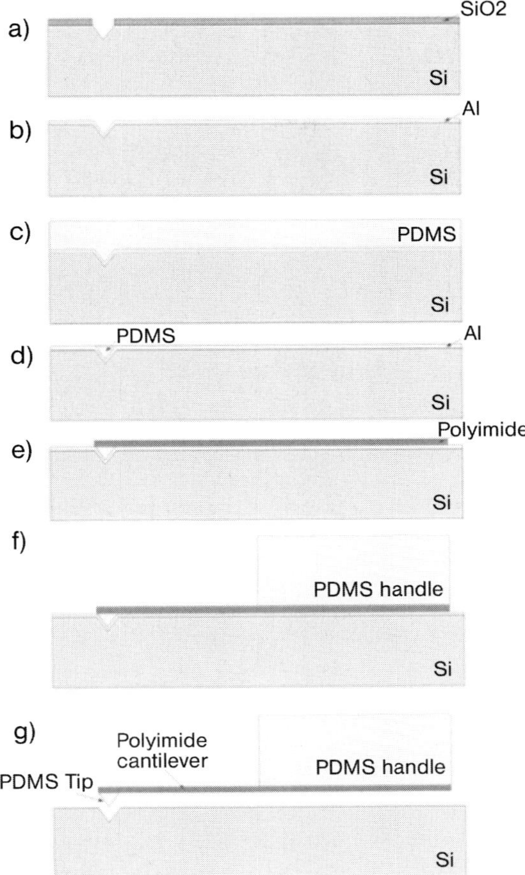

Figure 4.8. Fabrication process of the cantilever with an integrated elastomeric tip [77]. The sequence is described in detail in the text.

A wear-resistant probe for contact-mode AFM techniques was fabricated using ultra-nanocrystalline diamond (UNCD) [83]. The all-diamond probe demonstrated molecular writing capability in DPN mode and potential use in scratch-based lithography. The probe was monolithically microfabricated with a pyramidal tip utilizing the molding technique on pits. Electroplating was utilized to form a handling body. The diamond was microfabricated to be either electrically conductive by nitrogen doping or nonconductive. The growth of diamond film was achieved by microwave plasma CVD (MPCVD) using a methane/argon mixture, which also contained nitrogen in the case of nitrogen-doped films [84]. This doped film permits AFM potentiometry and a large variety of conductive AFM techniques [85, 86] with an extended tip life cycle. Furthermore, diamond is chemically and biolog-

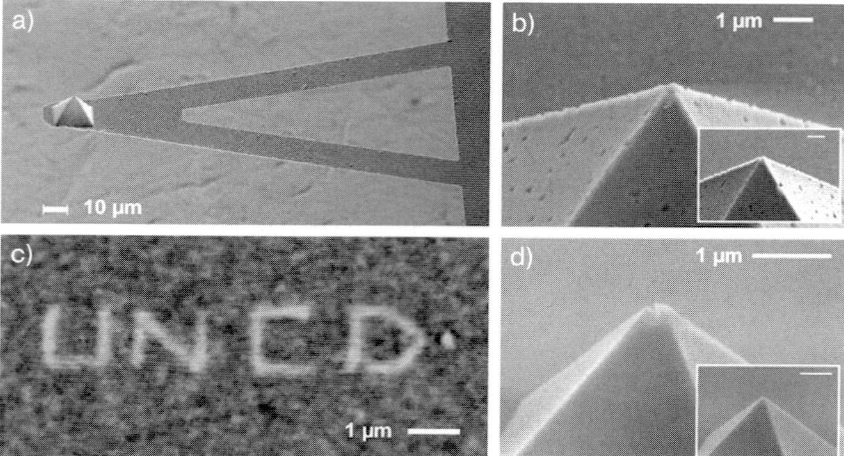

Figure 4.9. (a) SEM image of a UNCD cantilever with a tip. (b) SEM image of a UNCD tip after 1 h of scanning on a diamond substrate. Inset shows the tip before the scanning. (c) Frictional AFM image of alkanethiol monolayer patterned on a gold substrate by a UNCD tip. Patterning and imaging were performed by the same tip. (d) SEM image of a commercially available silicon nitride tip after 1 h of scanning with the same parameters used for the UNCD tip in (b). It shows damage at the tip apex; the inset shows the tip prior to the test.

ically inert, and forms an excellent electrode for electrochemical applications. It also has a tunable hydrophobicity that can be achieved by surface functionalization [82].

Two types of cantilevers were fabricated: V-shaped and arrow-shaped. The lengths and thicknesses of both cantilevers were 170 and 0.8–1.4 µm, respectively. The arm width of the V-shaped cantilever was 18.8 µm, whereas the rectangular one was 12 µm wide. The V-shape cantilevers had a stiffness of around 2 N m^{-1}, and were better suited for contact and tapping mode AFM techniques, while the rectangular cantilevers had a stiffness of around 1 N m^{-1} and performed well in lateral force imaging, well-suited for imaging DPN patterns. The fabrication started by forming of a thermal oxide on a (100) silicon wafer to be used as a mask in subsequent microfabrication steps and patterning square holes into it lithographically. Pyramidal pits were formed by KOH etching to be used as a molds for the future tips (Fig. 4.10a) and then sharpened by thermal oxidation utilizing the nonuniform growth of the oxide in the pit (Fig. 4.10b). The wafer was immersed in a diamond powder solution in methanol, followed by ultrasonication for uniform seeding. The growth of a diamond layer was achieved by MPCVD using a combination of argon and methane gases, with or without nitrogen gas (Fig. 4.10c). An aluminum layer was deposited by electron beam evaporation and then patterned to define cantilevers (Fig. 4.10d). The UNCD film was etched using the aluminum as a mask by oxygen RIE (Fig. 4.10e), followed by the removal of the aluminum mask (Fig. 4.10f). An etch window was patterned on the reverse side

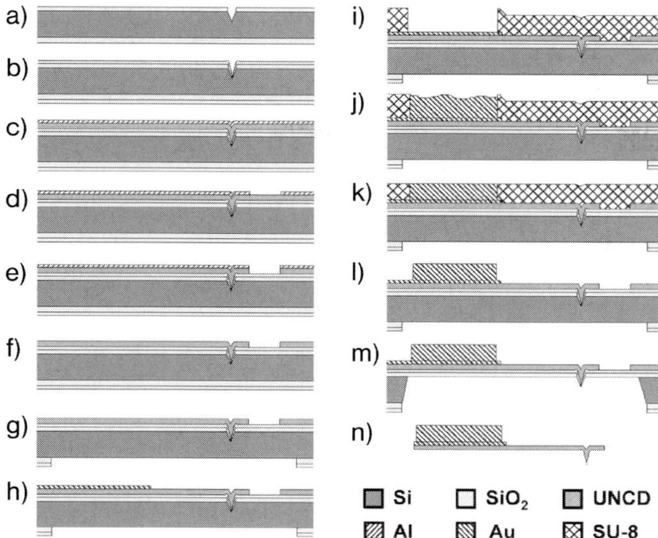

Figure 4.10. Microfabrication steps for UNCD probe [83]. Detailed steps are described in the text.

oxide layer for subsequent release of the cantilevers and chips (Fig. 4.10g). A seeding layer for electroplating was deposited and patterned on the front side (Fig. 4.10h). A thick photoresist (SU-8) was spin-coated and patterned to form a mold for the handling body of the chip (Fig. 4.10i). A layer of gold was then electroplated up to 300 μm thickness onto the seed layer to form the chip body (Fig. 4.10j). The top side of the gold was then flattened by polishing (Fig. 4.10k). After removing the photoresist (Fig. 4.10l), the chip was released in KOH solution (Fig. 4.10m).

The DPN compatibility of the diamond probe was tested. A commercial AFM (Digital Instruments; Dimension 3100) was used for the writing test using MHA as ink and gold substrates to write on. Lines produced were around 200 nm wide, whereas the smallest dots generated had a diameter of around 80 nm. In addition to the compatibility with the DPN technique, the UNCD probe demonstrated AFM imaging capabilities of topography and lateral force when used for scanning right after the patterning (Fig. 4.9c). Surface scratch testing was also performed, promising a favorable tool for scratch nanolithography. Furthermore, the scalability of the probes batch-fabricated with the integration of diamond films could lead to 1- and 2-D arrays of probes that could be used for massively parallel DPN-based fabrication of nanostructures. Such probes have the potential to perform DPN for long working times, virtually wear free.

We have described microfabrication aspects of simple microcantilevers in several applications related to the fabrication and manipulation of nanostructures and the detection of minute volumes of substances. Although microcantilevers are relatively simple structures, they have served as powerful tools for such purposes.

However, advanced applications demand more functionality from the devices. In the next section we discuss advanced microcantilevers that allow delivery of fluid materials and suspensions.

4.3
Cantilevers with Integrated Micro- and Nanofluidics

Material delivery in suspension has many potential applications, especially when the materials are macromolecules, such as biomolecules and polymers, nanoparticles, catalysts, and nanotubes. Controlled delivery of such materials would permit the fabrication of complicated nanostructures and nanoelectromechanical systems (NEMS). Also, in life sciences, it is oftentimes required to deliver large-size molecules and it is challenging to carry out scanning probe-based lithography with such molecules if the microcantilevers are not equipped with microfluidic systems. The implementation of micro- and nanofluidics into microcantilevers has been pursued to achieve material delivery in suspension as well as surface patterning with high resolution. Dispensing femtoliter or smaller volumes was achieved with a modified version of a conventional silicon nitride microcantilever [53]. Split-pin-type spotters for the fabrication of gene chips have been further miniaturized by microfabrication [87, 88]. Micro- and nanopipette-based AFM probes were used to deliver minute amounts of liquids [89]. Micropipette arrays with buried microchannels were micromachined for drug injection [90]. A nanofountain probe (NFP) was fabricated and tested, having an on-chip reservoir, cantilevers with embedded microchannels and high-resolution dispensing tips [91, 92].

In this section we describe microfabrication techniques for micro- and nanofluidic cantilevers and their applications along with their potential impact in life sciences.

4.3.1
Apertured Pyramidal Tips

The on-demand dispensing of single liquid attoliter droplets using AFM probes was demonstrated with modified tips containing a simple orifice. Meister and co-workers [18] created an aperture at the apex of a commercially available probe using focused ion beam (FIB) milling. The hollow reverse side of the pyramidal tip was used as a reservoir to store the liquid. Flow of the liquid was initiated simply by contacting the pipette with a surface and stopped by lifting it from the surface. Created features had dimensions of 100 nm and below. As the FIB process is serial in nature and precludes high-volume production, microfabrication techniques were employed to build directly apertured pyramidal tips integrated on cantilevers [53]. The device has potential applications in nanofabrication for the direct deposition of versatile materials, including biomolecules, catalysts, etch resists and nanoparticle suspensions. With the reservoir on the reverse side, extended writing was possible without reloading.

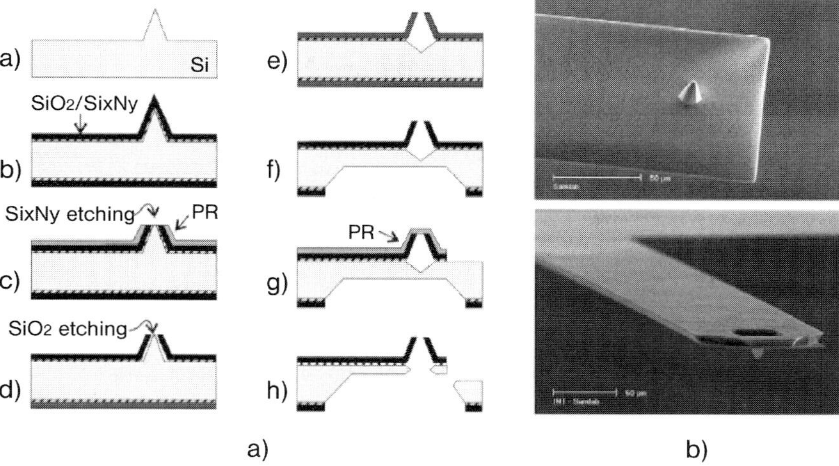

Figure 4.11. Microfabrication process for cantilevered probes with apertured tips [53]. Detailed steps are described in the text.

The fabrication process for the cantilevers with the apertured tips is illustrated in Fig. 4.11. The probes consisted of a hollow silicon nitride tip on a compound cantilever of silicon and silicon nitride. The molding technique on a pyramid was employed to integrate the tip on the cantilever. A molding tip was first fabricated on a silicon (110) wafer by underetching a precursor cap (Fig. 4.11A-a). The wafer was thermally oxidized, followed by LPCVD deposition of a silicon nitride layer to form a mold (Fig. 4.11A-b). Photoresist was spin-coated in such a manner that the tip apex was uncovered. With this opening at the tip apex, only that portion of the silicon nitride layer was etched by RIE (Fig. 4.11A-c). Through the opening of the silicon nitride layer, hydrofluoric acid etched the underlying silicon oxide layer (Fig. 4.11A-d) and, subsequently, KOH etching removed the underlying mold pyramid to produce an empty space surrounded by a pyramidal silicon nitride shell (Fig. 4.11A-e). A combination of photolithography and RIE patterned the silicon nitride and silicon oxide layers on the reverse side to open an etching window on the reverse side (Fig. 4.11A-f), while a cantilever was defined on the compound layer from the front side (Fig. 4.11A-g). Etching in KOH solution not only released the cantilever, but also opened a hole on the silicon portion of the hollow space of the tip to be used as a reservoir (Fig. 4.11A-h). The final thickness of the cantilever was 7–8 μm.

The patterning of the microfabricated probe was tested using alkanethiols to be deposited on gold substrates as well as Cy3 fluorescent dye in glycerol to be deposited on glass. The loading of the hollow probe was done by hand pipetting under an optical microscope. The probe had a pre-patterned loading area to avoid wetting the entire cantilever while loading the solution. A lower limit on the size of the deposition was given by the diameter of the aperture (around 100 nm to 1.5 mm). In addition, it was a challenge to routinely load ink into the reservoir island located at the end of the cantilever. Furthermore, evaporation constituted a major problem in this approach, preventing long-term writing.

4.3.2
Open-channel Cantilevered Microspotters

In microspotting technologies, a biochemical sample is loaded into a spotting pin by capillary action, and a small volume is transferred to a solid surface by physical contact between the pin and the solid substrate [93]. Other than microspotting pins, capillaries or tweezers can act as a printhead of biochemical samples. Printheads are moved by an *xyz* motion control system and brought into a contact with a surface to transfer pre-made substances. In an effort to fabricate massively parallel, high-density DNA, protein and cell chips, microfabricated cantilevers have been used as printheads (Fig. 4.12). Microcantilever spotters were used for depositing biological samples [87, 94] and nanoparticles [95].

Arrays of microcantilevers were also produced by microfabrication techniques. Deposition was achieved by direct contact between the cantilevers and the surface by capillary transport. An electrowetting technique for controlling surface tension was applied for the loading of the liquid. A passivated aluminum layer integrated on the cantilevers was employed as an electrode for the electrowetting. The sample loading required a droplet of a solution containing biological samples to be placed on a conductive solid substrate. After the end of cantilevers was dipped into a droplet of liquid to be deposited, an electric field was applied between the electrode and the substrate to increase the affinity of the liquid. The electric-field introduced a modified charge distribution that changed the free energy, causing the liquid to spread and wet the surface of the cantilevers [96]. As a result, the height of liquid rise on the cantilever surface was increased.

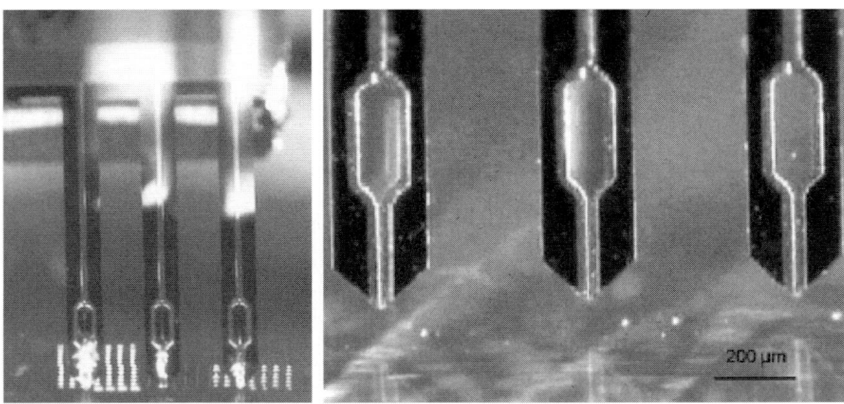

Figure 4.12. Optical photographs of microfabricated cantilever spotters [87, 95].

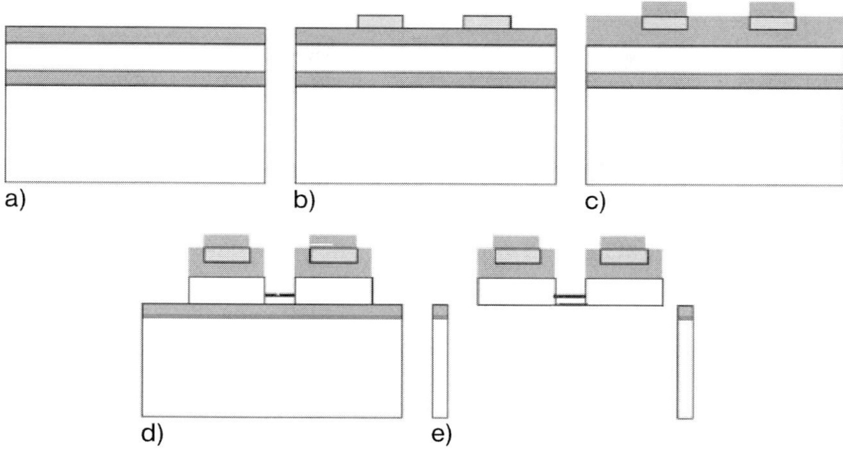

Figure 4.13. Microfabrication process of cantilevers [87]. Detailed steps are described in the text.

Figure 4.13 illustrates the fabrication steps of the cantilever microspotter as described by Belaubre and coworkers [87]. The microfabrication used a silicon-on-insulator (SOI) substrate with a 5-µm thick top silicon layer. First, a layer of low-temperature silicon oxide (LTO) was deposited by LPCVD (Fig. 4.13A). Aluminum electrodes were defined by a lift-off process (Fig. 4.13B). A second layer of LTO was deposited for passivation purposes (Fig. 4.13C). Cantilever shanks were defined by photolithography to be used as a mask for the following etching of the LTO layers by RIE. Subsequently, the top silicon layer of the SOI wafer was etched by another RIE step (Fig. 4.13D), in which microchannels and microreservoirs were formed on the cantilevers. Finally, the reverse side of the wafer was removed through etch windows in a deep RIE step. Etching was stopped by the buried oxide layer of the wafer. The silicon oxide layer was then etched from the reverse side by RIE to release the cantilevers (Fig. 4.13E). Arrays of cantilevers, with a spacing of 450 µm, were microfabricated, where each cantilever was 2 mm long, 210 µm wide and 5 µm thick.

The microfabricated cantilevers were used to pattern a glass slide with 1-pL volumes of a solution containing Cy3-labeled oligonucleotides (15mers) [94]. In addition, protein, anti-goat IgG (rabbit), microarrays were generated on a glass slide coated with dendrimer molecules as crosslinkers. With both molecules, 30-µm diameter spots were obtained. It was also demonstrated that two different biological samples were deposited with the same cantilevers using a cleaning procedure, in which no cross-contamination was observed.

Furthermore, microspotter cantilevers were used to deposit functionalized amorphous silica nanospheres on coated silicon surfaces by Leichle and coworkers [95]. Colloidal solutions, containing poly(ethylene glycol) (PEG)-600 and aminopropyltriethoxysilane (APTS) nanoparticles with diameters of around 300 and 150 nm,

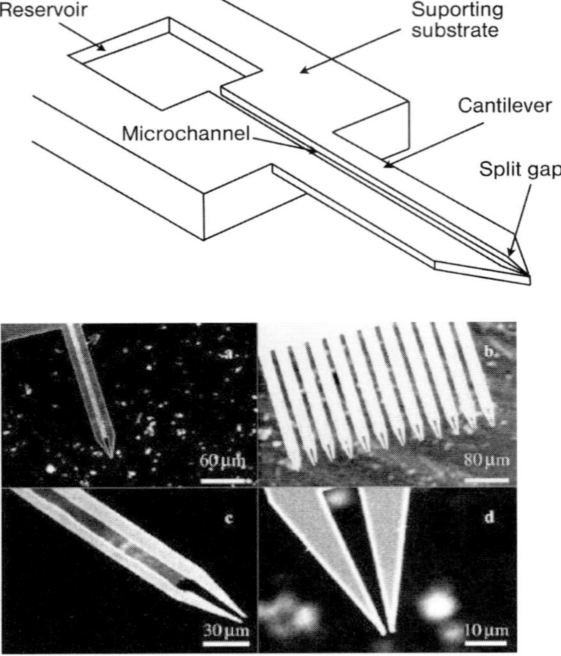

Figure 4.14. Schematic of the quill-type cantilever and SEM images of fabricated cantilever [88].

respectively, were deposited on surfaces to form spots of various diameters ranging from around 10 to more than 100 µm.

The minimum spot size writable with a common cantilever spotter is around 10 µm in diameter. To achieve higher resolution, cantilever microspotters were further miniaturized. A microcantilever-based tool with a 1-µm wide split gap at the end (Fig. 4.14) has been reported by Xu and coworkers [88]. Named the surface patterning tool (SPT), this device had integrated on-chip sample reservoirs and fluid transportation microchannels, which addressed limitations inherent in the use of the conventional AFM probes. An SPT consisted of a cantilever with a split gap at the end, a reservoir on the handling chip, and a 1-µm deep transportation microchannel connecting the gap and the reservoir. Sample loading was carried out by filling the reservoir with sample solutions as well as by dipping the cantilever end into sample fluid. These new designs, dedicated to biomolecular patterning, allowed reliable patterning of large molecular species and reduced reloading requirements.

The length of the SPT cantilevers ranged between 200 and 300 µm, and the width between 20 and 40 µm. The split gap was around 1 µm wide and around 40 µm long. At the fixed end of the cantilever, a 10-µm deep rectangular reservoir was located on the handling substrate. The depth and width of the microchannel

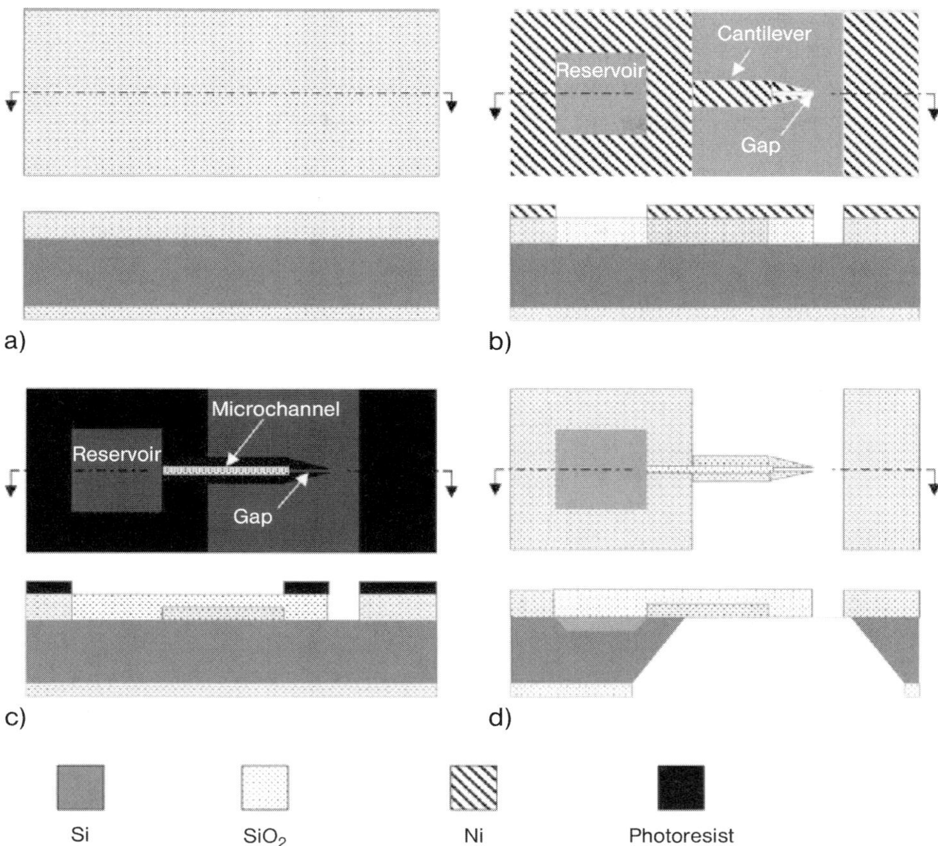

Figure 4.15. Schematic diagram of the microfabrication processes for the quill-type SPT [88].

were 1 and 1–10 µm, respectively. Silicon oxide was selected as the material for the cantilever due to its advantages in terms of mechanical properties, fabrication requirements and biocompatibility.

The fabrication steps of the SPT are shown in Fig. 4.15. A (100) double-side polished silicon wafer was thermally oxidized to have a thickness of 2–3 µm. The front side oxide layer was lithographically processed to define a cantilever shank, a split gap and a reservoir. In order to fabricate the cantilever with the 1-µm gap using conventional photolithography, a nickel layer was electroplated on top of the oxide to be used as mask, using a negative photoresist pattern. The oxide layer was then anisotropically etched by RIE with a gas mixture of CHF_3 and SF_6 (50:1 cm^2) and power of 50 W at 50 mTorr pressure. After patterning the cantilever and the gap, a 1-µm deep microchannel was etched by RIE. Finally, a window was patterned on the backside layer of SiO_2, followed by releasing the cantilever in KOH solution of concentration 35 wt%, at 80 °C.

Testing of the fabricated SPTs was performed using a dedicated commercial instrument called a NanoArrayer (BioForce Nanosciences, Ames, IA). This instrument was equipped with a precision motion control system and an environmental chamber. Although this instrument used an optical lever deflection scheme employed in an AFM, it did not scan or acquire images. SPTs were mounted to form a 12° angle with the deposition substrates such that only the tip end was in contact with the substrate. Patterning was demonstrated using Cy3–streptavidin in a standard protein patterning application [88]. The Cy3–streptavidin sample solution was loaded into the reservoir using a micropipette, prior to mounting the SPT into the-NanoArrayer. Spots were patterned on a dithiobis(succinimidyl undecanoate)(DSU) monolayer coated on a gold surface. Fluorescence microscopy was utilized to analyze the patterned features. A 10 × 10 array of spots with a diameter of 2–3 µm was routinely obtained. With a single loading, more than 3000 spots could be printed in about 1 h. It was also demonstrated that quantum dots conjugated to streptavidin could be deposited in patterns of lines and spots using the SPT [97]. Those features had line widths of around 150 nm to 7 µm and spot diameters of 3–5 µm.

Multiple cantilever-based SPTs for multiplexed biomolecular arrays were also reported by Xu and coworkers [98]. Their SPT featured a 1-D array of microcantilevers and a corresponding microfluidic network that was capable of transporting multiple fluid samples from macro-scale reservoirs located on the SPT substrate through micro-scale channels to the distal end of the cantilevers. Five cantilevers and five reservoirs were arranged in a chip so that multiple biological samples could be transferred from the reservoirs to the SPT cantilever array. The overall size of an SPT chip was 3 × 6 mm². Each cantilever was 250 µm long, 30 µm wide and 2 µm thick, and consecutive cantilevers were separated by a spacing of 50 µm. The microchannel on each cantilever was 15 µm wide and 1 µm deep.

The fabrication steps for the multiple-cantilever SPT are illustrated in Fig. 4.16. A double-side polished (100) silicon wafer was first thermally oxidized to form a 2.2-µm thick oxide film. The oxide layer on the front side was photolithographically patterned to define the cantilevers, reservoirs and channels, and etched by RIE (Fig. 4.16a). Subsequently, a 1-µm deep microchannel was etched on the cantilevers by overlay photolithography and follow-up RIE (Fig. 4.16b). In the next step, the oxide layer on the reverse side was patterned to form an etching window and the cantilevers were released by KOH etching (30 wt%). A wafer holder was used to protect the front side oxide pattern during the KOH step. Finally, once leakage began to occur through the wafer holder in the KOH etching bath, the wafer was released from the hold and dipped in a 40 wt% KOH solution to form the reservoirs and microchannels (Fig. 4.16c).

The multiple ink loading and patterning were tested using two types of fluorescent proteins: Cy2–donkey anti-goat IgG and Texas Red–donkey anti-rabbit IgG. These two proteins were diluted in phosphate-buffered saline (PBS) and alternatively loaded into the five reservoirs by hand pipetting. The solutions transferred from the reservoirs to the distal end of each channel by capillary action and the fluids were confined inside the microchannels without observed cross-contamination. A DSU/gold surface was patterned to generate 10 × 10 multiple

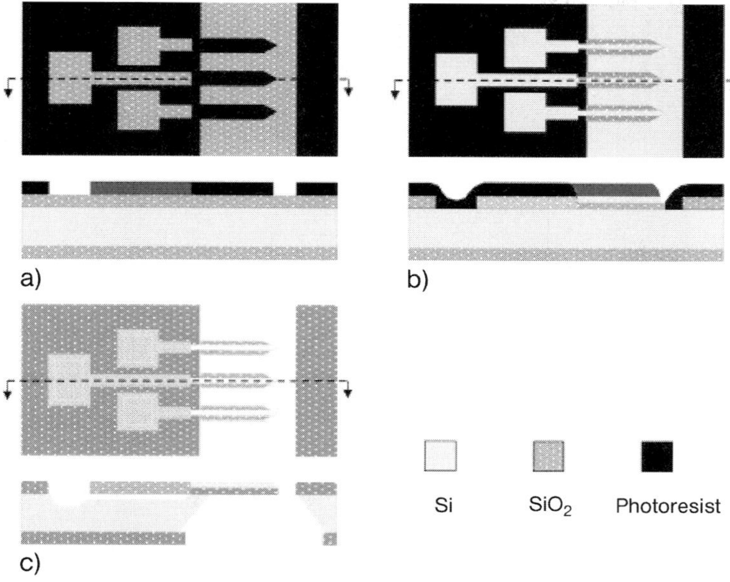

Figure 4.16. Fabrication steps for an array of SPT cantilevers [98].

ink dot arrays, with the mean spot diameter being about 12 µm. The SPTs generated biological arrays with a routine spot size of 2–3 µm. Several thousand spots could be printed without reloading. It was reported that the minimum spot size of the SPT was mainly limited by its gap width. The gap size can be further reduced with a higher-resolution lithography technique.

The SPTs have open microchannels integrated on cantilevers. This type of open channel has advantages of being clog-free, and allowing easy cleaning and simple microfabrication. However, such open microfluidic elements including microchannels and reservoirs are prone to cross-contamination via vapor by different types of samples, especially when loaded in arrays of cantilevers [99]. Also, evaporation may be critical in some applications, although its rate can be reduced with environmental conditioning. Enclosed microchannels are beneficial in such cases, although they are relatively difficult to microfabricate and the clogging issue needs to be addressed. Pipettes are conventionally used microfluidic devices with enclosed channels. Microneedles with embedded microchannels were also demonstrated to deliver liquid materials. In the following subsections, microcantilever devices with enclosed microchannels are described.

4.3.3
Closed-channel Cantilevered Nanopipettes

Pipettes are essential tools in biomedical applications in order to precisely manipulate and deliver fluidic samples, and they have been miniaturized for applications of local delivery of liquid/gaseous materials at the nanometer scale. It was reported that apertures as small as 3 nm at the tip with outer diameters at the tip of 10 nm were produced [89]. Such nanopipettes permit delivering or probing liquid samples in small volumes. In addition to delivering precise volumes of samples, accurate positioning is pivotal in biomedical applications. Nanopipettes were used in AFM for precise positioning to deliver liquid or gaseous materials on surfaces for localized chemistry [89, 100–104]. Nanopipette-based probes were used to deliver minute amounts of materials such as Cr etchant [89], photoresist [103], DNA [104], proteins [100] and enzymes [101, 102]. They are capable of continuous sample feeding through the capillary. In certain applications, nanopipettes were bent to have a cantilevered portion so that they could be used in AFM. The deflection, or contact force, of such cantilevered micropipettes was regulated by the AFM optical beam-deflection scheme, whereas straight nanopipettes, not compatible with the optical deflection scheme of AFM, employed other techniques such as shear-force feedback control [103] or an electrical current between the pipette and solution [104].

Protein solutions readily flowed through cantilevered nanopipettes to be directly deposited on a surface with dot diameters of around 200 nm [100]. A proteolytic enzyme was used to locally break a layer of bovine serum albumin (BSA) [101, 102]. Photoresist was delivered through the aperture formed at the end of a pulled nanopipette [103]. In this case, direct-write with photoresist replaced conventional photolithographic steps including spin-coating, mask alignment, exposure to UV light and developing. A nanopipette was used in a solution environment, whereas the tip–sample separation was controlled by the feedback signal of the ionic current established between two electrodes – one in the pipette and the other in the solution [104].

Production of cantilevered pipettes is based on pulling quartz capillary tubes by computerized systems called micropipette pullers. These pullers apply a controlled axial load on the capillaries, while heating them locally with a flame, electrical solenoid or, for better control, a laser beam. Two pipettes can be obtained after the pulled capillary breaks. A desired bending can be produced to make the probes usable in scanning probe systems [105, 106] (see Fig. 4.17).

The bending can be obtained close to the tip by heating the micropipette over a microflame. The operation can be performed in specialized tools called bevelers. Such cantilevered glass micropipettes can be coated with metals in order to allow the optical deflection detection scheme of AFM to function (Fig. 4.17). The diameter of the cantilevered portion of the micropipette can be around 12 μm and the cantilever length around 300 μm. Such dimensions of the glass cantilevers resulted in resonance frequencies up to 400 kHz and force constants ranging from tenths to tens of Newtons per meter. It was also reported that a filament made of metal thread was installed inside a pipette to mechanically strengthen the pulled tapered region [103]. The filament improves the wetting characteristics of the inner wall of the pipette such that the liquid can spontaneously fill the pipette lumen.

Taha and coworkers [100] reported micropipette printing of a yeast protein onto aldehyde-coated glass slides, in which an aldehyde group on the surface reacted

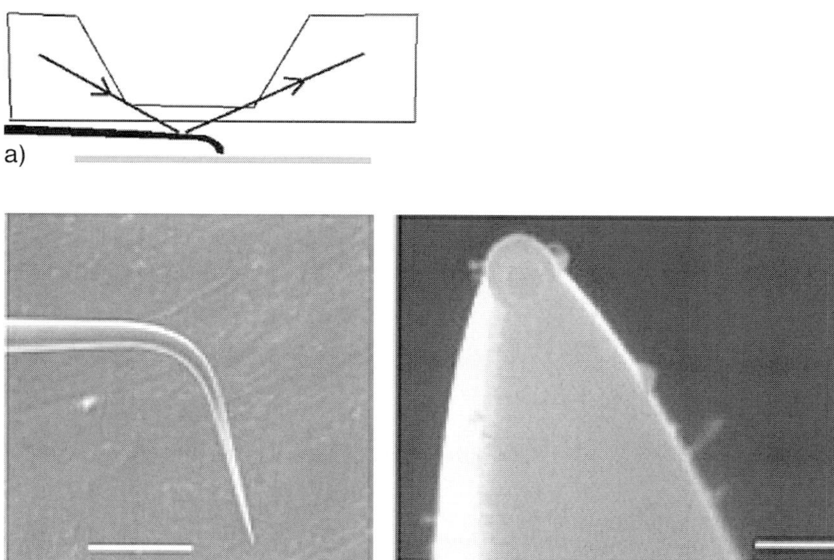

Figure 4.17. (a) Experimental setup of the cantilevered nanopipette in an AFM. (b) SEM image of a cantilevered nanopipet. Scale bar: 100 µm. (c) Close-up of the tapered end of a 50-nm aperture nanopipet. Scale bar: 500 nm [101].

preferentially with the primary amino group of the N-terminus of the protein for immobilization as described in Ref. [23]. A solution of the protein in PBS 100 µM with 10% glycerol to keep the proteins hydrated was prepared and manually loaded from the large end of the pipette with a syringe. The capillary action delivered the solution to the end of the pipette. Subsequently, the nanopipette deposited a line of the protein which was around 500 nm wide and around 40 nm thick. The experiment was performed at room temperature without humidity control. The same type of pipette was also used to pattern dots and lines of green fluorescent protein (GFP) on BSA-coated glass slides, resulting in dot diameters of around 250 nm and line widths of around 450 nm. This method demonstrated that the feature size can be made 1000 times smaller than with conventional spot arrayers.

A nanopipette filled with a solution of DNA or protein was utilized by Bruckbauer and coworkers [104] to pattern surfaces in an aqueous environment with a voltage applied between the pipette and surface. This method was based on scanning ion conductance microscopy (SICM) [107]. In this report, a pipette was filled with a solution of 100 nM DNA or protein and the pipette tip was inserted into a bath of ionic solution. The ion current was used to control the dispensing rate of molecules as well as the tip-to-surface distance. Spots of ssDNA labeled with Rhodamine Green were deposited on a streptavidin-coated glass surface and the measured full width at half-maximum (FWHM) was 830 ± 80 nm. Spots of Protein G were also patterned onto a positively charged glass surface utilizing electrostatic interaction as an immobilization mechanism. Measured feature sizes were 1.3 μm.

Although pulled micropipettes are capable of delivering small volumes of samples with precise positioning, it is a challenge to accomplish reproducibility and scalability when expansion to arrays is required, which is essential for tools in bio-nano applications, e.g. to fabricate microarrays or nanoarrays. Micromachined pipettes in the following section will give an idea how microcantilevers with channels can be microfabricated to address the reproducibility and scalability of conventionally produced micropipettes.

4.3.4
Micromachined Hypodermic Needle Arrays

Microcantilevers with closed microchannels were microfabricated to be used as micropipettes and microneedles for drug injection applications. The applications of micromachined cantilevered needles are different from the microcantilevers discussed in the preceding sections in terms of size (larger) and functionality (not for surface patterning). However, they are integrated with microchannels to effectively deliver biofluid into a body via injection and the microfabrication approach is worth discussing because it can be utilized to create embedded microchannels in microcantilevers. Microcantilever-based hypodermic needles are sometimes referred to as in-plane microneedles since they are fabricated in the plane of a silicon wafer as contrasted to out-of-plane needles, where short microneedles are microfabricated normal to the plane of the wafer surface [108, 109]. Microcantilever-based needles have been fabricated using several structural materials such as silicon [110, 111], polysilicon [112] and metals [90, 113].

Micromachined pipette arrays were reported by Papautsky and coworkers [90]. They have integrated microchannels on cantilevers as well as dispensing apertures at the end of the cantilevers. The cantilevers were 1.5 mm long and 400 μm wide, and the microchannels had an inner cross-sectional area of 400×30 μm^2 (W × H). For fabrication, a (100) silicon wafer was doped with boron to form a 4–6-μm p$^+$ layer, which later served as an etch stop. Then, a silicon nitride layer was deposited by plasma CVD to be used as a mask during the following anisotropic etching in KOH (Fig. 4.18a). Next, a palladium layer was selectively electroplated onto the silicon wafer to form the undersides of the needles (Fig. 4.18b). A thick photoresist layer (P4620) was deposited and photolithographically patterned as a sacrificial layer to define the inner lumen of the needles (Fig. 4.18c). A palladium layer was sputter-deposited to conformally cover the patterned thick photoresist layer, this was followed by the electrodeposition of a thin additional layer of palladium on top of the sputtered metal. The primary structural material was then electroformed to complete the top and side walls of the pipette (Fig. 4.18d). The sacrificial photoresist was dissolved by immersing the wafer in acetone (Fig. 4.18e). The p$^+$ membrane was removed by plasma etching with SF$_6$ gas from the reverse side to release the micropipettes (Fig. 4.18f).

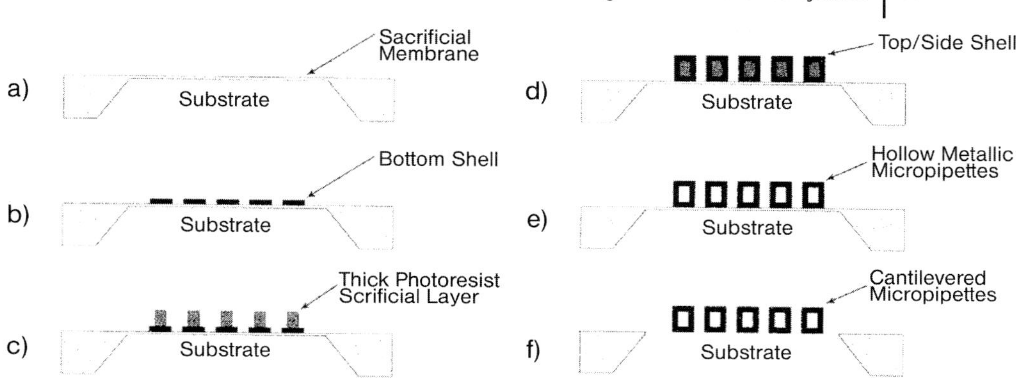

Figure 4.18. Fabrication steps for microneedle arrays [90].

The microfabricated microneedles were used for high lane density slab-gel electrophoresis. Capillary action was utilized to load samples into the microneedle. The samples were dispensed into wells of electrophoretic micro-gels using a syringe pump demonstrating a 2-fold increase in the number of theoretical plates and a 6-fold reduction in lane spacing compared to standard mini-gel separations. Although the micromachined pipettes were not intended for surface patterning, their closed microchannels demonstrated a means to effectively deliver biofluid by capillary action.

4.3.5
NFPs

NFPs can be defined as microfluidic dispensing probes that incorporate a reservoir, continuously feeding the ink to the tip, which can write with sub-100-nm resolution [91, 114]. On-chip reservoirs were met in several applications already described. Apertured pyramidal tips incorporated a small reservoir at the hollow back of each tip [53]. Cantilever microspotters were capable of containing liquid samples in the split at the end of the cantilever [87], while SPTs had on-chip reservoirs for storing larger volumes of samples [88]. However, the minimum resolution of all those tools was limited by the size of the aperture or the width of the gap. It seemed to be challenging to make patterns in the sub-100-nm region. Also, the open microfluidic components in the aforementioned devices resulted in restricted control of evaporation and possible cross-contamination when multiple tips/inks are used. From such a point of view, nanopipette-based probes showed advantages [89]. Since nanopipettes are fabricated by individual glass capillary pulling, difficulties arise in fabricating arrays of uniform probes. Also, the minimum feature size patterned by nanopipettes is limited since it depends critically on the aperture size, which typically leads to a larger dimension of the patterns due to the formation of an external liquid meniscus around the pipette tip during patterning.

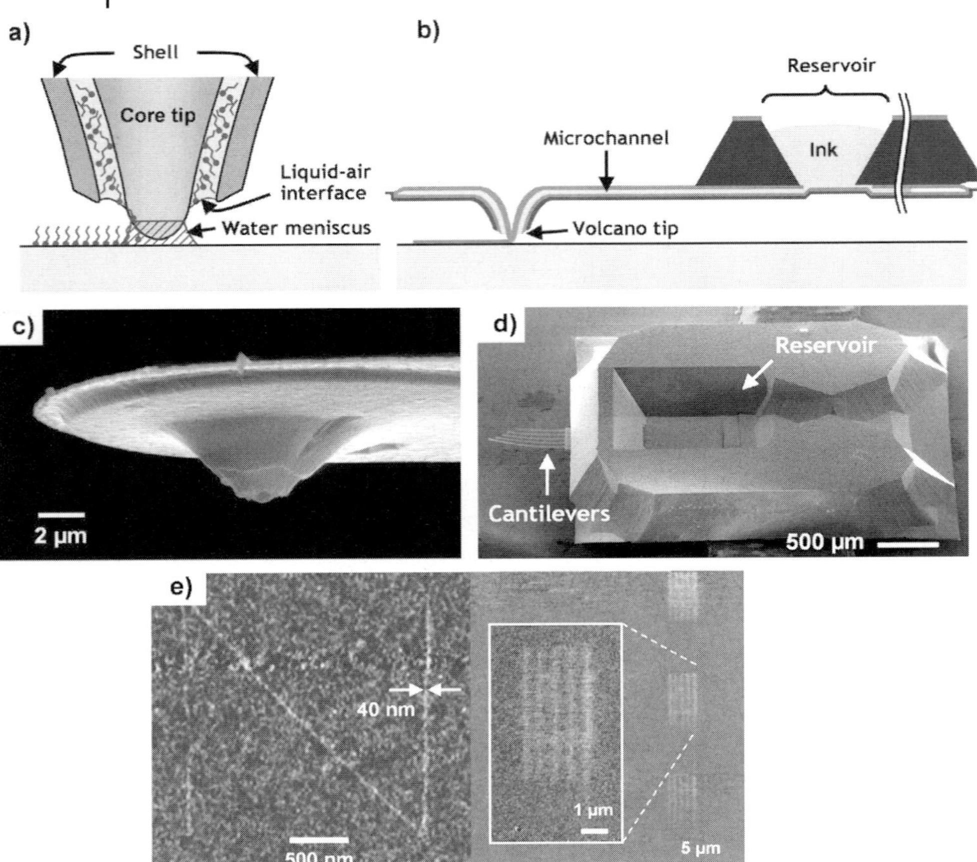

Figure 4.19. (a) Writing mechanism of the NFP device. A molecular ink drawn from an on-chip reservoir forms a liquid–air interface at the annular aperture of the volcano-like dispensing tip. Molecules are transferred by diffusion from the interface to a substrate and a water meniscus is formed by capillary condensation. (b) Liquid from the reservoir is delivered to the dispensing tip via capillary force. (c) SEM image of a dispensing tip. (d) SEM image of an NFP chip showing cantilevers and an on-chip reservoir. (f) Frictional AFM images of features patterned by an NFP. Patterns with line widths as small as 40 nm have been successfully generated.

Both high-resolution patterning and continuous sample feeding through closed microchannels were achieved by NFP. Figure 4.19 illustrates a first-generation microfabricated NFP device and its molecular-writing results. A volcano-like dispensing tip was configured to have a ring-shaped aperture through which samples were dispensed to generate sub-100-nm lines on a routine basis. Features with a line width as small as 40 nm were patterned, which experimentally demonstrated high-resolution fountain-pen-mode writing, i.e. a writing mode in which ink is transported from the on-chip reservoir to the substrate. An embedded microchannel and an on-chip reservoir continuously replenished the tip with liquid samples. Due to the unique volcano-like dispensing tip of the NFP, both continuous feeding and high-resolution writing were possible. The resolution of the NFP was controlled by the radius of the tip since it preserved the DPN-mode writing, compared to other techniques, where the aperture size or the gap width were considered as the dominant factors limiting the resolution. While the nanopipettes, apertured pyramidal tips and quill-type SPTs work by the formation of an outer meniscus between the probe and substrate, the NFP forms a meniscus between the ultra-sharp AFM-like tip and substrate, like traditional DPN.

The batch-fabrication processes developed for the NFP chips provided straightforward scaling-up to NFP arrays. Using standard microfabrication technologies, chips with five fountain-pen probes were batch-fabricated (Fig. 4.20). The fabrication steps started with the conventional tip formation process using a precursor cap defined by mask M1 (Fig. 4.20a). During the oxidation sharpening process, the precursor tip fell off (Fig. 4.20b). A silicon nitride film with a thickness of around 0.3 µm was deposited by LPCVD to form the floor of the channels. Mask M2 defined the space through which the on-chip reservoir would be connected to the channels. Layers of silicon oxide with a thickness of around 0.5 µm and silicon nitride with a thickness of 0.3–0.5 µm were deposited by LPCVD to form the sacri-

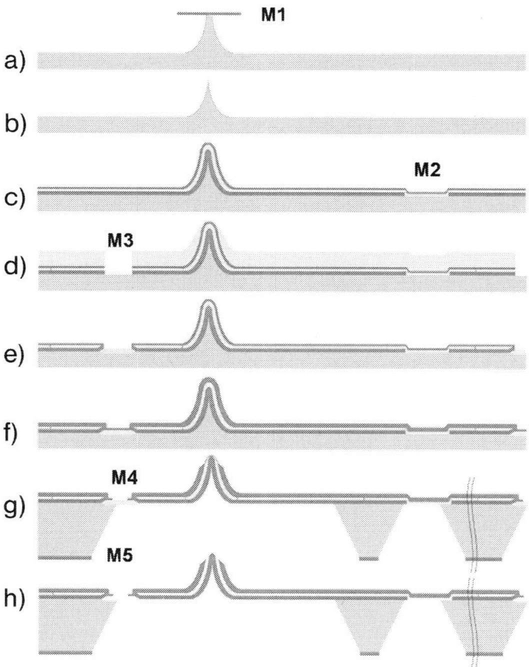

Figure 4.20. Microfabrication steps of NFP [114].

ficial layer and the ceiling layer of the microchannels, respectively (Fig. 4.20c). Lithography with mask M3 followed by RIE in CF_4 plasma defined the in-plane geometry of the channels, where the channels follow the edges of the pattern comprised in mask M3. A buffered oxide etching (BOE) solution was used for a controlled underetching of the structures, to provide the lumen of the microchannels (Fig. 4.20d). Subsequently, the lateral openings of the channels were closed by bird's beak oxidation (Fig. 4.20e). Sealing of the channels followed using either an LPCVD silicon nitride layer or an electron-beam evaporated gold layer (Fig. 4.20f). Lithography with mask M4 was conducted to pattern the sealing layer with the geometry of the cantilevers and chip boundaries. The reverse-side nitride was lithographically processed with mask M5 and subsequently etched by CF_4 RIE. Anisotropic etching in KOH solution formed the on-chip reservoir and the handling body (Fig. 4.20g). After the removal of the oxide, the chip remained suspended by small, easy-to-break silicon bridges, to provide good wafer-level maneuverability (Fig. 4.20h). After this step, a thin gold layer was deposited on the reverse side of the cantilevers to provide sufficient reflectivity for the optical beam-deflection sensing system of the AFM.

With the expansion to 1- or 2-D NFP arrays incorporating multiple on-chip reservoirs, simultaneous patterning of surfaces with multiple biological inks can be carried out. Potential applications include high-throughput manufacturing of nanosensors and nanoarrays. For example, in the DPN technique, a substrate was patterned with a MHA monolayer and subsequently antibodies were attached to be

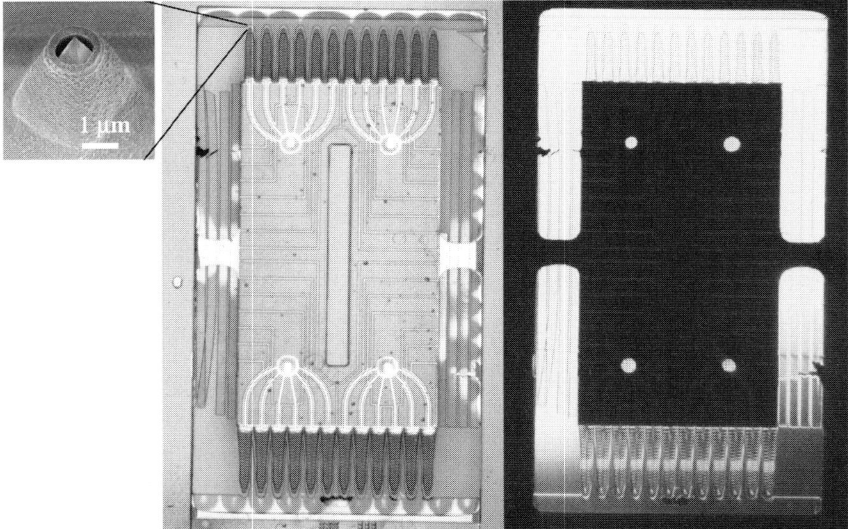

Figure 4.21. Second-generation NFP chip. Left: reflected light optical micrograph of the top side of the chip, showing two reservoirs feeding each six microchannels on each side of the chip. Right: transmitted light optical micrograph, showing the reservoir wells. Inset shows a SEM detail of the dispensing volcano-shape tip.

ready to sense antigens. With the NFP, most of the intermediate steps may not be necessary. It is also important to emphasize that NFP patterning does not require repeated dipping of the tip and specific treatment of the substrate for protein printing. A second-generation NFP device was recently fabricated by the same group, allowing two on-chip reservoirs, each feeding a linear array of six cantilevers with nanodispensing probes (Fig. 4.21). The new chip presents some advantages regarding channel sealing and fabrication, an increased uniformity and sharpness of the nanodispensing tips, a better control of the longitudinal and lateral bending stiffness of the cantilever and the possibility to add independent actuation of the probes and electrical contacts for conductive NFP lithography and voltamography.

4.4
Applications

We now describe applications of cantilevers for patterning biomolecules, including DNA, protein and viruses.

4.4.1
Patterning of DNA

Silicon nitride DPN tips created nanoscale patterns of oligonucleotides on both metallic and insulating substrates [14]. Hexanethiol-modified oligonucleotides were used as ink for DPN patterning on gold substrates with features ranging from 50 nm to several micrometers in size. The patterned surface was subsequently used to direct the assembly of complementary oligonucleotide-modified gold nanoparticles. The immobilization of the DNA was achieved by the covalent bonding between the hexanethiol group of the DNA and the gold surface. Prior to the patterning, the surface of the silicon nitride AFM cantilever was modified with $3'$-aminopropyltrimethoxysilane in order to improve the coating of the DNA on the surface of the tip. The modified tip surface was readily wetted by the DNA ink solution when dipped. In addition to utilizing the gold-thiol immobilization for DNA patterning, modified silicon oxide surfaces were demonstrated to work as a DPN substrate. Since gold surfaces prevented the study of electrical or optical characteristics of the DPN-patterned nanostructures due to the conductivity and quenching, respectively, of the surface, the use of silicon oxide opened the door to such studies. The silicon oxide surface was treated with $3'$-mercaptopropyltrimethoxysilane (MPTMS) and the oligonucleotides were modified with acrylamide groups in order to utilize the covalent link between the pendant thiol groups of the MPTMS and the acrylamide moieties of the DNA as an immobilization mechanism. The biological activity of the patterned oligonucleotides was confirmed in an ensuing hybridization process.

In DPN, functionalization of the tip surface is required to efficiently coat the DNA ink. However, the NFP did not need such tip modification due to the direct delivery of the DNA ink in suspension from the on-chip reservoir to the tip.

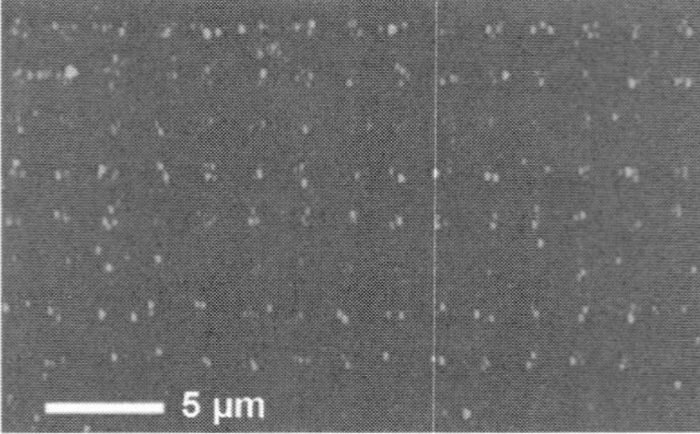

Figure 4.22. AFM topography image of an array of DNA dots patterned by fluidic spotting using an NFP probe.

Hexanethiol-modified oligonucleotides were deposited on a gold substrate in a fountain-pen fashion using the device presented in Fig. 4.21. An array of dots was patterned and verified by AFM topography scanned with a commercially available silicon nitride tip as shown in Fig. 4.22.

4.4.2
Patterning of Proteins

DPN was used to fabricate a nanoarray for the HIV-1 immunoassay as a proof-of-concept DPN-based biodetection [15]. DPN-generated dots of MHA on a thin gold film were used to immobilize anti-HIV-1 p24 antibodies. When immersed into a sample containing HIV-1 p24, the nanoarray reacted with the antigens leading to a height difference of the MHA features due to the bound antigens. This height difference is not easily detectable by AFM. Hence, the height difference was amplified by gold nanoparticles that were functionalized with the same antibody. The DPN-generated nanoarray with the double-sandwich immunoassay demonstrated the capability to detect and measure antigens with unprecedented sensitivity. The important feature of this approach is the high number and small size of individual dots, which reduces the time necessary for antigens to travel and bind to antibodies, and also provides the necessary redundancy for a correct statistics.

An enhanced tool was introduced by Lynch and coworkers [115] for fast and large-area patterning. One of the improvements consisted of replacing the AFM piezotube with a piezoelectric inchworm stage capable of 20-nm resolution over 25 mm of xy travel. Although the stage did not provide as high lateral resolution as the piezotube, it performed better in terms of repeatable movements over several centimeters. A second improvement was the shortening of the tip–surface contact time for patterning by applying precisely timed bursts of wet or dry air directly to the tip–surface interface. With this scheme, protein transfer could be achieved in less than 100 ms, which was an improvement upon the 250 s μm^{-1} in DPN [15]. For patterning, a commercially available AFM probe was mounted on the tool for building protein nanoarrays. In demonstrated experiments, an array of antigen (mouse IgG) was spotted with approximately 700-nm diameter spots on a pretreated substrate [115]. The specific binding of Cy3-labeled goat anti-mouse antibody was verified with minimal nonspecific background binding. It was demonstrated that nanoarray assays could be produced in a rapid fashion and the miniaturized size would lead to a reduced sample volume of the protein for bioanalysis. For example, each 1-μm dot in the nanoarray would cover less than 1/1000 of the surface area of a conventional microarray dot while maintaining enough antibodies to provide a useful dynamic range [115].

4.4.3
Patterning of Viruses

AFM was utilized to identify viruses utilizing type-specific immunocapture and the morphological properties of the capture viruses by Nettikadan and coworkers [116, 117]. Multiple virus-specific antibody capture domains were constructed on a chip using ink-jet protein arraying technology. The chip, termed ViriChip, was constructed with 600-μm diameter antibody domains. The ViriChips were individually exposed to each of six group B coxsackieviruses [116]. Each of the six group B coxsackievirus types bound extensively to their specific ViriChip with little or no binding observed on the nonspecific chips. Counting of the number of bound viruses was performed using AFM inspection.

Substrates for the chips were prepared from diced silicon wafers (4×4 mm^2). After ultrasonic cleaning in water and ethanol, metal films of 5-nm chromium and 10-nm gold were sequentially sputter coated. Target areas (600 μm diameter) were created using copper electron microscopy grids as masks during sputtering. A SAM layer was coated by immersing the substrate in an alkanethiolate solution. The target areas were covered with recombinant Protein A/G followed by blocking the unreacted alkanethiol groups. The chip was completed by placing 1 μl of antiviral antibody on the Protein A/G domain of a substrate, followed by incubation and rinsing. Each virus sample (1 μl) of the six group B coxsackieviruses was applied onto the antibody-coated domains. AFM scanning operated in tapping mode was used to obtain morphological and counting information on the viruses.

It was inferred that a reduction of the antibody domain size to 2–5 μm should increase the sensitivity of the assay [116] based on the previous study on the phenomenon of analyte harvesting [23, 118]. Further, it was verified that the efficiency of virus capture increases dramatically with decreased sample volume. With microfluidic delivery of samples to the capture domain, the capture efficiency would likely be increased.

4.5
Conclusions and Outlook

Since they have been used in AFM for imaging, microcantilevers have evolved to meet the requirements from many life sciences applications. Microcantilevers are fundamental tools for biopatterning and biosensing in SPM-based techniques. Different materials targeting the biochemical applications were integrated and used for fabricating the tips and cantilevers. Massive parallel arrays are being pursued in an attempt to increase the throughput. Efforts have been made to integrate microfluidics into microcantilevers, which include apertured pyramidal tips, miniaturized spotters with nib-type reservoirs, open and closed microchannels, on-chip reservoirs, and complex systems with several of the previously mentioned features. Microcantilevers may serve to create ultra-miniaturized bioanalytical devices and testing methodologies in which the efficiency relies on the higher detection speed, higher accuracy and increased statistical significance of the results.

References

1 BINNIG G., QUATE C. F., GERBER C. Atomic force microscope, *Phys. Rev. Lett.* **1986**, *56*, 930–933.

2 ALBRECHT T. R., QUATE C. F. Atomic resolution imaging of a nonconductor by atomic force microscopy, *J. Appl. Phys.* **1987**, *62*, 2599–2602.

3 PETERSEN K. E. Silicon as a mechanical material, *Proc. IEEE* **1982**, *70*, 39–76.

4 MADOU M. J. *Fundamentals of Microfabrication*. CRC Press, Boca Raton, FL, **2002**.

5 BLANCHARD A. P., KAISER R. J., HOOD L. E. High-density oligonucleotide arrays, *Biosens. Bioelectron.* **1996**, *11*, 687–690.

6 BIETSCH A., HEGNER M., LANG H. P., GERBER C. Inkjet deposition of alkanethiolate monolayers and DNA oligonucleotides on gold: evaluation of spot uniformity by wet etching, *Langmuir* **2004**, *20*, 5119–5122.

7 FODOR S. P. A. DNA sequencing – massively parallel genomics, *Science* **1997**, *277*, 393–395.

8 JAMES C. D., DAVIS R. C., KAM L., CRAIGHEAD H. G., ISAACSON M., TURNER J. N., SHAIN W. Patterned protein layers on solid substrates by thin stamp microcontact printing, *Langmuir* **1998**, *14*, 741–744.

9 BERNARD A., DELAMARCHE E., SCHMID H., MICHEL B., BOSSHARD H. R., BIEBUYCK H. Printing patterns of proteins, *Langmuir* **1998**, *14*, 2225–2229.

10 LANGE S. A., BENES V., KERN D. P., HORBER J. K. H., BERNARD A. Micro-contact printing of DNA molecules, *Anal. Chem.* **2004**, *76*, 1641–1647.

11 DELAMARCHE E., BERNARD A., SCHMID H., MICHEL B., BIEBUYCK H. Patterned delivery of immunoglobulins to surfaces using microfluidic networks, *Science* **1997**, *276*, 779–781.

12 BERNARD A., MICHEL B., DELAMARCHE E. Micromosaic immunoassay, *Anal. Chem.* **2001**, *73*, 8–12.

13 PINER R. D., ZHU J., XU F., HONG S., MIRKIN C. A. "Dip-pen" nanolithography, *Science* **1999**, *283*, 661–663.

14 DEMERS L. M., GINGER D. S., PARK S.-J., LI Z., CHUNG S.-W., MIRKIN C. A. Direct patterning of modified oligonucleotides on metals and insulators by dip-pen nanolithography, *Science* **2002**, *296*, 1836–1838.

15 LEE K.-B., PARK S.-J., MIRKIN C. A., SMITH J. C., MRKSICH M. Protein nanoarrays generated by dip-pen nanolithography, *Science* **2002**, *295*, 1702–1705.

16 MINH P. N., TAKAHITO O., MASAYOSHI E. *Fabrication of Silicon Microprobes for Optical Near-Field Applications*. CRC Press, Boca Raton, FL, **2002**.

17 LANG H. P., BERGER R., BATTISTON F., RAMSEYER J. P., MEYER E., ANDREOLI C., BRUGGER J., VETTIGER P., DESPONT M., MEZZACASA T., SCANDELLA L., GUNTHERODT H. J., GERBER C., GIMZEWSKI J. K. A chemical sensor based on a micromechanical cantilever array for the identification of gases and vapors, *Appl. Phys. A* **1998**, *66*, S61–S64.

18 MEISTER A., LILEY M., BRUGGER J., PUGIN R., HEINZELMANN H. Nanodispenser for attoliter volume deposition using atomic force microscopy probes modified by focused-ion-beam milling, *Appl. Phys. Lett.* **2004**, *85*, 6260–6262.

19 BULLEN D., WANG X., ZOU J., HONG S., CHUNG S., RYU K., FAN Z., MIRKIN C., LIU C. Micromachined arrayed dip pen nanolithography probes for sub-100 nm direct chemistry patterning. Presented at: *16th IEEE International Micro Electro Mechanical Systems Conference*, Kyoto, **2003**.

20 ZHANG M., BULLEN D., CHUNG S.-W., HONG S., RYU K. S., FAN Z., MIRKIN C. A., LIU C. A MEMS nanoplotter with high-density parallel dip-pen nanolithography probe arrays, *Nanotechnology* **2002**, *13*, 212–217.

21 VETTIGER P., DESPONT M., DRECHSLER U., DURIG U., HABERLE W., LUTWYCHE M. I., ROTHUIZEN H. E., STUTZ R., WIDMER R., BINNIG G. K. The "Millipede" – more than one thousand tips for future AFM data storage, *IBM J. Res. Dev.* **2000**, *44*, 323–340.

22 RAMSAY G. DNA chips: state-of-the-art, *Nat. Biotechnol.* **1998**, *16*, 40–44.

23 MACBEATH G., SCHREIBER S. L. Printing proteins as microarrays for high-throughput function determination, *Science* **2000**, *289*, 1760–1763.

24 LANGE D., BRAND O., BALTES H. *CMOS Cantilever Sensor Systems*. Springer, Berlin, **2002**.

25 MINNE S. C., MANALIS S. R., QUATE C. F. *Bringing Scanning Probe Microscopy Up to Speed*. Kluwer, Boston, MA, **1999**.

26 ALESSANDRINI A., FACCI P. AFM: a versatile tool in biophysics, *Meas. Sci. Technol.* **2005**, *16*, R65–R92.

27 SAMORI P. Scanning probe microscopies beyond imaging, *J. Mater. Chem.* **2004**, *14*, 1353–1366.

28 GREENE M. E., KINSER C. R., KRAMER D. E., PINGREE L. S. C., HERSAM M. C. Application of scanning probe microscopy to the characterization and fabrication of hybrid nanomaterials, *Microsc. Res. Tech.* **2004**, *64*, 415–434.

29 WOUTERS D., SCHUBERT U. S. Nanolithography and nanochemistry: probe-related patterning techniques and chemical modification for nanometer-sized devices, *Angew. Chem. Int. Ed. Engl.* **2004**, *43*, 2480–2495.

30 GINGER D. S., ZHANG H., MIRKIN C. A. The evolution of dip-pen nanolithography, *Angew. Chem. Int. Ed. Engl.* **2004**, *43*, 30–45.

31 WOLTER O., BAYER T., GRESCHNER J. Micromachined silicon sensors for scanning force microscopy, *J. Vac. Sci. Technol. B* **1991**, *9*, 1353–1357.

32 RAITERI R., GRATTAROLA M., BUTT H.-J., SKLÁDAL P. Micromechanical cantilever-based biosensors, *Sens. Actuators B* **2001**, *79*, 115–126.

33 BALLER M. K., LANG H. P., FRITZ J., GERBER C., GIMZEWSKI J. K., DRECHSLER U., ROTHUIZEN H., DESPONT M., VETTIGER P., BATTISTON F. M., RAMSEYER J. P., FORNARO P., MEYER E., GUNTHERODT H. J. A cantilever array-based artificial nose, *Ultramicroscopy* **2000**, *82*, 1–9.

34 BAIN C. D., TROUGHTON E. B., TAO Y. T., EVALL J., WHITESIDES G. M., NUZZO R. G. Formation of monolayer films by the spontaneous assembly of organic thiols from solution onto gold, *J. Am. Chem. Soc.* **1989**, *111*, 321–335.

35 BERGER R., DELAMARCHE E., LANG H. P., GERBER C., GIMZEWSKI J. K., MEYER E., GUNTHERODT H. J. Surface stress in the self-assembly of alkanethiols on gold, *Science* **1997**, *276*, 2021–2024.

36 FRITZ J., BALLER M. K., LANG H. P., ROTHUIZEN H., VETTIGER P., MEYER E., GÜNTHERODT H.-J., GERBER C., GIMZEWSKI J. K. Translating biomolecular recognition into nanomechanics, *Science* **2000**, *288*, 316–318.

37 WU G., JI H., HANSEN K., THUNDAT T., DATAR R., COTE R., HAGAN M. F., CHAKRABORTY A. K., MAJUMDAR A. Origin of nanomechanical cantilever motion generated from biomolecular interactions, *Proc. Natl Acad. Sci. USA* **2001**, *98*, 1560–1564.

38 WU G., DATAR R. H., HANSEN K. M., THUNDAT T., COTE R. J., MAJUMDAR A. Bioassay of prostate-specific antigen

(PSA) using microcantilevers, *Nat. Biotechnol.* **2001**, *19*, 856–860.

39 ARNTZ Y., SEELIG J. D., LANG H. P., ZHANG J., HUNZIKER P., RAMSEYER J. P., MEYER E., HEGNER M., GERBER C. Label-free protein assay based on a nanomechanical cantilever array, *Nanotechnology* **2003**, *14*, 86–90.

40 THUNDAT T., WACHTER E. A., SHARP S. L., WARMACK R. J. Detection of mercury-vapor using resonating microcantilevers, *Appl. Phys. Lett.* **1995**, *66*, 1695–1697.

41 THUNDAT T., CHEN G. Y., WARMACK R. J., ALLISON D. P., WACHTER E. A. Vapor detection using resonating microcantilevers, *Anal. Chem.* **1995**, *67*, 519–521.

42 WACHTER E. A., THUNDAT T. Micromechanical sensors for chemical and physical measurements, *Rev. Sci. Instrum.* **1995**, *66*, 3662–3667.

43 BATTISTON F. M., RAMSEYER J. P., LANG H. P., BALLER M. K., GERBER C., GIMZEWSKI J. K., MEYER E., GUNTHERODT H. J. A chemical sensor based on a microfabricated cantilever array with simultaneous resonance-frequency and bending readout, *Sens. Actuators B* **2001**, *77*, 122–131.

44 SHU W., LIU D., WATARI M., RIENER C. K., STRUNZ T., WELLAND M. E., BALASUBRAMANIAN S., McKENDRY R. A. DNA molecular motor driven micromechanical cantilever arrays, *J. Am. Chem. Soc.* **2005**, *127*, 17054–17060.

45 ALBRECHT T. R., AKAMINE S., CARVER T. E., QUATE C. F. Microfabrication of cantilever styli for the atomic force microscope, *J. Vac. Sci. Technol. A* **1990**, *8*, 3386–3396.

46 WATANABE S., FUJII T. Microfabricated piezoelectric cantilever for atomic force microscopy, *Rev. Sci. Instrum.* **1996**, *67*, 3898–3903.

47 YAMAMOTO T., SUZUKI Y., MIYASHITA M., SUGIMURA H., NAKAGIRI N. Development of a metal patterned cantilever for scanning capacitance microscopy and its application to the observation of semiconductor devices, *J. Vac. Sci. Technol. B* **1997**, *15*, 1547–1550.

48 RASMUSSEN J. P., TANG P. T., SANDER C., HANSEN O., MOLLER P. Fabrication of an all-metal atomic force microscope probe. In: *Proceedings of the International Conference on Solid-State Sensors and Actuators*, Chicago, IL, **1997**, pp. 463–466.

49 AKAMINE S., QUATE C. F. Low temperature thermal oxidation sharpening of microcast tips, *J. Vac. Sci. Technol. B* **1992**, *10*, 2307–2310.

50 RAVI T. S., MARCUS R. B., LIU D. Oxidation sharpening of silicon tips, *J. Vac. Sci. Technol. B* **1991**, *9*, 2733–2737.

51 MARCUS R. B., RAVI T. S., GMITTER T., CHIN K., LIU D., ORVIS W. J., CIARLO D. R., HUNT C. E., TRUJILLO J. Formation of silicon tips with <1 nm radius, *Appl. Phys. Lett.* **1990**, *56*, 236–238.

52 BOISEN A., HANSENYAN O., BOUWSTRA S. AFM probes with directly fabricated tips, *J. Micromech. Microeng.* **1996**, *6*, 58–62.

53 MEISTER A., JENEY S., LILEY M., AKIYAMA T., STAUFER U., DE ROOIJ N. F., HEINZELMANN H. Nanoscale dispensing of liquids through cantilevered probes, *Microelectron. Eng.* **2003**, *67–68*, 644–650.

54 FAROOQUI M. M., EVANS A. G. R., STEDMAN M., HAYCOCKS J. Micromachined silicon sensors for atomic force microscopy, *Nanotechnology* **1992**, *3*, 91–97.

55 FAROOQUI M. M., EVANS A. G. R. Silicon sensors with integral tips for atomic force microscopy: a novel single-mask fabrication process, *J. Micromech. Microeng.* **1993**, *3*, 8–12.

56 ITOH J., TOHMA Y., SHIMIZU S., KA K. Fabrication of an ultrasharp and high-aspect-ratio microprobe with a silicon-on-insulator wafer for scanning force microscopy, *J. Vac. Sci. Technol. B* **1995**, *13*, 331–334.

57 FOLCH A., WRIGHTON M. S., SCHMIDT M. A. Microfabrication of oxidation-sharpened silicon tips on silicon nitride cantilevers for atomic force microscopy, *J. Microelectromech. Syst.* **1997**, *6*, 303–306.

58 FLORIN E. L., MOY V. T., GAUB H. E. Adhesion forces between individual ligand–receptor pairs, *Science* **1994**, *264*, 415–417.

59 Lee G. U., Kidwell D. A., Colton R. J. Sensing discrete streptavidin biotin interactions with atomic-force microscopy, *Langmuir* **1994**, *10*, 354–357.

60 Krautbauer R., Rief M., Gaub H. E. Unzipping DNA oligomers, *Nano Lett.* **2003**, *3*, 493–496.

61 Rief M., Oesterhelt F., Heymann B., Gaub H. E. Single molecule force spectroscopy on polysaccharides by atomic force microscopy, *Science* **1997**, *275*, 1295–1297.

62 Rief M., Gautel M., Oesterhelt F., Fernandez J. M., Gaub H. E. Reversible unfolding of individual titin immunoglobulin domains by AFM, *Science* **1997**, *276*, 1109–1112.

63 Fisher T. E., Oberhauser A. F., Carrion-Vazquez M., Marszalek P. E., Fernandez J. M. The study of protein mechanics with the atomic force microscope, *Trends Biochem. Sci.* **1999**, *24*, 379–384.

64 Wilson D. L., Martin R., Hong S., Cronin-Golomb M., Mirkin C. A., Kaplan D. L. Surface organization and nanopatterning of collagen by dip-pen nanolithography, *Proc. Natl Acad. Sci. USA* **2001**, *98*, 13660–13664.

65 Maynor B. W., Filocamo S. F., Grinstaff M. W., Liu J. Direct-writing of polymer nanostructures: poly(thiophene) nanowires on semiconducting and insulating surfaces, *J. Am. Chem. Soc.* **2002**, *124*, 522–523.

66 Ivanisevic A., Mirkin C. A. "Dip-pen" nanolithography on semi-conductor surfaces, *J. Am. Chem. Soc.* **2001**, *123*, 7887–7889.

67 Su M., Dravid V. P. Colored ink dip-pen nanolithography, *Appl. Phys. Lett.* **2002**, *80*, 4434–4436.

68 Li Y., Maynor B. W., Liu J. Electro-chemical AFM "dip-pen" nano-lithography, *J. Am. Chem. Soc.* **2001**, *123*, 2105–2106.

69 Lim J.-H., Ginger D. S., Lee K.-B., Heo J., Nam J.-M., Mirkin C. A. Direct-write dip-pen nanolithography of proteins on modified silicon oxide surfaces, *Angew. Chem. Int. Ed. Engl.* **2003**, *42*, 2309–2312.

70 Lee K.-B., Lim J.-H., Mirkin C. A. Protein nanostructures formed via direct-write dip-pen nanolithography, *J. Am. Chem. Soc.* **2003**, *125*, 5588–5589.

71 Agarwal G., Sowards L. A., Naik R. R., Stone M. O. Dip-pen nanolithography in tapping mode, *J. Am. Chem. Soc.* **2003**, *125*, 580–583.

72 Agarwal G., Naik R. R., Stone M. O. Immobilization of histidine-tagged proteins on nickel by electrochemical dip pen nanolithography, *J. Am. Chem. Soc.* **2003**, *125*, 7408–7412.

73 Weeks B. L., Vaughn M. W., DeYoreo J. J. Direct imaging of meniscus formation in atomic force microscopy using environmental scanning electron microscopy, *Langmuir* **2005**, *21*, 8096–8098.

74 Hong S., Mirkin C. A. A nanoplotter with both parallel and serial writing capabilities, *Science* **2000**, *288*, 1808–1811.

75 Salaita K., Lee S. W., Wang X., Huang L., Dellinger T. M., Liu C., Mirkin C. A. Sub-100 nm, centimeter-scale, parallel dip-pen nanolithography, *Small* **2005**, *1*, 940–945.

76 Lee S. W., Sanedrin R. G., Oh B.-K., Mirkin C. A. Nanostructured polyelectrolyte multilayer organic thin films generated via parallel dip-pen nanolithography (p NA), *Adv. Mater.* **2005**, *17*, 2749–2753.

77 Wang X., Ryu K. S., Bullen D. A., Zou J., Zhang H., Mirkin C. A., Liu C. Scanning probe contact printing, *Langmuir* **2003**, *19*, 8951–8955.

78 Xia Y. N., Whitesides G. M. Soft lithography, *Angew. Chem. Int. Ed. Engl.* **1998**, *37*, 551–575.

79 Zhang H., Elghanian R., Amro N. A., Disawal S., Eby R. Dip pen nanolithography stamp tip, *Nano Lett.* **2004**, *4*, 1649–1655.

80 Xia Y. N., Rogers J. A., Paul K. E., Whitesides G. M. Unconventional methods for fabricating and patterning nanostructures, *Chem. Rev.* **1999**, *99*, 1823–1848.

81 CHUNG S. W., GINGER D. S., MORALES M. W., ZHANG Z. F., CHANDRASEKHAR V., RATNER M. A., MIRKIN C. A. Top-down meets bottom-up: dip-pen nanolithography and DNA-directed assembly of nanoscale electrical circuits, *Small* **2005**, *1*, 64–69.

82 JIAN W., FIRESTONE M. A., AUCIELLO O., CARLISLE J. A. Surface functionalization of ultrananocrystalline diamond films by electrochemical reduction of aryldiazonium salts, *Langmuir* **2004**, *20*, 11450–11456.

83 KIM K.-H., MOLDOVAN N., KE C., ESPINOSA H. D., XIAO X., CARLISLE J. A., AUCIELLO O. Novel ultrananocrystalline diamond probes for high-resolution low-wear nanolithographic techniques, *Small* **2005**, *1*, 866–874.

84 KRAUSS A. R., AUCIELLO O., GRUEN D. M., JAYATISSA A., SUMANT A., TUCEK J., MANCINI D. C., MOLDOVAN N., ERDEMIR A., ERSOY D., GARDOS M. N., BUSMANN H. G., MEYER E. M., DING M. Q. Ultrananocrystalline diamond thin films for MEMS and moving mechanical assembly devices, *Diamond Rel. Mater.* **2001**, *10*, 1952–1961.

85 HERSAM M. C., HOOLE A. C. F., O'SHEA S. J., WELLAND M. E. Potentiometry and repair of electrically stressed nanowires using atomic force microscopy, *Appl. Phys. Lett.* **1998**, *72*, 915–917.

86 TRENKLER T., STEPHENSON R., JANSEN P., VANDERVORST W., HELLEMANS L. New aspects of nanopotentiometry for complementary metal-oxide-semiconductor transistors, *J. Vac Sci. Technol. B* **2000**, *18*, 586–594.

87 BELAUBRE P., GUIRARDEL M., LEBERRE V., POURCIEL J. B., BERGAUD C. Cantilever-based microsystem for contact and non-contact deposition of picoliter biological samples, *Sens. Actuators A* **2004**, *110*, 130–135.

88 XU J., LYNCH M., HUFF J. L., MOSHER C., VENGASANDRA S., DING G., HENDERSON E. Microfabricated quill-type surface patterning tools for the creation of biological micro/nano arrays, *Biomed. Microdevices* **2004**, *6*, 117–123.

89 LEWIS A., KHEIFETZ Y., SHAMBRODT E., RADKO A., KHATCHATRYAN E., SUKENIK C. Fountain pen nano-chemistry: atomic force control of chrome etching, *Appl. Phys. Lett.* **1999**, *75*, 2689–2691.

90 PAPAUTSKY I., BRAZZLE J., SWERDLOW H., WEISS R., FRAZIER A. B. Micromachined pipette arrays, *IEEE Trans. Biomed. Eng.* **2000**, *47*, 812–819.

91 KIM K.-H., MOLDOVAN N., ESPINOSA H. D. A nanofountain probe with sub-100 nm molecular writing resolution, *Small* **2005**, *1*, 632–635.

92 MOLDOVAN N., KIM K.-H., ESPINOSA H. D. Multi-ink linear array of nanofountain probes, submitted to *J. Micromech. Microeng.* **2006**.

93 SCHENA M., HELLER R. A., THERIAULT T. P., KONRAD K., LACHENMEIER E., DAVIS R. W. Microarrays: biotechnology's discovery platform for functional genomics, *Trends Biotechnol.* **1998**, *16*, 301–306.

94 BELAUBRE P., GUIRARDEL M., GARCIA G., POURCIEL J. B., LEBERRE V., DAGKESSAMANSKAIA A., TREVISIOL E., FRANCOIS J. M., BERGAUD C. Fabrication of biological microarrays using microcantilevers, *Appl. Phys. Lett.* **2003**, *82*, 3122–3124.

95 LEICHLE T., SILVAN M. M., BELAUBRE P., VALSESIA A., CECCONE G., ROSSI F., SAYA D., POURCIEL J. B., NICU L., BERGAUD C. Nanostructuring surfaces with conjugated silica colloids deposited using silicon-based microcantilevers, *Nanotechnology* **2005**, *16*, 525–531.

96 LEE J., MOON H., FOWLER J., SCHOELLHAMMER T., KIM C. J. Electrowetting and electrowetting-on-dielectric for microscale liquid handling, *Sens. Actuators A* **2002**, *95*, 259–268.

97 VENGASANDRA S. G., LYNCH M., XU J., HENDERSON E., Microfluidic ultra-microscale deposition and patterning of quantum dots, *Nanotechnology* **2005**, *16*, 2052–2055.

98 XU J., LYNCH M., NETTIKADAN S., MOSHER C., VEGASANDRA S., HENDERSON E. Microfabricated "biomolecular ink cartridges" –

surface patterning tools (SPTs) for the printing of multiplexed biomolecular arrays, *Sens. Actuators B* **2005**, *113*, 1034–1041.

99 BIETSCH A., ZHANG J. Y., HEGNER M., LANG H. P., GERBER C. Rapid functionalization of cantilever array sensors by inkjet printing, *Nanotechnology* **2004**, *15*, 873–880.

100 TAHA H., MARKS R. S., GHEBER L. A., ROUSSO I., NEWMAN J., SUKENIK C., LEWIS A. Protein printing with an atomic force sensing nanofountainpen, *Appl. Phys. Lett.* **2003**, *83*, 1041–1043.

101 IONESCU R. E., MARKS R. S., GHEBER L. A. Nanolithography using protease etching of protein surfaces, *Nano Lett.* **2003**, *3*, 1639–1642.

102 IONESCU R. E., MARKS R. S., GHEBER L. A. Manufacturing of nanochannels with controlled dimensions using protease nanolithography, *Nano Lett.* **2005**, *5*, 821–827.

103 HONG M.-H., KIM K. H., BAE J., JHE W. Scanning nanolithography using a material-filled nanopipette, *Appl. Phys. Lett.* **2000**, *77*, 2604–2606.

104 BRUCKBAUER A., YING L. M., ROTHERY A. M., ZHOU D. J., SHEVCHUK A. I., ABELL C., KORCHEV Y. E., KLENERMAN D. Writing with DNA and protein using a nanopipet for controlled delivery, *J. Am. Chem. Soc.* **2002**, *124*, 8810–8811.

105 SHALOM S., LIEBERMAN K., LEWIS A., COHEN S. R. A micropipette force probe suitable for near-field scanning optical microscopy, *Rev. Sci. Instrum.* **1992**, *63*, 4061–4065.

106 LIEBERMAN K., LEWIS A., FISH G., SHALOM S., JOVIN T. M., SCHAPER A., COHEN S. R. Multifunctional, micro-pipette based force cantilevers for scanned probe microscopy, *Appl. Phys. Lett.* **1994**, *65*, 648–650.

107 HANSMA P. K., DRAKE B., MARTI O., GOULD S. A. C., PRATER C. B. The scanning ion-conductance microscope, *Science* **1989**, *243*, 641–643.

108 MCALLISTER D. V., ALLEN M. G., PRAUSNITZ M. R. Microfabricated microneedles for gene and drug delivery, *Annu. Rev. Biomed. Eng.* **2000**, *2*, 289–313.

109 ZAHN J. D., DESHMUKH A., PISANO A. P., LIEPMANN D. Continuous on-chip micropumping for microneedle enhanced drug delivery, *Biomed. Microdevices* **2004**, *6*, 183–190.

110 LIN L. W., PISANO A. P. Silicon-processed microneedles. *J. Micro-electromech. Syst.* **1999**, *8*, 78–84.

111 PAIK S. J., BYUN A., LIM J. M., PARK Y., LEE A., CHUNG S., CHANG J. K., CHUN K., CHO D. D. In-plane single-crystal-silicon microneedles for minimally invasive microfluid systems, *Sens. Actuators A* **2004**, *114*, 276–284.

112 ZAHN J. D., TALBOT N. H., LIEPMANN D., PISANO A. P. Microfabricated polysilicon microneedles for minimally invasive biomedical devices, *Biomed. Microdevices* **2000**, *2*, 295–303.

113 CHANDRASEKARAN S., BRAZZLE J. D., FRAZIER A. B. Surface micromachined metallic microneedles, *J. Micro-electromech. Syst.* **2003**, *12*, 281–288.

114 MOLDOVAN N., KIM K.-H., ESPINOSA H. D. Design and fabrication of a novel microfluidic nanoprobe, *J. Microelectromech. Syst.* **2006**, *15*, 204–213.

115 LYNCH M., MOSHER C., HUFF J., NETTIKADAN S., JOHNSON J., HENDERSON E. Functional protein nanoarrays for biomarker profiling, *Proteomics* **2004**, *4*, 1695–1702.

116 NETTIKADAN S. R., JOHNSON J. C., VENGASANDRA S. G., MUYS J., HENDERSON E. ViriChip: a solid phase assay for detection and identification of viruses by atomic force microscopy, *Nanotechnology* **2004**, *15*, 383–389.

117 NETTIKADAN S. R., JOHNSON J. C., MOSHER C., HENDERSON E. Virus particle detection by solid phase immunocapture and atomic force microscopy, *Biochem. Biophys. Res. Commun.* **2003**, *311*, 540–545.

118 MACBEATH G. Protein microarrays and proteomics, *Nat. Genet.* **2002**, *32*, 526–532.

5
Nanobioelectronics

Ross Rinaldi and Giuseppe Maruccio

5.1
Introduction

Nanobiotechnology has been boosted by the advancements in fabrication technologies, which enable the construction of artificial structures in the size range of biological entities. The fascinating world of the bio-self-assembly provides new opportunities and directions for future electronics, opening the way to a new generation of computational systems based on biomolecules and biostructures. Here, we review the advances in the field of nanobiomolecular electronics, starting from the description of a few selected bio-units up to the implementation of hybrid devices. Possible applications of bioelectronic devices to standard electronics, such as rectification, amplification and storage, as well as novel functionalities will be discussed on the basis of recent concept studies.

The chapter is organized as follows: Section 5.2 is dedicated to the motivations of nanobioelectronics research, Section 5.3 summarizes the fundamentals of bio-building blocks for nanoelectronics (DNA and proteins), and Section 5.4 deals with how to interconnect biomolecules and exploit their self-assembly properties for the implementation of nanobiodevices. Sections 5.5 and 5.6 are dedicated, respectively, to the investigation of DNA and proteins for nanobioelectronics applications. Finally, some conclusions are drawn and possible future research directions are forecast in Section 5.7.

5.2
Bio-self-assembly and Motivation

Moore's Law (Fig. 5.1), the 1965 prediction by Intel co-founder Gordon Moore that manufacturers would double the number of transistors on a chip every 18 months, with resulting declining prices and increasing performances, has been fulfilled for four decades by the semiconductor industry. However, the latest editions of the annual International Technology Roadmap for Semiconductors [1] – a joint effort of worldwide semiconductor industry associations predicting the main trends in the

Nanotechnologies for the Life Sciences Vol. 4
Nanodevices for the Life Sciences. Edited by Challa S. S. R. Kumar
Copyright © 2006 WILEY-VCH Verlag GmbH & Co. KGaA, Weinheim
ISBN: 3-527-31384-2

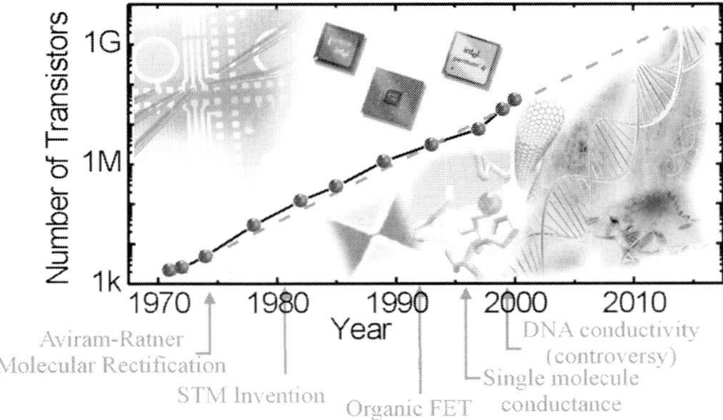

Figure 5.1. Moore's law – number of transistors as a function of time (points refer to the various processors introduced by Intel: 4004, 8008, 8080, 8086, 286, 386, 486, Pentium, Pentium II, Pentium III and Pentium 4, respectively). The concomitant occurrence of some milestones of molecular electronics is also indicated. In 1965, Gordon Moore, co-founder of Intel, first observed and then predicted that the number of transistors per integrated circuit grew exponentially. So far, this prediction has been fulfilled thanks to the continuous advances in miniaturization driven by the progress in lithographic techniques. However, it now seems to be in serious danger and modern electronics has to face the restrictions dictated by the laws of physics when the minimum feature size of a chip approaches 100 nm, at least with current technologies. Thus, gaining the nanometer scale and/or further enhancing the computational capabilities requires a turning point, a change in architecture and the development of conceptually new devices exploiting spin, quantum mechanics and/or molecular building blocks. (Adapted from Ref. [2].)

semiconductor industry spanning 15 years into the future – lists reasons for thinking that this may soon change.

The Roadmap explores "technology nodes" – advances needed to keep shrinking the so-called DRAM half-pitch, i.e. half the spacing between cells in memory chips. Currently, the industry is moving to a DRAM half-pitch of 90 nm, about 2/1000 the width of the proverbial human hair. Analyzing various aspects of chipmaking, the Roadmap forecasts that researchers should lower that figure to 35 nm by 2014, continuing doubling the number of transistors per integrated circuit. This miniaturization trend is expected to continue for another 15–20 years, but the downscaling with the usual top-down approach (i.e. thanks to the improvement of the lithographic techniques) is becoming increasingly difficult because of physical limitations including size of atoms, wavelengths of radiation used for lithography, interconnect schemes, etc. Three fundamental limits exist [2]: (a) the energy needed to write a bit must be bigger than the average energy of the thermal fluctuations (kT) to avoid bit errors (thermal limit), (b) the energy necessary to read or write a bit and the frequency of the circuits are limited by the uncertainty principle (quantum limit), and (c) a maximum tolerable power density exists (power dissipation limit).

Although minimum feature sizes in the few nanometers range have been demonstrated in various laboratories and a number of new lithographic solutions [such as extreme UV, X-ray proximity, and electron (EBL)- and ion-beam lithography (IBL)] have been proposed as post-optical techniques, to a large extent these tools are not suitable for mass production and/or capable of meeting the requirements of the Roadmap, and no known solutions currently exist for these problems.

Thus, the Roadmap underlines the pressing need for developing beyond-CMOS devices and post-optical lithographies, together with the necessity to engineer manufacturable interconnection schemes compatible with the new materials and processes recently proposed. Gaining the nanometer scale and/or further enhancing computational capabilities is expected to require a turning point, a change in architecture and the development of conceptually new devices [2–5].

One of the potential roadblocks to continue the scaling beyond the 50-nm node is molecular electronics. Having the final goal of using interconnected molecules to perform the basic functions of digital electronics, molecular electronics (for a review see Ref. [2]) has recently attracted much interest as a different approach to maintain the historical trend of reducing the cost/function ratio by around 25–30% per year. The concept of molecular electronics can be traced back to the Aviram–Ratner prediction [6] that molecules having a donor–spacer–acceptor structure would exhibit rectifying properties when placed between two electrodes.

A bottom-up approach is promising for building nanodevices from molecular building blocks instead of carving lithographically bigger pieces of matter into smaller and smaller chunks, since it provides a new solution to miniaturization. In fact, although it needs a cross-disciplinary effort (merging chemical, biological and physical expertise), the molecular approach offers important advantages of high reproducibility and small size of the molecular building blocks (naturally identical and with well-defined sizes and electronic levels), along with their thermodynamically driven self-assembly and self-recognition properties [7]. From this perspective, the fascinating world of biomolecules (accurately optimized by nature over billions of years of evolution) provides new opportunities and directions for further miniaturization. While engineers and scientists have long been aspiring to manipulate structures controllably and specifically at the micro- and nanometer scale, nature has been performing these tasks with great accuracy and high efficiency using highly specific biological molecules such as DNA and proteins. One of the biggest drivers behind nanotechnology's enthusiasm for biological systems revolves around the organism's impressive ability to manufacture complex molecules (like DNA and proteins) with atomic precision. However, the strongest motivation for using biomolecules as building blocks for the construction of artificial computational systems lies in their self-assembly properties, which open the way for a new generation of devices and fabrication strategies [8]. The self-assembly properties of biological units can be defined as "the self-organization of one or more entities without any external source of information about the structure to be formed as the total energy of the system is minimized to result in a more stable state" [9]. This process inherently implies (a) some mechanism driving movement of entities like diffusion, electric fields, etc., and (b) the concept of "recognition" between different elements, or "bio-linkers", that drives the self-assembly and results in binding of elements dictated by forces (electrical, covalent, ionic, hydrogen bonding, van der Waals, etc.), such that the final physical placement of the entities originates a state of lowest energy [10].

Researchers are now looking at biology as a source of inspiration, and working to join biology and nanotechnology, fusing useful biomolecules to man-made structures in order to fabricate devices that mimic nature in sensing [11], performing complex computational tasks [5], transferring electrons [12], relaying, processing and storing information and producing energy (e.g. biofuel cells exploiting enzymes that extract energy from compounds such as glucose to power life) [13], as well in creating multifunctional systems and molecular-scale motors [14] capable of performing mechanical movements. The result is a merger that attempts to blend biology's ability to assemble complex structures with the nanoscientist's capacity to build useful devices or the chemist's ability to synthesize complex systems (up to hundreds of atoms in size) controlling the position of every atom without too much trouble or fusing useful biomolecules to chemically synthesized nanoclusters in arrangements that do everything from emitting light to storing tiny bits of magnetic data.

Over the last decade there have been dramatic advances toward the realization of molecular-scale devices and integrated computers at the molecular scale [2, 15]. First pioneering experiments were performed demonstrating that individual molecules can serve as nanorectifiers [16] and switches [17, 18], 1000 times smaller than those on conventional microchips. Very recently, the assembly of tiny computer logic circuits built from such molecular-scale devices has been demonstrated [19, 20]. Moreover, molecular electronics is expected to also develop nanodevices exhibiting new functionalities, which exploit peculiar properties of engineered molecules [21] or of biomolecules capable of performing efficiently a lot of different functions.

5.3
Fundamentals of the Bio-building Blocks

5.3.1
DNA

DNA is the basic building block of life. It has a double-stranded helical structure. The famous double-helix structure discovered by Watson and Crick in 1953 [22] consists of two strands of DNA wound around each other (Fig. 5.2). Each strand is about 2 nm wide, and has a long polymer backbone built from repeating sugar molecules and phosphate groups. Each sugar group is attached to one of four "bases". The four bases – guanine (G), cytosine (C), adenine (A) and thymine (T) – form the genetic alphabet of the DNA and their order or "sequence" along the

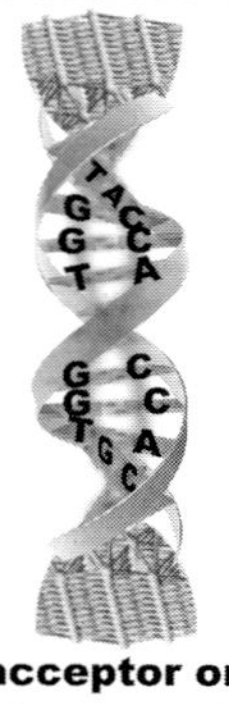

Figure 5.2. (a) Diagram of the DNA double-stranded helical structure. (b) Possible interconnection of DNA fragment with two electrodes or an acceptor and a donor group for transport studies.

molecule constitutes the genetic code. Two single-strand (ss) DNA molecules can join together through hydrogen bonding to form a double-strand (ds) DNA. This process is called hybridization. The chemical bonding is such that an A can only pair with a T, while a G is always paired with a C. The base pairs look like the rungs of a helical ladder. Since the phosphate groups on the backbone are negatively charged, the DNA is usually surrounded by positive "counterions". The negative charges produce electrostatic repulsion of the two strands, so that positive ions are needed to neutralize the negative charges and keep them together. DNA molecules have been proposed as molecular wires because some of the electron orbitals belonging to the bases overlap quite well with each other along the long axis of the molecule. Such so-called stacking interactions also underlie many one-dimensional (1-D) molecular conductors, including one of the most widely studied organic conductors – tetrathiafulvalene-tetracyano-p-quinodimethane (TTF-TCNQ).

5.3.2
Proteins

Proteins have a vital role in all living systems. The biological function of proteins is determined by the interaction with other molecules. The ability of proteins to interact with the outside world is based on their 3-D structure and conformational flexibility. Proteins are encoded by DNA. The primary structure of the protein determines how a protein will fold into a functional 3-D shape. A scheme concerning

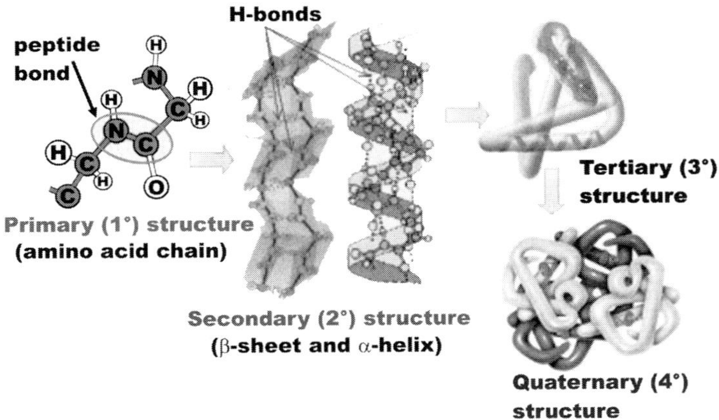

Figure 5.3. Fundamentals of protein structure.

the fundamentals of protein structure is reported in Fig. 5.3. The folded (native-state) protein consists of hydrophilic (polar) residues on the inside and hydrophobic (nonpolar) residues on the outside. In solution, the polar residues are usually in contact with the solvent, whereas the nonpolar residues are protected from solvent in a hydrophobic core. The hydration shell in aqueous solution, where the H_2O molecules forms hydrogen bonds onto the protein surface, solvates the protein.

Proteins play a fundamental role in all biological processes, as they may have different functionalities such as enzymatic catalysis, ionic and electron transfer, coordinated moving, mechanical support, immunization, production and transmission of neuronal pulses, growth and transformation control, etc. Typical examples are hemoglobin that transports oxygen in the erythrocytes and myoglobin that plays the same role in the muscles, whereas iron is carried in the blood by the transferrin and is deposited in the liver by means of a complex with a different protein, ferritin.

In particular, proteins capable of charge transport can most likely be used for building bioelectronic devices. For example, blue copper proteins like Azurin from *Pseudomonas aeruginosa* and poplar plastocyanin are involved in the bacterial respiratory phosphorylation and in some steps on the photosynthesis of green plants, respectively. They show little differences in molecular mass (14 000–10 500) and in the structure of the copper active site (distorted trigonal bipyramid for Azurin and distorted tetrahedral for plastocyanin), which plays the role of an electron reservoir in the electron transfer process. Both of them are redox proteins whose functionality in physiological conditions consists of performing electron transfer between other metalloproteins [cytochrome c_{551} and nitrite reductase for Azurin and cytochrome *bf* and photosystem I (PS I) for plastocyanin].

5.4
Interconnection, Self-assembly and Device Implementation

To date, a fully biological device, presumably operating in a living (liquid) environment, is far from being realistic. An intermediate and necessary step is to fabricate hybrid devices in which the functionality of the biological molecules is exploited through the interconnection with a more conventional solid-state inorganic device. This is the case, for instance, of a field-effect transistor (FET) whose gate channel consists of a self-assembled biomolecular layer.

Although at the early stage, such technology is very fundamental for the understanding of biomolecular systems and for the exploration of their potentialities. The processes for the implementation of hybrid biodevices are based on two fundamental steps:

- The nanofabrication of a pattern which interconnects the bio-entity to the external world (load, power supply, circuit, etc.), following a typical "top-down" lithographic approach. This is a general issue concerning all molecular electronics (see Section 5.4.1).
- The self-assembling of the biomolecules ("bottom-up" fabrication) to immobilize the biosystem in the pattern. A possible systematic approach to build complex biomolecular devices is to start from well-characterized clean substrates [such as Si(100), SiO_2 or Au(111) single crystal surfaces] and deposit submonolayers of spatially isolated molecules or biomolecular functional units. The self-assembly of the biological units on the substrate can follow different pathways like (a) bulk (polymer) entrapment, i.e. absorption, (b) surface absorption (physisorbtion), (c) nondirected covalent binding to the surface (chemisorption), (d) electrostatic adsorption, (e) covalent binding at defined (molecularly engineered) sites of the biocompound or (f) binding by bio-specific interactions (Fig. 5.4) [23]. See Section 5.4.2 for further details.

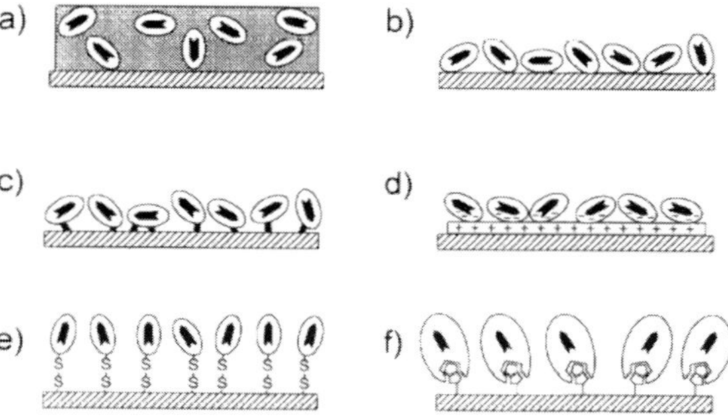

Figure 5.4. Schematic of the different self-assembling mechanisms of biological units onto a substrate. (reprinted with permission from W. Göpel, Biosensor and Bioelectronics 1995, 10, 35–39. Copyright 1995 Elsevier Science Ltd.)

5.4.1
Interconnecting Molecules

A major challenge in molecular electronics is how to interconnect single molecules and probe molecular conductivity in real devices working at the nanoscale (both two- and three-terminal devices). This issue requires the fabrication of nanometer-spaced electrodes. Since optical lithography suffers from physical limitations (mainly related to diffraction effects), new techniques are needed for patterning below 100 nm. Even though none of the proposed methods equals the advantages of photolithography for low cost and high throughput, a number of lithographic solutions (extreme UV lithography [24], X-ray proximity lithography, and EBL and IBL) are available, and new techniques have been developed to produce very close electrodes. For instance, Reed and coworkers [25] have investigated charge transport through molecules of benzene-1,4-dithiol self-assembled onto the two facing gold electrodes of a mechanically controllable break junction fabricated by the gentle fracture of an electrode by means of mechanical deformation. Another possibility is to constrain molecules in free-standing junctions, which consist of suspended electrodes used to electrostatically trap molecules and/or nanoparticles by monitoring the current and interrupting the trapping fields at the first increase in conductivity. Among these various alternatives, planar nanojunctions – consisting of two facing metallic electrodes separated by an insulating medium and fabricated by EBL and lift-off processes – are ideal to implement molecular devices since they have a high process yield (typically more than 90% of the nanojunctions are good with an open-circuit resistance around 1 TΩ) and provide the opportunity to perform transport experiments on molecules and/or nanostructures self-assembled at the surface of a solid, as well as to fabricate field-effect devices by adding a third electrode at the bottom of the device. Recently, Kervennic and coworkers [26] succeeded in producing pairs of platinum electrodes with a separation down to 3.5 nm by combining EBL and chemical electrodeposition. The separation between EBL-fabricated electrodes is reduced to atomic separation by monitoring the interelectrode conductance and stopping the electrodeposition process at predefined conductance values. In the particular case of DNA, the conductance has been measured using nanofabricated electrodes, atomic force microscopy (AFM), scanning tunneling microscopy (STM) and low-energy electron point source (LEEPS) microscopy. See Tab. 5.1.

Tab. 5.1. Transport characterization techniques.

Scanning probe techniques	STM	Conductive-probe AFM	Mercury drop electrodes
Nanojunctions for devices	break junctions	free standing junctions	planar junctions

5.4.2
Delivering Molecules

Molecular-scale assembly of individual components in working devices is another challenge. In the last few years, great attention has been focused on the development of techniques capable of fabricating molecular layers and patterns [evaporation, spin-coating, Langmuir–Blodgett techniques, thermodynamically driven self-assembly (TDSA) and soft lithography] or to deliver molecules to specific locations (electrostatic trapping, dip-pen and related techniques, and optical tweezers). All these strategies have their own advantages and disadvantages. For example, evaporation usually is not suitable for delicate biomolecules, while spin-coating is fast and cheap, but films are strongly anisotropic and inhomogeneous. TDSA is particularly attractive for bioelectronics applications since it opens the way for the fabrication of engineered highly ordered layers, taking advantage of the specific reactivity of biomolecules having functional groups with affinity for specific surface atoms and/or molecular sites [7]. Although a detailed knowledge of the positions, identities and affinities of the functional groups is required in order to develop a reasonable linkage strategy, TDSA allows us to fully exploit all the intrinsic advantages of the biomolecular approach. Recently, soft lithographies (consisting of different fabrication techniques embracing imprint and stamping methods) have also been attracting considerable attention since they are biocompatible and allow us to create patterns on a surface of interest using a pattern-transfer element (a rigid stamp) made of an elastomer, a thermoplastic polymer network that deforms under an applied force and recovers its shape when the force is released.

In addition to these techniques (parallel and static), a number of scanning methods (serial or parallel) have been developed to place molecules in specific, well-established locations (e.g. in the gap between two nanoelectrodes). "Dip-pen" nanolithography (DPN), proposed by Mirkin's group [27], is a powerful resistless nanopatterning technique allowing the delivery of molecules to a suitable substrate from a solvent meniscus. The concept of DPN is quite simple. A conventional silicon nitride AFM tip literally draws molecules of virtually any material from the solvent meniscus to a surface of interest ("paper"). The molecular "ink" consists of a chemically reactive material placed on the tip. The fine controls of the tip position and the chemisorption of ink molecules as the tip is scanned across the substrate enables the deposition of organic molecules in a nanometer-scale region. Using DPN it is currently possible to achieve high-resolution patterns (linewidths as small as 15 nm and less than 10-nm spatial resolution). The main parameters affecting the nanopatterning results are the diffusion and the relative humidity. When deposited onto solid surfaces, molecules diffuse out, therefore dramatically influencing the spatial resolution. The relative humidity controls the dimension of the naturally formed water droplet between the substrate and the AFM tip, and thus the effective contact area. Another limiting parameter is the tip radius of curvature. Due to the high resolution and the possibility to use the same apparatus to image and to write a pattern, DPN is a very powerful tool for molecular electronics. More recently, Mirkin's group has also developed an eight-pen nanoplotter capable of performing parallel DPN [28] using a dense array of AFM tips and has employed DPN to generate DNA-based templates capable of guiding the assembly of nanoscale building blocks [29]. This technique opens the way to placing molecules and/or nanostructures at established positions. Another interesting tool for molecular delivery is the optical tweezer where light is used to move matter and manipulate nanoscopic objects, exploiting photons momentum. For examples, nanoparticles can be trapped in the focus of a laser beam and serially moved. The force felt by the trapped particle is primarily given by the dipole and by the gradient force due to the interaction of a dielectric object with light and lies in the piconewton range, therefore allowing the manipulation of very fragile structures. Recently, MacDonald and coworkers have constructed 3-D trapped structures within an optical tweezers setup using an interferometric pattern between two laser beams. These results point towards the creation of extended 3-D crystalline structures [30]. Assembling functional molecular devices with this serial manipulation approach is, however, extremely slow, and thus technologically unattractive and not suitable for mass production. An alternative, promising strategy to produce large-scale functional circuits at the molecular level is so-called "molecular lithography" directly exploiting self-recognition and self-assembly of biomolecules on surfaces (a parallel process), as described in Section 5.5.

In particular, in order to attach single DNA molecules to metal electrodes, DNA oligomers have been functionalized with thiol groups (SH) [31] or other sulfur–gold interactions have been exploited [32–34], such as derivatizing oligonucleotides with disulfide groups. Electrostatic trapping has been also employed [35, 36] to align DNA molecules between the electrodes. See Tab. 5.2.

Tab. 5.2. Nanopatterning and molecular delivery methods.

Local/scanning	Static/large area	Type
Optical tweezers	extreme UV and X-ray proximity lithography	light-based
Electrostatic trapping; EBL and IBL		electromagnetic
AFM nanolithography, AFM storage, "Millipede"	nanoimprint lithography	thermomechanical
AFM nanostencil, membrane nanostencil	evaporation, spin-coating, Langmuir–Blodgett technique	deposition
DPN and parallel DPN	soft lithography	molecular delivery
Molecular lithography	self-assembled monolayers	self-assembly/self recognition

5.5
Devices Based on DNA and DNA Bases

DNA plays a crucial role in biology as the carrier of genetic information in all living species. Recently, however, physicists and chemists have become increasingly interested in the electronic properties of the "molecule of life", as demonstrated by the large number of fundamental and applied works published in the recent years, mainly concerning its charge transfer properties. Such considerable interest is largely explained by the number of possible applications ranging from electronic devices to long-range detection of DNA damage.

Despite the current hot debate, the subject is far from new. Soon after the Watson-Crick discovery of the double-helix structure of DNA in 1953, Eley and Spivey suggested that DNA could serve as an electronic conductor [37], as the result of interbase hybridization and π–π interactions in the stacked base pairs of double-stranded DNA. If z is the direction of the DNA helical axis (i.e. perpendicular to the base plane), atomic p_z orbital can originate delocalized π bonding and π^* antibonding orbitals, enabling charge transport along the helical axis (similar stacked aromatic crystals, such as the Bechgaard salt, are indeed metallic). (Note that although DNA bases are also aromatic, there are important differences such as the nonperiodic nature of biological DNA and the expectation of Anderson localization for the electronic states. Moreover, the chemical surroundings and molecular vibrations have also to be taken into account in the case of DNA.)

The advent of measurements on single DNA molecules recently revived the field. In particular, measuring the fluorescence produced by an excited molecule, the Barton's group at the California Institute of Technology [38] found that it no longer emitted light if attached to a DNA molecule and ascribed this fluorescence quenching to the charge transfer from the excited donor molecule to a nearby acceptor molecule along the DNA strand. In other words, DNA would act as a conducting molecular wire, mediating long-range transport on fast timescales. Moreover, they reported a profound sensitivity of long-range charge transfer to stacking and a significant reduction of electron migration as the result of the presence of base mismatches or other stacking perturbations [38].

It is worth noting how these results could have deep implications in the fundamental understanding of important biological processes such as the mechanisms for sensing DNA damage in living organisms, which unlike the DNA repairing processes still present various open-points. For example, in response to DNA-damaging agents, cells induce the expression of DNA-repair enzymes, activate cell-cycle checkpoints, inhibit DNA replication and mitosis and promotes DNA repair, recombination, or, under some circumstances, undergo apoptosis (in the case of fatally damaged DNA, cells are triggered to self-destruct so that they cannot cause cancer). However, the mechanisms by which these cellular responses are activated are not completely understood. Proteins moving along the DNA could be responsible for DNA damage recognition, but in such case the screening of the genome would be quite inefficient since this process is expected to be highly time-consuming and very slow along, for example, the 2 m of DNA in the human genome. However, if DNA is a conducting molecular wire, as reported by Barton and coworkers, the presence of damaged regions could be electronically detected exploiting DNA-mediated long-range charge transport, by two proteins (a transmitter and a receiver) working as a donor group at one end and an acceptor at the other end. If a damaged region intervenes between them, a significant reduction of charge transfer is expected and the transmitter could simply move along the DNA until the identification of the site of damage, which is then marked for repair [39].

After much initial controversy, the chemists studying charge transfer in DNA are now moving towards a common view [40, 41] and it is generally accepted that charge carriers can hop along the DNA over distances of few nanometres. On the contrary, the understanding of physicists investigating transport through DNA molecules is much less clear and the issue of DNA conductivity still presents some open-points, since the experimental results of different worldwide groups are surprisingly contrasting and range from an insulating [31–33, 42], semiconducting [35, 43] or ohmic [34, 36, 43–46] up to a proximity-induced superconducting behavior [47].

5.5.1
Charge Transfer in DNA

The process of charge transfer (intermolecular from one molecule to another or intramolecular from one end to the other of the same molecule) is one of the most fundamental in chemistry and materials science and has been extensively studied in extended molecules, like DNA, with a donor group at one end and an acceptor at the other end. Experimental and theoretical studies have shown that the electron-transfer reactions within such a single molecule can occur by two main mechanisms. The first consists of a single-step quantum mechanical tunnelling (QMT) from the donor to the acceptor, a coherent process since no energy is exchanged between the electron and the molecule during the transfer and the electron is never localized. QMT is quantitatively described by the Marcus theory [48] and characterized by a rate decreasing exponentially with the distance. As a consequence, it is possible only for short distances events, while for very long distances, it is expected to be insignificant. The second possible mechanism is thermal hopping, an incoherent process involving several uncorrelated events resulting in a random multi-step walk. In this case, the electron is localized on the molecule and exchanges energy with it. Hopping is predominant for long-distance electron transfer.

Shortly, the main factors influencing the electron transfer (ET) rate are [49]: (1) the distance between the donor and acceptor (in QMT the ET rate decreases exponentially with distance, whereas in hopping it is inversely related to the distance); (2) the nature of the micro-environment separating the donor and the acceptor (which mediates the virtual state or provides intermediate states, respectively); (3) the reorganization energy λ, i.e. the energy required for all structural adjustments (in the reactants and in the surrounding molecules) that are needed for the transfer of the electron [50]; (4) the driving force.

In the past, electron transfer in DNA was largely studied by means of various techniques. However, it was the important work of the Barton's group at Caltech [38] and the Turro's group at Columbia University [51] to revive the field, investigating its distance dependency and reporting the apparent coherent transfer over distances as long as 4 nm. As a result, DNA became a new paradigm for electron transfer, with possible applications as a molecular wire. In this respect, the extensive chemical variability of DNA (including differences in sequence, structures – kinks, bends, bulges and distortions – as well as the polyelectrolyte character of the double helix with the possible flow of positively charged counterions along the negatively charged phosphate backbone) opens large possibilities for tuning electron transport in DNA. On the other hand, the same variability is at the origin of the different results reported in DNA charge-transfer experiments. However, a common view about the basic mechanisms (coherent tunnelling and thermal hopping) involved in DNA charge-transfer was laboriously achieved in the last years thanks to extensive experimental and theoretical work.

To understand why the DNA sequence makes a difference, we have to compare the relative energies of the G-C and A-T base pairs obtained from computational models, photoemission experiments and electrochemical measurements. Since the thermal energy of the charge carrier is substantially smaller than the energy difference between these two base pairs, a hole will localize on G-C pairs while the A-T pairs act as barriers. Hole transfer is thus possible by tunnelling or hopping between G-C sites, with the second mechanism dominant for large distances due to the different distance dependencies.

Both the charge-tunnelling and the thermal-hopping mechanisms have been verified in experiments, notably by Bernd Giese and co-workers [52, 53]. In their experiment (Fig. 5.5), the charge transfer between G bases, separated by adenine-thymine $(A \bullet T)_n$ bridges of various lengths, in double DNA strands was measured using gel electrophoresis. They found that from being a coherent superexchange charge transfer (tunnelling) process at short distances, the mechanism becomes a thermally induced hopping process for long $(A \bullet T)_n$ sequences, where the adenines are involved as charge carriers (A-hopping). A switch between these reaction mechanisms occurs because the tunnelling rates decrease considerably as the distance increases. Therefore, in DNA strands where the guanines are separated by long $(A \bullet T)_n$ sequences (more than three), endothermic transfer of the positive charge from a guanine radical cation (G^+) to an adjacent adenine becomes faster than the direct transfer of this charge to the distant guanine. The subsequent migration of the positive charge between the adenines (A-hopping) is so rapid that the length of the $(A \bullet T)_n$ sequence plays only a minor role. Note that charge transfer (and transport) in DNA strongly depends on the base-stacking characteristics, the dynamic structural distortion of the DNA, the base sequence and the different redox potentials of the bases (Fig. 5.6). In this scenario, having the smallest oxidation potential between the DNA bases, guanines (G) or sequences of guanines are easily oxidizable sites. Thus, once holes are created on the DNA chain, then the charge transport can occur among discrete G sites or delocalized GGG domains (e.g., polaron). This means also that they represent easy targets for many oxidizing

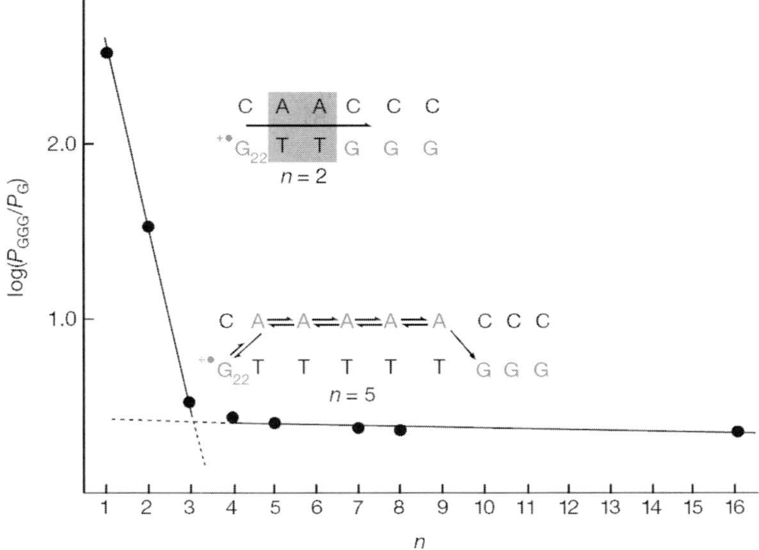

Figure 5.5. Electron transfer between a GGG sequence and a guanine radical cation, separated by $(AT)_n$ base pairs. The ratio PGGG/PG is a measure of the efficiency of the electron transfer. (reprinted with permission from B Giese et al., Nature 2001, 412, 318. Copyright 2001 Macmillan Publishers Ltd.)

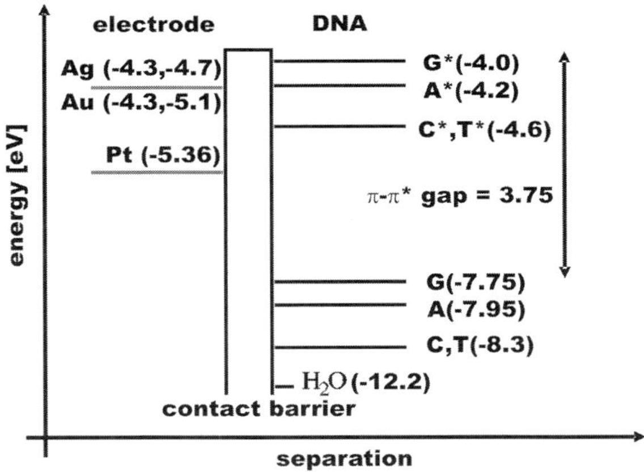

Figure 5.6. Energy levels for DNA bases and some metal work functions. (reprinted with permission from R. G. Endres et al., Rev. Mod. Phys. 2004, 76, 195. Copyright 2004 American Physical Society)

agents and guanine-rich regions of DNA could serve as "cathodic" protectors against oxidation for genes which are in electrical contact with them [54].

Meanwhile, Lewis and coworkers, at Northwestern University, have directly observed both thermal hopping and coherent transfer [55], while Barton and Zewail, at Caltech, also explained their fluorescence-quenching experiments using these two mechanisms [56]. For a more detailed description of the theory of charge transfer, see Ref. [57]. Most of the reported examples of charge transfer in DNA are in liquid environment.

5.5.2
DNA Conductivity

While chemists are now converging towards a common view, among physicists there is still a hot debate about the issue of DNA conductivity due to the conflicting results yielded in direct electrical measurements by different groups worldwide. This is a quite new field which in the last years took advantage of the nanotechnology tools (such as electron beam lithography and scanning probe microscopy) to image samples at the molecular level and probe conduction in single DNA molecules by interconnecting them between two metal electrodes.

5.5.2.1 Near-ohmic Behavior (Activated Hopping Conductor)

In 1999, Fink and Schönenberger at the Basel University [44] carried out the first direct electrical measurements on small bundles of DNA developing a special high-vacuum low-energy electron microscope able to image thin free-standing bundles of DNA stretched across a hole in a membrane. The molecules were placed onto a regular array of 2-µm holes in a carbon foil by cast deposition of a drop of water containing 0.3 µg/ml of λ-DNA onto the sample holder. This procedure usually resulted in λ-DNA networks spanning the sample-holder holes with DNA strands associated into a rope and only very occasionally individual DNA molecules were obtained. The DNA conductance was then measured in a vacuum environment ($\sim 10^{-7}$ mbar) by touching the DNA bundles with an additional tip (Fig. 5.7) placed at an electrical potential U with respect to the grounded sample holder and used to achieve mechanical contact to a specific site of the DNA rope, to break the ropes at a certain distance from the rim of the hole, and to apply a voltage bias to one end of the ropes (while the other one is grounded). These manipulations were performed *in situ* while observing the projection images of the molecules on a TV monitor and the sample holder was covered with a gold layer to improve the electrical contact to the molecules. The result of this experiment was quite unexpected: a single DNA rope of 600 nm in length was found to behave like a ohmic conductor with a resistivity of the order of 1 mΩcm, while two DNA ropes simultaneously interconnected behaved like two resistors in parallel (with the overall measured resistance reduced by a factor of 2). Contrarily, in the simplest picture one expects DNA to behave like a semiconductor with a large energy gap between the valence and conduction bands.

Results from the groups of Tran, Rakitin, Cai, Hartzell and Yoo [45, 43, 36, 34,

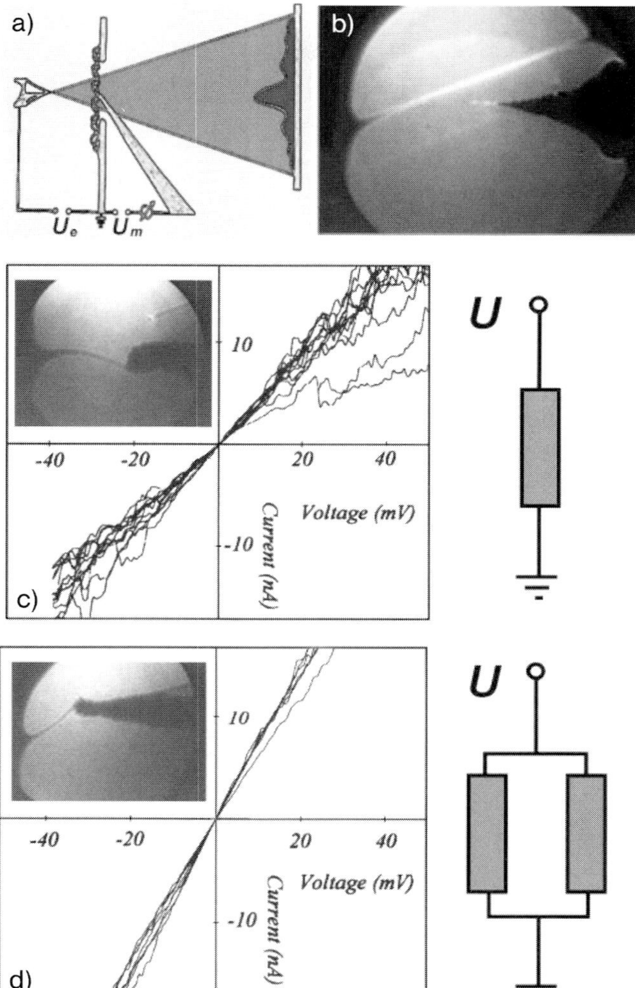

Figure 5.7. (a) The experimental setup used by Fink and Schonenberger to probe the conductivity of DNA. (b) A 2-mm hole spanned by a single DNA rope and the shadow image of the manipulation tip. (c and d) $I–V$ characteristics of an individual DNA rope and two DNA ropes in parallel. (reprinted with permission from H. W. Fink et al., Nature 1999, 398, 407. Copyright 1999 Macmillan Publishers Ltd.)

46] also suggested almost ohmic conductance at room temperature by means of very different experimental techniques. In all these studies DNA bundles, networks of bundles or supercoiled samples were investigated (see Tab. 5.3).

In particular, due to the numerous open points and difficulties concerning the contact effects and charge injection into DNA, Tran and coworkers measured the

Tab. 5.3. Summary of recent measurements of DNA conductivity (adapted with permission from Ref. [57], Copyright 2004 American Physical Society).

Class	Group	DNA sample	Result	Electrodes	Method	Ions
1. Anderson insulator	Storm et al. (2001)	single λ-DNA/ poly(G)–poly(C)	insulating (at room temperature) (DNA height: 0.5 nm)	Pt/Au	on SiO$_2$, mica surface	Mg^{2+}
	Braun et al. (1998)	single λ-DNA	insulating (at room temperature)	Au	free hanging (gluing technique)	Na$^+$
	Zhang et al. (2002) de Pablo et al. (2000)		(conducting if doped)		SFM, on mica	
2. Bandgap insulator	Porath et al. (2000)	single poly(G)– poly(C) (only 30 base pairs)	wide bandgap semiconductor (at room temperature)	Pt	free hanging	Na$^+$
	Rakitin et al. (2001)	single, short oligomer-λ-DNA	(at room temperature)	Au	(gluing technique)	
3. Activated hopping conductor	Rakitin et al. (2001)	bundles of λ-DNA	narrow "bandgap"	Au	free hanging	Na$^+$
	Yoo et al. (2001)	supercoiled poly(G)–poly(C)/ poly(A)–poly(T)	linear ohmic at room temperature insulating at low temperature	Au/Ti	on SiO$_2$	
	Cai et al. (2000)	networks of bundles poly(G)– poly(C)/poly(A)– poly(T)	linear ohmic (at room temperature)	Au	SFM, on mica	
	Tran et al. (2000)	supercoiled dry and wet λ-DNA	hopping conductivity	None	microwave absorption	
	Fink and Schonenberger (1999)	bundles of λ-DNA	conducting (doped) (at room temperature)	Au	free hanging	
4. Conductor	Kasumov et al. (2001)	few λ-DNA molecules	induced superconductivity ($T < 1$ K)	Re/C	on mica	Mg^{2+}

temperature dependence of the AC conductivity in a contact-less configuration by evaluating the absorption in a microwave resonant cavity [45]. The measured conductivity was found to be only weakly frequency dependent and influenced by the environment of the double helix (with larger values in buffer). They interpreted the observed temperature dependence by means of a crossover between two transport

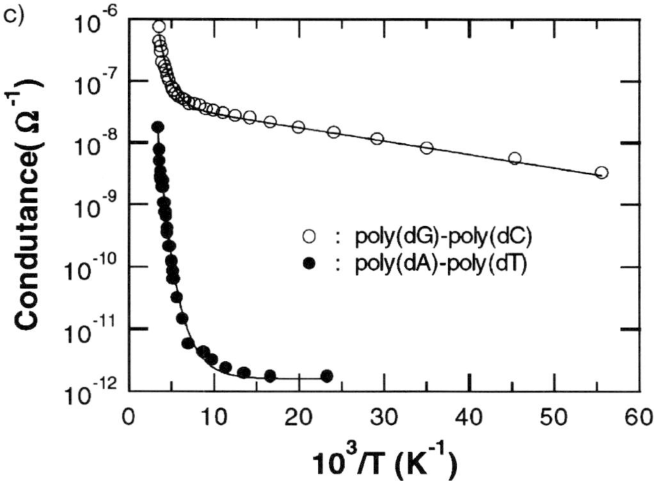

Figure 5.8. Temperature dependence of the conductance for poly(dA)–poly(dT) and poly(dG)–poly(dC). (reprinted with permission from Yoo et al., Phys. Rev. Lett. 2001, 87, 198102. Copyright 2001 American Physical Society)

mechanisms: ionic conduction due to the counterions at low temperatures and carrier excitations across single particle gaps or temperature-driven hopping transport processes at high temperatures. Notably, the same temperature dependence was observed by Yoo and coworkers (Fig. 5.8 [46]), which proposed an explanation in terms of a small polaron hopping model over the whole temperature range (the same model predicting a dependence of $I \propto \sinh(bV)$ also fits their I–V data very well).

Yoo and coworkers investigated transport in poly(dA)–poly(dT) and poly(dG)–poly(dC) DNA molecules and reported room-temperature resistances of 100 and 1.3 MΩ, respectively. Other interesting differences were observed like the much weaker temperature dependence for poly(dG)–poly(dC) and its behavior as a p-type semiconductor, while poly(dA)–poly(dT) acts as a n-type semiconductor under the effect of a gate voltage. Tran and coworkers reported a resistivity of 0.005 Ωcm for the dry λ-DNA, while Yoo and coworkers estimated a value of 0.025 Ωcm in the case of poly(dG)–poly(dC).

Other experimental results came from the Kawai group at Osaka University [36]. They used scanning probe microscopy and a conductive tip to inject current in DNA bundles connected to a metal strip at different positions in order to determine the length dependence of the current. The resistance was found to increase exponentially with the distance, from 10^9 ohm to 10^{12} ohm (Fig. 5.9). Moreover, DNA made of C and G bases were more conductive than DNA with A and T bases [36] (estimated resistivity for poly(dG)-poly(dC) around 1 Ωcm). The increasing of the resistance with the length was attributed to the number of defect in the molecules, while they proposed that the more compact base-stacking structure of

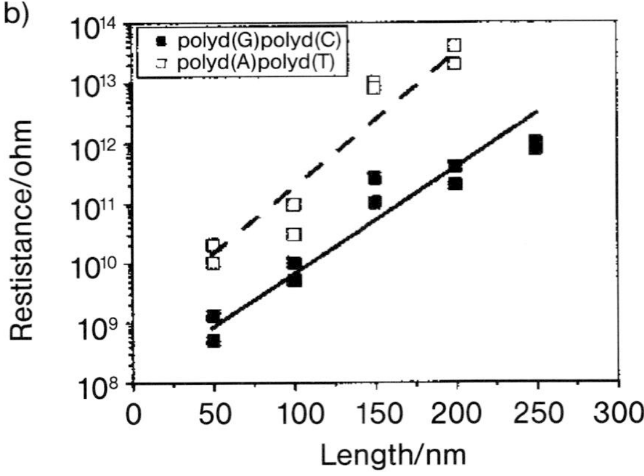

Figure 5.9. Resistance versus DNA length for poly(dA)–poly(dT) and poly(dG)–poly(dC). (reprinted with permission from Cai et al., Appl. Phys. Lett. 2000, 77, 3105. Copyright 2000 American Institute of Physics)

poly(dG)-poly(dC) (crystal structure data indicates that the helical rises of residues are 2.88 Å and 3.22 Å for poly(dG)-poly(dC) and for poly(dA)-poly(dT), respectively [36]) can also contribute to its better conductance as compared to poly(dA)-poly(dT). In the same group, in a four probes current voltage measurement [58] the effect of oxygen doping was tested on poly(dG)–poly(dC) DNA and poly(dA)-poly(dT) DNA samples. The authors found that the conductance of the DNA samples increases by orders of magnitudes due to oxygen hole doping and that the poly(dG)–poly(dC) DNA samples behave like a p-type semiconductor, whereas the poly(dA)–poly(dT) ones are more likely n-type semiconductors. This difference was attributed to the lower oxidation potential of dG with respect to dA. In this experiment the spacing between the electrodes was varied from 100 to 200 nm.

Finally, Hartzell and coworkers [34] compared the conduction in nicked and repaired λ-DNA, showing a gap up to ±3 V and almost ohmic behavior (it is worth noting, once again, the important possible implications in DNA damage recognition). Moreover, Kasumov and coworkers [47] reported proximity induced superconductivity in long λ-DNA, which in their experiment exhibits metallic behavior down to very low temperature (this would be the proof of the existence of true extended states). However, this result has not been independently reproduced and will not be further discussed here.

5.5.2.2 Semiconducting (Bandgap) Behavior

The role of the DNA sequence on transport was investigated at the nanoscale for the first time by Porath and co-workers at Delft University, who measured very short DNA molecules consisting of two strands containing only G and C bases, respectively, in order to probe the importance of G-C pairs in hole transfer through

DNA [35]. Using electrostatic trapping from a dilute aqueous buffer containing about one molecule per $(100 \text{ nm})^3$, Porath and co-workers placed individual double-stranded poly(G)-poly(C) DNA molecules between two nanoelectrodes only 8 nm apart. The DNA molecule was 10.4 nm long to span the two closely spaced metal nanoelectrodes as a linear, nearly stiff structure. After a DNA molecule was trapped from the solution, the device was dried in a flow of nitrogen and electrical transport was measured. On applying a voltage to the device, they first observed an insulating gap (i.e. no current) at low voltage followed by a conduction onset above 1 V. The low-current region widened as the temperature increased. Two different peaks were present in the differential conductance curves and suggested a molecular-band-mediated transport mechanism and a wide-gap semiconductor behavior, as expected for short DNA molecules if we assume that DNA bases have a rather large highest occupied–lowest unoccupied molecular orbital (HOMO–LUMO) gap (4 eV) and the metal work function lies in the gap. In this case, in fact, a very large electric field is needed to drive transport and allow charge carriers to enter the DNA by aligning the molecular energy bands with the energy levels in the electrodes. The reported dI/dV curve is thus a measure of the DOS, while the voltage gap (1–2 eV) can be related to the energy difference between the metal work function and the nearest molecular level available for conduction (either the guanine HOMO or the cytosine LUMO). Porath and coworkers [35] interpreted their results as reasonable evidence of the existence of coherent electronic states extended across the molecule. It is worth noting that, in this experiment, transport through DNA molecules only a few base pairs long is probed: the persistence length in dsDNA is about 100 base pairs at room temperature, thus the investigated molecules are reasonably free of the kinks and defects that would lead to an interruption of the π–π interactions [57]. This experimental finding was confirmed by the model developed by Cuniberti and coworkers [59] on a similar DNA short sequence.

More recently, Watananbe et al. at Xerox in Japan [60] performed other experiments on single DNA molecules by means of a triple probe atomic force microscope in which two conductive carbon nanotube (CNT) electrodes were employed as source and drain electrodes to interconnect a DNA molecule and a third CNT (fixed to the AFM probe) served as the gate electrode. Due to the small drain-source distance (25 nm), also in this case only few base pairs were probed. Experiments were carried out under nitrogen atmosphere at room temperature and non linear I_{DS}-V_{DS} characteristics with a non-conducting gap decreasing from 2 to 0.2 V with increasing V_G from 0 to 5 V were observed. Electrical measurements with drain-source distance of only 5 nm showed equally spaced steps in the I_{DS}-V_{DS} curve with a period of 80 meV and a gap of 400 meV at $V_G = 2$ V.

5.5.2.3 **Insulating Behavior**

The electronic conduction at length scales of 40 nm and longer was measured at Delft University by Storm and co-workers with different electrode shape [31]. They obtained clear AFM images of DNA molecules interconnecting the nanoelectrodes, but did not observed any conduction for such long lengths of DNA with applied bias voltages up to 10 V and also with an additional gate voltage in the range between -50 V and 50 V. They used both individual DNA molecules and small DNA bundles (mixed sequence and poly(dG)-poly(dC)) and employed a cast deposition technique to deposit the molecules between the electrodes. Although under their conditions DNA self-assembles into networks similar to those obtained by Cai and coworkers, Storm and colleagues measured a higher resistance in all their experiments of around 1 TΩ, which was observed to increase in a flow of dry nitrogen gas ($>10^{13}$ Ω, i.e. 10 TΩ). Thus, they suggested that the observed conduction was due to the thin water layer on the hydrophilic mica. These findings clearly contrast with the Basel results (and other high-conductance reports), but they confirm earlier measurements by Braun and coworkers at Technion in Israel [32], who also observed insulating behavior for DNA over a scale length of about 16 μm in length, and results of de Pablo and coworkers [42], who measured the DNA conductivity using a scanning force microscopy (SFM) tip covered by gold (they reported a lower bound of 10^6 Ωcm for the resistivity). The discrepancy between the Storm lower value (1 TΩ) and the conductivity reported by Cai and coworkers [about 200 GΩ for bundles of poly(dG)–poly(dC) at a length scale of 200 nm] is, however, not so large.

de Pablo and coworkers [42] also presented "first principles" electronics structures calculations for a double helix of infinite length in acidic dry conditions, which supported the picture of an insulating behavior for the λ-DNA. They started by considering a poly(C)–poly(G) sequence and then analyzed the effect of swapping guanine and cytosine bases in one of every 11 base pairs of their unit cell. This resulted in a dramatic change of the electronic structure of the chain and in a cut in the HOMO-state channel (more or less the same happens for the unoccupied band), as shown in Fig. 5.10. Thus, Anderson localization (over very few base pairs) and an exponential decay of the conductance with the length are expected in biological DNA, due to the nonperiodic nature of its base sequence. Residual conduction by hopping mechanisms (polaronic or not) is still possible, but it should exhibit a marked dependence on temperature and frequency [42]. Taking into account the electronic structure obtained for poly(C)–poly(G), de Pablo and coworker's results could be consistent with those of Porath, but are strongly in contrast with those of Fink and Schonenberger (whose resistivity is 10 orders of magnitude lower). An insulating behavior for a λ-DNA molecule on the micron scale has been also reported by Zhang and coworkers [33] (resistivity higher than 10^6 Ωcm).

5.5.2.4 Discussion of DNA Conductivity

Despite the large worldwide efforts, the issue of DNA conductivity is, thus, still subject of intense debate and rather unsettled. Assessing and understanding DNA transport properties, with all their facets, it is crucial to gain a deep insight inside important life mechanisms (like DNA damage recognition and reparation) as well as for a variety of applications in molecular electronics and biotechnology, e.g. the implementation of DNA chips with an electronic readout. Microarrays for the de-

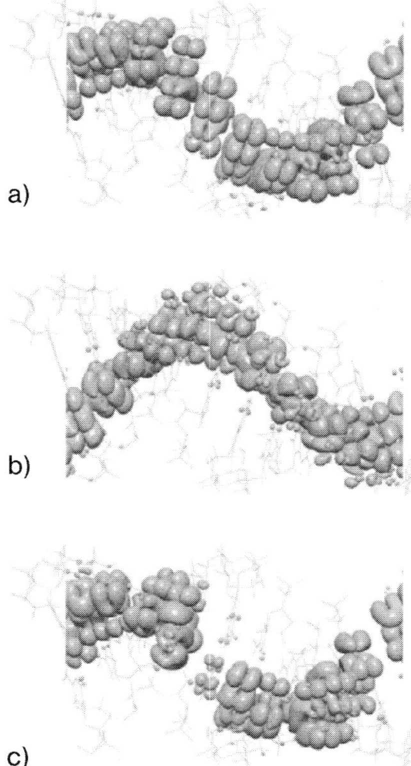

Figure 5.10. Isosurfaces of constant density for (a and b) the 11 highest occupied and lowest unoccupied states of poly(G)–poly(C), and (c) the 11 highest occupied states when swapping a guanine and a cytosine base in one of every 11 base pairs of the unit cell. (reprinted with permission from de Pablo et al., Phys. Rev. Lett. 2000, 85, 4992. Copyright 2000 American Physical Society)

tection of nucleic acids (DNA chips) are widely used for DNA sequencing, disease screening and gene expression analysis since they allow us to obtain information on nucleic acid sequences in a manner that is faster, simpler and cheaper than traditional methods [61]. Currently, the readout schemes for such devices are optical and involve the use of fluorescent dyes, but the exploitation of DNA conductivity (if large enough, after hybridization, in well-stacked dsDNA as compared to floppy, insulating single strands of DNA) or of its self-assembly properties (for the delivery of conductive elements between pairs of electrodes in large arrays) could enable in the near future the development of an electronic readout which is expected to allow a significant increase in the number of different probe sites (whose density would be no more resolution limited) per unit area, and thus a potentially faster and more efficient sequence analysis.

The main difficulties in addressing the DNA conductivity lie in the complexity of the system and in making clean (i.e. fully controllable) experiments, due to three main issues:

(a) Differences in the DNA molecules: sequence, length, character (single molecules or ropes).
(b) Influence of the molecular environment (hardly controllable and possibly preparation dependent): water molecules, counterions, substrate, physisorbed/chemisorbed DNA. For example, the number of water molecules per base is believed to influence the DNA structure [62].
(c) The detection scheme and the role (nature and quality) of contacts between electrodes and molecules. It is very difficult to have an ideal ohmic molecule–metal contact and the organic–inorganic interfaces usually produce tunnel barriers that can strongly influence the results. Only the metal work function is typically known.

Even theoretically, the situation is quite confused with the large unit cell limiting computational approaches [57], and a number of proposed transport models involving electrons, holes [63], polarons [64] and solitons [65].

However, recently, some progress in the analysis and understanding of the wide range of experimental results has been done [57]. We will try to provide a more unitary as possible discussion of the literature. As noted above, there has been only one group reporting proximity-induced superconductivity in long λ-DNA, thus these results will not be discussed here (Zhang and coworkers [33] proposed contamination from carbon or rhenium to explain such results).

Although the contact-less detection scheme of Tran and coworkers could detect very short dissipative regions embedded in an insulating DNA molecule [33] and de Pablo and coworkers [42] demonstrated that in Fink and Schonenberger's experiment [44] the measured DNA conductivity was affected by the irradiation with the imaging electron beam of the LEEPS microscope (resulting in a contamination layer with subsequent increase of conduction), the different experimental evidence of conducting DNA molecules needs an explanation in a comprehensive framework, which also takes into account the observed semiconducting and insulating behaviors. In this respect, it is worth noting how all reports of near-ohmic behavior concern DNA bundles, networks of bundles or supercoiled samples, while insulating behavior was observed in single molecules and short molecules were found to be semiconducting.

The possible stabilization of floppy single DNA molecules by the bundles or condensed water and counterions trapped between the DNA molecules (leading to a different pathway for charge transport) could explain the observed near-ohmic behavior [57] and also the dependence of the conductance on humidity (the contribution of the alternative pathway to conduction would be of course strongly influenced by the water content, but the more regular structure of DNA at high humidity could also contribute) as well as the lack of anisotropy seen in films of oriented DNA molecules [62].

A 1-D pathway through stacked base pairs could be argued from the weak sequence dependence reported by Yoo and coworkers; however, this could also be related to the different helical rises of poly(G)–poly(C) and poly(A)–poly(T) DNA, which could lead to a different contribution from water and counterions. In this scenario, small molecules could easily exhibit a semiconductor behavior, as observed by Porath and coworkers. The effects of the water solvent on the electronic states are accurately discussed in [57].

5.5.2.5 **Other Applications of DNA in Molecular Electronics**

The interest in DNA for molecular electronics, however, does not stop at its conduction properties, but also lies in its self-assembly ability. For example, in Rinaldi's group at the National Nanotechnology Laboratory, a single modified DNA base, the lipophilic deoxyguanosine, was adopted to fabricate electronic devices. Guanosine was chosen due to the lowest oxidation potential among the DNA bases, which favors transport, and its self-assembling properties related to its peculiar sequence of hydrogen-bond donor and acceptor groups [66]. Using this approach, Maruccio and coworkers [67], from the same group, succeeded in the fabrication of a prototype FET based on this modified DNA base (Fig. 5.11) with a maximum voltage gain of 0.76. Unlike other molecular electronics devices based on CNTs, this prototype FET is based on ordered and self-assembled layers, instead of a single molecule with tremendous interconnection problems [68]. The transistor was fabricated by cast deposition of a droplet of the deoxyguanosine derivative in chloroform solution between the source and drain contacts consisting of two EBL-fabricated chromium/gold electrodes separated by a distance of 20–100 nm, while the control electrode was a layer of silver deposited on the reverse of the Si/SiO$_2$ substrate. The guanosine-based FET operated at room temperature and ambient pressure, and exhibited a voltage threshold for the conduction which could be modulated by the gate voltage. The electrical characteristics were explained in terms of resonant tunneling, the threshold voltage being defined by the alignment between the molecular minibands and the Fermi level in the electrodes.

DNA exhibits a surprising range of structural forms and possible modifications. One interesting example is the replacement of certain hydrogen atoms in the base pairs of the DNA with metal ions (Zn^{2+}) as reported by Rakitin and co-workers at Brown University in the US [43], in collaboration with researchers at the University of Saskatchewan in Canada. Specifically, they prepared and investigated four types of λ-DNA samples:

(1) *B*-DNA in standard buffer at pH 7.5.
(2) *M*-DNA where the imino-proton of each base pair was substituted with a metal ion (Zn^{2+}).
(3) *B*-DNA samples in which the sticky ends of the DNA were attached to surface-bound oligomers.
(4) *B*-DNA at pH 7.5 with 0.1 mM of Zn^{2+}, since at this pH *M*-DNA does not form but the contribution of DNA surface-bound Zn^{21} ions on the measured electrical characteristics can be determined.

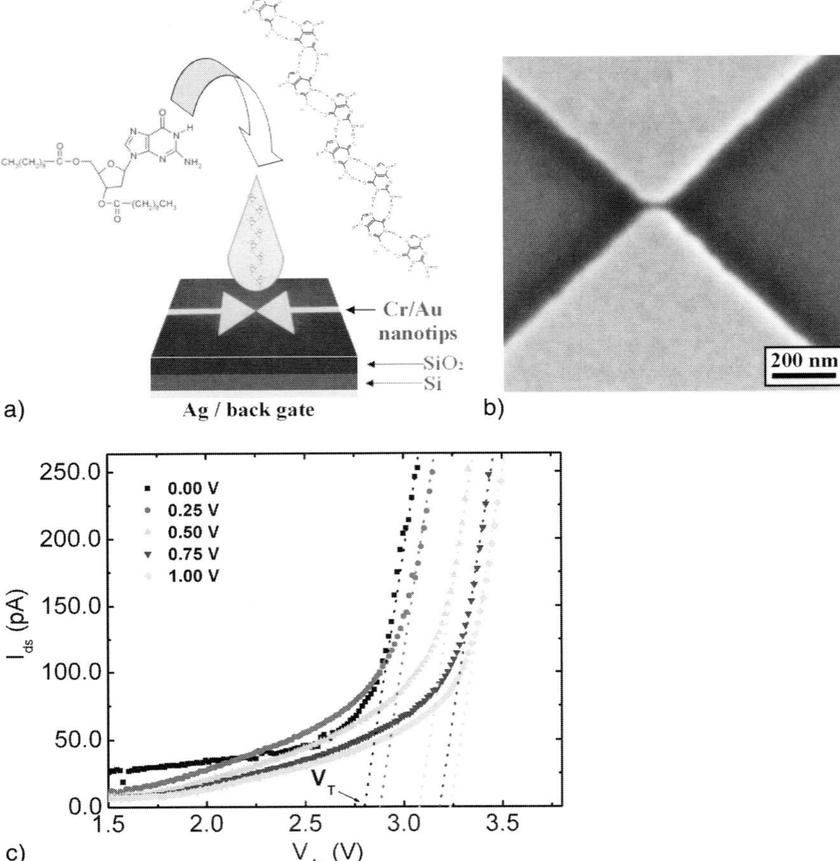

Figure 5.11. (a) Self-assembly and cast deposition of $dG(C_{10})_2$ on the three-terminal device, consisting of two arrow-shaped Cr/Au electrodes on a SiO_2 substrate and a third Ag back electrode (not to scale). (b) High-magnification SEM image of two Cr/Au nanotips with a separation of 20 nm. (c) Characteristics of the FET at different gate voltages (V_G). The dashed lines extrapolate the voltage threshold V_T for any V_G value. By changing the gate voltage, V_T can be modulated since the alignment condition for resonant tunneling is modified due to a shift in the energy of the molecular band. (reproduced with permission from G. Maruccio et al., *Nanoletters*, 2003, 3, 479–483. Copyright 2003 American Chemical Society)

The subsequent exposure of the samples to "freeze-dry" methods could not rule out the ionic contribution because of the existence of salt bridges formed on the substrate surface between electrodes. To solve this problem a new design for the contact was implemented by placing DNA between two electrodes separated by a physical gap of width 1–30 μm and practically infinite depth. A metallic-like conduction was observed in 15 μm long *M*-DNA, while a semiconducting behavior with a few hundred meV band gap was obtained in the case of *B*-DNA before conversion into *M*-DNA. Finally, a decrease in the zero bias conductance of about three orders of magnitude with respect to M-DNA was found in sample (4). The possibility to achieve such drastic change in conductivity could have important applications in molecular electronics.

An entirely different use of DNA was proposed by Braun and co-workers at the Technion and is based on the exploitation of its unrivaled assembly properties [32]. In fact, a strategy for the assembly of integrated circuits is still missing in molecular electronics and in this respect the highly specific binding between DNA strands may provide a key tool to control the geometry and connectivity of future electronic circuits without the use of destructive lithography techniques. Inspired by this, the Braun group firstly assembled DNA between two electrodes adding sticky ends to DNA fragments and then replaced the counterions with silver ions demonstrating that DNA can be used as a linear template to grow a thin metallic wire. The same approach was followed by Richter at Dresden to build up highly conductive palladium nanowires on a DNA template [69]. Moreover, the group of Erez Braun developed also the so-called sequence-specific molecular lithography [8], an alternative and promising strategy to produce large-scale functional circuits at the molecular level exploiting self-recognition and self-assembly of DNA (a parallel process). The information encoded in the DNA molecules replaces the masks used in conventional lithography, while a RecA protein serves as the resist. This technique enables high resolution, the fabrication of three-way junctions (branchpoints) and the sequence-specific positioning of molecular objects. More recently, Keren and coworkers [70] reported the fabrication of a carbon nanotube FET, self-assembled using a DNA scaffold molecule to provide the address for precise localization of the nanotube as well as the template for the extended metallic wires contacting it (Fig. 5.12). Besides DNA and proteins, genetically engineered viruses have also been employed to order nanostructures [71].

Recently Williams and coworkers at Delft joined the conducting properties of single-walled carbon nanotubes (SWNTs) with the specific molecular-recognition features of DNA by coupling SWNTs to peptide nucleic acid (PNA, an uncharged DNA analogue) and hybridizing these macromolecular wires with complementary DNA [72]. The oligonucleotide adducts imparted recognition properties, used to programme the attachment of SWNTs to each other and to the electrodes. This general approach can be exploited in the field of biosensors.

DNA can be also combined with specific chemical side groups, providing the basis for new functional devices and accurate biosensors. For example, nanocomposite structures made of gold nanoparticles covered with single-strand DNA fragments were prepared by the Alivisatos's group at Berkeley [73] and Mirkin and coworkers at Northwestern [74] with the spacing between the nanoparticles controlled statically (by using different sticky ends) or dynamically (by exploiting the melting properties of DNA with changes in temperature or solvent). Moreover, DNA has the potential for assembling intricate spatial structures and networks with a variety of geometries, as those produced by Seeman and co-workers at New

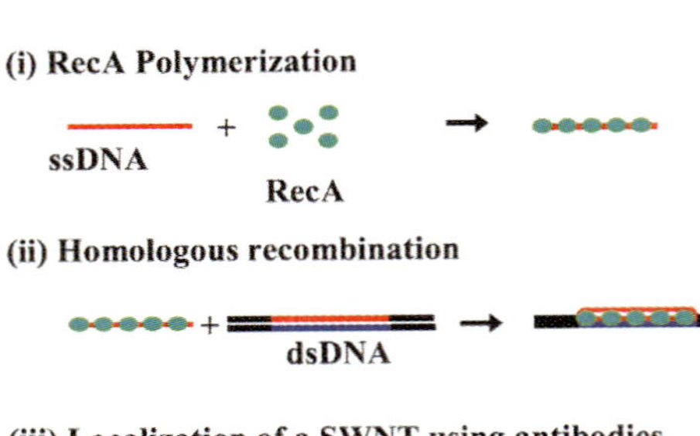

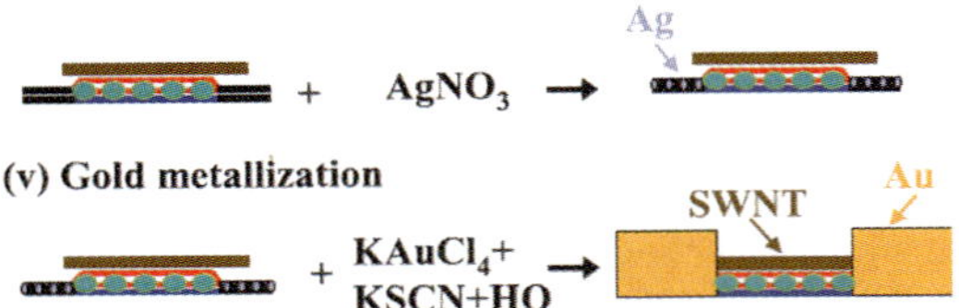

Figure 5.12. Assembly of a DNA-templated FET by molecular lithography. (i) Polymerization of RecA monomers on a ssDNA molecule, resulting in the formation of a nucleoprotein filament. (ii) Binding of the nucleoprotein filament at a desired address by homologous recombination. (iii) Delivery of a streptavidin-functionalized SWNT on the DNA-bound RecA by means of a primary antibody to RecA and a biotin-conjugated secondary antibody. (iv) Incubation in an $AgNO_3$ solution and formation of silver clusters on segments unprotected by RecA. (v) Formation of two DNA-templated gold wires to contact the SWNT by electroless gold deposition (the silver clusters are used as nucleation centers). (reproduced with permission from K. Keren et al., *Science*, 2003, 302, 1380–1382. Copyright 2003 American Association for the Advancement of Science)

York University, including loops, knots, one- and two-dimensional arrays, three-dimensional cubes and nanolattices based on synthetic DNA structures [75].

As previously discussed, an important application of the assembly properties of DNA are the DNA chips which make use of many parallel DNA (single strands) probes to check whether certain genetic codes are present in a given specimen of DNA. Their read out schemes are currently optical, but an electronic read-out possibly exploiting the electron-transfer properties of DNA or using single-strands with sticky ends attached to electrically active molecular elements (such as metal clusters, fullerenes or certain molecular switches) could enable further miniaturization. The different electrochemical responses of single- and double-strand DNA molecules that attach to a surface can be also used in this frame [76].

Finally, concerning the applications of DNA self-assembly, we have of course to mention the DNA computation and its proof-of-concept by Adleman's group in 1994 [5], which demonstrated that the recombinant properties of DNA can be exploited to solve problems appropriately encoded into single DNA strands, using five simple operations: (a) synthesis of a large numbers of oligonucleotides, (b) annealing/hybridization of oligonucleotides to produce dsDNA molecules, (c) extraction of molecules containing a given sequence of bases, (d) detection of DNA molecules and (e) amplification of DNA molecules. The massive parallelism intrinsic in this approach allowed Adleman's group to solve a problem that had resisted conventional methods. The problem-specific technique of Adleman's group has been recently extended theoretically to more general Boolean operations.

In conclusion, although the advantages to use DNA as a molecular wire in electronic devices have probably to be evaluated in more detail, its unique assembly and recognition properties seem destined to be a major tool in molecular electronics.

5.6
Devices Based on Proteins

Besides DNA, proteins have been also intensively investigated in the last years for application in molecular electronics, molecular motors and biosensors, since their nature-tailored functions presents clear advantages for a number of specific applications. For example, the potential of photochromism has generated great interest in the synthesis of new photosensitive materials and in the development of new techniques for their use. Some biological systems, like bacteriorhodopsin and the green fluorescent protein (GFP) of the *Aequorea Victoria* jelly fish, offer naturally evolved optimized structures with unique properties that can further be tailored for specific applications by genetic engineering. The green fluorescent protein (GFP) is very interesting for optoelectronic applications since it exists in two distinct configurations (bright and dark) and has a very efficient fluorescence emission that makes possible also single-molecule detection. Moreover, genetic engineering allows producing mutants with modified spectral characteristics, enhanced brightness, photostability, quantum yield, and other properties tailored for specific applications. In particular, two GFP mutants – EGFP (enhanced GFP) and E^2GFP (obtained by a single point mutation T203Y of EGFP) – were recently investigated by the Beltram group [77] who achieved optical control of transition between the bright and dark states in E^2GFP by means of two laser beams with different wavelengths ($\lambda = 476$ nm and 350 nm). Both the investigated proteins exhibited in solution absorption peaks at 400 and 515 nm, ascribed respectively to the protonated neutral and deprotonated anionic forms (states *A* and *B*) of the chromophore. A weak emission was observed for both mutants after excitation of state A and was ascribed to excited-state photoconversion from A to B. Moreover, the authors observed a reversible turning on and off (blinking) of the emission and its ultimate switching off (photobleaching) into a long-lasting dark state (*C*) within a few seconds after excitation at 476 nm. In the case of E^2GFP mutants, it was possible to achieve a reversible photoconversion from state C back into state B by irradiation at 350 nm. Since unlimited optically controllable cycles between the bright and dark configurations were reported, these results open the way to the use of the E^2GFP mutant in memory devices by employing the states B and C to encode a $(0, 1)$ bit. In fact, information can be stored and manipulated and the basic required operations (write, read and erase data) are performed at the single molecule level. According to the authors, one possible implementation is to use photoconversion from the dark to bright state by irradiation at 350 nm as WRITE step, fluorescence emission following weak excitation at 476 nm as READ step and photobleaching as ERASE step.

As discussed above, another interesting protein is the Bacteriorhodopsin (BR), a trans-membrane protein found in the cellular membrane of *Halobacterium salinarium*, which functions as a light-driven proton pump and thus is expected to be useful as photonic material. Due to its peculiar properties, its use as the active component in memory devices was investigated (for a recent review on Bacteriorhodopsin-based applications see [78]), in particular in holographic associative memories and branched-photocycle three-dimensional optical memories. More in detail, the proposed holographic associative memories were based on a Fourier transform optical loop and the real-time holographic properties of a BR films. On the other hand, in the branched-photocycle three-dimensional optical memories, parallel write, read, and erase processes were performed exploiting a sequential multiphoton process and an unusual branching reaction that creates a long-lived photoproduct. A very broad range of applications was proposed, from electronics to optoelectronics and computing including random access thin film memories, neural-type logic gates, photon counters and photovoltaic converters, artificial retinas, picosecond photodetectors, multilevel logic gates optical computing, and different kinds of memories (see [78] for further details).

Then, very recently, Yasutomi and coworkers at the Kyoto University [79] reported the fabrication of a molecular photodiode that can switch photocurrent direction by changing the wavelength of an irradiating light (Fig. 5.13). Their device was based on bicomponent SAMs of two helical peptides carrying different chromophores (thus selectively activable) and having opposite direction of dipole moments (when immobilized on gold). Since at a certain range of the applied potentials the direction of the photocurrent is determined by the direction of the dipole moment, photocurrent can be switched from anodic to cathodic alternating photoirradiation at 351 nm and 459 nm. The optimum voltage for photocurrent switching was determined by investigating the dependence of photocurrents on applied potential for the unicomponent SAMs.

The Greenbaum group reported on the generation of exogenous photovoltages by the Photosystem I (PSI) reaction centers, nanometer-size robust supramolecular structures that can be isolated and purified from green plants. A diode laser at $\lambda = 670$ nm was used to illuminate heterostructures composed of PSI and organosulfur molecules onto gold substrates, while the photovoltage was measured by Kelvin force probe microscopy. Under illumination, the potentials of the central region of the PSIs were found to be typically more positive than the periphery by 6–9 kT, where kT is the Boltzmann energy at room temperature [80]. Possible applications of the PSI to the implementation of artificial retina are currently under exper-

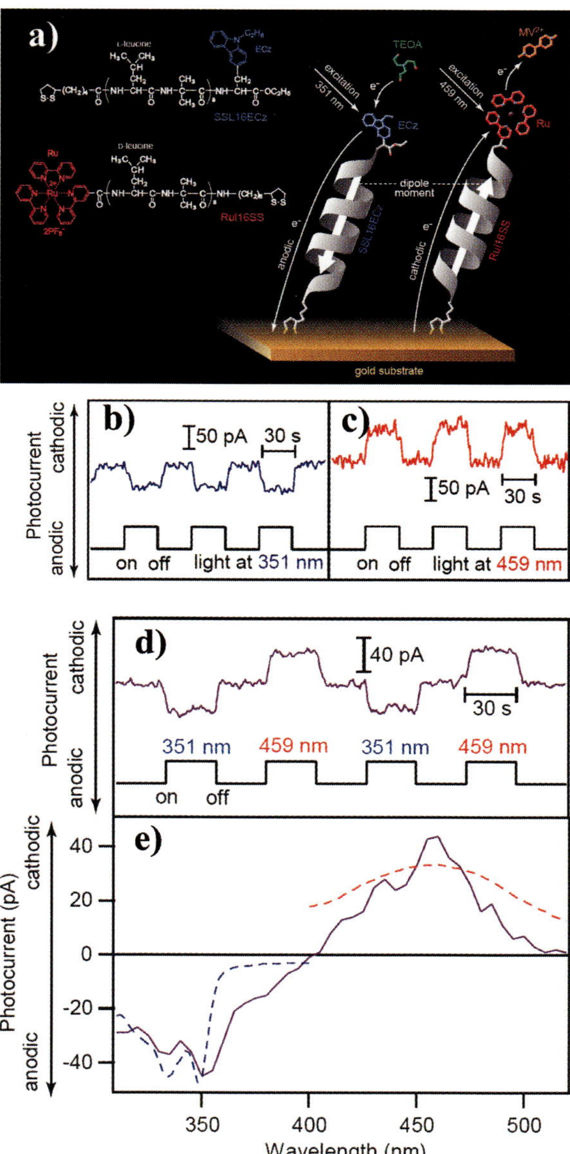

Figure 5.13. (a) Molecular structures of the two kinds of hexadecapeptides employed (SSL16ECz and Ru16SS) whose dipole moments (with opposite directions) determine the photocurrent switching from anodic to cathodic under selective photoexcitation of the sensitizer (an ECz or Ru group respectively) when the peptides are coassembled in a highly-ordered bicomponent SAM on a gold substrate via an Au–S linkage. (b–c) Periodic photocurrent generation by the SSL16ECz (Ru16SS) SAM upon photoirradiation at 351 nm (459 nm) in a 50 mM TEOA (MV²⁺) aqueous solution. (d) Time course of photocurrent switching upon alternating photoirradiation at 351 and 459 nm. (e) The action spectra (purple solid line) consists of anodic and cathodic photocurrent regions, which agree with the absorption spectra of SSL16ECz (blue dashed line) and Ru16SS (red dashed line) in ethanol, respectively. (reproduced with permission from S. Yasutomi et al. Science, 2004, 304, 1944. Copyright 2004 American Association for the Advancement of Science)

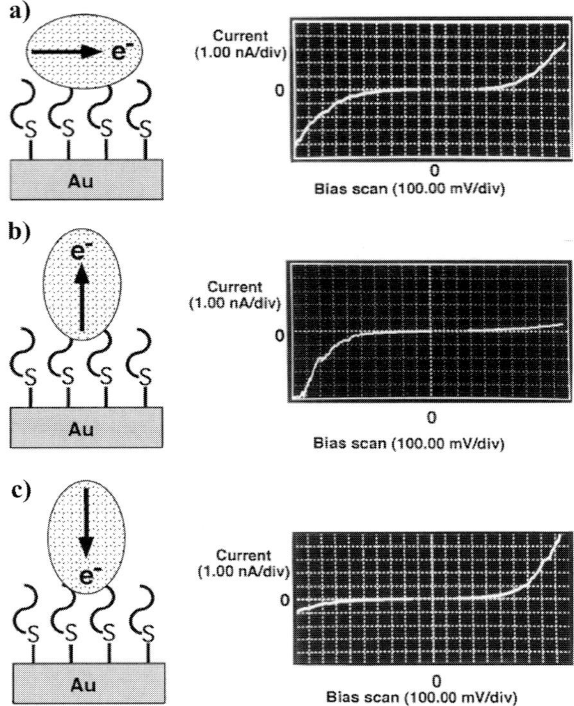

Figure 5.14. Orientation-dependent STS characteristics of individual PS I reaction centers. (a) If a PS I is oriented parallel to the electrode surface, a semiconductor-like behavior with a bandgap of around 1.8 eV can be observed. (b and c) If a PS I is anchored perpendicular to the gold surface, a diode-like (current rectification) *I–V* curve can be observed. (reproduced with permission from I. Lee et al., *Phys. Rev. Lett.* 1997, 79, 3294–97. Copyright 1997 American Physical Society)

imentation in the same group. Moreover, they also explored two-dimensional vectorial arrays of functional PSI reaction centers (prepared on atomically flat derivatized gold surfaces) by STS [12]. The nature and extent of PSI orientation were controlled by chemical modification of the surface derivative and checked by STS. When gold electrodes where treated with mercaptoacetic acid, 83% of the electron transport vectors were parallel to the electrode surface and a semiconductor-like I-V curve with a band gap of ≈ 1.8 eV (corresponding to the first excited singlet state of chlorophyll) was observed (Fig. 5.14a). On the other hand, using 2-mercaptoethanol, 70% were oriented perpendicularly in the "up" position and only 2% were in the "down" position and current-rectification (i.e. diode-like behavior) was observed (Fig. 5.14b–c). Authors ascribed the asymmetry of the I-V characteristics to either a difference in electronic energy (in analogy with a solid-state p-n junction) or a difference in the tunneling distance for each end.

A novel example of protein-based active electronic device has been recently demonstrated by the group of Rinaldi and coworkers at NNL, taking advantage of the redox properties and of the functional groups of the blue copper protein Azurin, that, *in vitro*, is able to mediate electron transfer (ET) from cytochrome c_{551} to nitrite reductase from the same organism [81]. Azurin from *P. aeruginosa* is a 14.6 kDa blue-copper protein existing in two stable configurations – Cu(I) and Cu(II) – and its ET capability depends on the equilibrium between these two oxidation states by means of the reversible redox reaction $Cu^{2+} + e^- \leftrightarrows Cu^{1+}$, which converts continuously the Cu(II) copper oxidized state into the Cu(I) reduced state and *vice versa*. A disulfide bridge (Cys-3-Cys-26) located at a distance of $\approx$ 2.6 nm from the copper site [82] allows the chemisorption of Azurin in oriented monolayers onto crystalline gold or other suitably functionalized surfaces [83, 84]. Since protein adsorption on surfaces may lead to denaturation, in their recent publication, Maruccio and co-workers [49] examined the integrity of proteins in dry monolayers after immobilization by non-contact atomic-force microscopy and by intrinsic fluorescence (the fluorescence of aromatic residues in proteins is strongly influenced by their microenvironment) concluding that neither the covalent binding to the functionalized SiO_2 surface nor the application of strong electric fields induce gross denaturation and/or conformational transitions in Azurin. Their prototype device consists of a planar metal-insulator-metal nanojunction, fabricated by EBL [85] and connected by the self-assembled protein monolayer, and a silver back gate electrode in a field-effect transistor configuration. An oriented protein monolayer was formed in a two-step procedure involving (a) the self-assembly of 3-mercaptopropyltrimethoxysilane (3-MPTS) and (b) the reaction of the free thiol groups of 3-MPTS with the unique surface disulfide bridge of Azurin. The current-voltage characteristic (I_{ds}-V_{ds}) exhibited a low-current plateau at low field and then rises up to hundreds of pA (see Fig. 5.15). The transfer characteristic exhibited a pronounced resonance centered at $V_g = 1.25$ V. This feature gradually disappeared after some gate sweeps. From an electronic viewpoint, their device switches from a n-MOS FET behavior before resonance to a p-MOS FET after resonance. Although among all the fabricated nanodevices, only a limited group exhibited a clear gate effect over a number of gate sweeps (whereas the others failed during the first few sweeps – the ageing of nanodevices is a general issue of molecular electronics [16, 85, 87]), this is a key result because it would allow to exploit the advantages of a complementary logic, fabricating both p-type and n-type devices on the same chip.

The authors ascribed these results to the unique transport mechanism of their biomolecular devices. The transport of electrons through systems containing redox sites may occur via physical displacement of the redox molecules and/or electron hopping from one reduced molecule to an adjacent oxidized molecule [88]. The authors discarded the first mechanism since the proteins were covalently bound to the substrate. Thus, they ascribed electron transport in the Pro-FET to hopping from one reduced (Cu(I)) protein to an adjacent oxidized (Cu(II)) protein, which behave as a redox pair. For electrons to flow, therefore, both reduced and oxidized Azurins must be present and their relative proportion determines the ET rate. The authors proposed that the Azurin redox state is regulated by V_g: the higher is

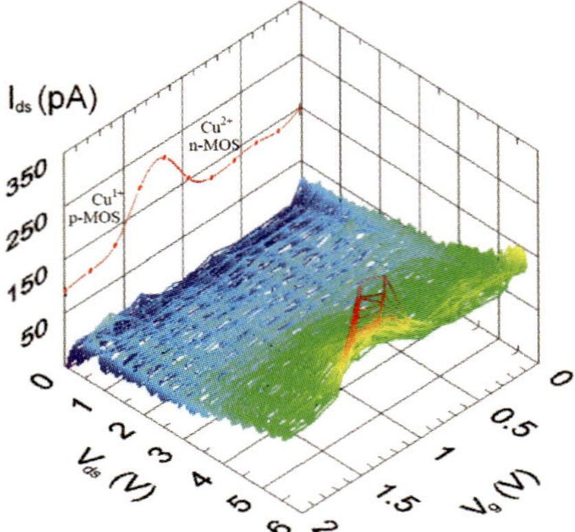

Figure 5.15. Characteristic of the Pro-FET: 3-D plot of the drain–source current as a function of the drain–source bias (V_{DS}) and gate bias (V_G) measured in the dark and at room temperature. No leakage current was observed to flow between the planar electrodes and the back-gate (values as low as few picoamperes and a negligible variation with V_G up to 8 V). A pronounced resonance centered at $V_G = 1.25$ V is present (see also the transfer characteristic at $V_{DS} = 5.5$ V in the projection). (Reproduced with permission from G. Maruccio et al., Towards Protein Field-Effect Transistors: Report and Model of a Prototype, Adv. Mater., 2005, 17, 816–822)

V_g, the greater is the fraction of reduced Azurins. At a particular value of V_g the fraction of oxidized molecules will equal that of reduced molecules and therefore ET will be maximal. In their model the change in the protein oxidation state was ascribed to the influence of the applied electric field on the redox energy levels of the proteins. To verify whether such model based on a hopping mechanism between neighbouring proteins was compatible with their experimental findings, they performed numerical simulations of the current flowing in the device as a function of the applied potentials. The basic features of the experimental curves were exhibited also in the results of their simulation. Moreover, the proposed model is also consistent with the interpretation of the redox peak in cyclic voltammetry curves and in electrochemical STM experiments [84, 89] performed on Azurins chemisorbed on Au(111) substrates. The key role of the copper atom in electron transfer was further supported by a comparison with the current-voltage curves measured in devices implemented with two Azurin variants (the first with the Cu atom replaced by a Zn atom, the second without metal atom), where the flowing currents were significantly lower and no modulation was observed between -6 and 6 volts.

The same metalloprotein has been investigated also at Oxford, where Davis and coworkers acquired current–voltage characteristics by means of conductive-probe AFM and studied the dependence of the conductance from the force load, finding that forces larger than 3 nN were necessary to achieve a reliable contact to the protein [90].

Before concluding, we would also like to mention the work of Yu and coworkers, who succeeded in the construction of ordered neuronal networks by positioning neurons on a biolectronic chips by means of a negative dielectrophoretic force [91].

5.7
Conclusions

We have briefly discussed the present status of the research in the field of nano-bioelectronics, with special emphasis on DNA- and protein-based devices.

Although this field is rather young and very open, the worldwide results obtained so far are promising, and deserve further studies to determine the actual potentiality of biodevices and to further investigate a number of important issues, such as:

- Interconnection of biomolecules and inorganic devices
- Molecular engineering and self-organization
- Transport mechanisms
- Reproducibility and ageing of biomolecular devices
- Biochemical driven implementation of fully biomolecular nanodevices

On the basis of recent progress in the field of nanobioelectronics we fell as though we are just at the beginning of a great journey that will have many unexpected findings. There is no turning back and we cannot even envisage the future developments of this field and where it will take us.

There is much work under way actively pursuing molecular and biomolecular engineering, and the building of composite materials at the nanoscale to be joined together, but there is still an enormous challenge ahead. Building materials from the bottom up requires a multidisciplinary approach. This arena is unquestionably in the nano-dimension, where all fields of science and engineering meet. New ideas will be fostered from collaboration among scientists with diverse backgrounds. As Francis Crick put it, "In nature, hybrid species are usually sterile, but in science the reverse is often true. Hybrid subjects are often astonishingly fertile, whereas if a scientific discipline remains too pure it usually wilts".

Acknowledgments

We are thankful for the invaluable support and exciting collaboration by various colleagues. We would like to thank Elisa Molinari, Rosa Di Felice, Francesca De Rienzo, Stefano Corni and Paolo Facci at S^3-CNR-INFM research center in Modena (Italy), Gerard Canters and Martin Verbeet at Leiden University (The Netherlands), Salvatore Masiero, Tatiana Giorgi, Gianpiero Spada and Giovanni Gottarelli at University of Bologna (Italy), Roberto Cingolani, Valentina Arima, Adriana Biasco, Alessandro Bramanti, Franco Calabi, Stefano D'Amico, Eliana

D'Amone, Antonio Della Torre, Pier Paolo Pompa and Paolo Visconti at NNL-CNR-INFM research center in Lecce (Italy). Financial support by NNL-CNR-INFM, EC through SAMBA project and Italian MIUR (FIRB molecular devices) is gratefully acknowledged.

References

1 The Internet Technology Roadmap for Semiconductors (ITRS) is available at http://public.itrs.net.

2 MARUCCIO, G., CINGOLANI, R., RINALDI, R. Projecting the nanoworld: concepts, results and perspectives of molecular electronics, *J. Mater. Chem.* **2004**, *14*, 542–554 and references therein.

3 WOLF, S. A., AWSCHALOM, D. D., BUHRMAN, R. A., DAUGHTON, J. M., VON MOLNAR, S., ROUKES, M. L., CHTCHELKANOVA, A. Y., TREGER, D. M. SPINTRONICS: a spin-based electronics vision for the future, *Science* **2001**, *294*, 1488–1495.

4 (a) STEANE, A. Quantum computing, *Rep. Prog. Phys.* **1998**, *61*, 117–173; (b) DE MARTINI, F., BUZEK, V., SCIARRINO, F., SIAS, C. Experimental realization of the quantum universal NOT gate, *Nature* **2003**, *419*, 815–818.

5 (a) ADLEMAN, L. M. Molecular computation to solutions of combinatorial problems, *Science* **1994**, *266*, 1021; (b) BRAICH, R. S., CHELYAPOV, N., JOHNSON, C., ROTHEMUND, P. W. K., ADLEMAN, L. Solution of a 20-variable 3-SAT problem on a DNA computer, *Science* **2002**, *296*, 499–502.

6 AVIRAM, A., RATNER, M. A. Molecular rectifiers, *Chem. Phys. Lett.* **1974**, *29*, 277–283.

7 LEHN, J. M. *Supramolecular Chemistry – Concept and Perspectives.* VCH, Weinheim, **1995**.

8 KEREN, K., KRUEGER, M., GILAD, R., BEN-YOSEPH, G., SIVAN, U., BRAUN, E. Sequence-specific molecular lithography on single DNA molecules, *Science* **2002**, *297*, 72–75.

9 MARUCCIO, G., VISCONTI, P., BIASCO, A., BRAMANTI, A., DELLA TORRE, A., POMPA, P. P., FRASCERRA, V., ARIMA, V., D'AMONE, E., CINGOLANI, R., RINALDI, R. Nano-scaled biomolecular field-effect transistors: prototypes and evaluations, *Electroanalysis* **2004**, *16*, 1853–1862.

10 BASHIR, R. DNA-mediated artificial nanobiostructures: state of the art and future directions, *Superlattices Microstruct.* **2001**, *29*, 1–16.

11 LOOGER, L. L., DWYER, M. A., SMITH, J. J., HELLINGA, H. W. Computational design of receptor and sensor proteins with novel functions, *Nature* **2003**, *423*, 185–190.

12 LEE, I., LEE, J. W., GREENBAUM, E. Biomolecular electronics: vectorial arrays of photosynthetic reaction centers, *Phys. Rev. Lett.* **1997**, *79*, 3294–3297.

13 CHEN, T., CALABRESE BARTON, S., BINYAMIN, G., GAO, Z., ZHANG, Y., KIM, H., HELLER, A. A miniature biofuel cell, *J. Am. Chem. Soc.* **2001**, *123*, 8630–8631.

14 GIMZEWSKI, J. K., JOACHIM, C., SCHLITTLER, R. R., LANGLAIS, V., TANG, H., JOHANNSEN, I. Rotation of a single molecule within a supramolecular bearing, *Science* **1998**, *281*, 531–533.

15 JOACHIM, C., GIMZEWSKI, J. K., AVIRAM, A. Electronics using hybrid-molecular and mono-molecular devices, *Nature* **2000**, *408*, 541.

16 METZGER, R. M., CHEN, B., HOPFNER, U., LAKSHMIKANTHAM, M. V., VUILLAUME, D., KAWAI, T., WU, X., TACHIBANA, H., HUGHES, T. V., SAKURAI, H., BALDWIN, J. W., HOSCH, C., CAVA, M. P., BREHMER, L., ASHWELL, G. J. Unimolecular electrical rectification in hexadecylquinolinium tricyanoquinodimethanide, *J. Am. Chem. Soc.* **1997**, *119*, 10455–10466.

17 REED, M. A., CHEN, J., RAWLETT, A. M., PRICE, D. W., TOUR, J. M. Molecular random access memory cell, *Appl. Phys. Lett.* **2001**, *78*, 3735.

18 ROTH, S., JOACHIM, C. *Atomic and Molecular Wires*. Kluwer, Dordrecht, **1997**.

19 BACHTOLD, A., HADLEY, P., NAKANISHI, T., DEKKER, C. Logic circuits with carbon nanotube transistors, *Science* **2001**, *294*, 1317–1320.

20 HUANG, Y., DUAN, X., CUI, Y., LAUHON, L. J., KIM, K., LIEBER, C. M. Logic gates and computation from assembled nanowire building blocks, *Science* **2001**, *294*, 1313–1317.

21 ROCHA, A. R., GARCIA-SUAREZ, V. M., BAILEY, S. W., LAMBERT, C. J., FERRER, J., SANVITO, S. Towards molecular spintronics, *Nat. Mater.* **2005**, *4*, 335.

22 WATSON, J., CRICK, F. A structure for deoxyribose nucleic acid, *Nature* **1953**, *171*, 737.

23 GÖPEL, W. Controlled signal transduction across interfaces of "intelligent" molecular systems, *Biosens. Bioelectron.* **1995**, *10*, 35–59.

24 GWYN, C. W., STULEN, R., SWEENEY, D., ATTWOOD, D. Extreme ultraviolet lithography, *J. Vac. Sci. Technol. B* **1998**, *16*, 3142–3149.

25 REED, M. A., ZHOU, C., MULLER, C. J., BURGIN, T. P., TOUR, J. M. Conductance of a molecular junction, *Science* **1997**, *278*, 252–254.

26 KERVENNIC, Y. V., VAN DER ZANT, H. S. J., MORPURGO, A. F., GUREVICH, L., KOUWENHOVEN, L. P. Nanometer-spaced electrodes with calibrated separation, *Appl. Phys. Lett.* **2002**, *80*, 321–323.

27 PINER, D., ZHU, J., XU, F., HONG, S., MIRKIN, C. A. Dip-Pen nanolithography, *Science* **1999**, *283*, 661–663.

28 (a) HONG, S., ZHU, J., MIRKIN, C. A. Multiple ink nanolithography: towards a multiple-pen nanoplotter, *Science* **1999**, *286*, 523–525; (b) HONG, S., MIRKIN, C. A. A nanoplotter for soft lithography with both parallel and serial writing capabilities, *Science* **2000**, *288*, 1808–1811.

29 DEMERS, L. M., GINGER, D. S., PARK, S. J., LI, Z., CHUNG, S. W., MIRKIN, C. A. Direct patterning of modified oligonucleotides on metals and insulators by dippen nanolithography, *Science* **2002**, *296*, 1836–1838.

30 MACDONALD, M. P., PATERSON, L., VOLKE-SEPULVEDA, K., ARLT, J., SIBBETT, W., DHOLAKIA, K. Creation and manipulation of three-dimensional optically trapped structures, *Science* **2002**, *296*, 1101–1103.

31 STORM, A. J., VAN NOORT, J., DE VRIES, S., DEKKER, C. Insulating behavior for DNA molecules between nanoelectrodes at the 100 nm length scale, *Appl. Phys. Lett.* **2001**, *79*, 3881.

32 BRAUN, E., EICHEN, Y., SIVAN, U., BEN-YOSEPH, G. DNA-templated assembly and electrode attachment of a conducting silver wire, *Nature* **1998**, *391*, 775.

33 ZHANG, Y., AUSTIN, R. H., KRAEFT, J., COX, E. C., ONG, N. P. Insulating behavior of λ-DNA on the micron scale, *Phys. Rev. Lett.* **2002**, *89*, 198102.

34 (a) HARTZELL, B., MCCORD, B., ASARE, D., CHEN, H., HEREMANS, J. J., SOGHOMONIAN, V. Comparative current–voltage characteristics of nicked and repaired λ-DNA, *Appl. Phys. Lett.* **2003**, *82*, 4800. (b) HARTZELL, B., MCCORD, B., ASARE, D., CHEN, H., HEREMANS, J. J., SOGHOMONIAN, V. Current–voltage characteristics of diversely disulfide terminated λ-deoxyribonucleic acid molecules, *J. Appl. Phys.* **2003**, *94*, 2764.

35 PORATH, D., BEZRYADIN, A., DE VRIES, S., DEKKER, C. Direct measurement of electrical transport through DNA molecules, *Nature* **2000**, *403*, 635.

36 CAI, L., TABATA, H., KAWAI, T. Self-assembled DNA networks and their electrical conductivity, *Appl. Phys. Lett.* **2000**, *77*, 3105.

37 ELEY, D. D., METCALFE, E., WHITE, M. P., Semiconductivity of organic substances. Part 17. – Effects of ultraviolet and visible light on the conductivity of the sodium salt of deoxyribonucleic acid, *J. Chem. Soc., Faraday Trans. 1*, **1975**, *71*, 955.

38 (a) HALL, D. B., HOLMLIN, R. E., BARTON, J. K. Oxidative DNA damage through long-range electron transfer, *Nature* **1996**, *382*, 731; (b) DANDLIKER, P. J., HOLMLIN, R. E., BARTON, J. K. Oxidative thymine dimer repair in the DNA helix, *Science* **1997**, *275*, 1465; (c) KELLEY, S. O., HOLMLIN, R. E., STEMP, E. D. A., BARTON, J. K. Photoinduced electron transfer in ethidium-modified DNA duplexes: dependence on distance and base stacking, *J. Am. Chem. Soc.* **1997**, *119*, 9861; (d) KELLEY, S. O., BARTON, J. K. Electron transfer between bases in double helical DNA, *Science* **1999**, *283*, 375.

39 RAJSKI, S. R., JACKSON, B. A., BARTON, J. K. DNA repair: models for damage and mismatch recognition, *Mutat. Res.* **2000**, *447*, 49.

40 DEKKER, C., RATNER, M. A. Electronic properties of DNA, *Phys. World* **2001**, *14*, 29.

41 E. K. WILSON, DNA Charge Migration: No Longer An Issue, *Chemical & Engineering News* **2001**, *79*, 1.

42 DE PABLO, P. J., MORENO-HERRERO, F., COLCHERO, J., GOMEZ HERRERO, J., HERRERO, P., BAR, A. M., ORDEJON, P., SOLER, J. M., ARTACHO, E. Absence of dc-conductivity in lambda-DNA, *Phys. Rev. Lett.* **2000**, *85*, 4992.

43 RAKITIN, A., AICH, P., PAPAPDOPOULOS, C., KOBZAR, Yu., VEDENEEV, A. S., LEE, J. S., XU, J. M. Metallic conduction through engineered DNA: DNA nanoelectronic building blocks, *Phys. Rev. Lett.* **2001**, *86*, 3670.

44 FINK, H. W., SCHÖNENBERGER, C. Electrical conduction through DNA molecules, *Nature* **1999**, *398*, 407.

45 TRAN, P., ALAVI, B., GRUNER, G. Charge transport along the lambda-DNA double helix, *Phys. Rev. Lett.* **2000**, *85*, 1564.

46 YOO, K. H., HA, D. H., LEE, J. O., PARK, J. W., KIM, J., KIM, J. J., LEE, H. Y., KAWAI, T., CHOI, H. Y. Electrical conduction through poly(dA)–poly(dT) and poly(dG)–poly(dC) DNA molecules, *Phys. Rev. Lett.* **2001**, *87*, 198102.

47 KASUMOV, A. Y., KOCIAK, M., GUERON, S., REULET, B., VOLKOV, V. T. Proximity-induced superconductivity in DNA, *Science* **2001**, *291*, 280.

48 (a) MARCUS, R. A. On the theory of oxidation–reduction reactions involving electron transfer: I, *J. Chem. Phys.* **1956**, *24*, 966; (b) MARCUS, R. A. Electrostatic free energy and other properties of states having nonequilibrium polarization. I, *J. Chem. Phys.* **1956**, *24*, 979; (c) MARCUS, R. A. Electron transfer reactions in chemistry. Theory and experiment, *Rev. Mod. Phys.* **1993**, *65*, 599; (d) MARCUS, R. A. Ion pairing and electron transfer, *J. Phys. Chem. B* **1998**, *102*, 10071.

49 MARUCCIO, G., BIASCO, A., VISCONTI, P., BRAMANTI, A., POMPA, P. P., CALABI, F., CINGOLANI, R., RINALDI, R., CORNI, S., DI FELICE, R., MOLINARI, E., VERBEET, M. P., CANTERS, G. W. Towards protein field-effect transistors: report and model of a prototype, *Adv. Mater.* **2005**, *17*, 816–822.

50 *IUPAC Compendium of Chemical Terminology*, 2nd edn. IUPAC, Research Triangle Park, NC, **1997**.

51 ARKIN, M. R., STEMP, E. D. A., TURRO, C., TURRO, N. J., BARTON, J. K. Luminescence quenching in supramolecular systems: a comparison of DNA- and SDS micelle-mediated photoinduced electron transfer between metal complexes, *J. Am. Chem. Soc.* **1996**, *118*, 2267.

52 GIESE, B., AMAUDRUT, J., KÖHLER, A. K., SPORMANN, M., WESSELY, S. Direct observation of hole transfer through DNA by hopping between adenine bases and by tunneling, *Nature* **2001**, *412*, 318.

53 GIESE, B., WESSELEY, S., SPORMANN, M., LINDEMANN, U., MEGGERS, E., MICHEL-BEYERLE, M. E. On the mechanism of long-range electron transfer through DNA, *Angew. Chem. Int. Ed. Engl.* **1999**, *38*, 996.

54 HELLER, A. Spiers Memorial Lecture. On the hypothesis of cathodic protection of genes, *Faraday Discuss.* **2000**, *116*, 1.

55 (a) LEWIS, F. D., WU, T. F., ZHANG, Y. F., LETSINGER, R. L., GREENFIELD, S. R., WASIELENWSKI, M. R. Distance-dependent electron transfer in DNA hairpins, *Science* **1997**, *277*, 673; (b) Lewis, F. D., Wu, T. F., Liu, X. Y., Letsinger, R. L., Greenfield, S. R., Miller, S. E., Wasielenwski, M. R. Dynamics of photoinduced charge separation and charge recombination in synthetic DNA hairpins with stilbenedicarboxamide linkers, *J. Am. Chem. Soc.* **2000**, *122*, 2889.

56 Fiebig, T., Wan, C. Z., Kelley, S. O., Barton, J. K., Zewail, A. H. Femtosecond dynamics of the DNA intercalator ethidium and electron transfer with mononucleotides in water, *Proc. Natl Acad. Sci. USA* **1999**, *96*, 1187.

57 Endres, R. G., Cox, D. L., Singh, R. R. P. Colloquium: the quest for high-conductance DNA, *Rev. Mod. Phys.* **2004**, *76*, 195.

58 Lee, H., Tanaka, H., Otsuka, Y., Yoo, K., Lee, J., Kawai, T. Control of electrical conduction in DNA using oxygen hole doping, *Appl. Phys. Lett.* **2002**, *80*, 1670.

59 Cuniberti, G., Craco, L., Porath, D., Dekker, C. Backbone-induced semiconducting behavior in short DNA wires, *Phys. Rev. B* **2002**, *65*, 241314.

60 Watanabe, H., Manabe, C., Shigematsu, T., Shimotani, K. Single molecule DNA device measured with triple-probe atomic force microscope, *Appl. Phys. Lett.* **2001**, *79*, 2462.

61 Pirrung, M. C. How to make a DNA chip, *Angew. Chem. Int. Ed. Engl.* **2002**, *41*, 1276–1289.

62 Warman, J. M., de Haas, M. P., Rupprecht, A. DNA: a molecular wire?, *Chem. Phys. Lett.* **1996**, *249*, 319.

63 Beratan, D. N., Priyadarshy, S., Risser, S. M. DNA: insulator or wire?, *Chem. Biol.* **1997**, *4*, 3.

64 Conwell, E. M., Rakhmanova, S. V. Polarons in DNA, *Proc. Natl Acad. Sci. USA* **2000**, *97*, 4556.

65 Hermon, Z., Caspi, S., Ben-Jacob, E. Prediction of charge and dipole solitons in DNA molecules based on the behavior of phosphate bridges as tunnel elements, *Europhys. Lett.* **1998**, *43*, 482.

66 Gottarelli, G., Masiero, S., Mezzina, E., Pieraccini, S., Rabe, J. P., Samorì, P., Spada, G. P. The self-assembly of lipophilic guanosine derivatives in solution and on solid surfaces, *Chem. Eur. J.* **2000**, *6*, 3242.

67 Maruccio, G., Visconti, P., Arima, V., D'Amico, S., Biasco, A., D'Amone, E., Cingolani, R., Rinaldi, R., Masiero, S., Giorgi, T., Gottarelli, G. Field effect transistor based on a modified DNA base, *Nano Lett.* **2003**, *3*, 479–483.

68 Lefebvre, J., Lynch, J. F., Llaguno, M., Radosavljevic, M., Johnson, A. T. Single-wall carbon nanotube circuits assembled with an atomic force microscope, *Appl. Phys. Lett.* **2002**, *75*, 3014–3016.

69 Richter, J., Mertig, M., Pompe, W., Mönch, I., Schckert, H. K. Construction of highly conductive nanowires on a DNA template, *Appl. Phys. Lett.* **2001**, *78*, 536.

70 Keren, K., Berman, R. S., Buchstab, E., Sivan, U., Braun, E. DNA-templated carbon nanotube field-effect transistor, *Science* **2003**, *302*, 1380–1382.

71 Lee, S., Mao, C., Flynn, C. E., Belcher, A. M. Ordering of quantum dots using genetically engineered viruses, *Science* **2002**, *296*, 892–895.

72 Williams, K. A., Veenhuizen, P. T. M., de la Torre, B., Eritja, R., Dekker, C. Nanotechnology: carbon nanotubes with DNA recognition, *Nature* **2002**, *420*, 761.

73 Zanchet, D., Micheel, C. M., Parak, W. J., Gerion, D., Williams, S. C., Alivisatos, A. P. Electrophoretic and structural studies of DNA-directed Au nanoparticle groupings, *J. Phys. Chem. B* **2002**, *106*, 11758.

74 Taton, T. A., Mirkin, C. A., Letsinger, R. L. Scanometric DNA array detection with nanoparticle probes, *Science* **2000**, *289*, 1757.

75 Winfree, E., Liu, F. R., Wenzler, L. A., Seeman, N. C. Design and self-assembly of two-dimensional DNA crystals, *Nature* **1998**, *394*, 539.

76 Park, S. J., Taton, T. A., Mirkin, C. A. Array-based electrical detection of DNA with nanoparticle probes, *Science* **2002**, *295*, 1503.

77 Cinelli, R. A. G., Pellegrini, V., Ferrari, A., Faraci, P., Nifosi, R., Tyagi, M., Giacca, M., Beltram, F. Green fluorescent proteins as optically controllable elements in bioelectronics, *Appl. Phys. Lett.* **2001**, *79*, 3353.

78 Birge, R. R., Gillespie, N. B., Izaguirre, E. W., Kusnetzow, A., Lawrence, A. F., Singh, D., Song, Q. W., Schmidt, E., Stuart, J. A., Seetharaman, S., Wise, K. J. Biomolecular electronics: protein-based associative processors and volumetric memories, *J. Phys. Chem. B* **1999**, *103*, 10746.

79 Yasutomi, S., Morita, T., Imanishi, Y., Kimura, S. A molecular photodiode system that can switch photocurrent direction, *Science* **2004**, *304*, 1944.

80 Lee, I., Lee, J. W., Stubna, A., Greenbaum, E. Measurement of electrostatic potentials above oriented single photosynthetic reaction centers, *J. Phys. Chem. B* **2000**, *104*, 2439.

81 Vijgenboom, E., Busch, J. E., Canters, G. W. *In vivo* studies disprove an obligatory role of azurin in denitrification in *Pseudomonas aeruginosa* and show that azu expression is under control of rpoS and ANR, *Microbiology* **1997**, *143*, 2853.

82 Farver, O., Pecht, I., Long range intramolecular electron transfer in azurins, *J. Am. Chem. Soc.* **1992**, *114*, 5764.

83 Chi, Q., Zhang, J., Nielsen, J. U., Friis, E. P., Chorkendorff, I., Canters, G. W., Andersen, J. E. T., Ulstrup, J. Molecular monolayers and interfacial electron transfer of *Pseudomonas Aeruginosa* azurin on Au(111), *J. Am. Chem. Soc.* **2000**, *122*, 4047.

84 Alessandrini, A., Gerunda, M., Canters, G. W., Verbeet, M. P., Facci, P. Electron tunneling through azurin is mediated by the active site Cu ion, *Chem. Phys. Lett.* **2003**, *376*, 625.

85 Maruccio, G., Visconti, P., D'Amico, S., Calogiuri, P., D'Amone, E., Cingolani, R., Rinaldi, R. Planar nanotips as probes for transport experiments in molecules, *Microelectron. Eng.* **2003**, *67–68*, 838.

86 Park, J., Pasupathy, A. N., Goldsmith, J. I., Chang, C., Yaish, Y., Petta, J. R., Rinkoski, M., Sethna, J. P., Abruna, H. D., McEuen, P. L., Ralph, C. Coulomb blockade and the Kondo effect in single-atom transistors, *Nature* **2002**, *417*, 722.

87 (a) Lee, J., Lientschnig, G., Wiertz, F., Struijk, M., Janssen, R. A. J., Egberink, R., Reinhoudt, D. N., Hadley, P., Dekker, C. Absence of strong gate effects in electrical measurements on phenylene-based conjugated molecules, *Nano Lett.* **2003**, *3*, 113; (b) Kagan, C. R., Afzali, A., Martel, R., Gignac, L. M., Solomon, P. M., Schrott, A. G., Ek, B. Evaluations and considerations for self-assembled monolayer field-effect transistors, *Nano Lett.* **2003**, *3*, 119.

88 Blauch, D. N., Saveant, J. M. Dynamics of electron hopping in assemblies of redox centers. Percolation and diffusion, *J. Am. Chem. Soc.* **1992**, *114*, 3323.

89 Facci, P., Alliata, D., Cannistraro, S. Potential-induced resonant tunneling through a redox metalloprotein investigated by electrochemical scanning probe microscopy, *Ultramicroscopy* **2001**, *89*, 291.

90 Zhao, J., Davis, J. J., Sansom, M. S. P., Hung, A. Exploring the electronic and mechanical properties of protein using conducting atomic force microscopy, *J. Am. Chem. Soc.* **2004**, *126*, 5601–5609.

91 Yu, Z., Xiang, G., Pan, L., Huang, L., Yu, Z., Xing, W., Cheng, J. Negative dielectrophoretic force assisted construction of ordered neuronal networks on cell positioning bioelectronic chips, *Biomed. Microdevices* **2004**, *6*, 311–324.

6

DNA Nanodevices: Prototypes and Applications

Friedrich C. Simmel

6.1
Introduction

The unique biochemical and biophysical properties of DNA can be utilized to construct artificial, machine-like molecular structures which can perform simple mechanical or computational tasks, or both. This chapter gives an overview over the first prototypes of such "DNA nanodevices" as well as over recent developments towards functional DNA machines. In contrast to earlier review articles on this subject, particular emphasis is put on possible applications of DNA nanodevices in the life sciences – DNA devices interacting with proteins, as biosensors or as components of drug delivery systems.

The outline of this chapter is as follows. Section 6.2 gives a short introduction into the most important properties of DNA which make this molecule so interesting for nanoscale science and technology. Section 6.3 deals with "simple DNA devices" – prototype devices which can perform simple movements on the nanoscale. Section 6.4 focuses on more recent developments of "functional" nanodevices with an emphasis on possible applications in the life sciences. DNA-based molecular motors and automata will be discussed as well as the interaction of DNA devices with proteins. Section 6.5 deals with problems related to the construction of autonomous or "free-running" DNA devices and discusses a variety of different concepts to achieve such autonomous behavior. Section 6.5 closes the chapter with a few concluding remarks.

6.2
DNA as a Material for Nanotechnology

6.2.1
Nanoscale Science

One of the main goals of nanotechnology is the manipulation of matter at the level of single molecules or atoms. Such ultimate control could lead to the development

Nanotechnologies for the Life Sciences Vol. 4
Nanodevices for the Life Sciences. Edited by Challa S. S. R. Kumar
Copyright © 2006 WILEY-VCH Verlag GmbH & Co. KGaA, Weinheim
ISBN: 3-527-31384-2

of new functional materials and devices which obtain their functionality from the ordered assembly of components on the molecular scale. In the life sciences, such molecularly ordered systems could lead to extremely sensitive and even autonomous sensors and to intelligent drug delivery units.

One approach towards nanoscale control – often termed the "top-down" approach – is the manipulation of matter using macroscopic machines. Using scanning probe techniques, for example, the manipulation of single atoms or molecules has already been demonstrated during the last decade [1, 2]. Even though such examples show impressively the degree of control on the atomic level that has been achieved so far, it is not clear how scanning probe or other "top-down" techniques could be used for fast and efficient assembly of a large number of nanoscale components.

A different approach to molecular-scale engineering does not involve the "direct" manipulation of molecules and atoms with the help of a macroscopic device, but rather the utilization of self-assembly and self-organization principles. This approach is inspired by biological systems which often serve as "prototypes" for self-assembled nanotechnological structures.

In biological systems, complex and comparatively large structures are formed from smaller and simpler building blocks by "molecular recognition" events – usually the cooperative action of a multitude of weak bonds between two or more molecules which result in strong and highly specific interactions. Molecular construction is often assisted by "molecular machines" or enzymes – themselves self-assembled macromolecular complexes – which, among others, catalyze reactions, assist molecular transport or transduce signals. In addition, the molecules in a living cell are intimately linked within complex interaction networks which give rise to fascinating "emergent" properties.

Naturally, all these features would be highly desirable for artificial nanosystems. For these reasons, researchers in the field are trying to utilize the molecular recognition properties of biological molecules for the construction of nonbiological structures. Furthermore, first attempts have been made to construct artificial analogs of biological molecular machines and motors. One of the most interesting and most widely used biomolecules in this context is the famous molecule of heredity – DNA.

6.2.2
Biophysical and Biochemical Properties of Nucleic Acids

There are a number of good – and practical – reasons for the utilization of DNA in nanoscience (summarized in Fig. 6.1). The main feature of interest, of course, is the base-pairing interaction between complementary DNA bases which was discovered more than 50 years ago [3]. DNA molecules are composed of so-called nucleotides which themselves consist of a sugar unit (2′-deoxyribose) with a phosphate group and a "base" attached to it. The bases in DNA are the purines adenine (A) and guanine (G) and the pyrimidines thymine (T) and cytosine (C). In the Watson–Crick (WC) base-pairing scheme, the bases A and T bind together

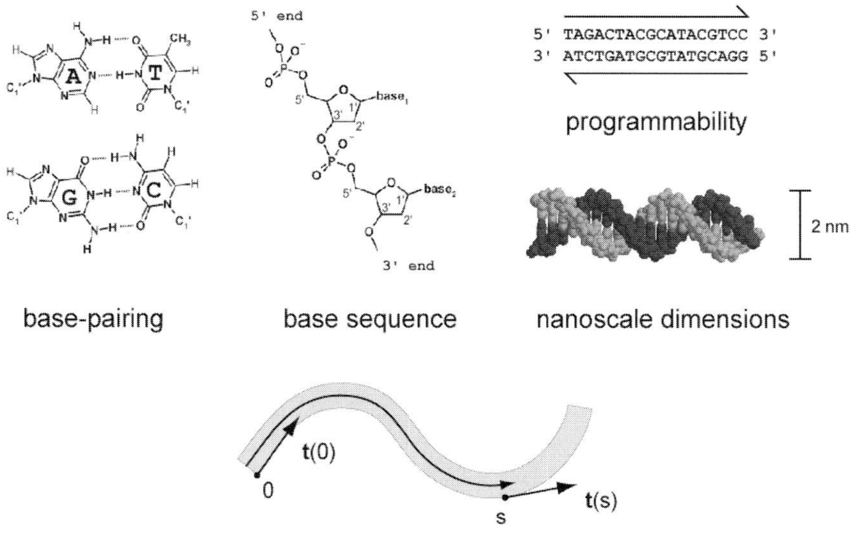

Figure 6.1. The most important features of DNA for nanoscience. *Base-pairing*: The specific binding between adenine (A) and thymine (T), on the one hand, and guanine (G) and cytosine (C), on the other hand, is the basis of the unique molecular recognition properties of DNA. *Base sequence*: In ssDNA, nucleotides which consist of deoxyribose, phosphate and a base are linked together. The bases can be chosen arbitrarily and therefore large numbers of distinct DNA molecules with different base sequences are possible. *Programmability*: If two DNA sequences are complementary, they may bind together to form a DNA duplex structure, i.e. sequence determines structure. *Nanoscale dimensions*: Under normal conditions DNA duplexes are right-handed double-helices with a diameter of 2 nm. *Mechanical properties*: On the nanoscale, DNA is a rigid molecule. The persistence length, i.e. the length over which the correlation between the tangential vector $t(s)$ decays, is of the order of 50 nm or 150 base pairs.

via two hydrogen bonds, whereas G and C are connected via three such bonds. In DNA, the nucleotides are linked together by the phosphate groups which connect the 3′ position of deoxyribose of one nucleotide with the 5′ position of the sugar ring of the following nucleotide. This linkage endows the DNA strand with a "direction", e.g. 5′-AG-3′ and 5′-GA-3′ (where the sequence is read from the 5′ end of the first nucleotide to the 3′ end of the last nucleotide) are two different molecules. If two DNA molecules with "complementary" sequences are brought together, they can bind to each other and form a DNA duplex. Here it is important that the two strands have opposite directionality, i.e. a DNA strand with sequence 5′-$N_1N_2\ldots N_{n-1}N_n$-3′ will bind to a strand with sequence 5′-$\underline{N}_n\underline{N}_{n-1}\ldots\underline{N}_2\underline{N}_1$-3′, where $\underline{N}_i$ denotes the WC complement of N_i. Under standard buffer conditions, the DNA duplex will adopt a double-helical structure with a diameter of 2 nm and a helix pitch of roughly 10.5 base pairs. The base pairs stack upon each other and are "hidden" within the double helix. The planes defined by the base pairs are oriented roughly perpendicularly to the axis of the helix, whereas the sugar units with the negatively charged phosphates point radially away from it. The vertical distance between two subsequent base pairs is approximately 0.34 nm. The phosphate linkages make DNA a highly negatively charged polyelectrolyte. Due to the double-helical structure and this high charge density, duplex DNA is a relatively rigid molecule. In terms of polymer science, the persistence length L_p of double-helical DNA is of the order of 50 nm or 150 base pairs. This is a measure for how far one has to follow the contour of a polymer until the original orientation is lost – essentially this is a measure of its flexibility. More technically, L_p is the correlation length of the tangent vector of the polymer. In the worm-like chain (WLC) polymer model, we have $\langle \vec{t}(s) \cdot \vec{t}(0) \rangle = \exp(-s/L_p)$, where s is the coordinate along the polymer and $\vec{t}(s)$ is the tangent vector at position s. An L_p of 50 nm makes DNA a considerably less flexible molecule than most synthetic polymers. For nano-construction, DNA can thus simply be regarded as a stiff molecular rod. At the same time, DNA is much more flexible than other biopolymers such as the cyto-skeletal F-actin or microtubuli. It is therefore not straightforward to produce or-dered micron-scale structures with DNA.

By the proper choice of DNA sequences, one can also *avoid* the formation of du-plex structures. To this end, sequences have to be chosen in such a way that they are not complementary to any other DNA molecule – or to itself – over stretches of more than a few bases. In contrast to double-stranded (ds) DNA, single-stranded (ss) DNA is a much more flexible polymer with a persistence length on the order of 1 nm. By "programming" DNA molecules to form partly double-stranded and partly single-stranded regions, one can therefore construct molecular networks consisting of rigid and flexible elements.

Unlike the folding of amino acid chains in proteins, the formation of secondary and tertiary structure in nucleic acids is hierarchical and sequential [4] – the binding energies for base-pairing are much larger than the energies for other inter-actions between remote residues, and therefore DNA and RNA folding are domi-nated by secondary structure formation. The prediction of the formation of base-paired regions within one DNA molecule or those formed between several DNA molecules therefore represents a major step in accurate structure prediction for nucleic acids structures. In the absence of significant three-dimensional (3-D) in-teractions between distant sections of a DNA or RNA molecule, it is already the final step. This property makes the relation between DNA sequence and structure much more transparent than in proteins and therefore a "rational" design of DNA structures can be achieved much easier.

This rational design is also facilitated by the wealth of thermodynamic data avail-able for nucleic acids structures. Based on these data, thermodynamic quantities such as the free energy of formation of DNA structures or useful experimental pa-rameters such as the melting temperature of a given DNA duplex (the temperature at which half of the duplexes are dissociated into single strands) can be calculated relatively reliably. In a given set of DNA molecules, strands may also hybridize to each other in several alternative configurations and here thermodynamic calculations aid in the design of structures which mainly self-assemble into one specific, desired structure.

Maybe one of the most important practical features of DNA, however, is the availability of automated synthesis methods which allow the production of reasonable quantities of DNA with virtually any desired base sequence. A designed DNA structure can therefore be readily translated from a sequence on your computer to a real molecule.

6.2.3
DNA Nanoconstruction

The unique properties of DNA summarized in the previous section have indeed already been utilized for the realization of a number of impressive supramolecular constructions. Starting with the synthesis of a DNA molecule with the topology of a cube by Seeman and coworkers in 1991 [5]. DNA has been used to construct molecules with the structure of a truncated octahedron [6], DNA catenanes [7], tetrahedra and octahedra [8, 9], and other geometrical objects [10]. DNA has also been used to construct 2-D molecular networks from DNA branched junctions [11, 12]. When these networks fold back to themselves, they can also form DNA "nanotubes" [13–15]. Recently, such networks have even been utilized to arrange nanoparticles and proteins in two dimensions [16–18]. Attempts to extend these structures into the third dimension are currently being made in a number of laboratories.

6.3
Simple DNA Devices

The biochemical and mechanical properties of DNA cannot only be used for the construction of static supramolecular structures, but they can also be utilized to achieve motion on the nanometer scale. So far, one can discern essentially two strategies. Some of the devices rely on conformational changes of the DNA molecules themselves which occur under certain buffer conditions. Other devices exploit the different rigidity of ssDNA and dsDNA, and switch back and forth between several structures by the addition or removal of DNA "fuel" or "set" strands.

6.3.1
Conformational Changes Induced by Small Molecules and Ions

The presence of multiply charged cations can lead to marked changes in DNA structure. DNA strands with a sequence of repeating CG residues, for example, may undergo a transition to the so-called Z form of DNA in the presence of cobalt hexammine. Z-DNA is a left-handed helical conformation of DNA and the transition from B form DNA (the "usual" right-handed double-helical DNA conformation) is therefore associated with a change in helicity of the molecule. In the "B–Z device" [19], the B–Z transition has been utilized to produce nanoscale rotatory motion. To this end, the DNA sequence $d(CG)_{10}$ which tends to undergo the B–Z transition, was flanked by two DNA supramolecular structures which do not. When the $d(CG)_{10}$ sequence was switched into the Z form, the two flanking parts rotated with respect to each other. Experimentally this could be proved by fluorescence resonance energy transfer (FRET) measurements between two strategically attached fluorescent dyes. In FRET, excitation energy is transferred from one donor fluorophore to another chromophore in close proximity. The strong distance dependence of this effect in the nanometer range makes it extremely valuable for the characterization of nanoscale motions. Consequently, it is the most commonly used experimental technique – apart from gel electrophoresis – to characterize the operation of DNA nanodevices.

Another example where the presence of cations influences DNA structure is related to the phenomenon of DNA condensation – even though duplex DNA is highly negatively charged, two DNA helices can overcome their mutual electrostatic repulsion in the presence of multivalent cations such as Mg^{2+}, putrescine, cobalt hexammine, spermidine or spermine [20]. In Ref. [21], magnesium ions were utilized to switch a network of biotinylated DNA molecules linked by streptavidin from an uncondensed into a condensed structure, in which neighboring DNA duplexes formed supercoils. This was accompanied by a movement of the nodes of the network with respect to each other.

The presence of protons may also induce a DNA conformational change. At low pH values the DNA bases adenine and cytosine, for example, are protonated at their N6 or N4 positions, respectively. Protonated cytidine ($pK_a = 4.2$) participates in a number of unconventional DNA secondary structures, e.g. the DNA "i-motif" and a DNA triplex structure. The i-motif [22] can form between DNA strands with repetitive stretches of cytosine residues and is held together by hemiprotonated cytosine pairs (Fig. 6.2a). The intramolecularly formed i-motif shown in Fig. 6.2(b)

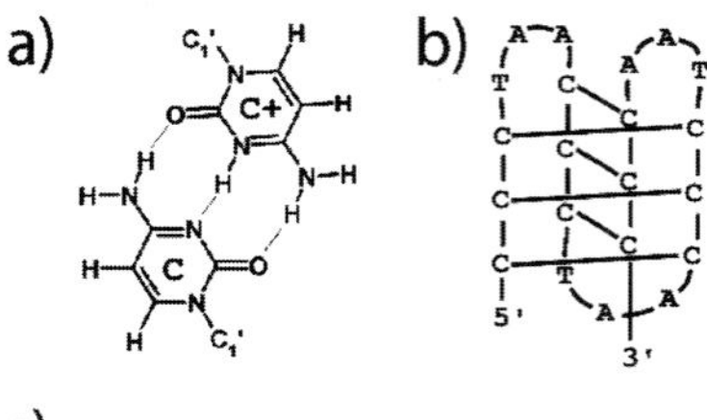

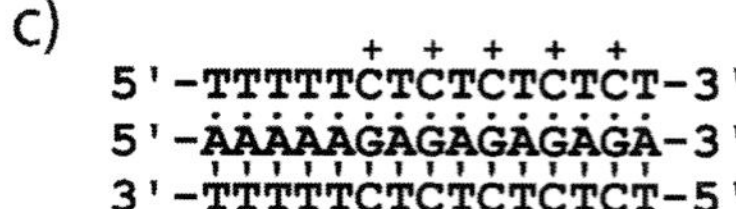

Figure 6.2. Unusual DNA structures which form with protonated cytosine residues. (a) A hemiprotonated CC+ base pair. (b) A C-rich strand of DNA forms the so-called "i-motif" at low pH values due to intramolecular CC+ base-pairing. (c) Protonated cytosine also plays a role in DNA triplex formation. Here, the lower two DNA strands bind to each other in standard WC mode, whereas the upper strand binds to the middle strand in the "Hoogsteen" mode. This is only possible when the cytosines are protonated.

has recently been utilized for a simple proton-driven DNA device [23]. Upon the addition of a small amount of hydrochloric acid to a solution of the DNA device strand 5′-CCCTAACCCTAACCCTAACCC-3′, the molecule folds into the i-motif. When the pH is subsequently raised by the addition of sodium hydroxide, the molecule unfolds again. At neutral pH, the DNA strand is available for hybridization and it can form a DNA duplex with its complement strand. During one operation cycle, the 5′ and 3′ ends of the device strand therefore move from a proximate position with a distance on the order of 1 nm to a position where they are separated by roughly 6 nm.

Under certain conditions DNA can also form triplex structures. In a triplex, two strands bind with each other in conventional WC mode while the third strand binds to the duplex via other base-pairing interactions. One such triplex structure is based on the binding of two homopyrimidine strands to one homopurine strand. In the example in Fig. 6.2(c), the homopurine strand contains the repetitive sequence $d(GA)_n$. One homopyrimidine strand [with the sequence $d(TC)_n$] binds to the hompurine strand by WC base-pairing while the other homopyrimidine strand binds via "Hoogsteen" base-pairing [24]. The latter binding mode is only possible when the cytosine residues are protonated. One can therefore switch between a duplex and a triplex DNA structure by lowering or raising the pH value [25]. This duplex–triplex transition has already been utilized to drive the simple DNA mechanical device depicted in Fig. 6.3 [26]. Here, a long DNA strand L has been hybridized to two shorter DNA strands F and S. The strands F and L form a struc-

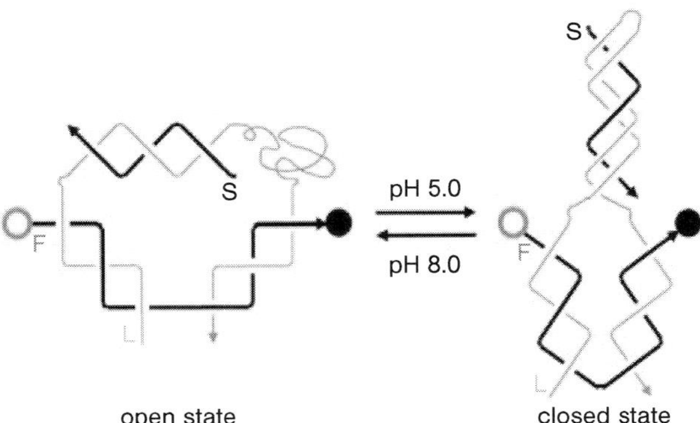

Figure 6.3. A proton-driven DNA nanodevice based on the duplex–triplex transition. The device consists of three DNA strands L, F and S. L and F form two rigid double-stranded "arms" connected by a flexible single-stranded hinge. Strand S is also hybridized to L and contains a DNA sequence which is able to form a triplex structure. The sequence which can bind to the duplex formed by S and L is contained in the unhybridized section schematically depicted as a "random coil" in the open state. At low pH, the cytosine residues in this coil region are protonated and this section of L can bind to the S–L duplex in Hoogsteen mode, thereby forming a triplex structure. This closes the arms of the structure. The structure can be cycled between the open, duplex state and the closed, triplex state by changing the pH value of the reaction solution between pH 8 and pH 5. (Reprinted with permission from Ref. [26], © 2004, Wiley-VCH.)

ture characterized by two rigid arms connected by a short flexible hinge (very similar to the structure of other, hybridization-driven devices to be discussed in the next section). Strand S also hybridizes to a section of L, while a part of L remains unhybridized at high pH values. S contains a hompurine sequence and can form a triplex structure with L at low pH values. As shown in Fig. 6.3, in the triplex configuration strand L folds back on itself and therefore moves the two arms of the device with respect to each other.

Another (maybe the earliest) example of a DNA nanomechanical device, demonstrated by the Seeman group, was based on the influence of the intercalating dye ethidium bromide on DNA structure [27]. Here, the position of base pairs in a cruciform DNA junction embedded into a circular dsDNA construct could be shifted by applying torsional stress to the DNA circle via the incorporation of the intercalator.

One great disadvantage of the devices based on conformational changes caused by small molecules or ions is the fact that here DNA only plays a structural role – the DNA code is only utilized for the construction of the devices, but not for their control. Hence all of the devices in a reaction volume are affected by a buffer change simultaneously and they cannot be addressed individually. This issue is, at least in principle, resolved in hybridization-driven devices.

6.3.2
Hybridization-driven Devices

A very different operation principle underlies hybridization-driven DNA devices. Here, the conformational change of a DNA supramolecular structure is brought about by the hybridization of "fuel strands" with single-stranded "motor" sections. "DNA tweezers" were the first prototype of this kind of device [28]. Their operational principle is shown in Fig. 6.4. In the open state, the DNA tweezers consist of three strands of DNA. One 40-nucleotide long strand is hybridized to two other, 42-nucleotide strands in such a way that together they form two 18-base pair (6-nm), rigid duplex "arms" connected by a 4-nucleotide, single-stranded flexible "hinge". In the open state, each of the 42-nucleotide strands still has 24-nucleotide (i.e. 42–18 nucleotides) single-stranded extensions available for hybridization. The addition of a 56-nucleotide "fuel" strand which is complementary to these extensions can then be used to close the tweezers structure, i.e. the two arms are forced together by the hybridization with the fuel strand. A "trick" is used to switch the device back to its original configuration. In the closed state, from the 56 nucleotides of the fuel strand, 8 nucleotides are deliberately left single-stranded. These 8 nucleotides serve as a "toehold" for another "anti-fuel" or "reset" strand which is exactly complementary to the fuel strand. A biochemical process known as "branch migration" unzips the structure, when fuel and anti-fuel try to bind with each other. After completion of this process, a stable 56-base pair "waste" duplex is formed and the DNA tweezers are in their open configuration once more. The device can be operated cyclically by the alternate addition of set and reset strands.

The same operation principle has since been used in many other DNA devices: a

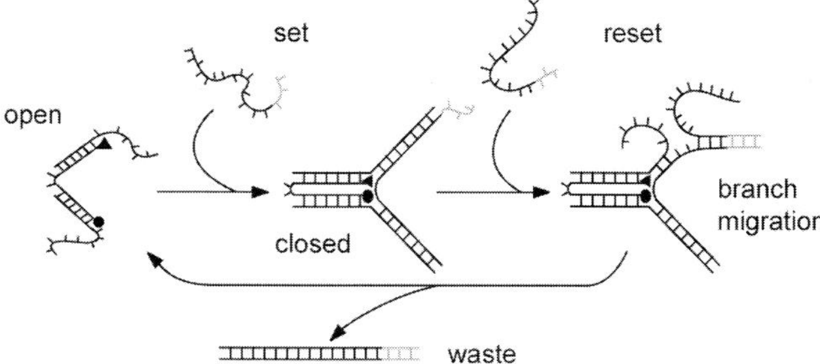

Figure 6.4. Operation cycle of the DNA tweezers (*cf.* Ref. [28]). In the open state, the DNA tweezers consist of three DNA strands which together form a "tweezers"-like structure consisting of two double-stranded arms connected by a flexible single-stranded hinge. The arms have two single-stranded extensions to which a "set" or "fuel" strand can attach. The hybridization of the fuel strand with these extensions closes the tweezers structure. Another fuel strand which is completely complementary to the first one can reset the tweezers to the open state. When the tweezers are in the closed state, the reset strand attaches to the single-stranded "toehold" region of the set strand (gray) and displaces the set strand from the tweezers by "branch migration". After completion of this process, a double-stranded waste product is formed by the set and reset strand, while the open state of the tweezers is restored.

simple variation of the DNA tweezers is the DNA actuator device [29, 30] shown in Fig. 6.5. Here, instead of two single-stranded extensions, the arms of the tweezers were connected by a single-stranded loop. Depending on the sequence of the fuel strands, the device could either be closed similarly to the tweezers or stretched into an elongated conformation. Whereas the intermediate configuration is a rather floppy structure (like the open tweezers), the closed and the stretched configurations are much more rigid. The distance between the fluorescent dyes attached to the arms of the actuator is switched from roughly 2 nm in the closed state to approximately 14 nm in the stretched state.

A rotary DNA device based on multiple crossover molecules (complex supramolecular DNA structures in which DNA strands are shared between two or more DNA duplexes) was demonstrated by Yan and coworkers [31]. Their DNA structure could be switched between a "paranemic crossover" (PX) conformation and a "juxtaposed" (JX_2) configuration (Fig. 6.6). The PX motif is a four-stranded DNA structure in which two neighboring duplexes exchange strands of the same directionality at every possible site [32]. If parts of this motif are removed and replaced by DNA sections without crossover, molecules such as the juxtaposed JX_2 structure result in which two helices are rotated with respect to the corresponding PX structure (Fig. 6.6). In the PX–JX_2 device, the central section of a supramolecular structure based on the PX motif was created with DNA molecules which could be removed by the same branch migration "trick" as used for the operation of the DNA tweezers. After removal, these molecules could be replaced by DNA strands which switched the device to a JX_2 configuration. Again using branch migration, the device could be switched back to the original PX structure.

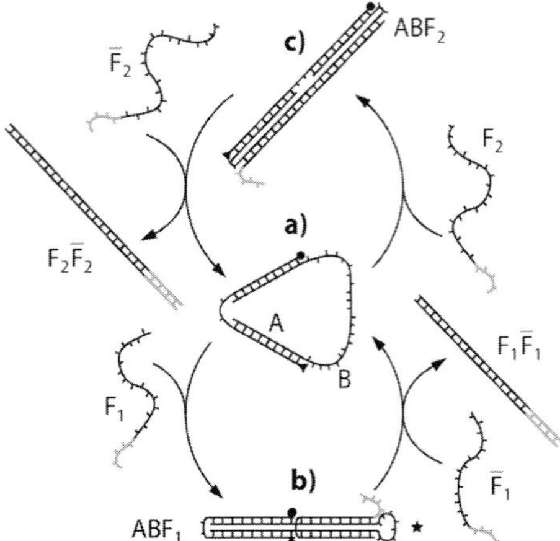

Figure 6.5. Operation cycle of a DNA actuator [30] switchable between three mechanical states. The actuator is similar to the tweezers and consists of two DNA strands (A and B) which hybridize together to form a circular structure as depicted in (a). By the addition of appropriate fuel or removal strands F_1, $\bar{F}_1$, F_2 or $\bar{F}_2$, the device can be switched between a relaxed state (a), a closed state (b) and a stretched state (c). Removal strands can bind to the toehold sections depicted in gray of the fuel strands and remove them by branch migration. The ends of the arms (circular and triangular symbols stand for fluorescent dyes which are used for characterization) are switched from a short distance of a few nanometers in the closed state (b) to almost 14 nm in the stretched state (c). (Reprinted with permission from Ref. [30], © 2002, American Institute of Physics.)

Other devices utilizing branch migration were based on unusual DNA structures such as guanine quartets ("G quartets"). In a G quartet, four guanine bases bind to each other into a cyclical configuration via non-WC base-pairing interactions. In the experiments described by Li and Tan [33] and by Alberti and Mergny [34], ssDNA structures formed by several stacked intramolecular G quartets were switched to an elongated duplex structure by the addition of a complementary DNA fuel strand. This transition could be reversed by removing the fuel strand using branch migration. The overall effect was a detectable change in distance between the 5′ and 3′ ends of the device strand from about 2 nm in the quadruplex state to 7 nm in the duplex state.

6.4
Towards Functional Devices

With the prototype devices described in the previous section it was established that it is possible to construct DNA structures which can perform movements on the nanometer scale. Except for their building material, these structures are nonbiological and completely "designed". Their realization can therefore be regarded as an

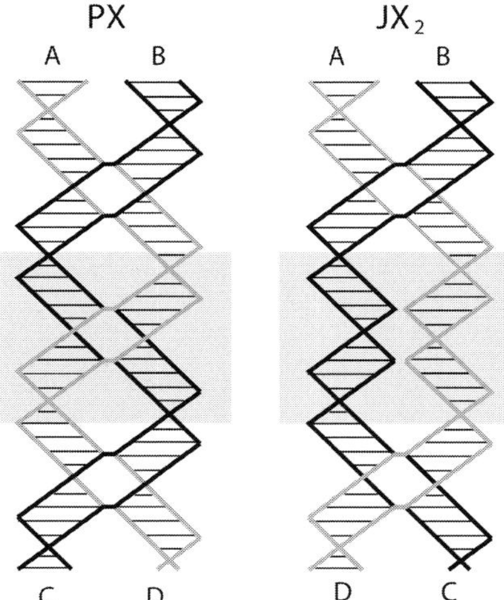

Figure 6.6. The transition between the PX and the JX$_2$ structure forms the basis of a nanomechanical device by Yan and coworkers [31]. Both structures are formed by two DNA duplexes which share common DNA strands. In the PX (paranemic crossover) structure, the two DNA double-helices exchange strands wherever it is possible, whereas in the JX$_2$ (juxtaposed) structure strands do not cross over in the middle part of the structure. Note that in the JX$_2$ structure, the helices C and D have a different position than in the PX structure. In Ref. [31], Yan and coworkers managed to switch the inner section of these structures (gray) using removable DNA strands and branch migration. This leads to a switchable transition between the PX and the JX$_2$ form. The corresponding "rotation" of the duplexes C and D with respect to each other can be observed.

important step towards the realization of complex artificial molecular machinery. However, the tasks these prototype devices could perform were not particularly "useful". In fact, their movements (rotation or stretching) were mostly idle, i.e. the motion was just an internal rearrangement of parts. To lift DNA devices to the next level of complexity, it is important to design structures that can perform specific tasks and interact with their environment. There are a number of different approaches towards such functional devices. One is to achieve directed motion rather than idle movements with the goal to produce artificial molecular motors. Another possibility for functional devices is to couple DNA devices to nanoscale structures like other biomolecules or nanoparticles. Such hybrid structures could be particularly interesting for applications in biotechnology and the life sciences.

6.4.1
Walk and Roll

Among the most fascinating biological molecular machines are molecular motors such as kinesin, dynein or the myosins [35]. Kinesin and dynein motors walk along cytoskeletal microtubuli, whereas myosin motors walk on supramolecular tracks made from actin. Kinesin, dynein and certain kinds of myosin help to actively transport vesicles within cells. Myosin motors are also responsible for other kinds of active movement in eukaryotic cells, including muscle contraction in higher eukaryotes. Many other enzymes like DNA or RNA polymerases can also be regarded as biological molecular motors. Clearly, the ability to actively transport molecular components or to exert forces on a molecular scale would also be highly desirable for nanotechnological applications or in biotechnology. However, building an artificial molecular motor is not an easy task.

Molecular motors work very differently from macroscopic machines as all movements on the molecular scale are dominated by Brownian motion. The energies available for the operation of molecular motors are only slightly higher than the thermal energy and they are therefore subjected to large fluctuations. Research on biological molecular motors has shown that these tiny machines do not work against Brownian motion, but rather utilize it for their operation. In many cases, a "Brownian ratchet" mechanism seems to be at work, where an irreversible chemical step is utilized to "rectify" the undirected Brownian motion [36]. The general features of such biological motors now serve as a guideline for the construction of artificial DNA motors.

Three different concepts to achieve directional motion in DNA-based systems are depicted schematically in Fig. 6.7. They correspond to the DNA walkers by Shin and Pierce [37] and Sherman and Seeman [38], the molecular "gears" by Tian and Mao [39], and the enzyme-assisted walker by Yin and coworkers [40]. The first simple DNA walker was introduced by Shin and Pierce [37]. A double-stranded DNA scaffold with ssDNA "docking positions" attached to it serves as a track for the DNA walker. The walker itself is a DNA duplex with two single-stranded "feet". As explained in Fig. 6.7(a), DNA fuel strands are used to connect the single-stranded feet to the docking positions on the track. The connector strands are equipped with single-stranded toehold sections and can be displaced from the device by removal strands via branch migration. The free DNA foot can then be connected to the next free position on the track. This can be repeated several times with the appropriate connector and removal strands to move the walker to arbitrary positions on the track.

A similar walking device based on more complex DNA structures was realized by Sherman and Seeman [38]. In their work, the walker consists of two DNA duplexes joined by flexible single-stranded linkers. The walker can be translocated along a track which consists of a triple crossover molecule ("TX", three double helices connected by shared DNA strands) which contains single-stranded position labels. Similar to the walker by Shin and Pierce, the duplexes of the Sherman and Seeman walker also contain single-stranded "feet" and can be attached to the docking positions on the TX track via DNA linker molecules. Again, these connections can be removed by branch migration and the free foot can be connected to the next position.

As shown in Fig. 6.7(b), the same principle can be applied to achieve a different kind of unidirectional motion. In the "molecular gears" system realized by Tian and Mao [39], two DNA "circles" roll against each other driven by hybridization

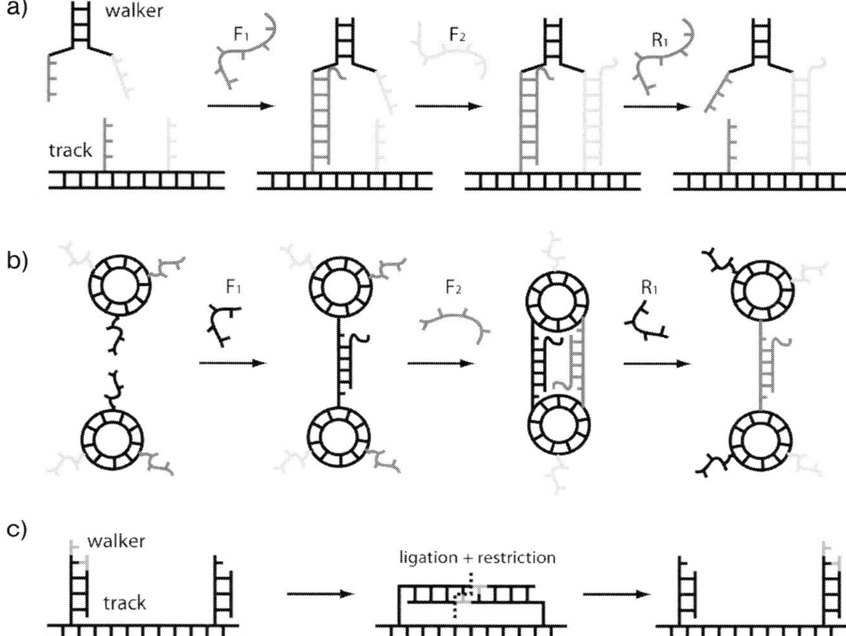

Figure 6.7. Three examples for unidirectional motion realized with DNA devices. (a) A simple walker system. A walker, which consists of a DNA duplex (or a more complex structure) and single-stranded "feet", can attach to a dsDNA track from which single-stranded binding positions or "footholds" extend. The attachment of the feet is achieved with DNA fuel strands F_1 and F_2, which are partly complementary to the footholds and partly complementary to the feet. The connection between the walker and track can be broken again with removal strands which can remove the fuel strands by branch migration. Using a different fuel strand the free foot can be attached to another position on the track. This walking principle has been applied in Refs. [37, 38]. (b) Essentially the same principle as in (a) can be used to make two DNA circles roll against each other. The two circles with single-stranded extensions can be connected to each other with one or two fuel strands. Using fuel and removal strands in the correct order, the circles can roll against each other (corresponds to the device in Ref. [39]). (c) An autonomous enzyme-driven "walker" (*cf.* Ref. [40]). The walker system consists of a dsDNA track with double-stranded position strands. Initially, the first two neighboring position strands have sticky ends which may hybridize with each other. The two position strands can be ligated together and cut with a restriction enzyme. This creates new sticky ends on the position strands. The net effect is the transfer of a few DNA bases from one position to the next (depicted in gray). Using two different restriction enzymes, transport of the bases can autonomously occur over a longer track with many positions.

and branch migration. The two circles consist of a closed single strand to which three strands are hybridized containing flexible hinges made of thymine residues and single-stranded extensions ("teeth"). The two circles can be connected by DNA linker strands which are partly complementary to one "tooth" of one circle and another tooth of the other circle. Due to the flexible hinges, two circles can be connected by two linker strands simultaneously. As in the case of the walkers, the linker strands contain single-stranded toehold sections at which removal strands can attach and displace the linker strands from the device. By the alternate addition of linker and removal strands in the correct order, the two circles can be made to roll against each other in one direction.

A different concept to achieve unidirectional motion in a DNA-based system was demonstrated by Yin and coworkers [40]. In contrast to the other walkers, their system needs the help of DNA-modifying enzymes, but it walks "autonomously". The basic idea is depicted in Fig. 6.7(c). Again, the walker (which now consists of only a few DNA bases) sits on a DNA track which consists of double-helical DNA with duplex position strands. These position strands have short, single-stranded ends. If the ends of neighboring strands are complementary, they can hybridize with each other and can then be ligated by a DNA ligase. The sequences of these duplexes are chosen such that in the ligated state the recognition sequence for a restriction endonuclease is formed. Upon the addition of the endonuclease, the connected neighboring position duplexes are cut and a new pair of sticky ends is produced. In this way, the DNA bases representing the "walker" are transferred from one to the next position. Using two different restriction enzymes, the system can be designed to unidirectionally transport the walker bases along a track without ever stepping back [40].

Quite recently, several improvements on these systems could be demonstrated. Turberfield and coworkers could realize an autonomous walker system similar to the system by Yin and coworkers, but now driven by the action of a nicking enzyme and branch migration [41]. In this new system, a longer DNA fragment is passed from one binding position to the next. In a different approach, Mao and coworkers could utilize the action of a DNA enzyme (*cf.* Section 6.5.1) to achieve autonomous unidirectional motion along a track [42].

The DNA walker systems demonstrated so far essentially belong to the "Brownian" type. The walkers diffusively find a binding position on a track and a chemical reaction (hybridization or enzymatic ligation, restriction, etc.) is used to fix the position. Directionality is introduced by the choice of the base sequence of the track positions and the walkers. So far, the walkers also move without load – so nothing can be said about the efficiency of these systems. It will be very interesting to see in the future how DNA walkers behave when they have to carry something, i.e. when they have to perform work.

6.4.2
Interaction with Proteins

The numerous advantages of DNA mentioned in Section 6.1.1, i.e. its controllability, programmability and stability, allow us to rationally construct artificial molecular devices which are (by their very nature) "automatically" biocompatible. The life sciences should therefore be a particularly interesting and fruitful area for the application of ideas developed in the context of DNA-based nanodevices: moving molecular components, sensing, information processing, etc.

Exciting possibilities arise, for example, when DNA nanodevices are made to interact with proteins. This interaction can be achieved in a variety of different ways: proteins can be coupled to DNA chemically or biochemically, proteins may be utilized which naturally bind to DNA, and proteins can bind to "aptamers" – nucleic acids which fold into a specific structure in which they strongly bind to a protein.

Many interesting examples for biochemically coupled protein–DNA conjugates come out of the Niemeyer lab. One example, not directly related to DNA nanomechanical devices, is the utilization of DNA recognition in the so-called immuno-polymerase chain reaction (PCR) technique [43]. Here biotinylated antibodies are linked to biotinylated DNA molecules via streptavidin. These DNA–protein constructs can then be used to detect a primary antibody–antigen recognition event by the amplification of the DNA fragment with the PCR. Another very interesting example for supramolecular constructs produced from biotinylated DNA and proteins is the assembly of bienzymic complexes from NADH:FMN oxidreductase and bacterial luciferase [44]. NADH:FMN oxidoreductase reduces flavin mononucleotide (FMN) to $FMNH_2$, which binds to luciferase. Luciferase in turn oxidizes an aldehyde to a carboxylic acid and emits blue light. As bacterial luciferase needs $FMNH_2$ for the reaction, holding the NADH:FMN oxidoreductase in close proximity greatly enhances its activity. Such a concept may allow one to tune biochemical activity by controlling the distance between enzymes in DNA-scaffolded multienzyme complexes.

There are also many proteins which naturally interact with DNA: nucleases, ligases, polymerases, recombination proteins, transcription factors, to name but a few. In many cases these DNA-binding proteins recognize specific DNA sequences and in some cases they also distort the structure of DNA. Seeman and coworkers demonstrated recently that DNA devices can be used to estimate the work performed by a DNA-binding protein when it distorts the double helix [45]. As protein, the *Escherichia coli* integration host factor (IHF) was used which bends the DNA double helix by more than 160°. The binding site for IHF was sandwiched between two DNA TX structures which were fluorescently labeled with a FRET pair (Fig. 6.8). The bending of the construct can be monitored with the FRET signal. To measure the work performed by the enzyme upon bending, cohesive ends of various lengths were also attached to the TX units. Upon bending the DNA, IHF also has to disrupt the duplex formed by these cohesive ends. The work can therefore be determined from the free energy of formation of the duplex. Interestingly, this concept is very similar to a recently demonstrated DNA-based force sensor [46], but here the "measuring device" itself is also on the molecular scale.

A different approach to DNA–protein hybrid devices is not to utilize proteins which bind to DNA, but rather to evolve specific DNA or RNA sequences which bind to proteins. Such so-called "aptamers" can be isolated in *in vitro* selection experiments (SELEX) for a large variety of proteins and also small ligands such as ATP [47]. Aptamers bind specifically to their targets and in many cases have comparable binding constants as antibodies with their antigens. The great advantage of aptamers as compared to antibodies is their simple preparation and their stability. In particular, once isolated, DNA aptamers can be easily synthesized using the

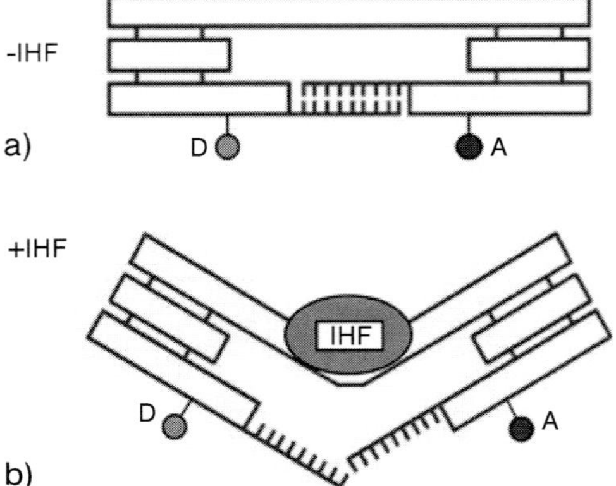

Figure 6.8. A DNA device that can be used to measure the work performed by a DNA-binding protein which distorts the double-helix. (a) The device consists of two connected triple-crossover (TX) constructs (left and right part; these are similar to the structures in Fig. 6.6, but with three interconnected double helices). The upper domains of the TX structures are connected by a DNA duplex which contains the recognition site for the DNA-binding protein IHF. The lower domains are connected by a cohesive part which consists of two complementary single-stranded sections of DNA. (b) When IHF binds to its recognition site it distorts the DNA duplex connecting the TX structures. However, to be able to do so it has to perform additional work to disrupt the cohesive tract connecting the lower domains of the device. Using cohesive tracts of varying lengths one can estimate the maximum work the protein is able to perform. (Reprinted with permission from Ref. [45], © 2004, Wiley-VCH.)

automated DNA synthesis methods mentioned before. In the context of DNA nanodevices, one additional advantage is that they can be incorporated into DNA nanomechanical devices or supramolecular structures.

Recently, Dittmer and coworkers [48] applied the concept of switching a DNA device between two conformations using branch migration to a DNA aptamer structure. The operation principle of this device is explained in Fig. 6.9. The nanodevice is based on a DNA aptamer selected to bind the human blood-clotting protein, thrombin. The 15-base DNA sequence 5′-GGTTGGTGTGGTTGG-3′ folds into a secondary structure consisting of two G-quartets, in the presence of potassium ions (Fig. 6.9). The device in its standard state binds thrombin. Release of the protein can be triggered by the addition of a single-stranded effector DNA containing a sequence that is complementary to a portion of the aptamer sequence, resulting in an unfolding of the aptamer to form a duplex, which is unable to bind the protein. The protein can be bound once again by the addition of another DNA strand, fully complementary to the previously added strand, which removes the effector by branch migration. The aptamer is thus allowed to return to its protein-binding quadruplex form. By the alternate addition of effector strands, the DNA device can repeatedly made to bind or release the protein – the device can therefore be thought of as a "nanohand". In principle, similar devices can be constructed to bind other proteins and ligands if an appropriate aptamer exists.

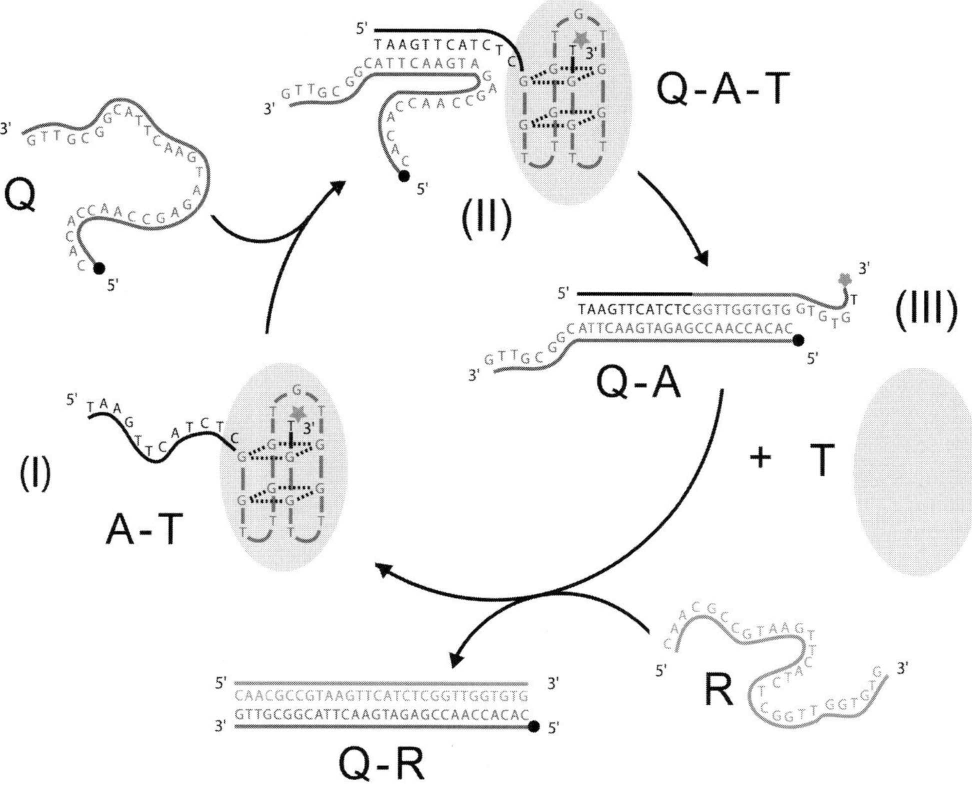

Figure 6.9. A DNA device based on a DNA aptamer. The aptamer sequence 5'-GGTTGGTGTGGTTGG-3' folds into a structure characterized by two stacked guanine quartets. In this structure, the aptamer binds strongly and specifically to the protein thrombin. To be able to switch this aptamer efficiently between a protein-binding and a non-binding conformation, it has been extended by a 12-nucelotide "toehold" section. In the folded state (I) the aptamer device A binds to the protein thrombin (T). In (II), the opening strand Q attaches to the 12-nucelotide toehold section of the device and displaces the protein. In the stretched duplex conformation (III), the device cannot bind to thrombin. The removal strand R can attach to a second toehold section in strand Q and displace Q from A by branch migration. Strand A then refolds and binds the protein thrombin again. Alternate addition of Q and R strands allows repeated release and binding of the protein. (Reprinted with permission from Ref. [48], © 2004, Wiley-VCH.)

The examples mentioned in this section should give an impression of the potential of hybrid systems composed of DNA nanodevices and proteins. Such systems can have many interesting applications in the life sciences. DNA structures can be used to arrange proteins into certain geometries to fine-tune or study their interactions. DNA nanodevices can also be used as sensors which measure the work performed by DNA-binding proteins. Finally, DNA aptamers can be combined with switchable DNA nanostructures to control the binding or release of molecules. This may find applications in biosensors or in intelligent drug delivery systems.

6.4.3
Information Processing

In biology DNA is used for the storage of genetic information. For artificial applications, this information-carrying nature of DNA has been utilized in the field of "DNA computing" for the solution of computational problems such as the "Hamiltonian path problem" [49], satisfiability problems [50–52] and many others. One more recent development in DNA computing is the construction of molecular automata which do not solve hard computational problems, but which can autonomously perform simple logical operations based on molecular inputs. This should be particularly interesting in the context of intelligent drug delivery systems. Here, one would like to construct autonomous molecular devices which can sense environmental information, perform a simple computation and decide which action to take, e.g. whether to release a specific molecule or not.

One approach towards autonomous information processing by DNA-based devices was taken by Benenson and coworkers [53–55]. In their work, the DNA-cleaving properties of the class IIS restriction endonuclease *Fok*I were utilized to build the molecular realization of the computer-theoretical concept of "finite state machines" or "finite automata". Finite automata are computing machines with a finite number of internal states which can undergo transitions between these states according to certain rules. A program for such an automaton consists of a series of such transitions. The operation of the automaton is based on the fact that the enzyme *Fok*I cleaves a dsDNA substrate 9 and 13 nucleotides offset from its recognition, creating a 4-nucleotide sticky end. Hence, the states and input symbols for the automaton are represented by different 4-nucleotide sequences. In Ref. [53], *Fok*I is used to sequentially cleave off pieces of a double-stranded program and thereby create a series of transitions between the various states of the automaton. Finite automata are related to Turing machines, but have limited computing power. In Ref. [53], Benenson and coworkers used their DNA automata to make simple decisions such as to determine whether a given input symbol occurs at least once. Keinan and coworkers recently used another class IIS endonuclease, *Bbv*I, to perform similar molecular computations in a chip-based approach [56].

Benenson and coworkers, on the other hand, were able to demonstrate how their *Fok*I-based DNA automaton could actually find application in autonomous diagnosis and drug delivery systems [57]. To this end, the transition molecules for a *Fok*I-based DNA automaton were designed to be active only in the presence of specific mRNA molecules. With these transition molecules, a molecular computer was realized which could decide whether the levels of certain mRNA molecules were high or low. Depending on the decision, a short DNA strand was released as a "drug". This DNA drug was chosen as an antisense molecule, which in principle could inhibit the synthesis of a cancer-related protein by binding to its corresponding mRNA.

A different approach towards autonomous computing with DNA is based on DNA enzymes. Similar to DNA aptamers, DNA enzymes are another example of DNA molecules which have functional properties in addition to their structural properties. They also fold into specific secondary and tertiary structures in which they exhibit catalytic activity. DNA enzymes (and their RNA counterparts, i.e. ribozymes) catalyze reactions such as RNA cleavage, ligation and even peptide bond formation. In Ref. [58], Stojanovic and coworkers realized several logic gates based on the action of the RNA cleaving deoxyribozymes E6 [59] and 8–17 [60, 61]. These DNA enzymes cleave RNA molecules or, alternatively, chimeric DNA/RNA hybrid molecules with an RNA base at the appropriate position, when these are bound to their substrate recognition sections. In Ref. [58], the recognition sequences are protected by self-hybridized stem–loop modules. By base-pairing with a DNA strand complementary to the loop sequence, the stem–loop modules can be opened. This makes the substrate recognition sequence accessible for the substrate, which is subsequently cleaved. By combining controlling and catalytic modules, Stojanovic and coworkers realized simple logic gates such as AND or XOR [58], a DNA half-adder [62] and a molecular computer which autonomously could play the game "Tic Tac Toe" [63]. A combination of autonomous DNA-based computers with aptamer-based molecular devices (as introduced above) seems natural. In fact, Kolpashchikov and Stojanovic recently introduced a molecular "robotic" system in which DNA logic gates controlled the binding state of a DNA aptamer for the chromophore malachite green [64].

6.4.4
Switchable Networks and Hybrid Materials

The properties of DNA nanomechanical devices may also be utilized to switch larger DNA nanostructures such as DNA-linked nanoparticle networks or DNA-based supramolecular networks between several states.

Yan and coworkers could show recently that the topology of large supramolecular DNA networks could be switched by nanomechanical actuators incorporated into these networks [65]. Again, switching between two different states was accomplished with DNA fuel strands and by using the branch migration mechanism. In a different publication, Yan and coworkers recently also demonstrated the incorporation of aptamer sequences into DNA supramolecular structures [18], which facilitated the spatial arrangement of proteins bound by the aptamers. Based on this work and ideas put forward above [44, 48], supramolecular structures are now conceivable on which several interacting proteins are held by different aptamer sequences and their distance (and hence their interactions) is regulated by actuating elements incorporated into the structures.

Similar structures can be realized which can switch the distance between inorganic nanoparticles rather than proteins. As a first example, Niemeyer and coworkers demonstrated a switchable network composed of gold nanoparticles and DNA molecules [66]. The experimental concept is displayed in Fig. 6.10. Two batches of gold nanoparticles are first tagged with thiolated oligonucleotides with

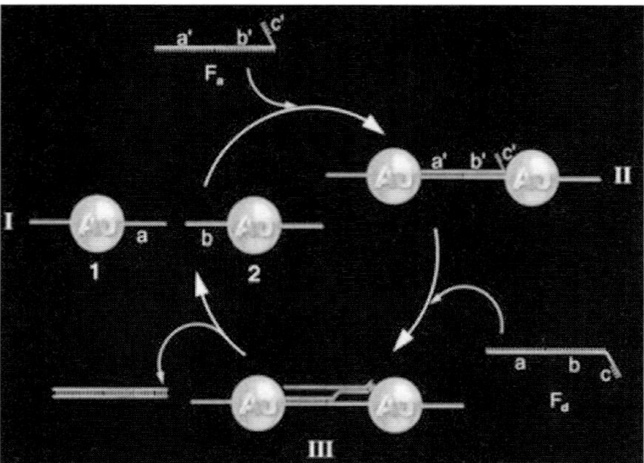

Figure 6.10. DNA can be used to reversibly control the aggregation of gold nanoparticles. Two species of gold nanoparticles (1 and 2) tagged with different single-stranded "address labels" can be connected using DNA strands F_a complementary to these labels. This leads to an aggregation of the nanoparticles which can be monitored, e.g. in the absorption characteristics of the sample. Removal strands F_d complementary to F_a can attach to the toehold section c' of F_a and remove the connecting strands from the aggregate by branch migration. This leads to a deaggregation of the DNA-linked nanoparticle network. By the alternate addition of F_a and F_d, the aggregation–deaggregation cycle can be performed many times. (Reprinted with permission from Ref. [66], © 2004, Wiley-VCH.)

different sequences. Two nanoparticles with different tags can then be connected by linker oligonucleotides which are partly complementary to the tags on one species of nanoparticle and partly complementary to the other. This leads to the formation of a crosslinked network of gold nanoparticles. To make the system switchable, the crosslinking oligonucleotides are equipped with a DNA "toehold" (as in the original "tweezers" system [28]) at which a DNA strand complementary to the crosslinker can attach. The crosslinker is then removed by branch migration and the nanoparticles are disconnected. By the alternate addition of crosslinkers and removal strands, the system can be driven through several operation cycles during which the mean distance between the particles is switched. In the case of metallic nanoparticles, the distance between the particles influences the coupling between surface plasmon excitations within the particles, and aggregation of the nanoparticles results in an observable shift in the absorbance of the reaction sample. This color change upon aggregation has already been applied in a variety of biosensors [67, 68] and such switchable networks may also find application in this area.

Apart from biosensing applications, switchable hybrid structures composed of DNA actuators and nanoparticles may also help to answer more fundamental questions regarding the coupling of excitations in nanoparticles, and to tune energy and charge transfer rates between them.

6.5
Autonomous Behavior

Most of the devices described so far could not operate autonomously, with the exception of the autonomous walkers by Yin and coworkers [40], Bath and coworkers [41], and Tian and coworkers [42], and the computing devices by Benenson and coworkers [53] and Stojanovic and coworkers [63]. Usually, an external operator had to intervene at each step of the operation cycle of the devices and add DNA "fuel" or "effector" strands, or change the buffer conditions. While this may be no hindrance in simple applications, in many cases it might be desired to have free-running devices rather than "clocked" devices. For example, for applications in intelligent drug delivery or in nanofabrication, the devices should respond to environmental stimuli by a complex series of actions without guidance by an external operator.

6.5.1
Driving Devices with Chemical Reactions

The successful concept of "fuel strands + branch migration" for the operation of DNA-based nanodevices has one drawback when it comes to autonomous operation: complementary fuel and removal strands cannot be added simultaneously to the reaction solution as they would immediately react with each other rather than driving a device through its states. A possible solution for this problem has been proposed by Turberfield and coworkers [69] based on the inhibition of hybridization between complementary strands by the formation of secondary structures. In Ref. [69] it is shown that the hybridization between two complementary strands can be slowed down considerably by forcing the strands into hairpin conformations using "protection strands". Hybridization can be sped up again by DNA "catalysts" which can open these hairpin structures. To construct free-running devices using this concept, such catalyst strands would have to be incorporated into DNA nanomechanical structures. These would then be continuously driven through their states by the consumption of the metastable protected "fuel" strands. These fuel molecules could be added simultaneously as now the reaction with the DNA device would be much faster than a direct reaction.

Another realization of a self-running device was demonstrated by Chen and coworkers with a machine consisting of a DNA actuator [29] modified with an RNA-cleaving DNA enzyme [70] similar to the DNA enzymes used for computing by Stojanovic and coworkers. In its standard state, the DNA enzyme is in a compact form and the actuator assumes its closed state (Fig. 6.11). In the presence of its substrate (a DNA/RNA chimera), the enzyme extends to its catalytically active structure as it forms a duplex with the substrate and the arms of the actuator are pushed apart from each other. When the substrate is cleaved by the DNA enzyme, two short fragments remain hybridized to the device instead of one long substrate strand. The two short duplexes are thermodynamically less stable and the cleaved

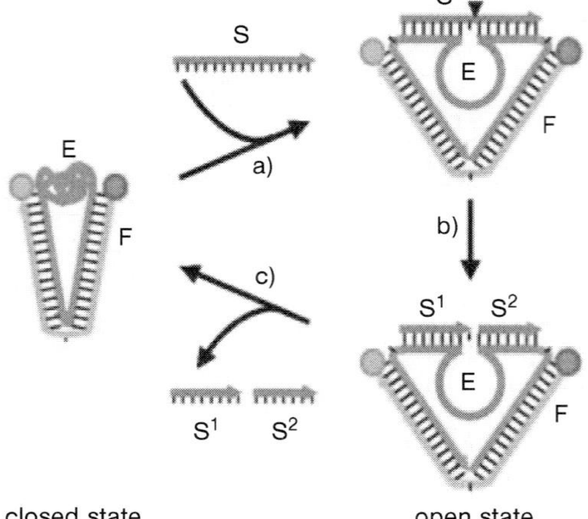

Figure 6.11. An autonomous DNA device based on the action of a DNA enzyme. The sequence for an RNA-cleaving DNA enzyme has been incorporated into a DNA actuator similar to the one shown in Fig. 6.5. When a chimeric RNA/DNA substrate S binds to the DNA enzyme section E, the two arms of the device are pushed apart from each other. After cleavage of the substrate the resulting shorter DNA fragments S^1 and S^2 unbind from E, and the device collapses into the closed state. This opening–closing motion can continue as long as substrate strands are available. (Reprinted with permission from Ref. [70], © 2004, Wiley-VCH.)

substrate is released from the device leading to the collapse of the enzyme section and thus an automatic closure of the actuator. The device will run through its operation cycle autonomously as long as the substrate is present. When using a noncleavable DNA substrate, the device can also be deliberately forced into a stalled configuration [71]. The recently demonstrated free-running walker by Tian and coworkers [42] is also based on this operation principle.

A different concept to autonomously operate buffer-driven DNA nanodevices has been demonstrated by Liedl and Simmel. It could be shown that the pH-sensitive conformational transition of a cytosine-rich DNA strand between a random coil conformation and the i-motif (Fig. 6.2b) could be driven by the oscillating proton concentration generated by a chemical oscillator [72]. In this system, the states of the DNA devices are synchronized by the dynamics of a nonlinear dynamical system rather than by an external operator.

6.5.2
Genetic Control

A very different possibility for the autonomous operation of DNA-based devices opens up when one considers the production of DNA or RNA effector molecules "on the run". RNA molecules are produced, of course, when a gene is transcribed into mRNA by an RNA polymerase. It is therefore quite natural to think about the utilization of genetic transcription for the operation of DNA nanomechanical devices.

It is well established that biochemical reaction networks have information-processing properties [73, 74]. This is particularly obvious in genetic networks which control the expression level of proteins through complex interactions between regulatory molecules and genes. The textbook example for a genetic switch is the *lac* operon in *E. coli* bacteria in which the expression of the protein β-galactosidase is regulated by the relative concentrations of lactose and glucose. Many other gene regulatory motifs with a variety of tasks and responsibilities have been found since the discovery of the *lac* operon and biochemical analogs of switches, oscillators, filters, components with memory functions and others have been identified [74].

Recently, the knowledge about these motifs has also been used to construct artificial genetic networks. By transferring circuit designs known from electrical engineering to artificial regulatory networks, such functions as a bistable genetic switch [75], a genetic oscillator [76], or a sender–receiver system [77] could be synthetically implemented in bacteria and similarly *in vitro* [78]. An overview of the developments in "synthetic biology" is given in Refs. [79, 80].

It is a tempting idea to use similar artificial genetic networks to drive DNA nanodevices through their mechanical states. One could imagine a genetic oscillator periodically producing RNA fuel strands which switch a DNA actuator from one conformation into another and back, possibly dependent on environmental variables which modulate the genetic activity. A first step in this direction was taken in Ref. [81] in which the DNA tweezers introduced in Section 6.2.2 were operated by an mRNA fuel strand which was transcribed from an artificial "fuel gene". More recently, the transcription of fuel strands was also put under the control of simple gene regulatory mechanisms [82]. One example is shown in Fig. 6.12, where the closure of DNA tweezers by RNA fuel is controlled by the Lac repressor and the inducer IPTG. In the presence of the repressor protein LacI, fuel transcription is suppressed and the DNA tweezers are in their open state. When the inducer isopropyl-D-thiogalactoside (IPTG) is added to the transcription solution, it binds to LacI, which in turn cannot bind to the DNA substrate in this form. Consequently, fuel transcription is switched on and the tweezers close. In Ref. [82], the transcription of an appropriate removal strand has also been put under the control of a second repressor (LexA). Closed tweezers are opened by RNA fuel only in the absence of this repressor. To achieve genetic control over a series of movements, in the future the temporal order of the transcription events will have to be regulated using transcriptional cascades or feedback loops.

These initial steps demonstrate that it is indeed possible to control the action of DNA-based nanomechanical devices by gene regulatory mechanisms. It should therefore be possible to program DNA devices to perform a complex series of tasks in response to certain environmental influences. It is also conceivable to construct similar DNA or RNA structures which act in response to naturally occurring mRNA molecules. This again could be interesting in the context of biosensing or drug delivery. For example, one could construct aptamer-based devices which

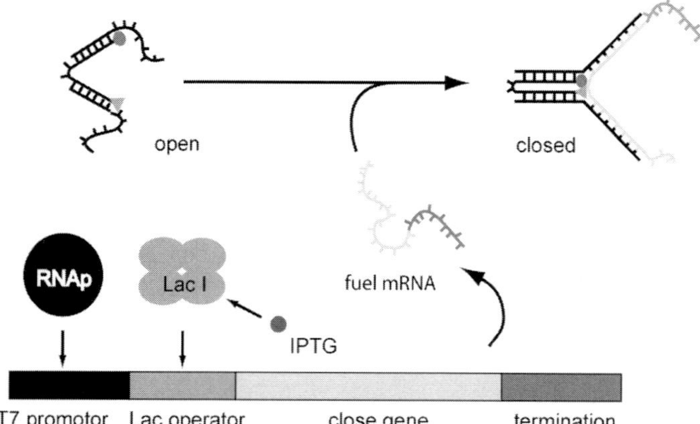

Figure 6.12. Gene regulatory control of DNA devices. A gene construct contains the instructions to close opened DNA tweezers (close gene). Its transcription will lead to the production of a fuel mRNA strand which can close the tweezers. In the construct, the promotor of the gene is put under the control of the Lac operator directly downstream of the promoter. When the repressor protein LacI binds to the operator, transcription extension by T7 polymerase is blocked. However, when the "inducer" IPTG binds to LacI, it cannot bind to the operator sequence. Then transcription is switched on, fuel mRNA is produced and the DNA tweezers are closed. (Reprinted with permission from Ref. [81], © 2005, Wiley-VCH.)

release their molecular load in response to the presence of specific mRNA molecules.

6.6
Conclusion

Apart from their original biological role, DNA molecules can be used as highly versatile building blocks for the construction of nanoscale structures and devices. The first simple prototype devices based on DNA could only perform simple movements such as stretching or rotation. Based on these prototypes and combining them with other concepts from nanotechnology and the life sciences, a variety of more functional DNA nanodevices has recently been demonstrated. The first examples of "DNA walkers" capable of unidirectional motion have been given; the interaction of DNA devices and proteins has been utilized for sensors, for the control of biochemical reaction rates and for the controlled binding and release of proteins; ideas from DNA computing have been integrated to logically control the action of DNA-based devices; and, finally, the action of DNA devices has been coupled to the genetic machinery.

These developments represent the first steps towards autonomous molecular-scale devices which can sense environmental information, perform computations and act independently as molecular motors, drug reservoirs or signal transducers.

As such, they are expected to have a significant impact on the development of advanced biosensors or as components for intelligent drug delivery systems.

Acknowledgments

The author gratefully acknowledges funding by the Deutsche Forschungsgemeinschaft through its Emmy Noether program.

References

1 M. F. Crommie, C. P. Lutz, D. M. Eigler. Confinement of electrons to quantum corrals on a metal-surface, *Science* **1993**, *262*, 218–220.

2 S. W. Hla, L. Bartels, G. Meyer, K. H. Rieder. Inducing all steps of a chemical reaction with the scanning tunneling microscope tip: towards single molecule engineering, *Phys. Rev. Lett.* **2000**, *85*, 2777–2780.

3 C. Dennis, P. Campbell. The eternal molecule, *Nature* **2003**, *421*, 396–396.

4 I. Tinoco, C. Bustamante. How RNA folds, *J. Mol. Biol.* **1999**, *293*, 271–281.

5 J. H. Chen, N. C. Seeman. Synthesis from DNA of a molecule with the connectivity of a cube, *Nature* **1991**, *350*, 631–633.

6 Y. W. Zhang, N. C. Seeman. Construction of a DNA-truncated octahedron, *J. Am. Chem. Soc.* **1994**, *116*, 1661–1669.

7 C. D. Mao, W. Q. Sun, N. C. Seeman. Assembly of Borromean rings from DNA, *Nature* **1997**, *386*, 137–138.

8 R. P. Goodman, R. M. Berry, A. J. Turberfield. The single-step synthesis of a DNA tetrahedron, *Chem. Commun.* **2004**, 1372–1373.

9 W. M. Shih, J. D. Quispe, G. F. Joyce. A 1.7-kilobase single-stranded DNA that folds into a nanoscale octahedron, *Nature* **2004**, *427*, 618–621.

10 M. Scheffler, A. Dorenbeck, S. Jordan, M. Wüstefeld, G. v. Kiedrowski. Self-assembly of trisoligonucleotidyls: the case for nano-acetylene and nano-cyclobutadiene, *Angew. Chem. Int. Ed. Engl.* **1999**, *38*, 3311–3315.

11 E. Winfree, F. R. Liu, L. A. Wenzler, N. C. Seeman. Design and self-assembly of two-dimensional DNA crystals, *Nature* **1998**, *394*, 539–544.

12 T. H. LaBean, H. Yan, J. Kopatsch, F. R. Liu, E. Winfree, J. H. Reif, N. C. Seeman. Construction, analysis, ligation, and self-assembly of DNA triple crossover complexes, *J. Am. Chem. Soc.* **2000**, *122*, 1848–1860.

13 D. Liu, S. H. Park, J. H. Reif, T. H. LaBean. DNA nanotubes self-assembled from triple-crossover tiles as templates for conductive nanowires, *Proc. Natl Acad. Sci. USA* **2004**, *101*, 717–722.

14 J. C. Mitchell, J. R. Harris, J. Malo, J. Bath, A. J. Turberfield. Self-assembly of chiral DNA nanotubes, *J. Am. Chem. Soc.* **2004**, *126*, 16342–16343.

15 P. W. K. Rothemund, A. Ekani-Nkodo, N. Papadakis, A. Kumar, D. K. Fygenson, E. Winfree. Design and characterization of programmable DNA nanotubes, *J. Am. Chem. Soc.* **2004**, *126*, 16344–16352.

16 H. Yan, S. H. Park, G. Finkelstein, J. H. Reif, T. H. LaBean. DNA-templated self-assembly of protein arrays and highly conductive nanowires, *Science* **2003**, *301*, 1882–1884.

17 J. Malo, J. C. Mitchell, C. Venien-Bryan, J. R. Harris, H. Wille, D. J. Sherratt, A. J. Turberfield. Engineering a 2D protein–DNA

crystal, *Angew. Chem. Int. Ed. Engl.* **2005**, *44*, 3057–3061.

18 Y. Liu, C. X. Lin, H. Y. Li, H. Yan. Protein nanoarrays – aptamer-directed self-assembly of protein arrays on a DNA nanostructure, *Angew. Chem. Int. Ed. Engl.* **2005**, *44*, 4333–4338.

19 C. D. Mao, W. Q. Sun, Z. Y. Shen, N. C. Seeman. A nanomechanical device based on the B–Z transition of DNA, *Nature* **1999**, *397*, 144–146.

20 W. M. Gelbart, R. F. Bruinsma, P. A. Pincus, V. A. Parsegian. DNA-inspired electrostatics, *Phys. Today* **2000**, *53*, 38–44.

21 C. M. Niemeyer, M. Adler, S. Lenhert, S. Gao, H. Fuchs, L. F. Chi. Nucleic acid supercoiling as a means for ionic switching of DNA–nanoparticle networks, *ChemBioChem* **2001**, *2*, 260–264.

22 K. Gehring, J. L. Leroy, M. Gueron. A tetrameric DNA-structure with protonated cytosine.cytosine base-pairs, *Nature* **1993**, *363*, 561–565.

23 D. S. Liu, S. Balasubramanian. A proton-fuelled DNA nanomachine, *Angew. Chem. Int. Ed. Engl.* **2003**, *42*, 5734–5736.

24 V. A. Bloomfield, D. M. Crothers, I. Tinoco, Jr. *Nucleic Acids*. University Science Books, Sausalito, CA, **2000**.

25 M. Brucale, G. Zuccheri, B. Samori. The dynamic properties of an intramolecular transition from DNA duplex to cytosine–thymine motif triplex, *Org. Biomol. Chem.* **2005**, *3*, 575–577.

26 Y. Chen, S. H. Lee, C. D. Mao. A DNA Nanomachine based on a duplex–triplex transition, *Angew. Chem. Int. Ed. Engl.* **2004**, *43*, 5335–5338.

27 X. P. Yang, A. V. Vologodskii, B. Liu, B. Kemper, N. C. Seeman. Torsional control of double-stranded DNA branch migration, *Biopolymers* **1998**, *45*, 69–83.

28 B. Yurke, A. J. Turberfield, A. P. Mills, F. C. Simmel, J. L. Neumann. A DNA-fuelled molecular machine made of DNA, *Nature* **2000**, *406*, 605–608.

29 F. C. Simmel, B. Yurke. Using DNA to construct and power a nanoactuator, *Phys. Rev. E* **2001**, *6304*, 041913.

30 F. C. Simmel, B. Yurke. A DNA-based molecular device switchable between three distinct mechanical states, *Appl. Phys. Lett.* **2002**, *80*, 883–885.

31 H. Yan, X. P. Zhang, Z. Y. Shen, N. C. Seeman. A robust DNA mechanical device controlled by hybridization topology, *Nature* **2002**, *415*, 62–65.

32 Z. Y. Shen, H. Yan, T. Wang, N. C. Seeman. Paranemic crossover DNA: A generalized Holliday structure with applications in nanotechnology, *J. Am. Chem. Soc.* **2004**, *126*, 1666–1674.

33 J. W. J. Li, W. H. Tan. A single DNA molecule nanomotor, *Nano Lett.* **2002**, *2*, 315–318.

34 P. Alberti, J. L. Mergny. DNA duplex–quadruplex exchange as the basis for a nanomolecular machine, *Proc. Natl Acad. Sci. USA* **2003**, *100*, 1569–1573.

35 J. Howard. *Mechanics of Motor Proteins and the Cytoskeleton*, 1st edn. Sinauer, Sunderland, MA, **2001**.

36 C. Bustamante, D. Keller, G. Oster. The physics of molecular motors, *Acc. Chem. Res.* **2001**, *34*, 412–420.

37 J. S. Shin, N. A. Pierce. A synthetic DNA walker for molecular transport, *J. Am. Chem. Soc.* **2004**, *126*, 10834–10835.

38 W. B. Sherman, N. C. Seeman. A precisely controlled DNA biped walking device, *Nano Lett.* **2004**, *4*, 1203–1207.

39 Y. Tian, C. D. Mao. Molecular gears: a pair of DNA circles continuously rolls against each other, *J. Am. Chem. Soc.* **2004**, *126*, 11410–11411.

40 P. Yin, H. Yan, D. X. G., A. J. Turberfield, J. H. Reif. A unidirectional DNA walker that moves autonomously along a track, *Angew. Chem. Int. Ed. Engl.* **2004**, *43*, 4906–4911.

41 J. Bath, S. J. Green, A. J. Turberfield. A free-running DNA motor powered by a nicking enzyme, *Angew. Chem. Int. Ed. Engl.* **2005**, *44*, 4358–4361.

42 Y. Tian, Y. He, Y. Chen, P. Yin, C. D. Mao. Molecular devices – A DNAzyme that walks processively and autonomously along a one-dimensional track, *Angew. Chem. Int. Ed. Engl.* **2005**, *44*, 4355–4358.

43 C. M. NIEMEYER, M. ADLER, R. WACKER. Immuno-PCR: high sensitivity detection of proteins by nucleic acid amplification, *Trends Biotechnol* **2005**, *23*, 208–216.

44 C. M. NIEMEYER, J. KOEHLER, C. WUERDEMANN. DNA-directed assembly of bienzymic complexes from *in vivo* biotinylated NAD(P)H:FMN oxidoreductase and luciferase, *Chembiochem* **2002**, *3*, 242–245.

45 W. Q. SHEN, M. F. BRUIST, S. D. GOODMAN, N. C. SEEMAN. A protein-driven DNA device that measures the excess binding energy of proteins that distort DNA, *Angew. Chem. Int. Ed. Engl.* **2004**, *43*, 4750–4752.

46 C. ALBRECHT, K. BLANK, M. LALIC-MULTHALER, S. HIRLER, T. MAI, I. GILBERT, S. SCHIFFMANN, T. BAYER, H. CLAUSEN-SCHAUMANN, H. E. GAUB. DNA: a programmable force sensor, *Science* **2003**, *301*, 367–370.

47 D. S. WILSON, J. W. SZOSTAK. *In vitro* selection of functional nucleic acids, *Annu. Rev. Biochem.* **1999**, *68*, 611–647.

48 W. U. DITTMER, A. REUTER, F. C. SIMMEL. A DNA-based machine that can cyclically bind and release thrombin, *Angew. Chem. Int. Ed. Engl.* **2004**, *43*, 3550–3553.

49 L. M. ADLEMAN. Molecular computation of solutions to combinatorial problems, *Science* **1994**, *266*, 1021–1024.

50 K. SAKAMOTO, H. GOUZU, K. KOMIYA, D. KIGA, S. YOKOYAMA, T. YOKOMORI, M. HAGIYA. Molecular computation by DNA hairpin formation, *Science* **2000**, *288*, 1223–1226.

51 D. FAULHAMMER, A. R. CUKRAS, R. J. LIPTON, L. F. LANDWEBER. Molecular computation: RNA solutions to chess problems, *Proc. Natl Acad. Sci. USA* **2000**, *97*, 1385–1389.

52 R. S. BRAICH, N. CHELYAPOV, C. JOHNSON, P. W. K. ROTHEMUND, L. ADLEMAN. Solution of a 20-variable 3-SAT problem on a DNA computer, *Science* **2002**, *296*, 499–502.

53 Y. BENENSON, T. PAZ-ELIZUR, R. ADAR, E. KEINAN, Z. LIVNEH, E. SHAPIRO. Programmable and autonomous computing machine made of biomolecules, *Nature* **2001**, *414*, 430–434.

54 Y. BENENSON, R. ADAR, T. PAZ-ELIZUR, Z. LIVNEH, E. SHAPIRO. DNA molecule provides a computing machine with both data and fuel, *Proc. Natl Acad. Sci. USA* **2003**, *100*, 2191–2196.

55 R. ADAR, Y. BENENSON, G. LINSHIZ, A. ROSNER, N. TISHBY, E. SHAPIRO. Stochastic computing with bio-molecular automata, *Proc. Natl Acad. Sci. USA* **2004**, *101*, 9960–9965.

56 M. SORENI, S. YOGEV, E. KOSSOY, Y. SHOHAM, E. KEINAN. Parallel bio-molecular computation on surfaces with advanced finite automata, *J. Am. Chem. Soc.* **2005**, *127*, 3935–3943.

57 Y. BENENSON, B. GIL, U. BEN-DOR, R. ADAR, E. SHAPIRO. An autonomous molecular computer for logical control of gene expression, *Nature* **2004**, *429*, 423–429.

58 M. N. STOJANOVIC, T. E. MITCHELL, D. STEFANOVIC. Deoxyribozyme-based logic gates, *J. Am. Chem. Soc.* **2002**, *124*, 3555–3561.

59 R. R. BREAKER, G. F. JOYCE. A DNA enzyme with Mg^{2+}-dependent RNA phosphoesterase activity, *Chem. Biol.* **1995**, *2*, 655–660.

60 S. W. SANTORO, G. F. JOYCE. A general purpose RNA-cleaving DNA enzyme, *Proc. Natl Acad. Sci. USA* **1997**, *94*, 4262–4266.

61 J. LI, W. C. ZHENG, A. H. KWON, Y. LU. *In vitro* selection and characterization of a highly efficient Zn(II)-dependent RNA-cleaving deoxyribozyme, *Nucleic Acids Res.* **2000**, *28*, 481–488.

62 M. N. STOJANOVIC, D. STEFANOVIC. Deoxyribozyme-based half-adder, *J. Am. Chem. Soc.* **2003**, *125*, 6673–6676.

63 M. N. STOJANOVIC, D. STEFANOVIC. A deoxyribozyme-based molecular automaton, *Nat. Biotech.* **2003**, *21*, 1069–1074.

64 D. M. KOLPASHCHIKOV, M. N. STOJANOVIC. Boolean control of aptamer binding states, *J. Am. Chem. Soc.* **2005**, *127*, 11348–11351.

65 L. P. FENG, S. H. PARK, J. H. REIF, H. YAN. A two-state DNA lattice switched by DNA nanoactuator, *Angew. Chem. Int. Ed. Engl.* **2003**, *42*, 4342–4346.

66 P. HAZARIKA, B. CEYHAN, C. M. NIEMEYER. Reversible switching of DNA-gold nanoparticle aggregation, *Angew. Chem. Int. Ed. Engl.* **2004**, *43*, 6469–6471.

67 J. J. STORHOFF, R. ELGHANIAN, R. C. MUCIC, C. A. MIRKIN, R. L. LETSINGER. One-pot colorimetric differentiation of polynucleotides with single base imperfections using gold nanoparticle probes, *J. Am. Chem. Soc.* **1998**, *120*, 1959–1964.

68 J. W. LIU, Y. LU. Adenosine-dependent assembly of aptazyme-functionalized gold nanoparticles and its application as a colorimetric biosensor, *Anal. Chem.* **2004**, *76*, 1627–1632.

69 A. J. TURBERFIELD, J. C. MITCHELL, B. YURKE, A. P. MILLS, M. I. BLAKEY, F. C. SIMMEL. DNA fuel for free-running nanomachines, *Phys. Rev. Lett.* **2003**, *90*.

70 Y. CHEN, M. S. WANG, C. D. MAO. An autonomous DNA nanomotor powered by a DNA enzyme, *Angew. Chem. Int. Ed. Engl.* **2004**, *43*, 3554–3557.

71 Y. CHEN, C. D. MAO. Putting a brake on an autonomous DNA nanomotor, *J. Am. Chem. Soc.* **2004**, *126*, 8626–8627.

72 T. LIEDL, F. C. SIMMEL. Switching the conformation of a DNA molecule with a chemical oscillator, *Nano Lett.* **2005**, *5*, 1894–1898.

73 A. ARKIN, J. ROSS. Computational functions in biochemical reaction networks, *Biophys. J.* **1994**, *67*, 560–578.

74 D. M. WOLF, A. P. ARKIN. Motifs, modules and games in bacteria, *Curr. Opin. Microbiol.* **2003**, *6*, 125–134.

75 T. S. GARDNER, C. R. CANTOR, J. J. COLLINS. Construction of a genetic toggle switch in *Escherichia coli*, *Nature* **2000**, *403*, 339–342.

76 M. B. ELOWITZ, S. LEIBLER. A synthetic oscillatory network of transcriptional regulators, *Nature* **2000**, *403*, 335–338.

77 R. WEISS, T. F. KNIGHT, JR. Engineering communications for microbial robotics, in: *DNA Computing: 6th International Workshop on DNA-Based Computers, Lecture Notes in Computer Science 2054*, A. E. CONDON, G. ROZENBERG (Eds.). Springer, Berlin, **2000**, pp. 1–16.

78 V. NOIREAUX, R. BAR-ZIV, A. LIBCHABER. Principles of cell-free genetic circuit assembly, *Proc. Natl Acad. Sci. USA* **2003**, *100*, 12672–12677.

79 J. HASTY, D. MCMILLEN, J. J. COLLINS. Engineered gene circuits, *Nature* **2002**, *420*, 224–230.

80 S. A. BENNER, A. M. SISMOUR. Synthetic biology, *Nat. Rev. Genet.* **2005**, *6*, 533–543.

81 W. U. DITTMER, F. C. SIMMEL. Transcriptional control of DNA-based nanomachines, *Nano Lett.* **2004**, *4*, 689–691.

82 W. U. DITTMER, S. KEMPTER, J. O. RÄDLER, F. C. SIMMEL. Using gene regulation to program DNA-based molecular devices, *Small* **2005**, *1*, 709–712.

7
Towards the Realization of Nanobiosensors Based on G-protein-coupled Receptors

Cecilia Pennetta, Vladimir Akimov, Eleonora Alfinito, Lino Reggiani,
Tatiana Gorojankina, Jasmina Minic, Edith Pajot-Augy,
Marie-Annick Persuy, Roland Salesse, Ignacio Casuso,
Abdelhamid Errachid, Gabriel Gomila, Oscar Ruiz, Josep Samitier,
Yanxia Hou, Nicole Jaffrezic, Giorgio Ferrari, Laura Fumagalli, and
Marco Sampietro

7.1
Introduction

Among membrane proteins, G-protein-coupled receptors (GPCRs) are of special importance because they form one of the widest groups of receptor proteins [1–6]. As they can be activated by a large variety of extracellular signals, such as light, odorant molecules, hormones, peptides, lipids, neurotransmitters and nucleotides, GPCRs mediate the sense of vision, smell, taste and pain [1–8], and are involved in an extraordinary number of physiological processes [1–8]. For the same reasons, GPCRs are also implicated in a number of pathologies and they constitute one of the most important classes of pharmacological targets [1–6]. Accordingly, many efforts in the field of nanobiotechnology are devoted to the realization of nanobiosensors based on a single or a few GPCRs and aimed at detecting the specific ligands [9–18]. In the cell, the detection and transduction process begins with the conformational change of the GPCRs associated with the capture of the specific ligand [1, 7]. This conformational transition then activates the so-called G-proteins, giving rise to a complex sequence of biological mechanisms which ends in the production of an electric pulse by neurons (action potential) [1, 7].

It must be noted that different ligands may induce and stabilize distinct conformational states among the various structures available, and thus promote different G protein activation and receptor desensitization/internalization [19–22]. Moreover, GPCRs can interact with other proteins through their C-terminal domains. They can also give rise to homodimers and heterodimers with other membrane-bound proteins involved in their function and pharmacology [23]. Of course, understanding these mechanisms is crucial for new drug design. However, the drug-discovery process and the research of novel GPCR-based therapeutics require a large body of data to allow the identification of new receptor ligands. Therefore,

Nanotechnologies for the Life Sciences Vol. 4
Nanodevices for the Life Sciences. Edited by Challa S. S. R. Kumar
Copyright © 2006 WILEY-VCH Verlag GmbH & Co. KGaA, Weinheim
ISBN: 3-527-31384-2

there is an increasing need to develop genetic and chemical high-throughput screening (HTS) methods for an efficient sorting of compounds with pharmaceutical potential [3, 24–26]. In this context, the production of nanobiosensors based on GPCRs can provide a very useful tool.

In nanobiosensors, the detection process bypasses the complicated sequence of biological events which follows G protein activation; rather, the goal is to achieve sensitive and reliable monitoring of the conformational change by direct analysis of the optical or the electrical response of the nanodevice itself.

The first difficulty to be faced in developing such chip-based sensors is to prepare and immobilize the receptors in such a manner as to preserve their function, i.e. their ability of undergoing the appropriate conformational transition. The second difficulty consists in achieving sensitive and reliable monitoring of the conformational change, by analysis of the optical or electrical response of the device. Of course, this analysis requires the development of suitable tools, both experimental and theoretical.

This development represents a true challenge. In particular, here we discuss the possibility of studying the detection process of a single receptor device by direct measurement of its electrical properties [16, 17]. Actually, most of efforts in the literature focus on optical or atomic force microscopy (AFM) techniques to monitor the ligand detection process (see Section 7.3). Only a few attempts [16–18] based on electrical measurements have been made until now, as a consequence of the difficulty arising from the very low values expected for currents, which requires very advanced amplification techniques. Thus, the method adopted here and illustrated in this chapter represents a very innovative approach. Concerning this point, we further underline the novelty of the theoretical model that, uniquely in the literature, uses an impedance network for the description of the electrical properties of a single protein device.

It is important to note that all the members of the GPCR superfamily [1] share a common molecular topology consisting of seven transmembrane helical domains (α-helices), connected by three intracellular and three extracellular loops (Fig. 7.1). This feature of the GPCR superfamily plays an important role in the development of theoretical and computational models of these receptors [1]. Actually, while the primary structure (amino acid sequence) is known for many GPCRs [27], the detailed atomic positions are available only in the case of rhodopsin, the rod cell photoreceptor that mediates light vision, where they can be measured by X-rays diffraction experiments performed in the protein crystalline state [27–31]. Indeed, the natural abundance and availability of rhodopsin, in contrast to other GPCRs, typically expressed at a very low level in the cell, results in the fact that only rhodopsin crystals have been obtained and studied at high resolution [27–33]. For all these reasons, rhodopsin is the best-studied GPCR, and it is often selected as a model protein for biological and biophysical studies of this kind of receptor [34–36]. Rhodopsin also serves as a pilot receptor for establishing biotechnological platforms for chip-based screening technologies, to characterize the function and interactions of GPCRs [37].

A three-dimensional structure for the activated state of rhodopsin, i.e. metarhodopsin II [19, 31, 37], was obtained at medium resolution by combining elements

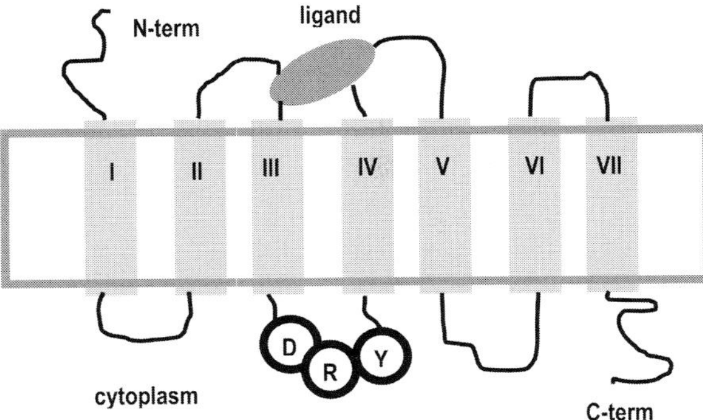

Figure 7.1. Schematic drawing of a GPCR. The big dark-grey rectangular box represents the cellular membrane which separates the extra-cellular region (above) from the cytoplasmic region (below). The seven grey vertical rectangles show the seven transmembrane α-helical domains of the receptor, while the loops and the two termini (N- and C-terminus) are shown by black lines. The three black circles represent the three subunits of the G-protein and, finally, the grey ellipse on the top shows a ligand captured by the receptor.

of secondary structure with experimental, long-range distance constraints inducing rigid body movements of secondary structures relative one to another. As a result, the metarhodopsin II state of the receptor exhibits a spatial organization different from the ground state rhodopsin [31].

In this chapter, we first briefly review some of the different techniques that have been previously developed to prepare and immobilize the receptors on the substrate. Then, we consider the detection process as characterized by the response of the device. In particular, we discuss the electrochemical impedance spectroscopy (EIS) technique, which is emerging as a very effective technique for the detection of biosensing events at the electrodes [38–42]. The largest part of this chapter is devoted to illustrating a recent theoretical model [17, 43–45] which studies the current response to an applied AC voltage of a nanodevice realized by a single GPCR embedded in its membrane and in contact with two functionalized metallic nano-electrodes.

This model, based on a coarse-grained approach [17, 46, 47], describes the protein as a network [48] of elementary impedances. The model starts with the construction of a time-independent (static) network, corresponding to an ideally frozen protein, in which all the atomic positions are fixed at the equilibrium values [17]. The nodes of this static network correspond to the positions of the α carbons (C_α) atoms of the rhodopsin amino acids, in the ground or in the activated state (meta-rhodopsin II), as taken from the protein data bank [27]. Then, an elementary impedance is associated with each link between a pair of amino acids, established according to a length cut-off criterion [46]. The elementary impedances, which mimic the electrical interactions among the protein amino acids, are taken dependent on the amino acid distance and on other parameters which account for the different physical and chemical properties of the amino acids [17]. The solution of the equivalent circuit determines the global impedance of the network itself. The results of this static-network model, applied to the case of rhodopsin and OR-I7 (rat olfactory receptor) [49], predict a detectable change of the global impedance associated with the conformational change of the receptor due to the sensing action [17, 45]. Afterwards, the frozen protein approximation is relaxed and the thermal fluctuations of the atomic positions are included in the model, first by adopting a single-force-constant approximation [43, 50, 51], as the simplest level of modeling, and then a two-force-constant approximation [44]. This last update in the model is introduced to describe the different flexibility of the atomic bonds within the α-helices and within the loops [1]. The protein thermal fluctuations result in an impedance noise which is calculated as a function of the temperature. The implications of this impedance noise on the ligand detection process are then discussed.

7.2
Preparation and Immobilization of GPCRs on Functionalized Surfaces

The first requirement to develop receptor-based nanobiosensors is the immobilization of receptors in such a manner as to preserve their function. Since GPCRs are composed of seven transmembrane α-helices [1], they are extremely hydrophobic and require a lipidic or detergent environment to keep their native conformation and function. Generally, membrane receptors are expressed in heterologous cells, solubilized, and then purified in an appropriate detergent before being reconstituted in proteoliposomes and immobilized on a sensor surface [52–54]. Although many such successful procedures have been reported, the method requires substantial effort for purification of GPCRs prior to analysis. This cannot be considered for the majority of GPCRs because of their very low expression level, even in recombinant systems. Furthermore, the receptor function can be influenced by its lipidic environment as shown for rhodopsin [55] and μ-opioid receptor [56]. Therefore, enormous care must be taken to avoid modification of the activity of the receptor during its solubilization and reconstruction.

The possibility of building bilayers harboring previously purified GPCRs on sensor surfaces has been demonstrated for rhodopsin [54, 57], β_2-adrenergic receptor [58, 59] and μ-opioid receptor [60, 61]. On-surface reconstitution of rhodopsin in lipidic membranes has also been achieved from mixed detergent/lipid micelles, with a high density of functional receptors [15]. However, one of the most promising strategies for the preparation of GPCRs seems to produce lipid vesicles by breaking membranes where GPCRs have been expressed [18, 42]. Indeed, this procedure bypasses the difficulties implied in the purification of GPCRs and in their subsequent insertion into the reconstituted membranes.

For what concerns the immobilization of GPCRs, the self-assembled multilayer technique appears to be the most suitable and effective for the construction of well-ordered and ultra-thin organic films, since it allows the required control at a molecular level. Actually, starting from the pioneering works of Nuzzo and Allara [62], the field of self-assembled monolayers (SAMs) has experienced an explosive growth [63]. The simplicity in the production of SAMs, their adaptability and the possibility of controlling the orientation of biomolecules on the surface ascribe to the SAM technique a central role in the construction of artificial biomolecular recognition surfaces, particularly in the development of biosensors [64–66]. The construction of self-assembled multilayers from biological components has been investigated intensively due to its potential application for biosensors [13, 14, 67, 68]. In particular, it has been found that the avidin–biotin system works very well as a bridge to anchor a bioreceptor, since the biotinylation of a biomolecule does not affect its biological activity. Furthermore, the noncovalent complex between avidin and biotin is characterized by a very high affinity constant of 10^{15} mol^{-1} L. Once formed, the bond is stable even if the pH of the solution is changed and it can easily resist multiple washings [69].

In recent studies, a mixed SAM formed by 16-mercaptohexadecanoic acid (MHDA) and biotinyl-PhosphoEthanolamine, inserted and bound to MHDA, has been produced on gold electrodes [18, 42]. Neutravidin was then used to anchor biotin-labeled, receptor-specific antibodies. Such a multilayer system using biotin/ avidin pairs acting as binding agents is of relevance for biosensor research, to immobilize a membrane receptor. Immobilization of the prototypical GPCR, rhodopsin, in its native membranes was thus achieved on functionalized surfaces [18, 42], which represents the first prerequisite for elaborating a GPCR-based biosensor. The rhodopsin membrane fraction preparations were obtained by simple sonication of the natural rod outer segments membranes. Negative-staining electron microscopy of the sample showed that the fragments are circularized into microsomes, although unclosed membrane fragments are still present and the suspension is mostly constituted of vesicles with diameters of 40–100 nm. This procedure provides samples of rather uniform size, which fit the geometrical requirements of the nanobiosensor supports. Moreover, as the receptor remains at all times in its native environment, this procedure offers the advantage of avoiding the risks of altering or loosing receptor activity as a consequence of the purification and reconstitution of the GPCR in artificial membranes or liposomes. Furthermore, this specific immobilization methodology can be very useful to develop molecular arrays of other GPCRs that, in contrast to rhodopsin, are usually present in a low amount in the membrane fraction. Therefore, this form of *in situ* purification and immobilization on biostructured solid supports can overcome the difficulties usually encountered in studying most of the GPCRs and that arise from a low expression level.

7.3
Signal Techniques

Different techniques have been developed in recent years. A particularly important role in this field is played by optical techniques which, in most cases, make use of visible fluorescent proteins (VFPs) [70] and of confocal microscopy analysis, as in fluorescence resonance energy transfer (FRET) and fluorescence lifetime imaging (FLIM) techniques [71–73].

Other important tools for the investigation of this kind of system are offered by surface plasmon resonance (SPR) [18, 54, 58] and AFM. This last technique, in particular, represents a very powerful tool for structural biology studies since it gives access to the molecular architecture [74, 75], and it can also be used to follow and characterize receptor immobilization on the selected biostructured surfaces.

EIS has become one of the most effective electrochemical techniques for the characterization of biomaterial-functionalized electrodes and of biocatalytic transformations at electrode surfaces; specifically, for the detection of biosensing events at electrodes [38–42]. Compared to other electrochemical techniques, one of the great advantages of EIS is the small amplitude of the perturbation from the steady state, which makes it possible to treat the response theoretically by linearized or otherwise simplified current–potential characteristics [38–42]. Therefore, EIS has been performed extensively to characterize the fabrication of biosensors and to monitor biomolecular recognition [12, 38–41, 76]. Very recently, electrochemical impedance measurements have demonstrated the sensitivity and selectivity of self-assembled multilayer systems for the specific grafting of the rhodopsin membrane fraction [18, 42].

7.4
Theoretical Approach

The theoretical study of the electrical response to an external voltage of a GPCR-based device is a hard task. Thus, we limit ourselves to consider a "simple" two-terminal device, made by a single receptor embedded in a small portion of its native membrane and inserted between two ohmic electrical contacts (functionalized metallic layers) through which an AC voltage is applied. The device is placed within a physiological buffer solution, necessary to keep the protein in the correct conformation. This device is sketched in Fig. 7.2, where, without loss of generality,

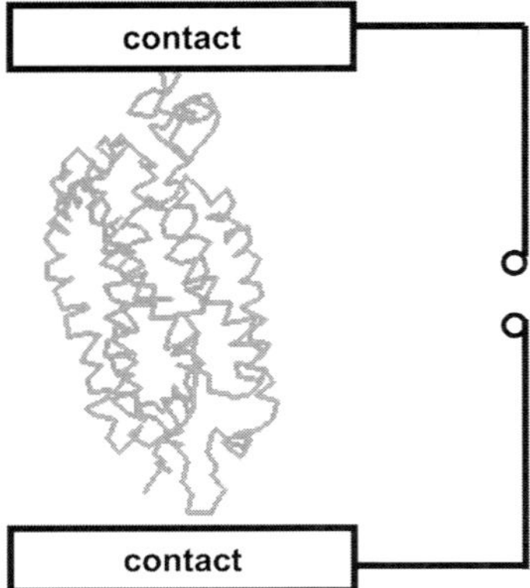

Figure 7.2. Representation of a single receptor device: the receptor is contacted between two functionalized electrodes through which an external AC voltage is applied.

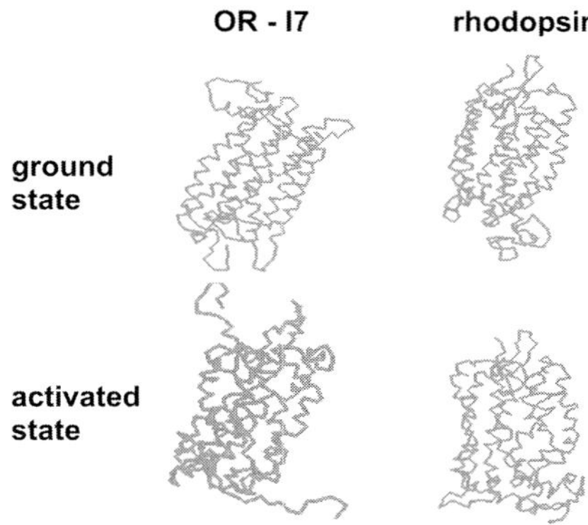

Figure 7.3. Schematic representation of the ground and activated states of OR-I7 and rhodopsin.

a vertical configuration of the electrodes has been assumed (other configurations can also be considered).

As discussed before, the main mechanism by which GPCRs perform their biological functions is associated with the conformational transition undergone as consequence of the interaction and capture of the ligands [1, 9]. Thus, the main task of a theoretical study of the electrical properties of a GPCR-based device consists in estimating the magnitude of the difference of the electrical response to the external voltage in the two receptor conformations. The GPCR conformational transitions have been intensively studied [27, 31, 32, 46, 77–82], especially by experiments [27, 31, 32, 77–80], and it has been realized that the activation of the receptor involves a release of constraints in the transmembrane helix bundle, resulting in the opening of a cleft at its cytoplasmic end [19, 37], and in a net volume change of the protein [79]. In particular, a lot of information is available concerning the most stable light-activated state of rhodopsin, known as metarhodopsin II [27, 31, 32], where the available information also includes the coordinates of the atomic positions. To illustrate qualitatively the conformational change of GPCRs, Fig. 7.3 shows a schematic representation of the ground and activated states of rhodopsin and of the rat OR-I7. Apart from these large-scale conformational transitions, it is well known that proteins are not rigid, but instead they sample a variety of conformations in the neighborhood of their native conformation [82]. These fluctuations within different conformations, all near to the equilibrium one, are called equilibrium fluctuations [82].

Both the complicated relationship between structure and functions of the proteins and the details of the molecular motion in the folded state can, in principle, be studied by molecular dynamics (MD) simulations and normal-mode analysis, by using classical MD or *"ab initio"* quantum MD, or even hybrid schemes [81, 83–85]. These last mix a quantum treatment of a limited number of atoms and a classical treatment of the remaining atoms, thus allowing the study of large biological molecules [81, 83–85]. However, the use of these atomic approaches becomes computationally very heavy and inefficient with increasing protein size [46].

Recently, alternative methods have been proposed, based on coarse-grained models of proteins and simplified force fields [50, 51]. Successively, many other studies have shown the success and the effectiveness of these kinds of models in describing protein dynamics, particularly for the case of large proteins composed by more than several thousands of amino acids [46, 47].

Therefore, in developing a theoretical model of the electrical properties of a GPCR-based device we have also followed a coarse-grained approach, by formulating it in terms of an electric network of impedances [17, 43–45] instead of the usual elastic network of springs, considered in the above cited works [46, 47, 50, 51]. In the following we illustrate in detail the procedure and we discuss the results for the case of rhodopsin. Moreover, we also show results concerning OR-I7 [49]. However, it is important to note that, in addition its computational simplicity, an important advantage of this coarse-grained network approach [17, 46, 47, 50, 51] is given by the fact that, by focusing mainly on protein topology, it represents an appropriate tool for providing predictions for the other GPCRs, for which detailed information on the atomic positions is missing.

7.5
The Impedance Network Model

We start by considering an ideally frozen bovine rhodopsin molecule, in which no fluctuation occurs and all the atoms occupy the equilibrium positions [17]. We describe this protein as a static network made of elementary impedances (in this context, a static network means a network made of time-independent elements). The network is built in the following manner [17].

First, we have to choose the nodes of the networks, i.e. we have to choose a reference position within each amino acid (residue). By following a choice largely adopted in the literature [46, 47, 51], we assume the C_α atoms present in any amino acid as nodes of the network. Accordingly, we have taken the atomic coordinates of bovine rhodopsin from the Protein Data Bank (PDB) [27], where several sets of data from independent experiments are present in the standard PDB file format. We have used PDB with IDs 1F88 [28] and 1JFP [32] for the rhodopsin ground state, and the data set II (PDB ID 1LN6) [31] (engineered data) for meta-rhodopsin II. Then, we have extracted the spatial coordinates of the C_α atoms of each amino acid (348 for bovine rhodopsin) from the PDB file.

We assume that the amino acids interact electrically between each other and that charge transfers between neighboring residues [78, 86] and/or changes of their electronic polarization [87] affect these interactions. Accordingly, a link is drawn between any pair of nodes, i and j separated by a distance $l_{i,j}$ less than a given cut-off value, $d = 2R_a$, where R_a is an electrical interaction radius (Fig. 7.4). Moreover, we introduce two extra nodes (contact nodes) which mimic the electrodes. These contact nodes are linked to a given set of amino acids, depending on the particular geometry of the contacts in the real device (each electrode is linked at least to one amino acid). A representation of this interaction network for the case

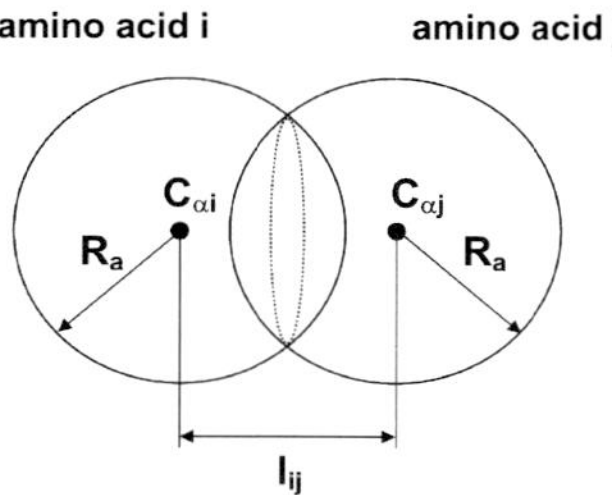

Figure 7.4. A link is drawn between the nodes *i* and *j*, which stay at the positions of the α carbon atoms $C_{\alpha i}$ and $C_{\alpha j}$ of the *i*th and *j*th amino acid, when their relative distance is less than twice the interaction radius R_a.

of a hypothetical protein made of 18 residues is shown in Fig. 7.5. We note that the solution and the membrane are not directly taken into account at this stage of the model (however, as the main effect of the membrane is to keep the protein in the folded state, this effect is implicitly accounted for by taking the coordinates of the C_α corresponding to a given folded conformation of the protein).

The next step of the model consists of associating an elementary impedance $Z_{i,j}$ with the link between the nodes *i* and *j* [17]. We take this elementary impedance as the impedance of a RC parallel circuit (the most usual passive AC circuit) and we denote as $R_{i,j}$ and $C_{i,j}$, respectively, the resistance and the capacitance of the link between the nodes *i* and *j*. Different expressions can be adopted for $Z_{i,j}$. Here, we discuss three possibilities, corresponding to an increasing level of complexity. The first possibility, model (i), is the simplest one: all the impedances are taken to be equal: $Z_{i,j} = Z_0$. The second possibility, model (ii), consists of assuming that $R_{i,j}$ and $C_{i,j}$ are, respectively, the resistance of a simple ohmic resistor and the capacitance of a planar homogeneous capacitor. Thus, $R_{i,j} \propto l_{i,j}$ and $C_{i,j} \propto l_{i,j}$. Consequently, $Z_{i,j}$ takes the expression [17]:

$$Z_{i,j} = l_{i,j} \frac{\rho}{A(1 + i\rho\omega\varepsilon\varepsilon_0)} \tag{1}$$

Figure 7.5. Interaction network associated with a hypothetical protein made of 18 residues: the full circles show the nodes positioned at the α carbon atom of each amino acid and the lines represent the links arising from electrical interactions between a pair of amino acids with a relative distance shorter than a cut-off value. The open circles represent two extra nodes associated with the electrodes.

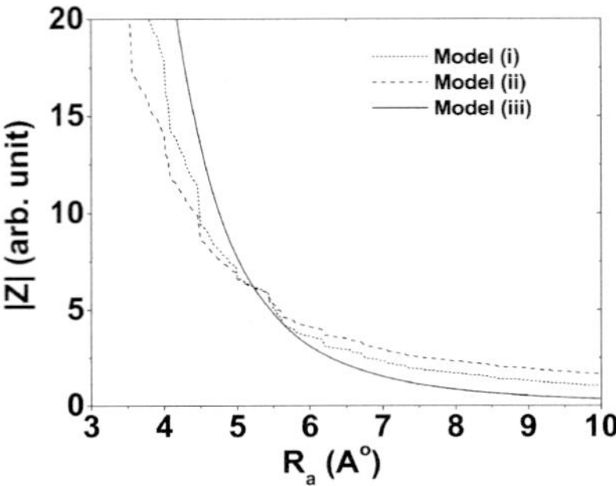

Figure 7.6. Modulus of the network impedance as a function of the interaction radius: the dotted curve is obtained from model (i), the dashed curve from model (ii) and the solid curve from model (iii). The impedance is expressed in arbitrary units and the interaction radius is in Ångstroms.

where A is the cross-sectional area of the capacitor and of the resistor, ρ is the resistivity, ε the relative dielectric constant, ε_0 is the vacuum dielectric permittivity, ω is the angular frequency of the external AC voltage. The third choice, model (iii), consists of taking the cross-sectional area of the resistor and of the capacitor equal to the area of the cross-section defined by the overlap of the two spheres pertaining to the given amino acids, as shown in Fig. 7.4. Thus:

$$A = \pi \left[R_{\mathrm{a}}^2 - \frac{l_{i,j}^2}{4} \right] \tag{2}$$

To compare these three models, we plot in Fig. 7.6 the modulus of the total network impedance, $|Z|$, calculated by using models (i), (ii) and (iii) as a function of the interaction radius R_{a}. The total network impedance Z is calculated by solving Kirchhoff's node equations. We note that in the case of an irregular network, with complex topology [48], the solution of the circuit by node equations is particularly convenient with respect to the use of loop equations. The systematic decrease of $|Z|$ shown in Fig. 7.6 at increasing values of R_{a} reflects the increasing importance of parallel connections with respect to series connections. One can see that the curves of the first two models show a step-like behavior related to the sharp discontinuity in the value of $|Z|$ when R_{a} becomes equal to $l_{i,j}$. However, the curve obtained by using model (iii) shows a continuous behavior. Furthermore, model (iii) appears to be more sensitive to a variation of the number of links in the networks.

Of course, an improvement in the description of the protein is expected if the physical and chemical properties of the different amino acids are accounted for in the expression of $Z_{i,j}$. At a simplest level, this can be done by considering ε and/or ρ as dependent on the indices i and j. For this purpose, we have taken [17] the following expression $\varepsilon_{i,j} = 1 + g(\bar{a} - 1)$ for the dielectric constant of the capacitor associated with the pair i and j of amino acids, where $\bar{a} \equiv (a_i + a_j)/2$ is the average intrinsic polarizability of the pair, and α_i and α_j are the intrinsic polarizabilities of the ith and jth amino acid, respectively. Actually, by describing a protein in solution as a set of polarizable dipoles embedded in a dielectric medium of solvent molecules, Song [87] has calculated a set of values for the intrinsic polarizabilities of the 20 amino acids, portable for all the proteins in nature. Finally, we have chosen the value of g in the expression of $\varepsilon_{i,j}$ in such a manner as to obtain the values of $\varepsilon_{i,j}$ distributed between 1 and 80 (vacuum and water) proportionally to $\bar{a}$. Thus, with the last update, model (iii) provides for $Z_{i,j}$ the expression [17]:

$$Z_{i,j} = \frac{l_{i,j}}{\pi(R_a^2 - l_{i,j}^2/4)} \cdot \frac{1}{(\rho^{-1} + i\omega\varepsilon_{i,j}\varepsilon_0)} \tag{3}$$

In the following we will show and discuss results obtained by using this expression of $Z_{i,j}$.

An important feature characterizing the topology of a network is represented by the degree distribution, i.e. the distribution function of the node connectivity [48]. Figure 7.7 shows the degree distribution of the interaction networks of rhodopsin and OR-I7 obtained for two values of the interaction radius, $R_a = 4$ and 5 Å. The degree distribution is found to be only roughly described by a Poissonian curve. Indeed, secondary peaks are present for both rhodopsin and OR-I7. This behavior of the degree distribution characterizes a pseudo-random network [48]. We note that for the values of R_a considered in Fig. 7.7 the width of the distribution increases by increasing the interaction radius. The contrary will occur at the highest values of R_a. In particular, above a given maximum value of the radius, $R_{a,\max}$, of the order of the size of the receptor, each node will be connected with all the other nodes and the degree (number of links) of all the nodes will be become $(N_{am} - 1)$,

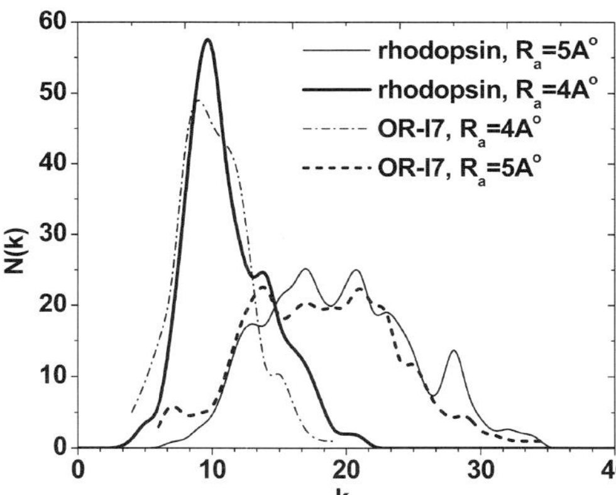

Figure 7.7. Degree distribution (distribution function of the node connectivities) of the interaction network of rhodopsin and OR-I7. The thin and thick solid curves are obtained in the case of rhodopsin by taking $R_a = 4$ and 5 Å, respectively; the dot-dashed and the dashed curves are obtained in the case of OR-I7 by taking $R_a = 4$ and 5 Å.

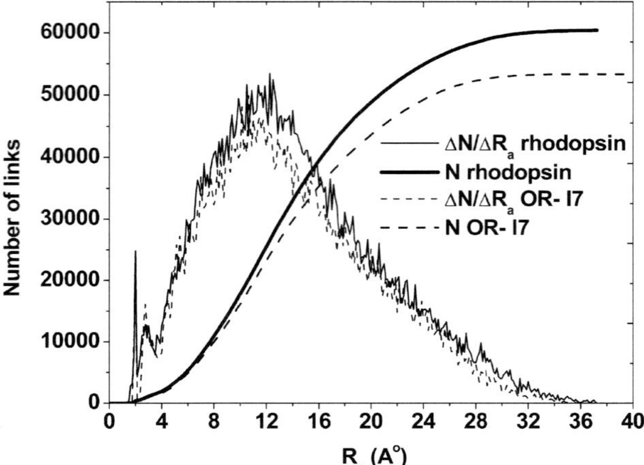

Figure 7.8. Total number of links in the network, N, and relative increment as a function of the interaction radius. The thick solid and the grey dashed curves show N for the interaction network corresponding to the basic state of rhodopsin and OR-I7. The thin solid and the grey short-dashed curves show the relative increment of N. The interaction radius is expressed in Ångstroms.

where N_{am} is the number of amino acids. Consequently, for $R_a \geq R_{a,max}$ the degree distribution will become a δ function independently of R_a.

Figure 7.8 reports the total number of links, N, existing in the network, as a function of the interaction radius. The data concerning the ground state of rhodopsin and OR-I7 are shown by the thick solid and the grey dashed curves, respectively. We can see that for any value of R_a, the number of links in the interaction network of rhodopsin is higher than that existing in the case of OR-I7. Moreover, the difference in the link number is rather significant (about 10%) for values of $R_a \geq 8\,\text{Å}$. Figure 7.8 also displays as a function of R_a the behavior of the relative increment of the number of links, $\Delta N/\Delta R_a$, for the ground state of rhodopsin (black solid curve) and OR-I7 (grey short-dashed curve). Thus, from Fig. 7.8 we can conclude that when $R_a \sim 8\text{–}18\,\text{Å}$ the connectivity of the network is very sensitive to the value of the interaction radius. On the other hand, by comparing the behavior versus R_a of N and $\Delta N/\Delta R_a$ in the ground and in the activated state of rhodopsin, it has been found [17] that the range of values $R_a \sim 8\text{–}18\,\text{Å}$ corresponds to the maximum sensitivity of the network to conformational changes.

This statement is also confirmed by the analysis of the dependence of the total network impedance, Z, on the interaction radius in the case of rhodopsin and metarhodopsin, as shown in Fig. 7.9. Precisely, in Fig. 7.9 we report the difference of the impedance modulus, $|Z|$, calculated in the case of rhodopsin and metarhodopsin. This difference has been normalized to the impedance modulus of the rhodopsin network. We can see that when $R_a \sim 8\text{–}18\,\text{Å}$ the total network impedance depends strongly on the conformation of the protein. It must be noted that this

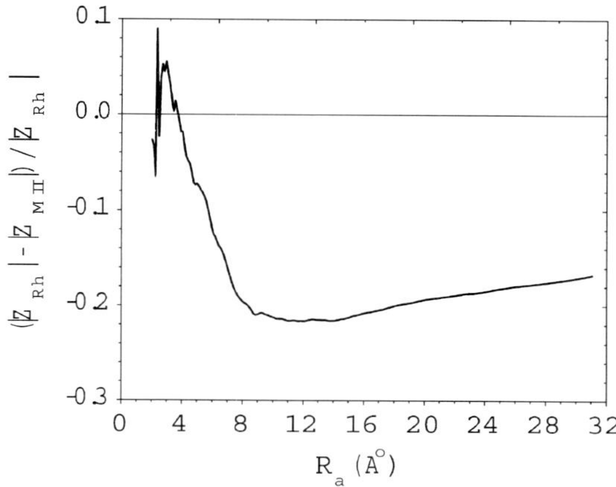

Figure 7.9. Relative difference of the values of the impedance modulus in the ground state (Rh) and in the activated state (MII) of rhodopsin as a function of the interaction radius (R_a expressed in Ångstroms).

sensitivity of the total network impedance to the conformational changes arises not only from the above discussed sensitivity of N to the protein conformation, but also from the fact that the elementary impedances $Z_{i,j}$ are assumed to be functions of the distances $l_{i,j}$.

To provide a prediction of the expected electrical response of the GPCR-based nanodevice and to test our model in a direct comparison with the results of EIS measurements [38–42], we have calculated the network impedance as a function of the frequency of the applied AC voltage. Figure 7.10 shows the Nyquist plot of the network impedance calculated in the case of rhodopsin. Precisely, Fig. 7.10 displays the opposite of the imaginary part of the network impedance versus the real part, where both these quantities are calculated in the frequency range $0 \doteq 1$ kHz. The three curves reported in Fig. 7.10 are obtained for $R_a = 2\,\text{Å}$ (short-dashed curve), $R_a = 5\,\text{Å}$ (long-dashed curve) and $R_a = 12.5\,\text{Å}$ (solid curve). In all cases, we have taken $\rho = 10^9\ \Omega\text{m}$ while the amplitude of the applied voltage, V_0, is $V_0 = 1$ V. Moreover, the real and the imaginary parts of the impedance have been normalized to the static value of the real part, $\text{Re}[Z(\omega = 0)]$, which takes the values $\text{Re}[Z(\omega = 0)] = 302$ GΩ, 11.2 GΩ and 1.67 MΩ, respectively, for $R_a = 2, 5$ and $12.5\,\text{Å}$. As a general trend, when $R_a \geq 5\,\text{Å}$, the shape of the Nyquist plot is indistinguishable from that corresponding to a single RC parallel circuit (semicircle). In contrast, when $R_a < 5\,\text{Å}$, the Nyquist plot deviates from this behavior. In particular, when $R_a = 2\,\text{Å}$, the degree of most of the nodes of the network is 2 (Fig. 7.7) and the series combination of elementary impedances $Z_{i,j}$ becomes dominant in the network structure. In other terms, changes in the shape of the Nyquist plot are only detected near to the sequential limit. Of course, the value of $\text{Re}[Z(\omega = 0)]$ depends on both R_a and ρ. Therefore, impedance spectroscopy measurements

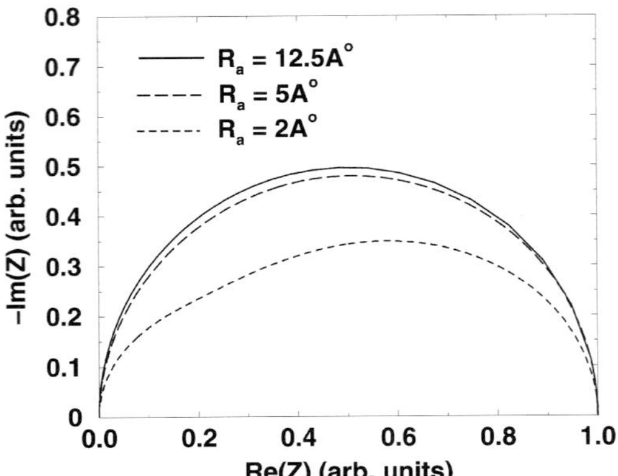

Figure 7.10. Nyquist plot of the network impedance: the short-dashed curve corresponds to an interaction radius equal to 2 Å, the long-dashed curve to a radius of 5 Å and the solid curve to a radius of 12.5 Å. The real and the imaginary parts of the impedance are normalized to the static value of the real part, which is 302 GΩ, 11.2 GΩ, and 1.67 MΩ, respectively, when R_a is 2, 5 and 12.5 Å. All curves correspond to the basic state of rhodopsin and are obtained by taking a resistivity of 1 GΩm.

could be a method which allows the identification of the values of the parameters to be used in the modelization of the receptor.

Figure 7.11 displays a comparison between the Nyquist plot corresponding to the rhodopsin interaction network and that corresponding to the activated state metarhodopsin II. Precisely, the solid curve is obtained for rhodopsin while the

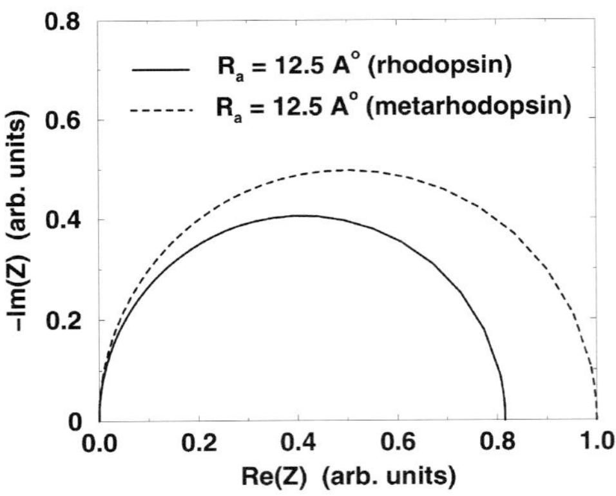

Figure 7.11. Nyquist plot of the network impedance: the solid curve is obtained for rhodopsin by taking the interaction radius equal to 12.5 Å and a resistivity of 1 GΩm, the dot-dashed curve is obtained for metarhodopsin by taking the same values of the parameters. The real and the imaginary parts of the impedance are normalized to the static value of the real part of the network impedance in the metarhodopsin state which takes the value of 2.04 MΩ.

dot-dashed curve for metarhodopsin II. In both cases we have taken $R_a = 12.5\,\text{Å}$, $\rho = 10^9\,\Omega\text{m}$ and $V_0 = 1\,\text{V}$. The real and the imaginary parts of the impedance are normalized to the static value of the real part of the impedance in the metarhodopsin state: $\text{Re}[Z(\omega = 0)] = 2.04\,\text{M}\Omega$. Figure 7.11 shows that the conformational change of the receptor due to the detection of a photon implies a significant variation in the Nyquist curve, in principle detectable by impedance spectroscopy measurements. This result is of particular relevance also in view of the application of the model to other GPCRs.

7.6
Equilibrium Fluctuations

The above results have been obtained by considering an ideally frozen protein, in which no fluctuation occurs and all the atomic positions are fixed at the equilibrium values. However, as discussed in a previous section, proteins are not rigid, but they sample a variety of conformations in the neighborhood of the equilibrium conformation [46, 50, 51, 80]. These fluctuations of the atomic positions result in an intrinsic impedance noise which is present during an electrical measurement performed on such a nanodevice. The level of this noise turns out to be crucial to the actual detection of the ligand capture by the GPCR when compared with the impedance variation due to the conformational transition and with the electrode/amplifier noise.

Therefore, we have extended the previous model by relaxing the frozen protein approximation and by studying the effect of the fluctuations of the atomic positions on the electrical response to an AC field [17, 43–45]. For this purpose, we allow the nodes of the network to fluctuate around their equilibrium positions with an amplitude depending on the temperature [43, 44]. For the sake of simplicity and to get a first qualitative estimation, we describe the system of fluctuating nodes as a set of independent, isotropic, harmonic quantum oscillators, with common values of the force constant γ and of the proper frequency, ω_0. The oscillators are assumed to be in contact with a thermal bath at temperature T. We denote with $\vec{r}_{n,\text{eq}}$ and $\delta\vec{r}_n = \vec{r}_n - \vec{r}_{n,\text{eq}}$, respectively, the equilibrium position and the displacement from the equilibrium at the time t of the nth oscillator, and with $\langle(\delta\vec{r})^2\rangle$ its mean square displacement. It can be easily seen that $\langle(\delta\vec{r})^2\rangle$ is given by:

$$\langle(\delta\vec{r})^2\rangle = \frac{3}{2}\cdot\frac{k_B\theta}{\gamma} + \frac{k_B\theta}{\gamma}\cdot\frac{1}{\exp[\theta/T]-1} \tag{4}$$

where $\theta = \hbar\omega_0/k_B$. Thus, when $T \gg \theta$, Eq. (4) simplifies as:

$$\langle(\delta\vec{r})^2\rangle \approx k_B T/\gamma \tag{5}$$

At an arbitrary temperature the wave function of the oscillator is a superposition of several excited states and its probability density cannot be expressed in a simple form. Again, for simplicity, we take the following expression for the probability density of presence of the nth oscillator around its equilibrium position:

$$P(\delta\vec{r}_n) = \frac{3^{3/2}}{[2\pi\langle(\delta\vec{r})^2\rangle]^{3/2}} \cdot \exp\left[-\frac{3}{2}\cdot\frac{(\delta\vec{r}_n)^2}{\langle(\delta\vec{r})^2\rangle}\right] \tag{6}$$

where $\langle(\delta\vec{r})^2\rangle = 3\langle(\delta x)^2\rangle$ and $\langle(\delta x)^2\rangle = \langle(\delta y)^2\rangle = \langle(\delta z)^2\rangle$. The fluctuations of the network impedance have been calculated by a Monte Carlo simulation which generates the position of all the nodes at each step according to the space probability given by Eq. (6). The calculations have been performed by choosing the value $\gamma = 2.5$ kJ mol^{-1} Å^{-2} for the oscillator force constant, according to the literature [46, 50, 51]. This choice of γ provides $\theta = 12$ K. Thus, the condition $T \gg \theta$ is well satisfied at room temperature. The choice of a unique force constant for all the amino acids of the protein is a frequently adopted assumption [46, 50, 51], as in the previously mentioned Gaussian Network Models [46, 50].

However, it is also well known that α-helices are much more stable under thermal fluctuations than loops and termini [1–4]. For this reason, we have also calculated the impedance noise by assuming two different force constants, γ_1/γ_2 and $\gamma_2 < \gamma_1$ for α-helices and loops/termini, respectively [44]. The values of γ_1 and γ_2 have been chosen by taking the mean value of the force constant $\bar{\gamma} = 2.5$ kJ mol^{-1} Å^{-2} and then by considering different values of the ratio γ_1/γ_2. The greater the value of this ratio, the more flexible the loops and termini are with respect to helices.

Figure 7.12 displays the values of the impedance modulus calculated as a function of the time for the case of rhodopsin. Similar data obtained for metarhodopsin

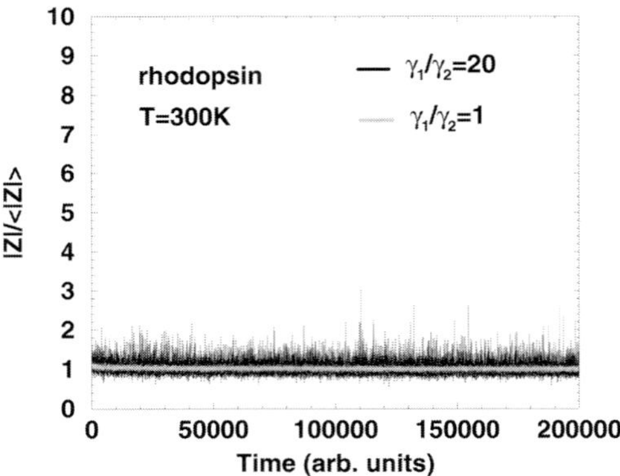

Figure 7.12. Calculated values of the impedance modulus versus time for rhodopsin. The impedance modulus has been normalized to its average value and the time is expressed in simulation steps. The grey and the black curves are calculated by taking the ratio $\gamma_1/\gamma_2 = 1$ and 20, respectively.

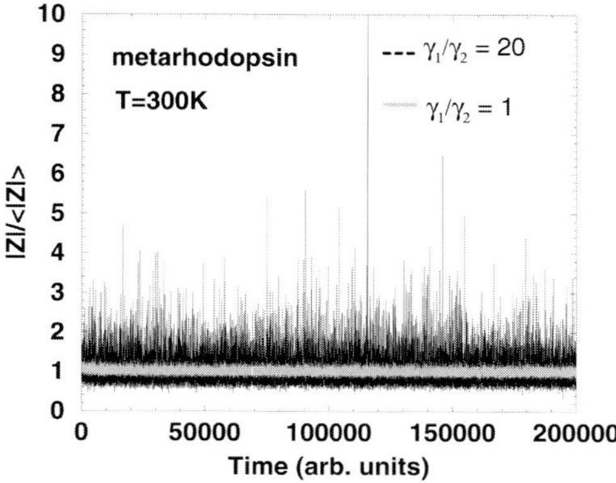

Figure 7.13. Calculated values of the impedance modulus versus time for metarhodopsin. The impedance modulus has been normalized to its average value and the time is expressed in simulation steps. The grey and the black curves have the same meanings as in Fig. 7.12.

II are reported in Fig. 7.13. All the data in Figs. 7.12 and 7.13 have been normalized to their respective average values and refer to simulations performed at room temperature. In Figs. 7.12 and 7.13, the curves corresponding to $\gamma_1/\gamma_2 = 1$ and 20 are reported, respectively, in grey and black. The comparison of the data shows that the metarhodopsin state is characterized by impedance fluctuations significantly wider than those in the rhodopsin state. This is particularly true when the force constant of the α-helices is 20 times greater than the force constant of loops/termini. This result reflects the minor thermal stability of metarhodopsin with respect to rhodopsin, probably due to an expansion of the entire structure consequent to the conformational transition [35].

This conclusion is supported also by Figs. 7.14 and 7.15, which report the probability distribution function, ϕ, of the impedance fluctuations for rhodopsin and metarhodopsin calculated at room temperature by taking the ratio $\gamma_1/\gamma_2 = 1$ (open diamonds) and 20 (full circles). The probability densities have been obtained by analyzing $Z(t)$ signals made of about 5×10^5 data. Here, σ is the root mean square deviation from the average value of the impedance modulus $\langle|Z|\rangle$. This normalized representation has been adopted for convenience because it makes the distribution independent of its first and second moments [88]. In particular, Fig. 7.15 shows that for metarhodopsin the impedance fluctuations are non-Gaussian even when $\gamma_1/\gamma_2 = 1$ (in contrast, in the case of rhodopsin the fluctuations are Gaussian). Moreover, strong non-Gaussian tails are present when $\gamma_1/\gamma_2 = 20$ and, in particular, for metarhodopsin. This strong non-Gaussianity at room temperature for $\gamma_1 \gg \gamma_2$ (and, in particular, for metarhodopsin) is due to the presence of big spikes corresponding to the loss of links that are crucial to ensure the connectivity of the network [48, 88].

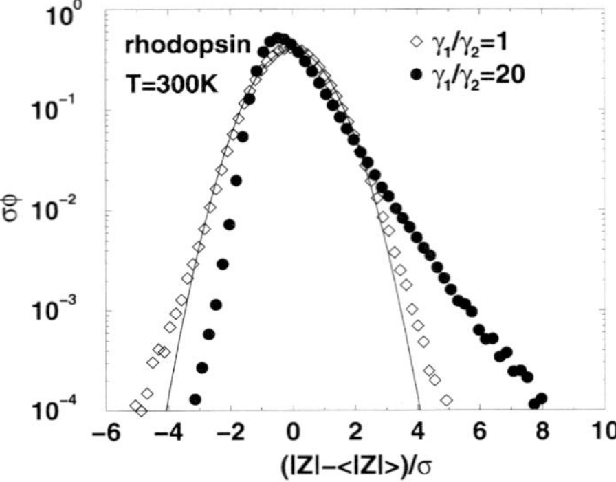

Figure 7.14. Normalized PDF of impedance modulus fluctuations at $T = 300$ K for rhodopsin. Diamonds and full circles correspond to simulations performed by taking the ratio $\gamma_1/\gamma_2 = 1$ and 20, respectively; σ is the root mean square deviation from the average modulus. The solid curve represents the Gaussian distribution.

Finally, Fig. 7.16 reports the values of the relative root mean square fluctuation of the impedance modulus, $\sigma/\langle|Z|\rangle$, as a function of the temperature. We can see that once the higher flexibility of loops and termini is accounted for by taking two different values for the force constants γ_1 and γ_2, the relative fluctuation of the impedance becomes strongly sensitive to the temperature. In particular, for $\gamma_1/\gamma_2 = 20$, the relative root mean square fluctuation of the impedance modulus for rhodopsin is about 15%, while for metarhodopsin it is about 35%. Such high levels of noise would make it problematic to detect the conformational change in terms of variation of the impedance of the nanodevice, being this variation roughly

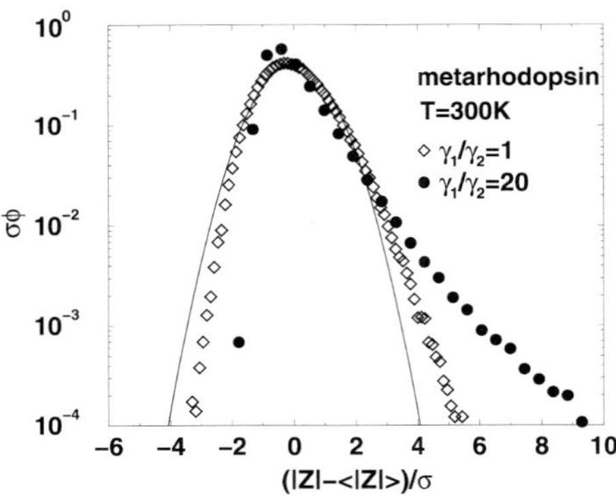

Figure 7.15. Normalized PDF of impedance modulus fluctuations at $T = 300$ K for metarhodopsin. The solid curve, the diamonds and full circles have the same meanings as in Fig. 7.14. σ is the root mean square deviation from the average modulus.

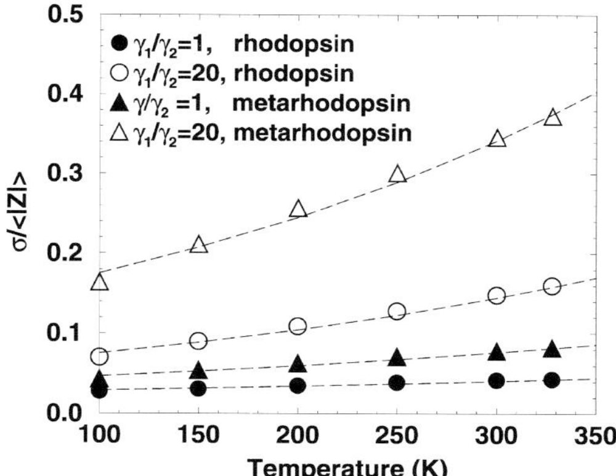

Figure 7.16. Relative root mean square deviation of the impedance modulus as a function of temperature for rhodopsin and metarhodopsin. The data shown by full and open symbols are calculated by taking the ratio $\gamma_1/\gamma_2 = 1$ and 20, respectively. The dashed lines represent the best fit of the data with an exponential law. The temperature is expressed in Kelvin.

of the same order [17]. However, the increase of the impedance noise itself in the rhodopsin $\rightarrow$ metarhodopsin transition (more than a factor of 2 at room temperature) could in principle provide a way to detect this transition.

7.7
Conclusions

We have considered the possibility of realizing nanobiosensors based on GPRCs. First, we discussed some of the main difficulties to be faced in developing such chip-based sensors, i.e. (a) the preparation and the immobilization of the GPCRs on functionalized surfaces in a manner suitable to preserve their function (to enable the conformational transition associated with the capture of the ligands) and (b) the signal techniques. In this regard, the problem consists of developing experimental and theoretical tools allowing a sensitive and reliable monitoring of the ligand capture and/or of the conformational transition by an analysis of the optical or the electrical response of the nanodevice. Focusing on this last approach, we discussed the EIS technique, which is emerging as a very effective technique for the detection of biosensing events at the electrodes [38–42]. Then, the largest part of this chapter was devoted to illustrating a theoretical model recently proposed for the study of the current response to an external AC voltage of a single-protein device. In particular, we have considered, as a prototype, a two-terminal device consisting of a GPRC (rhodopsin or rat OR-I7) embedded in a lipid bilayer and contacted by two ohmic electrodes. A coarse-grain approach has been developed for the description of the electrical properties of the receptor which is modeled as a network of elementary impedances. The conformational changes of the receptor induced by the capture of the ligand (photon or odorant molecule) are then translated into a variation of the network impedance. The role played by the different model parameters on the network structure and on the impedance spectral properties was analyzed. Furthermore, the study of the impedance noise associated with the equilibrium fluctuations of the protein showed that the conformational transition is followed by a significant increase of the impedance noise level, which, by itself, would offers an interesting possibility to detect the transition. Thus, the results obtained look promising for the practical realization of a GPCR-based nanobiosensor able to perform its sensing functions based on a change of its electrical properties.

Acknowledgments

This work has been performed within the SPOT-NOSED project IST-2001-38899 of the EC (http://www.nanobiolab.pcb.ub.es/projectes/Spotnosed/). Partial support from the cofin-03 project "Modelli e misure di rumore in nanostrutture" financed by Italian MIUR is also gratefully acknowledged.

References

1 LEFKOWITZ R. J. The superfamily of heptahelical receptors, *Nat. Cell Biol.* **2002**, *2*, E133–E136.

2 LAMEH J., CONE R. I., MAEDA S., PHILIP M., CORBANI M., NADASDI L., RAMACHANDRAN J., SMITH G. M., SADÈE W. Structure and function of G protein coupled receptors, *Pharmac. Res.* **1990**, *7*, 1213–1221.

3 VAIDEHI N., FLORIANO W. B., TRABANINO R., HALL S. E., FREDDOLINO P., CHOI E. J., ZAMANAKOS G., GODDARD W. A. Prediction of structure and function of G protein coupled receptors, *Proc. Natl Acad. Sci. USA* **2002**, *99*, 12622–12627.

4 GETHER U., KOBILKA B. K. G protein-coupled receptors, *J. Biol. Chem.* **1998**, *273*, 17979–17982.

5 MINIC J., SAUTEL M., SALESSE R., PAJOT-AUGY E. Yeast system as a screening tool for pharmacological assessment of G-protein-coupled receptors, *Curr. Med. Chem.* **2005**, *12*, 961–969.

6 BOCKAERT J., PIN J. P. Molecular tinkering of G protein-coupled receptors: an evolutionary success, *EMBO J.* **1999**, *18*, 1723–1729.

7 FIRESTEIN S. How the olfactory system makes sense of scents, *Nature* **2001**, *413*, 211–218.

8 CRASTO C., SINGER M. S., SHEPHERD G. M. The olfactory receptor family album, *Genome Biol.* **2001**, *2*, 10271–4.

9 BAYLEY H., CREMER P. S. Stochastic sensors inspired by biology, *Nature* **2001**, *413*, 226–230.

10 WU T. Z. A piezoelectric biosensor as an olfactory receptor for odour detection: electronic nose, *Biosens. Bioelectron.* **1999**, *14*, 9–18.

11 JOACHIM C., GIMZEWSKI J. K., AVIRAM A. Electronics using hybrid-molecular and mono-molecular devices, *Nature* **2000**, *408*, 541–548.

12 ALFONTA L., BARDEA A., KHERSONSKY O., KATZ E., WILLNER I. Chronopotentiometry and faradaic impedance spectroscopy as signal transduction methods for the biocatalytic precipitation of an insoluble product on electrode supports: routes for enzyme sensors, immunosensors and DNA

sensors, *Biosens. Bioelectron.* **2001**, *16*, 675–687.

13 BOOZER C., YU Q., CHEN S., LEE C., HOMOLA J., YEE S. S., JIANG S. Surface functionalization for self-referencing surface plasmon resonance (SPR) biosensors by multistep self-assembly, *Sens. Actuat. B* **2003**, *90*, 22–30.

14 CUI X., PEI R., WANG Z., YANG F., MA Y., DONG S., YANG X. Layer-by layer assembly of multilayer films composed of avidin and biotin-labeled antibody for immunosensing. *Biosens. Bioelectron.* **2003**, *18*, 59–67.

15 KARLSSON O. P., LOFAS S. Flow-mediated on-surface reconstitution of G-protein coupled receptors for applications in surface plasmon resonance biosensors, *Anal. Biochem.* **2002**, *300*, 132–138.

16 PATOLSKY F., ZHENG G., HAYDEN O., LAKADAMYALI M., ZHUANG X., LIEBER C. M. Electrical detection of single viruses, *Proc. Natl Acad. Sci. USA* **2004**, *101*, 14017–14022.

17 PENNETTA C., AKIMOV V., ALFINITO E., REGGIANI L., GOMILA G. Fluctuations of complex networks: electrical properties of single protein nano-devices, *Proc. SPIE* **2004**, *5472*, 172–182.

18 MINIC J., GROSCLAUDE J., AIOUN J., PERSUY M. A., GOROJANKINA T., SALESSE R., PAJOT-AUGY E., HOU Y., HELALI S., JAFFREZIC-RENAULT N., BESSUEILLE F., ERRACHID A., GOMILA G., RUIZ O., SAMITIER, J. Immobiliza-tion of native membrane-bound rho-dopsin on biosensor surfaces, *Biochem. Biophys. Acta* **2005**, *3*, 324–332.

19 OKADA T., ERNST O. P., PALCZEWSKI K., HOFMAN K. P. Activation of rho-dopsin: new insights from structural and biochemical studies, *Trends Bio-chem. Sci.* **2001**, *26*, 318–324.

20 SCARAMELLINI C., LEFF P. A three-state receptor model: predictions of multiple agonist pharmacology for the same receptor type, *Ann. NY Acad. Sci.* **1998**, *861*, 97–103.

21 KENAKIN T. Ligand-selective receptor conformations revisited: the promise and the problem, *Trends Pharmacol. Sci.* **2003**, *24*, 346–354.

22 SWAMINATH G., XIANG Y., LEE T. W., STEENHUIS J., PARNOT C., KOBILKA B. Sequential binding of agonists to the beta2-adrenoceptor, *J. Biol. Chem.* **2004**, *279*, 686–691.

23 ROCHEVILLE M., LANGE D. C., KUMAR U., PATEL S. C., PATEL R. C., PATEL Y. C. Receptors for dopamine and somatostatin: formation of hetero-oligomers with enhanced functional activity, *Science* **2000**, *288*, 154–157.

24 BISSANTZ C., BERNANRD P., HIBERT M., ROGNAN D. Protein-based virtual screening of chemical databases. II. Are homology models of G-protein coupled receptors suitable targets?, *Proteins* **2003**, *50*, 5–15.

25 EVERS A., GOHLKE H., KLEBE G. Ligand-supported homology modelling of protein binding-sites using knowledge-based potentials, *J. Mol. Biol.* **2003**, *334*, 327–345.

26 BLEICHER K. H., GREEN L. G., MARTIN R. E., ROGERS-EVANS M. Ligand identification for G-protein-coupled receptors: a lead generation perspec-tive, *Curr. Opin. Chem. Biol.* **2004**, *8*, 287–296.

27 BERMAN H. M., WESTBROOK J., FENG Z., GILLILAND G., BHAT T. N., WEISSING H., SHINDYALOV I. N., BOURNE P. E. Protein Data Bank, *Nucleic Acids Res.* **2000**, *28*, 235–242.

28 PALCZEWSKI K., HORI T., BEHNKE C. A., MOTOSHIMA H., FOX B. A., LE TRONG I., TELLER D. C., OKADA T., STENKAMP R. E., YAMAMOTO M., MIYANO M. Crystal structure of rhodopsin: a G protein-coupled receptor, *Science* **2000**, *289*, 739–745.

29 TELLER D. C., OKADA T., BEHNKE C. A., PALCZEWSKI K., STENKAMP R. E. Advances in determination of a high-resolution three-dimensional structure of rhodopsin, a model of G-protein-coupled receptors (GPCRs), *Biochemistry* **2001**, *40*, 7761–7772.

30 LI J., EDWARDS P. C., BURGHAMMER M., VILLA C., SCHERTLER G. F. Struc-ture of bovine rhodopsin in a trigonal crystal form, *J. Mol. Biol.* **2004**, *343*, 1409–1438.

31 CHOI G., LANDIN J., GALAN J. F., BIRGE R. R., ALBERT A. D., YEAGLE

P. L. Structural studies on meta-rhodopsin II, the activated form of the G protein-coupled receptor rhodopsin, *Biochemistry* **2002**, *41*, 7318–7324.

32 YEAGLE P. L., CHOI G., ALBERT A. D. Studies on the structure of G protein-coupled receptor rhodopsin including the putative G protein binding site in unactivated and activated forms, *Biochemistry* **2001**, *40*, 11932–11937.

33 SAKMAR T. P., MENON S. T., MARIN E. P., AWAD E. S. Rhodopsin: insights from recent structural studies, *Ann. Biomol. Struct.* **2002**, *31*, 443–482.

34 YAN E. C., KAZMI M. A., GANIM Z., HOU J. M., PAN D., CHANG B. S., SAKMAR T. P., MATHIES R. A. Retinal counterion switch in the photoactivation of the G protein-coupled receptor rhodopsin, *Proc. Natl Acad. Sci. USA* **2003**, *100*, 9262–9267.

35 MENON S. T., HAN M., SAKMAR T. P. Rhodopsin: structural basis of molecular physiology, *Physiol. Rev.* **2001**, *81*, 1659–1688.

36 BALLESTEROS J. A., SHI L., JAVITCH J. A. Structural mimicry in G protein coupled receptors: implications of the high-resolution structure of rhodopsin for structure–function analysis of rhodopsin-like receptors, *Mol. Pharmacol.* **2001**, *60*, 1–19.

37 MENG E. C., BOURNE H. R. Receptor activation: what does the rhodopsin structure tell us?, *Trends Biochem. Sci.* **2001**, *22*, 587–593.

38 GUAN J., MIAO Y., ZHANG Q. Impedimetric biosensors, *J. Biosci. Bioeng.* **2004**, *97*, 219–226.

39 KATZ E., WILLNER I. Probing biomolecular interaction at conductive and semiconductive surfaces by impedance spectroscopy: routes to impedimetric immunosensors, DNA-sensors, and enzyme biosensors, *Electroanalysis* **2003**, *15*, 913–947.

40 PEI R., CHENG Z., WANG E., YANG X. Amplification of antigen-antibody interactions based on biotin labeled protein–streptavidin network complex using impedance spectroscopy, *Biosens. Bioelectron.* **2001**, *16*, 355–361.

41 RICKERT J., GÖPEL W., BECK W., JUNG G., HEIDUSCHKA P. A "mixed" self-assembled monolayer for an impedimetric immunosensor, *Biosens. Bioelectron.* **1996**, *11*, 757–768.

42 HOU Y., HELALI S., ZHANG A., JAFFREZIC-RENAULT N., MARTELET C., MINIC J., GOROJANKINA T., PERSUY M. A., PAJOT-AUGY E., SALESSE R., BESSUEILLE F., SAMITIER J., ERRACHID A., AKINOV V., REGGIANI L., PENNETTA C., ALFINITO E. Immobilization of rhodopsin on a self-assembled multilayer and its detection by electrochemical impedance spectroscopy, *Biosens. Bioelectron.* **2006**, *21*, 1393–1402.

43 PENNETTA C., AKIMOV V., ALFINITO E., REGGIANI L., GOMILA G., FERRARI G., FUMAGALLI L., SAMPIETRO M. Modelization of thermal fluctuations in G protein-coupled receptors, in: *Proceedings of the 18th ICNF Conference*, Salamanca, **2005**, pp. 611–614.

44 ALFINITO E., AKIMOV V., PENNETTA C., REGGIANI L., GOMILA G. Thermal fluctuations of a GPCR: a two force constant model, in: *Proceedings of the 4th UPoN Conference*, Gallipoli, **2005**, pp. 381–387.

45 AKIMOV V., ALFINITO E., PENNETTA C., REGGIANI L., MINIC J., GOROJANKINA T., PAJOT-AUGY E., SALESSE R. An impedance network model for the electrical properties of a single protein nanodevice, in: *Proceedings of the 14th HCIS Conference*, Chicago, IL, **2005**, in press.

46 ATILGAN A. R., DURELL S. R., JERNIGAN R. L., DEMIREL M. C., KESKIN O., BAHAR I. Anisotropy of fluctuation dynamics of proteins with an elastic network model, *Biophys. J.* **2001**, *80*, 505–515.

47 LATTANZI G., MARITAN A. Force dependent transition rates in chemical kinetics models for motor proteins, *J. Chem. Phys.* **2002**, *117*, 10339–10349.

48 ALBERT R., BARABASI A. L. Statistical mechanics of complex networks, *Rev. Mod. Phys.* **2002**, *74*, 47–97.

49 HALL S. E., FLORIANO W. B., VAIDEHI N., GODDARD W. A. III. Predicted 3-D structures for mouse I7 and rat I7 olfactory receptors and comparison of predicted odor recognition profiles with experiment, *Chem. Sens.* **2004**, *29*, 595–616.

50 TIRION M. M. Large amplitude elastic motions in proteins from a single-parameter atomic analysis, *Phys. Rev. Lett.* **1996**, *77*, 1905–1908.

51 BAHAR I., ATILGAN A. R., ERMAN B. Direct evaluations of thermal fluctuations in proteins using a single parameter harmonic potential, *Fold. Des.* **1997**, *2*, 173–181.

52 REBOIS R. V., SCHUCK P., NORTHUP J. K. Elucidating kinetic and thermodynamic constants for interaction of G protein subunits and receptors by surface plasmon resonance spectroscopy, *Methods Enzymol.* **2002**, *344*, 15–42.

53 ALVES I. D., SALGADO G. F., SALAMON Z., BROWN M. F., TOLLIN G., HRUBY V. J. Phosphatidylethanolamine enhances rhodopsin photoactivation and transducin binding in a solid supported lipid bilayer as determined using plasmon-waveguide resonance spectroscopy, *Biophys. J.* **2005**, *88*, 198–210.

54 BIERI C., ERNST O. P., HEYSE S., HOFMANN K. P., VOGEL H. Micro-patterned immobilization of a G protein-coupled receptor and direct detection of G protein activation, *Nat. Biotechnol.* **1999**, *17*, 1105–1108.

55 GIBSON N. J., BROWN M. F. Lipid headgroup and acyl chain composition modulate the MI–MII equilibrium of rhodopsin in recombinant membranes, *Biochemistry* **1993**, *32*, 2438–2454.

56 LAGANE B., GAIBELET G., MEILHOC E., MASSON J. M., CEZANNE L., LOPEZ A. Role of sterols in modulating the human mu-opioid receptor function in *Saccharomyces cerevisiae*, *J. Biol. Chem.* **2000**, *275*, 33197–33200.

57 HEYSE S., ERNST O. P., DIENES Z., HOFMANN K. P., VOGEL H. Incorporation of rhodopsin in laterally structured supported membranes: observation of transducin activation with spatially and time-resolved surface plasmon resonance, *Biochemistry* **1998**, *37*, 507–522.

58 NEUMANN L., WOHLAND T., WHELAN R. J., ZARE R. N., KOBILKA B.

Functional immobilization of a ligand-activated G-protein-coupled receptor, *Chem. Biochem.* **2002**, *3*, 993–998.

59 DEVANATHAN S., YAO Z., SALAMON Z., KOBILKA B., TOLLIN G. Plasmon waveguide resonance studies of ligand binding to the human beta 2-adrenergic receptor, *Biochemistry* **2004**, *43*, 3280–3288.

60 ALVES I. D., CIANO K. A., BOGUSLAVSKI V., VARGA E., SALAMON Z., YAMAMURA H. I., HRUBY V. J., TOLLIN G. Selectivity, cooperativity, and reciprocity in the interactions between the delta-opioid receptor, its ligands, and G-proteins, *J. Biol. Chem.* **2004**, *279*, 44673–44682.

61 ALVES I. D., COWELL S. M., SALAMON Z., DEVANATHAN S., TOLLIN G., HRUBY V. J. Different structural states of the proteolipid membrane are produced by ligand binding to the human delta-opioid receptor as shown by plasmon-waveguide resonance spectroscopy, *Mol. Pharmacol.* **2004**, *65*, 1248–1257.

62 NUZZO R. G., ALLARA D. L. Adsorption of bifunctional organic disulfides on gold surfaces, *J. Am. Chem. Soc.* **1983**, *105*, 4481–4483.

63 ULMAN A. Formation and structure of self-assembled monolayers. *Chem. Rev.* **1996**, *96*, 1533–1554.

64 WINK T., ZUILEN S. J. V., BULT A., BENNEKOM W. P. V. Self-assembled monolayers for biosensors, *Analyst* **1997**, *122*, 43R–50R.

65 FERRETTI S., PAYNTER S., RUSSELL D. A., SAPSFORD K. E. Self-assembled monolayers: a versatile tool for the formulation of bio-surfaces, *Trends Anal. Chem.* **2000**, *19*, 530–540.

66 GOODING J. J., HIBBERT D. B. The application of alkanethiol self-assembled monolayers to enzyme electrodes, *Trends Anal. Chem.* **1999**, *18*, 525–533.

67 SPINKE J., LILEY M., GUDER H. J., ANGERMAIER L., KNOLL W. Molecular recognition at self-assembled monolayers: the construction of multi-component multilayers, *Langmuir* **1993**, *9*, 1821–1825.

68 LADD J., BOOZER C., YU Q., CHEN S.,

HOMOLA J., JIANG S. DNA-directed protein immobilization on mixed self-assembled monolayers via a streptavidin bridge, *Langmuir* **2004**, *20*, 8090–8095.

69 STORRI S., SANTONI T., MINUNNI M., MASCINI M. Surface modifications for the development of piezoimmunosensors, *Biosens. Bioelectron.* **1998**, *13*, 347–357.

70 LIDKE D. S., NAGY P., HEINTZMANN R., ARNDT-JOVIN D. J., POST J. N., GRECCO H. E., JARES-ERIJMAN E. A., JOVIN T. M. Quantum dots ligands provide new insights into erbB/HER receptor-mediated signal transduction, *Nat. Biotechnol.* **2004**, *22*, 198–203.

71 POMPA P. P., BIASCO A., FRASCERRA V., CALABI F., CINGOLANI R., RINALDI R., VERBEET M. P., DE WAAL E., CANTERS G. W. Solid state protein monolayers: morphological, conformational and functional properties, *J. Chem. Phys.* **2004**, *121*, 10325–10328.

72 WALLRABE H., ELANGOVAN M., BURCHARD A. FRET microscopy reveals clustered distribution of co-internalized receptor–ligand complexes in the apical recycling endosome of polarized epithelial MDCK cells, *Proc. SPIE* **2002**, *4620*, 64–72.

73 WALLRABE H., PERIASAMY A. Imaging protein molecules using FRET and FLIM microscopy, *Curr. Opin. Biotechnol.* **2005**, *16*, 19–27.

74 FOTIADIS D., SCHEURING S., MÜLLER S. A., ENGEL A., MÜLLER D. J. Imaging and manipulation of biological structures with the AFM, *Micron* **2002**, *33*, 385–397.

75 BIASCO A., MARUCCIO G., VISCONTI P., BRAMANTI A., CALOGIURI P., CINGOLANI R., RINALDI R. Self-chemisorption of azurin on functionalized oxide surfaces for the implementation of biomolecular devices, *Mater. Sci. Eng. C* **2004**, *24*, 563–567.

76 WANG M., WANG L., WANG G., JI X., BAI Y., LI T., GONG S., LI J. Application of impedance spectroscopy for monitoring colloid Au-enhanced antibody immobilization and antibody–antigen reactions, *Biosens. Bioelectron.* **2004**, *19*, 575–582.

77 KOBILKA B., GETHER U., SEIFERT M., LIN S., GHANOUNI P. Characterization of ligand-induced conformational states in the beta 2 adrenergic receptor, *J. Receptor Signal Transduct. Res.* **1999**, *19*, 293–300.

78 YANG H., LUO G., KARNCHANAPHANURACH P., LOUIE T. M., RECH I., COVA S., XUN L., XIE X. S. Protein conformational dynamics probed by single-molecule electron transfer, *Science* **2003**, *302*, 262–266.

79 HELMREICH, E. J. M., HOFMANN, K. P. Structure and function of proteins in G-protein-coupled signal transfer, *Biochim. Biophys. Acta* **1996**, *1286*, 285.

80 PARAK F. G. Physical aspects of protein dynamics, *Rep. Prog. Phys.* **2003**, *66*, 103–129.

81 BUDA F., DE GROOT H. J. M., BIFONE A. Charge localization and dynamics in rhodopsin, *Phys. Rev. Lett.* **1996**, *77*, 4474–4477.

82 FRAUENFELDER H., SLIGAR S. G., WOLYNES P. G. The energy landscapes and motions of proteins, *Science* **1991**, *254*, 1598–1603.

83 CARLONI P., ANDREONI W., PARRINELLO M. Self-assembled peptide nanotubes from first principles, *Phys. Rev. Lett.* **1997**, *79*, 761–764.

84 CARLONI P., ROTHLISBERGER U., PARRINELLO M. The role and perspective of ab initio molecular dynamics in the study of biological systems, *Acc. Chem. Res.* **2002**, *35*, 455.

85 KITAO A., GO N. Investigating protein dynamics in collective coordinate space, *Curr. Opin. Struct. Biol.* **1999**, *9*, 164–169.

86 XIE Q., ARCHONTIS G., SKOURTIS S. S. Protein electron transfer: a numerical study of tunneling through fluctuating bridges, *Chem. Phys. Lett.* **1999**, *312*, 237–246.

87 SONG X. An inhomogeneous model of protein dielectric properties: intrinsic polarizabilities of amino acids, *J. Chem. Phys.* **2002**, *116*, 9359–9383.

88 PENNETTA C., ALFINITO E., REGGIANI L., RUFFO S. Non-Gaussian resistance noise near breakdown, in granular materials, *Physica A* **2004**, *340*, 380–387.

8
Protein-based Nanotechnology: Kinesin–Microtubule-driven Systems for Bioanalytical Applications

William O. Hancock

8.1
Introduction

Protein machines carry out tasks critical to cell function, including DNA replication, intracellular transport, ion pumping and cell motility. They have evolved incredible diversity, specificity, efficiency and precision, and a considerable proportion of research in modern biology aims to uncover the fundamental mechanisms underlying their function [1]. The cytoskeletal motors, kinesins, dyneins and myosins, constitute a subset of these protein machines and are notable in being able to convert chemical energy directly to mechanical work. In cells, these motors generate the force that drives muscle contraction, they transport intracellular cargo throughout cells and they drive the critical movements that underlie cell division.

With the ability to engineer devices and systems at the micron and submicron scales, and to synthesize nanoparticles with novel and powerful functionalities, there is a growing need to transport and organize material at submicron dimensions. Because cytoskeletal motors have evolved specifically to transport and organize material at these size scales, there is a current effort to integrate these molecular motors and their cytoskeletal tracks into engineered devices [2–10]. This interdisciplinary research, which involves biologists, bioengineers, chemists, materials scientists, electrical engineers and others, is part of a larger effort to integrate proteins with highly evolved functions into nanoscale engineered systems. For instance, proteins and peptides are also being used as tools to drive self-assembly of inorganic materials such as semiconductors into functional materials [11, 12]. Another example of protein-based nanotechnology is the push to create electronic devices based on proteins or ion channels [13].

Compared to pressure-driven microfluidic flow, active transport by molecular motors in microfluidic channels offers a number of advantages as components of bioanalytical systems or biosensors. First, because of the small size of the motors and cytoskeletal filaments, the channels can be scaled down to dimensions below 100 nm. Second, motor-driven transport can occur up concentration gradients and against fluid flows. Third, because energy for transport (in the form of ATP) is delivered and consumed directly at the site of transport, dense arrays of multiplexed

Nanotechnologies for the Life Sciences Vol. 4
Nanodevices for the Life Sciences. Edited by Challa S. S. R. Kumar
Copyright © 2006 WILEY-VCH Verlag GmbH & Co. KGaA, Weinheim
ISBN: 3-527-31384-2

channels can be created without need for complex electrical connections. However, before biomolecular motors can be integrated into hybrid microscale and nanoscale engineered systems, there are a number of experimental hurdles that must be tackled. First, interfaces between the proteins and material surfaces must be optimized to attach proteins while maintaining their biochemical function. Second, the design of these microsystems needs to be optimized to best capitalize on the unique transport properties of these motors. Finally, to make analytical devices based on biomotors a reality, it is crucial to develop methods for attaching designated cargo to these proteins and to increase the stability of these proteins.

This chapter focuses on the integration of kinesin molecular motors and their microtubule tracks into microdevices for bioanalytical applications. A number of insightful reviews have been written on the molecular mechanism of motor proteins, in general [14, 15], and kinesin motors, in particular [16, 17]. There are also reviews on applications of biomolecular motors in nanotechnology [18, 19] and on applications of kinesin motors in microscale transport [4]. Finally, there is a parallel effort underway using actin and myosin for transport in microscale and nanoscale transport applications, [20–23], but that work is not discussed here.

This chapter is organized as follows. In Section 8.2, the relevant cell biology and biophysics of the kinesin–microtubule system is presented, including a description of the *in vitro* assays that have been developed to study kinesin function. Section 8.3 explores theoretical aspects of motor-driven microscale transport; in particular, the relationship between transport speeds and diffusion times for particles of various sizes. This analysis helps to frame applications in which kinesin-driven transport is best utilized. Section 8.4 discusses experimental approaches to interfacing motor proteins and microtubules with engineered surfaces. Section 8.5 reviews approaches that have been taken to control the direction of kinesin and microtubule movements. These studies provide the core work that needs to be accomplished towards integrating motor proteins into functional microscale devices. To create bioanalytical systems or biosensors driven by the kinesin–microtubule system, it is also crucial to develop strategies for attaching molecular or cellular cargo to microtubules or motors; these strategies are discussed in Section 8.6. Finally, Section 8.7 discusses higher-level design considerations for motor-driven devices, including ways to maximize the lifetime of motors and microtubules, and approaches for introducing minute samples into these devices and detecting low levels of analyte in microfluidic channels.

8.2
Kinesin and Microtubule Cell Biology and Biophysics

In eukaryotic cells, organelles, vesicles, chromosomes and protein complexes are actively transported throughout the cell by molecular motors moving along cytoskeletal filaments. This transport system consists of both kinesin and dynein motors moving along microtubule filaments, as well as myosin motors moving along

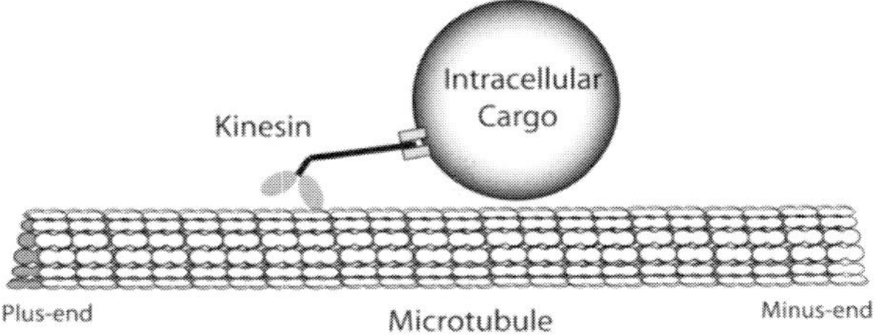

Figure 8.1. Structure and function of conventional kinesin. The two heads hydrolyze ATP and walk towards the "plus" end of microtubules. The kinesin tail binds to intracellular cargo. Microtubules, made from tubulin subunits, are 25 nm in diameter and tens of microns long.

actin filaments, but members of the kinesin family carry out the bulk of intracellular transport. Conventional kinesin, the founding member of the kinesin family, serves as a model protein for understanding the molecular basis of intracellular transport and for applications of molecular motors in nanotechnology. Conventional kinesin is a dimeric protein that contains three domains: the head or motor domain, the coiled-coil stalk that holds the two chains together and the tail that, along with two associated light chains, is responsible for binding cargo (Fig. 8.1) [24, 25]. Each motor domain contains both an ATP and a microtubule-binding site, and movement is achieved by a cycle in which each head alternately binds to the microtubule, undergoes a conformational change and releases from the track [16, 26–28]. Following the discovery of conventional kinesin [29], other members of the kinesin family were discovered based on sequence similarity in the motor head domain. The kinesin family can be divided into 14 classes based on sequence similarity and functional properties [30], and in the human genome there are 44 kinesin genes [31]. While almost all of the application work with kinesin motors has utilized conventional kinesin, because other kinesins have different motor properties, they may become useful in future applications.

Microtubules are cylindrical polymers of the protein tubulin that are 25 nm in diameter and up to tens of microns long. Tubulin dimers 8-nm long associate in a head-to-tail manner to make protofilaments, these protofilaments associate laterally to make sheets and the sheets close to make hollow cylinders that normally contain 14 protofilaments [32]. Because the subunits are asymmetric, microtubules have a structural polarity – the "minus" or slow-growing end is anchored near the center of the cell and the fast-growing "plus" ends extend to the perimeter of the cell. Experimentally, tubulin is normally isolated from cow or pig brains [33], which are large, inexpensive and rich in tubulin due to the neurons that require long-distance intracellular transport. Microtubules can be polymerized *in vitro* from purified tubulin, covalently modified with fluorophores or other functional groups and stabilized in polymer form with the drug taxol, making them stable for up to 1 week in normal buffers at room temperature [34].

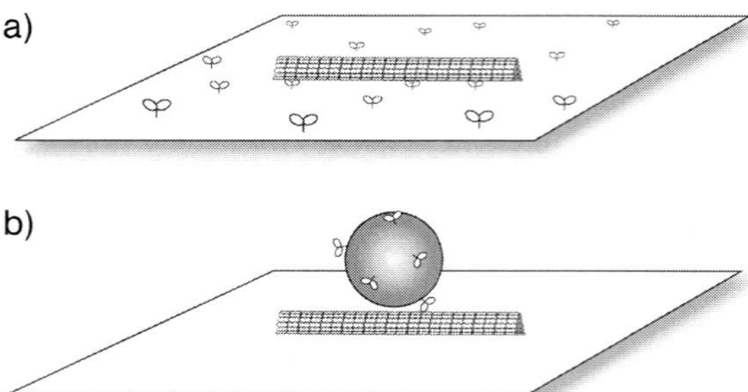

Figure 8.2. Assays for studying kinesin motor function. (A) Schematic of the microtubule gliding assay, in which motors are adsorbed to the surface and microtubules are transported across the surface. (B) In the bead assay, in which microtubules are immobilized on the surface, motors are adsorbed to micron-scale beads and the beads are transported along the immobilized microtubules.

8.2.1
Kinesin Motility Assays

Nearly two decades of biochemical and biophysical experiments on conventional kinesin have resulted in a solid quantitative characterization of this molecular motor. Conventional kinesin motors walk along microtubules at speeds of nearly 1 μm s^{-1}, taking 8-nm steps and hydrolyzing one ATP per step [27, 35, 36]. The motor speed decreases approximately linearly with applied load up to a single motor stall force of 5–7 pN [37, 38]. Kinesin movement *in vitro* is studied predominantly using two different assays – the microtubule gliding assay and the bead assay (Fig. 8.2) [39, 40]. In the microtubule gliding assay, motors are adsorbed to glass surfaces that have been treated with the blocking protein casein, and microtubules are observed landing on and moving over the motors, analogous to a rock star being passed over the hands of an eager crowd. Typically, these assays are performed in 20-μl flow cells constructed from a microscope slide, two pieces of double-sided tape and a cover glass, which enables facile solution exchange by simply pipeting solution in one side and wicking out the other side using filter paper or tissue [41]. Using this geometry, the motor concentration on the surface can be varied and different solutions can be introduced to optimize movement characteristics [42]. Microtubule movements are visualized by covalently labeling the microtubules with a fluorescent dye, observing them under a fluorescent microscope coupled to a sensitive CCD camera and recording the movements on videotape or computer [41]. As the assay is relatively easy to perform and the filaments are transported long distances along the surface, this geometry has generated the most attention for microscale transport applications of the kinesin–microtubule system.

In the other commonly used assay, the bead assay, microtubules are immobilized on glass surfaces, motors are adsorbed to micron-scale beads and the beads are transported along the microtubules. This system is analogous to the geometry found in cells, and one advantage is that optical tweezers can be used to grab the beads and measure displacements and forces generated by the motors to nanometer and piconewton precision [43]. For transport applications, the bead can in principle be replaced by functional nanoparticles or biomolecules like proteins or nucleotides. As the tail domain of kinesin motors can be deleted or significantly altered with no effect on the motor function [44], in theory antibody fragments, receptors, DNA-binding domains or other protein motifs can be fused to the motor tail and these motor–cargo complexes transported along microtubules.

Motor directionality is a key consideration for applications using either the gliding assay or the bead assay. In the gliding assay, microtubules diffuse out of solution and land on the motors, and the direction of microtubule transport is defined by the orientation of the filament (Fig. 8.3). Because the motors, which are immobilized, move to the microtubule "plus" end, the filaments move with their "minus" ends leading. The coiled-coil of conventional kinesin has a region of random coil that is thought to act like a swivel, enabling the heads to rotate freely and bind to filaments only in the proper stereospecific orientation [45]. As discussed below, a large portion of the work done to harness microtubule transport for nanoscale applications has involved finding ways to control the direction of microtubule transport. The bead geometry has its own directionality problems – the transport direction is determined by the orientation of the immobilized microtubules (Fig. 8.3). In theory, if a dense bundle of oriented microtubules could be immobilized on the surface, they would serve as ideal tracks to direct kinesin transport. However, as described in Section 8.5, achieving these oriented and aligned bundles has proven difficult [9, 46].

For applications using kinesin motors and microtubules, it is important to obtain sufficient amounts of protein and keep the proteins stable over time. As discussed above, tubulin can be purified from native sources following established protocols [33] or it can be bought from commercial sources (Cytoskeleton, Denver, CO). Conventional kinesin is generally bacterially expressed and purified [36, 44, 47], but it too can be purchased. One of the hurdles to using other motors in the kinesin family is that not all of them can be bacterially expressed [48, 49] and they can be more prone to denaturation or precipitation over time. Methods for extending the lifetime of kinesins and microtubules in engineered devices will be discussed below in Section 8.7. For more information and detailed protocols for kinesin and microtubule purification, and *in vitro* motility experiments, readers can consult the book *Kinesin Protocols* edited by Isabelle Vernos [50] or the Kinesin Home Page (http://www.proweb.org/kinesin/).

8.3
Theoretical Transport Issues for Device Integration

One of the main application goals for molecular motors to date has been active transport of analytes in microfluidic systems. There currently exist a number of ap-

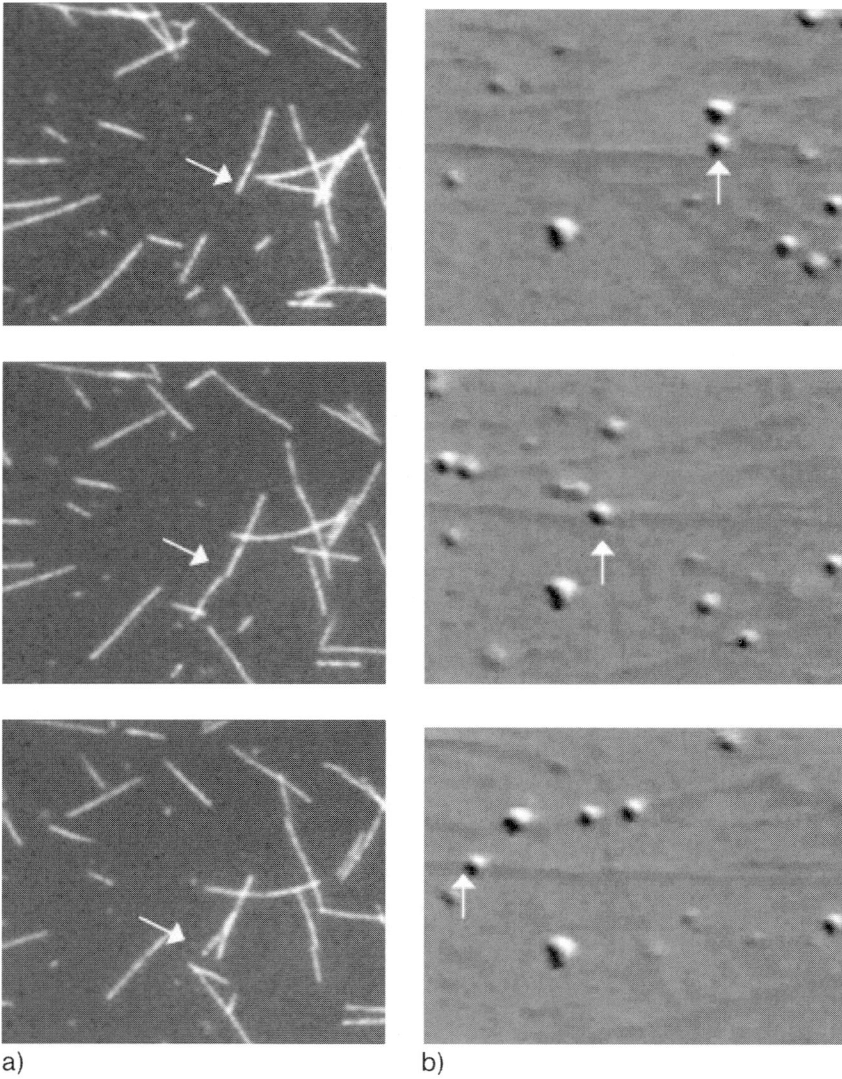

a) b)

Figure 8.3. Kinesin–microtubule *in vitro* motility assays. (A) Microtubule gliding assay, showing fluorescent microtubules moving over immobilized kinesin motors. Screens are 22 μm wide and images are 2 s apart. As can be seen, many microtubules move over the surface and their directions are determined by their orientation when they land on the surface. (B) Bead assay in which 0.2-μm diameter glass beads, to which many kinesin motors are adsorbed, are transported along surface-immobilized microtubules. Screens are 10 μm wide and images are 4 s apart. The direction of bead movement is determined by the orientation of the immobilized microtubules. Images in (A) are fluorescence and images in (B) are differential interference contrast microscopy.

proaches for moving fluids, analytes and particles through microscale geometries, such as pressure-driven convective flow, electrophoresis, dielectrophoresis and electro-osmotic flow. Rather than supplanting these methods, biomotor-driven transport should be thought of as a new approach that expands the toolbox. The best choice depends on the nature of the problem, and the hope is that in the future biomotor-driven transport will be combined with these other transport and separation approaches to make highly functional microscale devices.

It is informative to investigate from a theoretical perspective the range of particle sizes and transport distances where biomotor-driven transport is the most useful. Specifically, it is important consider the relative roles of diffusion (Brownian motion) versus motor-driven active transport. Small analytes (proteins and nucleotides) in aqueous solutions diffuse rapidly and any transport must overcome this diffusional mixing. For instance, as discussed below, an average protein can diffuse 1 μm in 5 ms and even for distances of 100 μm the average diffusion time is faster than the time it takes a kinesin motor moving at 1 μm s^{-1} to get there. It should be remembered that active transport is unidirectional, while diffusion is random, but the quantitative comparison serves as a useful guide.

8.3.1
Diffusion versus Transport Times

In Tab. 8.1, diffusion times and motor-driven transport times are compared for a range of potential analytes from proteins up to eukaryotic cells. The Einstein rela-

Tab. 8.1. Theoretical diffusional and transport properties for a range of biological analytes

Particle	Diameter	D (μm^2 s^{-1})	1 μm		100 μm		Transition distance (μm)
			$t_{Diffusion}$ (s)	$t_{Transport}$ (s)	$t_{Diffusion}$ (s)	$t_{Transport}$ (s)	
Protein	5 nm	100	0.005	1	50	100	200
Nanoparticle	20 nm	25	0.02	1	200	100	50
Virus	100 nm	5	0.1	1	1,000	100	10
Bacteria	1 μm	0.5	1	1	10,000	100	1
Cell	20 μm	0.02	20	1	2×10^5	100	0.05

Notes:
$D = k_B T / 6\pi\eta r$.
$k_B T = 4.1 \times 10^{-21}$ Nm at 25 °C.
Viscosity $\eta = 0.89$ cP at 25 °C $= 0.89 \times 10^{-3}$ Ns m^{-2}.
$t_{Diffusion} = x^2 / 2D$ (one-dimensional).
$t_{Transport} = x/v$, where $v = 1$ μm s^{-1} for kinesin.
Transition distance (when $t_{Transport} < t_{Diffusion}$) $= 2D/v$.

tion is used to derive the diffusion constant, D based on the drag coefficient of the particle γ:

$$D = k_{\mathrm{B}} T / \gamma \tag{1}$$

Where $k_{\mathrm{B}} T$ is Boltzman's constant (1.38×10^{-23} Nm) multiplied by the absolute temperature [51]. From Stokes' law, the drag coefficient of a spherical particle in low Reynold's number flow is defined as:

$$\gamma = 6\pi\eta r \tag{2}$$

where η is the solution viscosity (0.89×10^{-3} Ns m^{-2} for water at 25 °C) and r is the radius of the particle [51, 52]. Combining these equations, we get the equation for the diffusion constant for spherical objects:

$$D = k_{\mathrm{B}} T / 6\pi\eta r \tag{3}$$

For the particles in Tab. 8.1, approximate diameters are given and diffusion constants in aqueous buffers at 25 °C are calculated.

Using the diffusion constant, the average time it takes to diffuse a distance x in one dimension is defined as:

$$t_{\mathrm{Diffusion}} = x^2 / 2D \tag{4}$$

and the time required for kinesin-driven transport is defined as

$$t_{\mathrm{Transport}} = x / v \tag{5}$$

where $v = 1$ μm s^{-1} for conventional kinesin [36, 52]. In Tab. 8.1 these times are calculated for the different particles for distances of both 1 and 100 μm. As can be seen, small particles diffuse rapidly and motor-driven transport outpaces diffusion only for distances over a few hundred microns, while large particles like bacteria and eukaryotic cells diffuse very slowly and motor transport is much faster for all distances. A helpful way to understand the utility of kinesins in microscale transport applications is to ask: at what transition distance does motor-driven transport outpace diffusion ($t_{\mathrm{Transport}} < t_{\mathrm{Diffusion}}$)? This distance, which gives a rough value for where motor-driven transport becomes useful, is calculated in the final column of Tab. 8.1. The important result is that for proteins and small particles like nanoparticles in the range of tens of nanometers, diffusional mixing is sufficient to get particles where they need to go for distances less than a few hundred microns. However, for larger particles like bacteria and cells, diffusion is inadequate at virtually all length scales. Hence, there is significant potential for these nanoscale motors to drive and control movement of micron-scale biological objects. As discussed in Section 8.6, drag forces for these relatively large objects are negligible compared to the forces kinesin motors can exert.

While it is insightful to compare diffusion and transport times, it is important to appreciate that while diffusion occurs in all directions, motor-driven transport is directional. Hence, even for short distances this transport can establish and maintain concentration gradients provided the transport flux is sufficiently high. Second, because binding these small cargo pulls the particles out of solution and eliminates their random diffusional movement, it reduces the number of degrees of freedom and achieves a significant level of positional control. Furthermore, increasing the solution viscosity either by adding solutes or by creating permeable hydrogels should be able to significantly slow diffusion rates without necessarily slowing motor transport. Finally, as the channels and sorters are fabricated using versatile photolithography techniques [2, 6, 53], there is ample opportunity for optimizing the design of these systems to maximize the utility of motor-based transport and minimize diffusional mixing.

8.4
Interaction of Motor Proteins and Filaments with Synthetic Surfaces

A recurring technical hurdle in integrating functional proteins into engineered devices is the problem of interfacing synthetic materials with biological molecules. The problem of understanding and controlling protein–biomaterial interactions has been a major focus of the biomaterials field for decades [54, 55]. For instance, in the case of implanted biomaterials, the predominant approaches to maximizing biocompatibility are either to create surfaces that completely resist protein adsorption or surfaces to which proteins can adsorb, but not change their activities or otherwise produce an immune response. Due of the importance of this problem, there is a body of work on protein adsorption to biomaterials. However, there are no sure-fire techniques for (a) creating surfaces that bind proteins, but do not affect their function, or (b) creating surfaces that completely resist protein adsorption.

To harness the utility of motor proteins and their cytoskeletal tracks for nanotechnology and microscale transport applications, it is crucial to control the specific adsorption of these proteins to surfaces. Here, techniques for immobilizing motor proteins and cytoskeletal filaments are reviewed.

8.4.1
Motor Adsorption

It can be argued that the discovery of conventional kinesin's motor activity [29] was a result of this motor's ability to bind to glass surfaces and retain its function. With proper surface passivation to reduce motor denaturation, conventional kinesin binds functionally to many types of glass as well as various oxides and other hydrophilic surfaces [7, 41, 56, 57]. The most reliable surface passivation is pretreatment of surfaces with casein, a protein found in milk. In solution, casein forms heterogeneous aggregates with diameters of the order of 10–300 nm [58].

The aggregation state of unproteolyzed casein is heterogeneous and depends on the Ca^{2+} concentration, pH, degree of phosphorylation and other factors [59]. For kinesin experiments, commercially bought casein is dissolved in buffer and passed through a submicron filter and/or centrifuged to remove aggregates [41].

It is clear that pretreatment of glass surfaces with casein greatly increases the activity of kinesin motors adsorbed to the surface, but the precise mechanism of action is not clear [41]. The working model is that casein aggregates tens of nanometers in diameter pack on the surface, the tail domains of the kinesin motors bind between the casein particles, and the motor heads stick into solution and interact with microtubules. It cannot be ruled out that the kinesin motors bind directly to the surface-adsorbed casein, but the fact that a great excess of casein in the motor solution does not compete with the surface-adsorbed casein (there is no reduction in the concentration of functional motors on the surface) argues against this. Also, it has been reported that casein has a chaperone-like function in stabilizing proteins against denaturation [60]. It cannot be ruled out that part of the enhancement of kinesin function by casein is due to stabilizing the motor protein structure.

One of the problems with the casein pretreatment described above is that it does not work for every motor protein. Apart from conventional kinesin, there are many other kinesin motor proteins (44 total in the human genome) that vary both in their motor properties and in their intracellular cargo [31]. These different motor characteristics – direction of movement along the microtubule, affinity for the microtubule and speed of movement – provide a rich toolbox for engineering hybrid devices based on these motors. However, while there is considerable structural consistency in the head domain, there is great divergence structurally and functionally in the tail domains [31, 61–63]. The result of this diversity is that techniques for attaching one motor to a surface do not necessarily work for other motors. To date, three generalizable approaches have shown promise. The first is to adsorb antibodies to the surface and immobilize the motors through these antibodies [64]. This approach has the advantage that antibodies complementary to virtually any motor can be made or, by attaching a universal protein tag (such as a hexa-histidine tag) to recombinant motors, one reliable antibody to be used for a range of motors. A second approach for hexa-histidine tagged motors is to functionalize the surface with a surfactant terminated with nitrilotriacetic acid (NTA), which chelates nickel ion. His-tagged proteins bind tightly to the surface through the immobilized Ni-NTA and the motors remain functional on these surfaces [65]. A third method is to attach motors to surfaces through biotin–avidin chemistry [66]. Streptavidin (or its equivalents avidin or neutravidin) can be directly adsorbed to surfaces or specifically bound to biotin-functionalized surfaces and biotinylated motors can be attached to this immobilized streptavidin. Motors attached by this means have been shown to be completely functional and the bond strong enough to not be pulled off by motor forces [67–69].

While a range of hydrophilic surfaces support kinesin-driven microtubule movements, hydrophobic surfaces do not. When glass surfaces are treated with the hydrophobic silane octadecyl trichlorethylsilane or other hydrophobic surface treatments (with contact angles above 60°) prior to casein and kinesin treatment, no microtubule binding or movement is observed [57, 70]. The interpretation is that motors denature on these hydrophobic surfaces and even casein pretreatment cannot prevent this motor denaturation. This property of surface chemistry-dependent motor function has been used to define where on microfabricated surfaces microtubules will be transported and where they will not (discussed further in Section 8.5).

8.4.2
Microtubule Immobilization

For applications where microtubules are immobilized on surfaces and cargo-functionalized kinesins move along these filaments, a number of strategies have been developed for immobilizing microtubules. Microtubules will bind to clean glass or quartz microscope slides [71] and this immobilization is sufficient for some investigations into motor function. However, for transport applications, a more robust immobilization is generally required. One of the most common immobilization approaches is to functionalize glass surfaces with an amino silane compound (3-aminopropyltriethoxysilane) that confers a positive charge to the surface [39, 72]. Microtubules, which have a net negative charge at physiological pH, bind tightly to these surfaces and high microtubule densities can be achieved. By lithographically patterning silanes on silicon wafers, this approach has been used to pattern immobilized microtubules at microscale dimensions [73]. Poly-lysine-treated glass is another approach that uses electrostatic interactions to bind microtubules to surfaces [74].

Two other microtubule immobilization strategies have also been shown to work well. Microtubules can be biotinylated and attached to surfaces through streptavidin, either by adsorbing streptavidin directly to the surface or immobilizing another biotinylated protein (like biotinylated bovine serum albumin), and then using streptavidin as a glue between that protein and the biotinylated microtubule [37, 75]. In principle, self-assembled monolayers (SAMs) in which a portion of the groups are terminated with biotin could be patterned, and streptavidin and microtubules patterned by the underlying biotin. Other techniques for patterning biotin or streptavidin at micron or nanoscale dimensions that have been developed for DNA microarrays or nanoscale two-dimensional protein patterns could be applied to patterning biotinylated microtubules. The final immobilization strategy, in the "turning lemons into lemonade" category, involves using dysfunctional kinesin motors to immobilize microtubules [76, 77]. If kinesin protein is mishandled, e.g. stored at room temperature for days, or when problems arise during expression and purification of functional motors, the motors inactivate in such a way that they bind to microtubules, but do not move along them. When these "dead-heads" are adsorbed to surfaces, they act as an excellent adhesive to immobilize microtubules.

While the optimum immobilization approach depends upon the specific application, there are some general considerations that apply to all microtubule immobilization techniques. First, approaches that permit patterning microtubules on the micron or submicron scale are helpful only if filament orientation can be controlled. For instance, even with narrow strips of adhesive, microtubules can simply lay across the lines, negating the pattern. Finally, from experiences in our laboratory and others, it is clear that from the perspective of kinesins, not all microtubule immobilization strategies are created equal. For instance, when microtubules are immobilized through aminosilanes on glass, motors rarely interact with microtubules that are tightly adsorbed to the surface through their entire length, while they interact much more frequently with microtubules that are more loosely tacked down and have regions that are not directly attached to the surface. Whether these problems are due to deformation of the microtubules when they are tightly adsorbed or to unfavorable motor–surface interactions is not clear, but these observations emphasize that the success of any immobilization strategy must include an analysis of motor function as well.

8.5
Controlling the Direction and Distance of Microscale Transport

The key to harnessing the transport capabilities of molecular motors is controlling the direction of motion. For applications in microscale transport, it is this ability to direct the transport of a particle or analyte independent of fluid flows or concentration gradients that has generated the most interest in the field to date. Redirecting microtubule movements in the microtubule gliding assay has received the most attention because the filaments move long distances (detachment is not a concern), the filaments can be easily visualized by fluorescence microscopy and they can be functionalized to transport cargo. In this section, transport applications utilizing the filament gliding is covered first, followed by the opposite geometry, i.e. cargo-loaded motors moving along immobilized filaments.

8.5.1
Directing Kinesin-driven Microtubules

Building on the standard microtubule gliding assay, the first directed transport investigations showed that micron-scale grooves deposited on glass surfaces either by shearing polytetrafluoroethylene (PTFE) or by microfabrication guide microtubule motions parallel to the grooves [2, 4, 7, 78, 79]. This redirection arises from the nature of microtubule movements along surfaces – the front of the filament searches out new motors to bind to as it is propelled along the surface and physical barriers that reorient this free end act to reorient the direction of filament movement (Figs. 8.4 and 8.5). Building on these initial demonstrations, a number of studies using more sophisticated fabrication approaches have amply showed that microtubules can be guided and redirected using surface features. A key advance, first shown by Hiratsuka and coworkers [6], was the demonstration that microtubules moving in microfabricated channels can be redirected by arrowhead-shaped "rectifiers" built into the channels that pass filaments traveling one direction, and buckle and

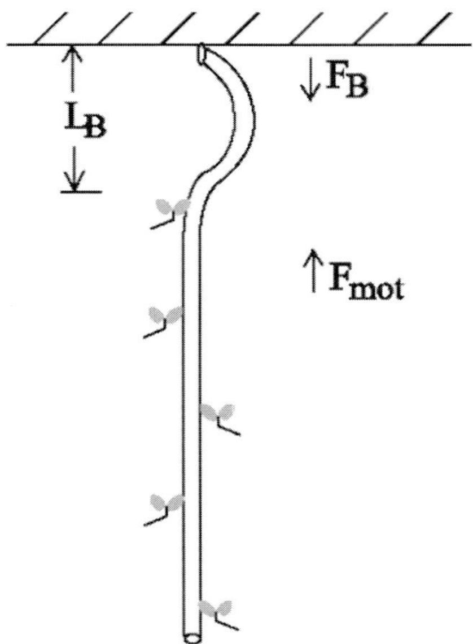

Figure 8.4. Physical model of a microtubule, propelled by immobilized kinesins, colliding with a wall and buckling. The length of filament that can buckle, L_B, is defined as the distance between the last motor and the wall. For the microtubule to buckle, the cumulative motor forces must be greater than the minimum buckling force for approach normal to the surface, but can be less than that if the incident angle is less than 90°. (Reprinted with permission from Ref. [2] © 2003 American Chemical Society.)

redirect filaments traveling in the opposite direction. Later investigations extended this result using different fabrication approaches and a range of rectifier shapes (Fig. 8.6) [53, 80]. An important advance in this study was to identify photoresist materials that prevent adsorption of functional kinesin motors, such that when the photoresist is patterned on glass, functional motors are only found on the glass surfaces in the bottom of the channel. In both the work by Hiratsuka and co-workers and in a related study by Moorjani and coworkers [2], the key to containing the movement in the channels was not preventing protein adsorption to the walls, but rather the fact that motors that adsorbed to the photoresist denatured or otherwise inactivated. When the photoresist walls were treated to enable functional motor adsorption, the microtubules simply crawled over the walls [2].

One goal in achieving useful transport by this approach is minimizing the number of filaments lost due to detachment from and diffusion away from the surface. If filaments are transported along the surface by a population of motors that each cyclically attach and detach from the filament, there is a low probability that all of the motors are detached at the same time and so long transport distances can be achieved. However, when filaments are redirected by walls or crawl over walls of the channels, there can be significant filament loss if the channels are not enclosed on all sides [2, 79, 81]. This filament loss can be minimized by undercutting the walls of the channels to create overhangs that tend to trap the filaments in the

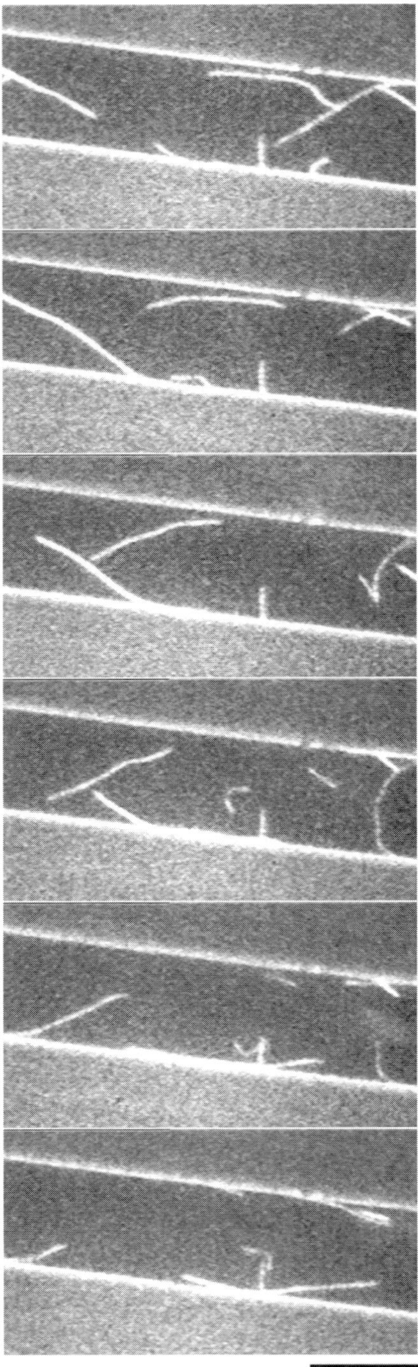

channels [3]. Also, by designing rectifying shapes and patterning gold on the bottom of the channel to maximize motor adsorption, and silanes on the sides of the channels to resist motor adsorption, high efficiencies of filament rectification have been achieved [80]. Experiments using this "open-top" configuration where motors are adsorbed to patterned and textured surfaces, and the top of the channels are left open to solution, have been instrumental in understanding the optimal channel design and surface chemistry for redirecting microtubule movements in microchannels. However, the best method for eliminating filament loss is to enclose these microchannels on all sides. For practical devices to emerge from this area of research, it is clear that enclosed microchannels will be required, together with apparatus for sample inputs and downstream detectors. As described below, there has been some initial progress toward these goals.

8.5.2
Movement in Enclosed Microchannels

Achieving directed movement in enclosed channels entails different design constraints than movement along topographically patterned surfaces. For instance, because microtubules are surrounded on four sides, detachment from the surface is less of a problem because subsequent rebinding is quite rapid. Furthermore, whereas guiding microtubules with open-top channels generally requires that the side walls are not populated with functional motors, in enclosed channels it is fine for microtubules to move along all four walls so long as they are moving down the channel in the desired direction. However, there are some important design constraints for creating enclosed channels for microtubule transport. First, because fluorescence microscopy is generally used to visualize microtubule movements in the channels, at least one side must be transparent and materials that autofluoresce (which includes many photoresists) need to be avoided. Second, as described in Section 8.7 below, polydimethylsiloxane (PDMS), which is regularly used in microfluidics work, can generally not be used because of its high oxygen permeability. Third, because the channels are initially filled and motor-functionalized by pressure-driven flow, the bond between the channels and the top enclosure must be sufficiently strong. Finally, the channel design must include fluidic connections for injecting motor and microtubule solutions into the channels.

Huang and coworkers recently succeeded in achieving kinesin-driven microtubule movements in enclosed microfabricated channels [57]. A number of channel designs were investigated, and a key breakthrough was creating hierarchical channels of decreasing dimensions such that solutions could be exchanged into small

Figure 8.5. A microtubule being redirected by a microfabricated channel wall. The channel is glass and the walls are 1.5-µm high SU-8 photoresist. The microtubule moves from left to right in the channel, bumps into the photoresist wall and is redirected back into the channel. The scale bar at right is 10 µm; images are 12 s apart. (Reprinted with permission from Ref. [2] © 2003 American Chemical Society.)

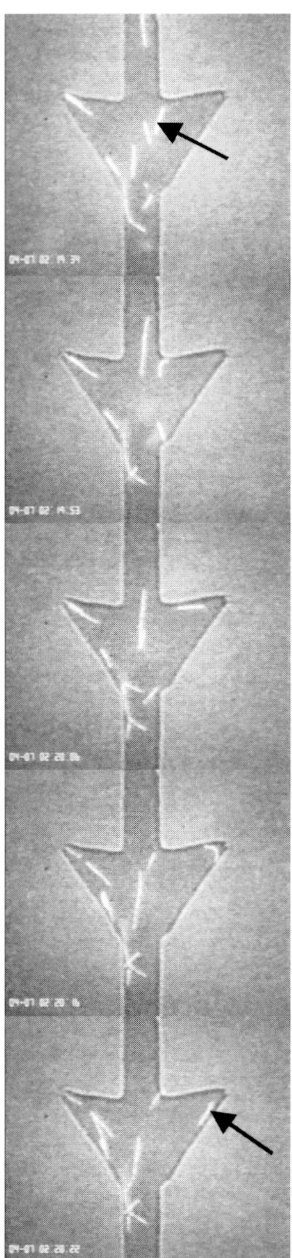

10 μm

channels 15 μm wide and 5 μm deep without large dead volumes. The channels were constructed in SU-8 photoresist patterned on glass, covered with dry film photoresist and topped with a silicon wafer or glass slide for mechanical stability. Microtubules moved long distances in the channels, indicating that the ATP fuel was not limiting in these enclosed volumes. Furthermore, the microtubules could move upstream in convective fluid flows, which means biomotor-driven transport can move against microfluidic transport [57]. Current work in this area aims to simplify fabrication processes and optimize the channel geometries to achieve ideal microtubule transport, concentration and redirection.

8.5.3
Immobilized Microtubule Arrays

While most studies investigating transport applications of kinesins have utilized the microtubule gliding configuration, there has also been progress on immobilizing aligned microtubules and transporting motor-functionalized cargo along these filaments. Conceptually, binding motors to analytes and transporting this cargo along immobilized filaments is the simpler geometry and because of the ease with which motors can be engineered to contain diverse cargo binding domains in place of their tail, it has significant potential. However, the key hurdle to making this geometry work is immobilizing parallel and *uniformly oriented* filaments on surfaces. While filaments can be aligned in fluid flows, they are of mixed orientation and hence are of little use in transport applications. Current progress in obtaining arrays of uniformly oriented immobilized filaments is described below.

Brown and Hancock created an array of uniformly oriented microtubules by immobilizing short microtubule seeds at defined locations, growing filaments selectively off of the "plus" ends of these seeds, aligning the newly polymerized filaments by fluid flow and then immobilizing them on a surface (Fig. 8.7A) [46]. This process created aligned filaments, but the arrays were not the density needed for useful transport and the method is difficult to adapt to microscale environments. Limberis and coworkers used an elegant technique to bind and align microtubules [82]. Antibodies to the α-tubulin subunit, which is exposed only at the "minus" ends of microtubules and not the "plus" ends, were immobilized on a surface and microtubules flowed in and allowed to bind end-on (Fig. 8.7B). Fluid flow was then used to push the filaments over and, although they were not immobilized in this study, a final immobilization step should not be difficult. A third technique, which may have the most potential, was demonstrated by Prots and coworkers,

Figure 8.6. Arrowhead-shaped structures in photoresist rectify microtubule movements. In this example, 1.5-μm high SU-8 photoresist walls on glass create a rectifier with inlet and outlet channels. A microtubule can be seen moving upward inside the arrowhead, bumping into the photoresist wall and being redirected back into the channel. Rhodamine-labeled microtubules and SU-8 channel walls were imaged using simultaneous fluorescence and differential interference contrast microscopy. (Reprinted from Ref. [53], with kind permission of Springer Science and Business Media.)

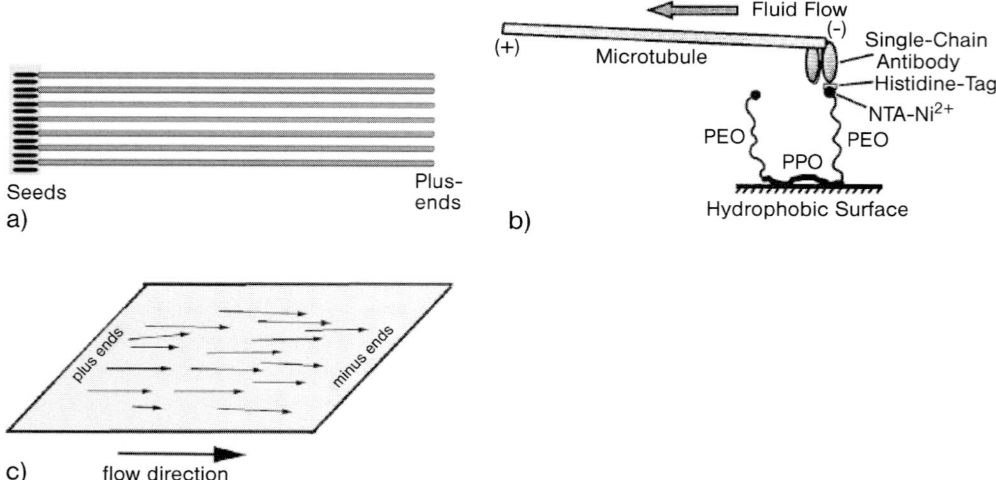

Figure 8.7. Three methods to create an array of uniformly oriented microtubules on a surface. (A) Brown and Hancock immobilized short microtubule seeds on an adhesive region, grew microtubules exclusively from the "plus" ends of these filaments, and then aligned and immobilized them. (B) Limberis and coworkers bound antibodies for α-tubulin to the surface, captured microtubule "minus" ends and then aligned the microtubules by fluid flow. (C) Prots and coworkers used fluid flow to align kinesin-driven microtubules and then immobilized these filaments on the surface. (Reprinted with permission from Refs. [46, 82, 83] © 2002 and 2001 American Chemical Society and 2003 Elsevier, respectively.)

who started with a field of microtubules moving over immobilized motors, aligned their direction of movement using fluid flow (over time all filaments move parallel and downstream to the flow) and then crosslinked the filaments to the motors using glutaraldehyde (Fig. 8.7C) [83]. Other experiments have shown that glutaraldehyde treatment does inhibit subsequent interactions of the microtubules with motors [84] and the authors showed that kinesin-functionalized beads move unidirectionally along these filament arrays [10].

One of the downsides of using kinesins moving over immobilized microtubules is that the transport distances are generally shorter than distances achieved by many immobilized motors moving long microtubules across surfaces. Single-molecule experiments have shown that during each encounter, conventional kinesin moves approximately 1 μm along a filament before detaching [39, 85]. These motors can then reattach and move further; however, for long distance transport, longer run lengths are desired. Somewhat longer transport distances can be achieved by modifying the motors to increase the amount of positive charge in the "neck" region adjacent to the head [86] or by binding multiple motors to the desired cargo such that dissociation of one motor does not lead to dissociation of the entire cargo. These dissociation problems highlight the fact that if immobilized microtubules are used, they not only need to be uniformly aligned, but they need to be immobilized at high densities to maximize the rate of motor reattachment. To make this geometry feasible for analyte transport applications, more work needs to be done to develop facile methods to create high density, uniformly oriented microtubule arrays.

8.6
Cargo Attachment

Another important hurdle for achieving useful transport from the kinesin–microtubule system is developing methods for selective and reversible attachment of cargo to the microtubules and/or motors. For microtubules moving in microchannels, the long-term vision is that the microtubules will pick up their cargo at one site and be transported along a channel, and the cargo will be deposited in a collection or analysis chamber. The question is: how is cargo attached to microtubules? Successful approaches and future ideas are reviewed here.

One problem with attaching cargo to microtubules is that, in contrast to motors, it is difficult to use the tools of molecular biology to modify tubulin sequences. Generally, tubulin for these applications is purified from cow or pig brain and it is currently prohibitively expensive to produce genetically engineered large animals for this purpose. *In vitro* expression of recombinant tubulin has been achieved [87], but it is technically demanding and yields are low. Hence, the primary route to modifying microtubules to enable cargo attachment has been to covalently link functional cargo attachment groups to the microtubules. By attaching reactive *N*-hydroxysuccinimidyl (NHS) groups to exposed amine residues on tubulin, microtubules have been covalently labeled with a number of fluorescent dyes and biotin [34]. Biotinylated microtubules have by far been the most widely used. Avidin (or its relatives streptavidin and neutravidin) contains four high-affinity biotin binding sites, enabling sandwich configurations where biotin-functionalized cargo can be attached to biotinylated microtubules [88, 89].

Proteins, single-stranded (ss) DNA or RNA molecules and cells represent three potential types of cargo that can be transported, separated and/or concentrated by the kinesin–microtubule system (Fig. 8.8). To date, all three types of cargo have been attached to microtubules through biotin–avidin chemistry. Biotinylated antibodies can be purchased from commercial sources or antibodies of interest can be biotinylated using amine-reactive biotin sold by Molecular Probes (Invitrogen,

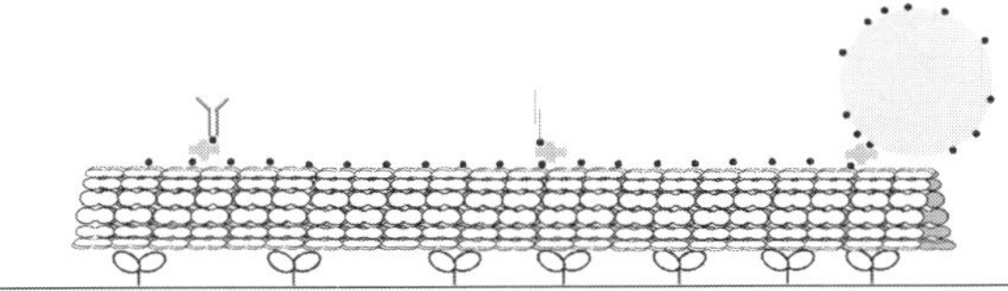

Figure 8.8. Attaching cargo to a microtubule through biotin–avidin chemistry. Biotin molecules (black dots) that are covalently linked to tubulin subunits bind streptavidin protein (which has four biotin binding sites). Biotin-functionalized antibodies (left), DNA oligonucleotides (center) or particles (right) can be linked through the streptavidin and transported by the immobilized kinesin motors.

Carlsbad, CA). These biotinylated antibodies can be attached to microtubules by incubating biotinylated microtubules first with neutravidin and then with the biotinylated antibody, but care must be taken not to overload the microtubule [89]. As the neutravidin binds to the same microtubule surface that the kinesin motors bind to, a portion of the tubulin subunits must remain unlabeled to enable motor binding. By attaching anti-GFP (green fluorescent protein) antibody to microtubules, we showed that microtubules can pick up and transport protein cargo along kinesin-functionalized surfaces [90]. This important proof of principle implies that microtubules could be modified to transport any protein for which a suitable antibody exists.

A second important cargo is nucleic acids. A possible application for a kinesin-based analytical device would be RNA analysis for examining gene expression in a small tissue sample or for detecting viruses. DNA microarrays are now widely used for expression profiling – determining what mRNA species are being expressed in a cell or tissue under certain conditions. However, if the sensitivity could be increased, smaller sample volumes could be used, ideally to the point where mRNA levels from individual cells could be analyzed. As this enhancement would remove any signal loss due to heterogeneity between different cells in a tissue biopsy, this is an active area of research. A kinesin-based transport device would be very valuable if it could bind selected mRNA from a cell lysate, transport these analytes down microscale channels, detect their presence and concentrate them in a collection chamber for later analysis such as sequencing or polymerase chain reaction (PCR) amplification (Fig. 8.9). To demonstrate the feasibility of nucleotide trans-

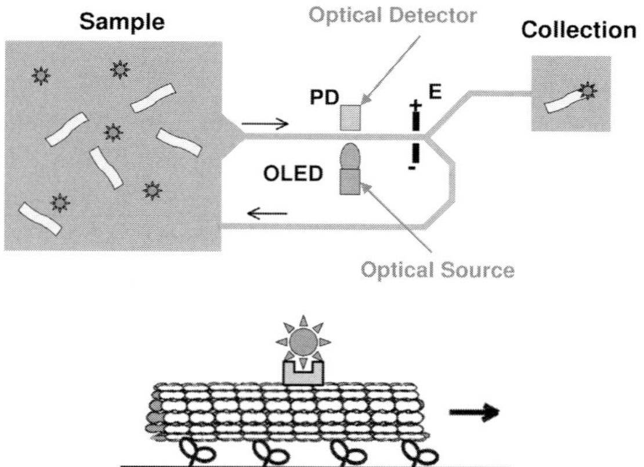

Figure 8.9. Diagram of a motor-based detection and purification scheme. Top panel shows an integrated system including a sample chamber containing microtubules and analyte, kinesin-functionalized microchannels through which cargo is transported, a LED excitation source and photodiode (PD) detector integrated into the channels, and electrodes (E) to direct cargo-laden microtubules to the collection chamber. Bottom panel shows that kinesin-based transport that will drive the system. (Reprinted from Ref. [53], with kind permission of Springer Science and Business Media.)

port by microtubules, Muthukrishnan and coworkers bound ssDNA oligonucleotides to microtubules through biotin–avidin chemistry and then bound fluorescently labeled complementary DNA oligonucleotides to these microtubules. Like the protein cargo-functionalized microtubules, these DNA cargo-functionalized microtubules were faithfully transported across kinesin functionalized surfaces [88]. Diez and coworkers showed that kinesins can also be used to transport and stretch large pieces of DNA. λ-phage DNA attached to microtubules through biotin–avidin chemistry was transported along surfaces by immobilized motors and, when bound at one end to the surface or another microtubule, could be stretched by the motor forces up to its maximum contour length [91].

Another cargo attachment strategy that has been investigated is using cyclodextrins attached to microtubules [92]. These versatile molecules are cyclic oligosaccharides that contain a hydrophobic cavity that can be engineered to bind small molecules and proteins [93]. As they can be biotinylated, they can be attached to biotinylated microtubules through a streptavidin crosslinker and transported by surface-immobilized kinesins. Kato and coworkers reported that cyclodextrin-functionalized microtubules moved more than 10-fold slower than normal microtubules, indicating the attachment unfavorably altered the surface of the microtubule [92]. Hence, work needs to be done to improve this strategy, but it has the potential to be of significant utility. Other cargo-binding approaches can be envisaged such as designing RNA aptamers through directed evolution [94], and attaching these to microtubules through biotinylated bases, but no work has been published to date in this area.

8.6.1
Maximum Cargo Size

An important question for kinesin-based transport is: what is the size limit for cargo transported by kinesins and microtubules? To investigate the potential utility of kinesins in microscale devices, Limberis and Stewart attached kinesins to $10 \times 10 \times 5\text{-}\mu\text{m}^3$ silicon microchips and found the chips moved at normal motor speeds along immobilized microtubules [56]. Similarly, Jia and coworkers attached clumps of gold nanowires 6 μm long to microtubules and found that they were transported at normal speeds along motor functionalized surfaces [53]. The fact that the viscous drag forces did not affect the motor transport speeds suggests that even larger objects can be transported. The theoretical maximum size of an object that can be attached to a microtubule and transported by kinesins is calculated below.

The theoretical size limit for spherical objects that can be attached to microtubules and transported by molecular motors can be calculated by comparing motor forces to viscous drag forces for a sphere in aqueous buffer. Individual motors can exert 5–8 pN of force [37, 38, 95]. Kinesin motors can be adsorbed on surfaces to densities of 1000 motors μm^{-2} or more, so it is reasonable to expect a long microtubule to be able to interact with hundreds of motors [42]. If motor forces sum (which has not been definitively shown, but is consistent with a number of experimental observations), then it is possible that motor forces pushing a microtubule could approach 1 nN. The drag force on a particle moving in aqueous solution is equal to

$$F = \gamma v \tag{6}$$

where γ is the drag coefficient and v is the velocity [52]. For a microtubule moving cargo at 1 μm s^{-1}, the drag coefficient, γ, that results in 100 pN of force, which would start to slow the microtubule, is 10^{-4} Ns m^{-1}. The drag coefficient of a spherical object can be calculated from Stokes' law, given in Eq. (2) above. Solving for the diameter of a sphere that would start to slow down the microtubule – the result is 1 cm! Hence, the range of objects that can in principle be transported by this system has only begun to be explored.

8.7
System Design Consideration

The development of biomotor-driven microscale devices has evolved in three phases: (a) recapitulating intracellular motility in *in vitro* experiments, (b) controlling microtubule transport using microfabrication, surface processing and other techniques, and (c) creating integrated functional devices. The first phase was completed by the early 1990s. The second phase has been the focus of considerable effort over the last five years and, although it will continue to advance, the progress will likely slow somewhat. It is now that researchers are turning to the third phase, i.e. device development. There are a number of issues that need to be resolved for this work to move forward, such as protein lifetimes, sample introduction, and analyte detection and processing. Progress to date in these directions and future efforts towards these goals are reviewed here.

8.7.1
Protein Stability and Lifetime

As a tool in nanotechnology, there is a trade-off between the exquisite functionality of biomolecular motors and their robustness. For their small size, kinesins are able to achieve long-range transport of microtubules at appreciable speeds and generate significant forces. Due to their small size, they have the potential to be integrated into very microscale and nanoscale device geometries, enabling redundancy, multiplexing, minute sample analysis and minimal reagent costs. However, because they are proteins, they are much more fragile than most nanoscale materials. To assess their potential utility when integrated into analytical devices, and to investigate alternate design approaches, it is necessary to determine the lifetime of kinesin motors and microtubules under normal operating conditions, and then to find conditions that extend these lifetimes.

Under normal laboratory conditions, taxol-stabilized microtubules are stable at room temperature for roughly 1 week [96, 97], and kinesin motors are stable on the order of hours to days depending on their concentration, storage buffer, temperature and other variables. For experiments on motor fundamentals, as long as the motors are stable for a working day, there is no need to put the work in to extend their lifetime because each morning a new aliquot of protein can be thawed from the $-80\,^{\circ}$C freezer. Hence, it has been the interest in device integration that has driven experiments to uncover the mechanisms that limit the working lifetimes of kinesin and microtubules, and to extend these lifetimes.

In the kinesin–microtubule field, the most convenient technique for observing motor function is fluorescence microscopy. Microtubules can be visualized by differential interference microscopy, but for applications where motors and/or cargo are to be observed fluorescence is much more versatile. In the microfluidics field, a standard approach for creating channels has been PDMS molding and bonding to glass [98] and in a perfect world these two approaches could be combined. However, workers studying fluorescent microtubule movements in PDMS channels quickly ran into problems with photodamage [99]. In the standard microtubule gliding assay, the microtubule cocktail includes buffering salts, ATP to drive motility, and an antifade consisting of an oxygen-scavenging system and reducing agent [41, 45]. In the absence of antifade, exposure of the fluorescent microtubules to the high-intensity mercury arc lamp used for fluorescence excitation results in rapid photobleaching, loss of movement and microtubule depolymerization due to oxygen free radical damage. With antifade, the movement of rhodamine-labeled microtubules can be visualized for roughly 10 min under continual high intensity illumination. Without illumination or by shuttering the light, microtubule movement lasts for around 10 h or longer [100]. However, when the flow cell is enclosed with PDMS instead of glass, illumination leads to inhibition of motion and microtubule breakage or depolymerization after approximately 1 min [99, 100]. The cause of failure was traced to the high oxygen permeability of this silicone polymer, which overwhelms the antifade's ability to remove all of the soluble oxygen. PMMA enclosures also showed this same property. While these results do not rule out these materials as device components, if they are to be used they must be combined with another barrier to oxygen diffusion, increasing the complexity of fabrication.

A central concern in the feasibility of kinesin–microtubule powered devices is long-term storage: if hybrid biological/synthetic devices are manufactured, how long can they be preserved before use? Interestingly, this is a similar concern in the field of tissue engineering, where products like artificial skin that include live cells integrated into a polymer scaffold must be made at a manufacturing site, stored and shipped, and then unpacked and used by a doctor. The obvious first approach is freezing; kinesin motors, tubulin and even taxol-stabilized microtubules can be frozen on liquid nitrogen and stored at $-80\,^{\circ}$C for years with little loss in functionality. In theory, hybrid devices could be simply dunked in liquid nitrogen and thawed before use, but there are a number of possible problems with this approach, including mechanical problems due to unequal expansion coefficients of the various materials and the requirement for a deep freeze at every point of use.

There has been some recent work on novel methods for extending kinesin and microtubule lifetimes that has yielded interesting results. With the goal of identifying material-processing approaches that could be used to created hybrid devices, Verma and coworkers tested the stability of casein, kinesin motors and microtubules when exposed to various solvents and strippers used in materials processing and nanofabrication. Interestingly, kinesin motors survived when exposed to either pure isopropyl alcohol or acetone – two solvents commonly used to remove photoresist [101]. Microtubules were less robust than motors, but they did survive exposure to 1:4 dilutions of these chemicals in aqueous buffer and these diluted solvents retained their ability to strip photoresist. While not directly impacting device lifetime, these studies provide new approaches for device manufacture and they suggest that these proteins are more stable than may be indicated by their lifetimes in buffer at room temperature. In other work, it was found that drying immobilized kinesins by wicking the solution out of flow cells and storing them under different conditions resulted in longer kinesin lifetimes [102]. Hence, it is possible that instead of freezing hybrid devices, simply desiccating them and then reconstituting them may provide a simple alternative. These findings should lessen the constraints to hybrid device design.

8.7.2
Sample Introduction and Detection

Two future hurdles to creating kinesin-driven analytical devices are sample introduction and analyte detection. To date there is scant work in this area, but because these requirements are universal to virtually all sensor systems, ideas and modules will undoubtedly be borrowed from other disciplines. With respect to sample introduction, one design constraint is that these biomotors work in aqueous environments. As discussed above, kinesins can survive being dried or exposed to solvents, but for their proper function they must remain in proper buffer solutions. Hence, to detect airborne particles (e.g. viruses or bacteria), the samples would need to be concentrated (i.e. filtered) from the atmosphere and then transferred to the aqueous state. While not impossible, there are clearly a number of engineering problems to be solved for this application to become feasible.

An area where the strengths of kinesin-powered devices are maximized is the detection of proteins, RNA or small molecules in minute aqueous samples. As discussed above, detection of specific molecules from single cells is a holy grail of sorts for bioanalytical detection. Delivering the contents of single cells to kinesin-powered transport and detection machinery is a task in itself. Both convective flows and dielectrophoresis have been used to transport cells in microfluidic systems [103, 104]. Additionally, both optical and electrical methods have been developed to rapidly lyse cells, enabling their intracellular contents to be delivered to the microanalysis machinery [105, 106].

8.7.3
Analyte Detection and Collection

As in all analytical systems, analyte detection will be a key component to future kinesin-powered hybrid devices. The first choice for analyte detection is fluorescence, because it is sensitive and is the standard technique for observing microtubules moving in channels. Building on work guiding microtubule transport in microfluidic channels, analyte detection could be achieved by binding the analyte to the microtubule and observing the change in fluorescence. If the analyte (cell, protein or nucleotide) is fluorescent and has a different emission peak than the fluorophore labeling the microtubule, analytes could be detected by observing the microtubule fluorescence at one wavelength and measuring the analyte fluorescence at a second wavelength. In this case, the microtubule acts to both concentrate and to localize the analyte.

While this detection could be achieved using epifluorescence microscopy, there is potential for achieving much more utility and sensitivity by integrating the optical components into the microfluidic chips containing the motors and analyte. For instance, if an LED and photodiode are integrated into the microchannel containing the microtubule and bound analyte, the detector could be within microns of the sample (Fig. 8.9). This placement should maximize photon capture and reduce background signal due to extraneous signal. One advantage of using the kinesin–microtubule transport system for analyte detection is that the movement speed is stereotyped, so for low signal levels the time average of fluorescence as the analyte moves across the detector could be used to increase the signal-to-noise ratio.

8.8
Conclusion

Due to the highly evolved properties of protein machines, there is incredible potential in integrating them into hybrid biological/synthetic devices. Their nanometer-size scales match well with emerging nanofabrication capabilities and novel functional nanoscale materials. A continuing struggle in nanotechnology is finding techniques to manipulate and organize molecules and nanoparticles in cases where self-assembly is not energetically favorable. As biomolecular motors have evolved to transport and organize biological materials on the nanoscale and microscale in cells, they have the potential to solve a number of these general problems in nanotechnology. This chapter reviewed the *in vitro* transport capabilities of kinesin motors, with an eye toward creating microfluidic devices for analyte detection. This application is one of the first envisaged for these particular protein machines and it is anticipated that as new approaches are developed for to gain ever more control over the kinesin–microtubule system, these biological machines will be applied to other nanoscale systems.

Acknowledgments

The author wishes to thank members of his laboratory, especially Maruti Uppalapati, Gayatri Muthukrishnan, Zach Donhauser and Samira Moorjani, as well as collaborators and their students, particularly Tom Jackson, Lili Jia, Ying-Ming Huang, Jeff Catchmark and Vivek Verma. The author's research in this area is supported by the NSF and NIH/NIBIB, and by the Penn State Center for Nanoscale Science (NSF MRSEC DMR0213623).

References

1 ALBERTS, B. The cell as a collection of protein machines: preparing the next generation of molecular biologists. *Cell* **1998**, *92*, 291–294.

2 MOORJANI, S. G., JIA, L., JACKSON, T. N., HANCOCK, W. O. Lithographically patterned channels spatially segregate kinesin motor activity and effectively guide microtubule movements. *Nano Lett.* **2003**, *3*, 633–637.

3 HESS, H., MATZKE, C. M., DOOT, R. K., CLEMMENS, J., BACHAND, G. D., BUNKER, B. C., VOGEL, V. Molecular shuttles operating undercover: a new photolithographic approach for the fabrication of structured surfaces supporting directed motility. *Nano Lett.* **2003**, *3*, 1651–1655.

4 HESS, H., BACHAND, G. D., VOGEL, V. Powering nanodevices with biomolecular motors. *Chemistry* **2004**, *10*, 2110–2116.

5 HESS, H., VOGEL, V. Molecular shuttles based on motor proteins: active transport in synthetic environments. *J. Biotechnol.* **2001**, *82*, 67–85.

6 HIRATSUKA, Y., TADA, T., OIWA, K., KANAYAMA, T., UYEDA, T. Q. Controlling the direction of kinesin-driven microtubule movements along microlithographic tracks. *Biophys. J.* **2001**, *81*, 1555–1561.

7 STRACKE, R., BOHM, K. J., BURGOLD, J., SCHACHT, H.-J., UNGER, E. Physical and technical parameters determining the functioning of a kinesin-based cell-free motor system. *Nanotechnology* **2000**, *11*, 52–56.

8 STRACKE, R., BOHM, K. J., WOLLWEBER, L., TUSZYNSKI, J. A., UNGER, E. Analysis of the migration behaviour of single microtubules in electric fields. *Biochem. Biophys. Res. Commun.* **2002**, *293*, 602–609.

9 BOHM, K. J., STRACKE, R., UNGER, E. Motor proteins and kinesin-based nanoactuatoric devices. *Tsitol. Genet.* **2003**, *37*, 11–21.

10 BÖHM, K. J., BEEG, J., MEYER ZUR HÖRSTE, G., STRACKE, R., UNGER, E., Kinesin-driven sorting machines on large scale microtubule arrays. *IEEE Adv. Packaging* **2005**, *28*, 571–576.

11 WHALEY, S. R., ENGLISH, D. S., HU, E. L., BARBARA, P. F., BELCHER, A. M. Selection of peptides with semiconductor binding specificity for directed nanocrystal assembly. *Nature* **2000**, *405*, 665–668.

12 SEEMAN, N. C., BELCHER, A. M. Emulating biology: building nanostructures from the bottom up. *Proc. Natl Acad. Sci. USA* **2002**, *99* *(Suppl. 2)*, 6451–6455.

13 ANRATHER, D., SMETAZKO, M., SABA, M., ALGUEL, Y., SCHALKHAMMER, T. Supported membrane nanodevices. *J. Nanosci. Nanotechnol.* **2004**, *4*, 1–22.

14 HOWARD, J. Molecular motors: structural adaptations to cellular functions. *Nature* **1997**, *389*, 561–567.

15 MAVROIDIS, C., DUBEY, A., YARMUSH, M. L. Molecular machines. *Annu. Rev. Biomed. Eng.* **2004**, *6*, 363–395.

16 VALE, R. D., MILLIGAN, R. A. The way things move: looking under the hood of molecular motor proteins. *Science* **2000**, *288*, 88–95.

17 HANCOCK, W. O., HOWARD, J. Kinesin: processivity and chemomechanical coupling. In: *Molecular Motors*, SCHLIWA, M. (Ed.). Wiley-VCH, Weinheim, **2003**, pp. 243–269.

18 KNOBLAUCH, M., PETERS, W. S. Biomimetic actuators: where technology and cell biology merge. *Cell. Mol. Life Sci.* **2004**, *61*, 2497–2509.

19 KINBARA, K., AIDA, T. Toward intelligent molecular machines: directed motions of biological and artificial molecules and assemblies. *Chem. Rev.* **2005**, *105*, 1377–400.

20 NICOLAU, D. V., SUZUKI, H., MASHIKO, S., TAGUCHI, T., YOSHIKAWA, S. Actin motion on microlithographically functionalized myosin surfaces and tracks. *Biophys. J.* **1999**, *77*, 1126–1134.

21 ASOKAN, S. B., JAWERTH, L., CARROLL, R. L., CHENEY, R. E., WASHBURN, S., SUPERFINE, R. Two-dimensional manipulation and orientation of actin-myosin systems with dielectrophoresis. *Nano Letters*. *3*, 431–437.

22 PATOLSKY, F., WEIZMANN, Y., WILLNER, I. Actin-based metallic nanowires as bio-nanotransporters. *Nat. Mater.* **2004**, *3*, 692–695.

23 MARTINEZ-NEIRA, R. K. M., NICOLAU, D., DOS REMEDIOS, C. G. A novel biosensor for mercuric ions based on motor proteins. *Biosens. Bioelectron.* **2005**, *20*, 1428–1432.

24 HIROKAWA, N., PFISTER, K. K., YORIFUJI, H., WAGNER, M. C., BRADY, S. T., BLOOM, G. S. Submolecular domains of bovine brain kinesin identified by electron microscopy and monoclonal antibody decoration. *Cell* **1989**, *56*, 867–878.

25 YANG, J. T., LAYMON, R. A., GOLDSTEIN, L. S. A three-domain structure of kinesin heavy chain revealed by DNA sequence and microtubule binding analyses. *Cell* **1989**, *56*, 879–889.

26 HACKNEY, D. D. Evidence for alternating head catalysis by kinesin during microtubule-stimulated ATP hydrolysis. *Proc. Natl Acad. Sci. USA* **1994**, *91*, 6865–6869.

27 HOWARD, J. The movement of kinesin along microtubules. *Annu. Rev. Physiol.* **1996**, *58*, 703–729.

28 SCHIEF, W. R., HOWARD, J. Conformational changes during kinesin motility. *Curr. Opin. Cell Biol.* **2001**, *13*, 19–28.

29 VALE, R. D., REESE, T. S., SHEETZ, M. P. Identification of a novel force-generating protein, kinesin, involved in microtubule-based motility. *Cell* **1985**, *42*, 39–50.

30 LAWRENCE, C. J., DAWE, R. K., CHRISTIE, K. R., CLEVELAND, D. W., DAWSON, S. C., ENDOW, S. A., GOLDSTEIN, L. S., GOODSON, H. V., HIROKAWA, N., HOWARD, J., MALMBERG, R. L., MCINTOSH, J. R., MIKI, H., MITCHISON, T. J., OKADA, Y., REDDY, A. S., SAXTON, W. M., SCHLIWA, M., SCHOLEY, J. M., VALE, R. D., WALCZAK, C. E., WORDEMAN, L. A standardized kinesin nomenclature. *J. Cell Biol.* **2004**, *167*, 19–22.

31 MIKI, H., SETOU, M., KANESHIRO, K., HIROKAWA, N. All kinesin superfamily protein, KIF, genes in mouse and human. *Proc. Natl Acad. Sci. USA* **2001**, *98*, 7004–7011.

32 DESAI, A., MITCHISON, T. J. Microtubule polymerization dynamics. *Annu. Rev. Cell Dev. Biol.* **1997**, *13*, 83–117.

33 WILLIAMS, R. C., JR., LEE, J. C. Preparation of tubulin from brain. *Methods Enzymol.* **1982**, *85(B)*, 376–385.

34 HYMAN, A., DRECHSEL, D., KELLOGG, D., SALSER, S., SAWIN, K., STEFFEN, P., WORDEMAN, L., MITCHISON, T. Preparation of modified tubulins. *Methods Enzymol.* **1991**, *196*, 478–485.

35 SVOBODA, K., SCHMIDT, C. F., SCHNAPP, B. J., BLOCK, S. M. Direct observation of kinesin stepping by optical trapping interferometry. *Nature* **1993**, *365*, 721–727.

36 COY, D. L., WAGENBACH, M., HOWARD, J. Kinesin takes one 8-nm step for each ATP that it hydrolyzes. *J. Biol. Chem.* **1999**, *274*, 3667–3671.

37 MEYHOFER, E., HOWARD, J. The force generated by a single kinesin molecule against an elastic load. *Proc. Natl Acad. Sci. USA* **1995**, *92*, 574–578.

38 SVOBODA, K., BLOCK, S. M. Force and velocity measured for single kinesin molecules. *Cell* **1994**, *77*, 773–784.

39 BLOCK, S. M., GOLDSTEIN, L. S., SCHNAPP, B. J. Bead movement by single kinesin molecules studied with optical tweezers. *Nature* **1990**, *348*, 348–352.

40 HOWARD, J., HUDSPETH, A. J., VALE, R. D. Movement of microtubules by single kinesin molecules. *Nature* **1989**, *342*, 154–158.

41 HOWARD, J., HUNT, A. J., BAEK, S. Assay of microtubule movement driven by single kinesin molecules. *Methods Cell Biol.* **1993**, *39*, 137–147.

42 HANCOCK, W. O., HOWARD, J. Processivity of the motor protein kinesin requires two heads. *J. Cell Biol.* **1998**, *140*, 1395–1405.

43 BLOCK, S. M. Making light work with optical tweezers. *Nature* **1992**, *360*, 493–495.

44 YANG, J. T., SAXTON, W. M., STEWART, R. J., RAFF, E. C., GOLDSTEIN, L. S. Evidence that the head of kinesin is sufficient for force generation and motility *in vitro*. *Science* **1990**, *249*, 42–47.

45 HUNT, A. J., HOWARD, J. Kinesin swivels to permit microtubule movement in any direction. *Proc. Natl Acad. Sci. USA* **1993**, *90*, 11653–11657.

46 BROWN, T. B., HANCOCK, W. O. A polarized microtubule array for kinesin-powered nanoscale assembly and force generation. *Nano Lett.* **2002**, *2*, 1131–1135.

47 HUANG, T. G., SUHAN, J., HACKNEY, D. D. Drosophila kinesin motor domain extending to amino acid position 392 is dimeric when expressed in *Escherichia coli*. *J. Biol. Chem.* **1994**, *269*, 16502–16507.

48 YAMAZAKI, H., NAKATA, T., OKADA, Y., HIROKAWA, N. KIF3A/B: a hetero-dimeric kinesin superfamily protein that works as a microtubule plus end-directed motor for membrane organelle transport. *J. Cell Biol.* **1995**, *130*, 1387–1399.

49 HUNTER, A. W., CAPLOW, M., COY, D. L., HANCOCK, W. O., DIEZ, S., WORDEMAN, L., HOWARD, J. The kinesin-related protein MCAK is a microtubule depolymerase that forms an ATP-hydrolyzing complex at microtubule ends. *Mol. Cell* **2003**, *11*, 445–457.

50 VERNOS, I. *Kinesin Protocols*. Humana Press, Totowa, NJ, **2000**.

51 HOWARD, J. *Mechanics of Motor Proteins and the Cytoskeleton*. Sinauer, Sunderland, MA, **2001**, pp. 58–61.

52 BERG, H. C. *Random Walks in Biology*. Princeton University Press, Princeton, NJ, **1993**, pp. 6–11, 53–58.

53 JIA, L., MOORJANI, S. G., JACKSON, T. N., HANCOCK, W. O. Microscale transport and sorting by kinesin molecular motors. *Biomed. Microdevices* **2004**, *6*, 67–74.

54 MRKSICH, M., WHITESIDES, G. M. Using self-assembled monolayers to understand the interactions of man-made surfaces with proteins and cells. *Annu. Rev. Biophys. Biomol. Struct.* **1996**, *25*, 55–78.

55 RATNER, B. D., BRYANT, S. J. Biomaterials: where we have been and where we are going. *Annu. Rev. Biomed. Eng.* **2004**, *6*, 41–75.

56 LIMBERIS, L., STEWART, R. J. Toward kinesin-powered microdevices. *Nanotechnology* **2000**, *11*, 47–51.

57 HUANG, Y. M., UPPALAPATI, M., HANCOCK, W. O., JACKSON, T. N. Microfabricated capped channels for biomolecular motor-based transport. *IEEE Adv. Packaging* **2005**, *28*, 564–570.

58 UDABAGE, P., McKINNON, I. R., AUGUSTIN, M. A. The use of sedimentation field flow fractionation and photon correlation spectroscopy in the characterization of casein micelles. *J. Dairy Res.* **2003**, *70*, 453–459.

59 WAUGH, D. F. *Milk Proteins: Chemistry and Molecular Biology*. Academic Press, New York, **1971**, pp. 10–12.

60 BHATTACHARYYA, J., DAS, K. P. Molecular chaperone-like properties of an unfolded protein, alpha(s)-casein. *J. Biol. Chem.* **1999**, *274*, 15505–15509.

61 KULL, F. J., SABLIN, E. P., LAU, R., FLETTERICK, R. J., VALE, R. D. Crystal structure of the kinesin motor domain reveals a structural similarity to myosin. *Nature* **1996**, *380*, 550–555.

62 SABLIN, E. P., KULL, F. J., COOKE, R., VALE, R. D., FLETTERICK, R. J. Crystal structure of the motor domain of the kinesin-related motor ncd. *Nature* **1996**, *380*, 555–559.

63 HIROKAWA, N. Kinesin and dynein superfamily proteins and the mechanism of organelle transport. *Science* **1998**, *279*, 519–526.

64 ZHANG, Y., HANCOCK, W. O. The two motor domains of KIF3A/B coordinate for processive motility and move at different speeds. *Biophys. J.* **2004**, *87*, 1795–1804.

65 DECASTRO, M. J., HO, C. H., STEWART, R. J. Motility of dimeric ncd on a metal-chelating surfactant: evidence that ncd is not processive. *Biochemistry* **1999**, *38*, 5076–5081.

66 BERLINER, E., MAHTANI, H. K., KARKI, S., CHU, L. F., CRONAN, J. E., JR., GELLES, J. Microtubule movement by a biotinated kinesin bound to streptavidin-coated surface. *J. Biol. Chem.* **1994**, *269*, 8610–8615.

67 HUA, W., YOUNG, E. C., FLEMING, M. L., GELLES, J. Coupling of kinesin steps to ATP hydrolysis. *Nature* **1997**, *388*, 390–393.

68 GELLES, J., BERLINER, E., YOUNG, E. C., MAHTANI, H. K., PEREZ-RAMIREZ, B., ANDERSON, K. Structural and functional features of one- and two-headed biotinated kinesin derivatives. *Biophys. J.* **1995**, *68*, 276S–281S; discussion 282S.

69 BERLINER, E., YOUNG, E. C., ANDERSON, K., MAHTANI, H. K., GELLES, J. Failure of a single-headed kinesin to track parallel to microtubule protofilaments. *Nature* **1995**, *373*, 718–721.

70 CHENG, L. J., KAO, M. T., MEYHÖFER, E., GUO, J. Highly efficient guiding of microtubule transport with imprinted CYTOP nanotracks. *Small* **2005**, *1*, 409–414.

71 LAKAMPER, S., KALLIPOLITOU, A., WOEHLKE, G., SCHLIWA, M., MEYHOFER, E. Single fungal kinesin motor molecules move processively along microtubules. *Biophys. J.* **2003**, *84*, 1833–1843.

72 HANCOCK, W. O., HOWARD, J. Kinesin's processivity results from mechanical and chemical coordination between the ATP hydrolysis cycles of the two motor domains. *Proc. Natl Acad. Sci. USA* **1999**, *96*, 13147–13152.

73 TURNER, D. C., CHANG, C., FANG, K., BRANDOW, S. L., MURPHY, D. B. Selective adhesion of functional microtubules to patterned silane surfaces. *Biophys. J.* **1995**, *69*, 2782–2789.

74 MALLIK, R., CARTER, B. C., LEX, S. A., KING, S. J., GROSS, S. P. Cytoplasmic dynein functions as a gear in response to load. *Nature* **2004**, *427*, 649–652.

75 BROUHARD, G. J., HUNT, A. J. Microtubule movements on the arms of mitotic chromosomes: polar ejection forces quantified *in vitro*. *Proc. Natl Acad. Sci. USA* **2005**, *102*, 13903–13908.

76 DESAI, A., VERMA, S., MITCHISON, T. J., WALCZAK, C. E. Kin I kinesins are microtubule-destabilizing enzymes. *Cell* **1999**, *96*, 69–78.

77 DESAI, A., WALCZAK, C. E. Assays for microtubule-destabilizing kinesins. *Methods Mol. Biol.* **2001**, *164*, 109–121.

78 DENNIS, J. R., HOWARD, J., VOGEL, V. Molecular shuttles: directed motion of microtubules along nanoscale kinesin tracks. *Nanotechnology* **1999**, *10*, 232–236.

79 HESS, H., CLEMMENS, J., QIN, D., HOWARD, J., VOGEL, V. Light-controlled molecular shuttles made from motor proteins carrying cargo on engineered surfaces. *Nano Lett.* **2001**, *1*, 235–239.

80 VAN DEN HEUVEL, M. G. L., BUTCHER, C. T., SMEETS, R. M. M., DIEZ, S., DEKKER, C. High rectifying efficiencies of microtubule motility on kinesin-coated gold nanostructures. *Nano Lett.* **2005**, *5*, 1117–1122.

81 CLEMMENS, J., HESS, H., HOWARD, J., VOGEL, V. Analysis of microtubule guidance in open microfabricated channels coated with the motor protein kinesin. *Langmuir* **2003**, *19*, 1738–1744.

82 LIMBERIS, L., MAGDA, J. J., STEWART,

R. J. Polarized alignment and surface immobilization of microtubules for kinesin-powered nanodevices. *Nano Lett.* **2001**, *1*, 277–280.

83 PROTS, I., STRACKE, R., UNGER, E., BOHM, K. J. Isopolar microtubule arrays as a tool to determine motor protein directionality. *Cell Biol. Int.* **2003**, *27*, 251–253.

84 TURNER, D., CHANG, C., FANG, K., CUOMO, P., MURPHY, D. Kinesin movement on glutaraldehyde-fixed microtubules. *Anal. Biochem.* **1996**, *242*, 20–25.

85 VALE, R. D., FUNATSU, T., PIERCE, D. W., ROMBERG, L., HARADA, Y., YANAGIDA, T. Direct observation of single kinesin molecules moving along microtubules. *Nature* **1996**, *380*, 451–453.

86 THORN, K. S., UBERSAX, J. A., VALE, R. D. Engineering the processive run length of the kinesin motor. *J. Cell Biol.* **2000**, *151*, 1093–1100.

87 SHAH, C., XU, C. Z., VICKERS, J., WILLIAMS, R. Properties of microtubules assembled from mammalian tubulin synthesized in *Escherichia coli*. *Biochemistry* **2001**, *40*, 4844–4852.

88 MUTHUKRISHNAN, G., ROBERTS, C. A., CHEN, Y.-C., ZAHN, J. D., HANCOCK, W. O. Patterning surface-bound microtubules through reversible DNA hybridization. *Nano Lett.* **2004**, *4*, 2127–2132.

89 BACHAND, G. D., RIVERA, S. B., BOAL, A. K., GAUDIOSO, J., LIU, J., BUNKER, B. C. Assembly and transport of nanocrystal CdSe quantum dot nanocomposites using microtubules and kinesin motor proteins. *Nano Lett.* **2004**, *4*, 817–821.

90 BASKAR, D., HANCOCK, W. O. Unpublished observations.

91 DIEZ, S., REUTHER, C., DINU, C., SEIDEL, R., MERTIG, M., POMPE, W., HOWARD, J. Stretching and transporting DNA molecules using motor proteins. *Nano Lett.* **2003**, *3*, 1251–1254.

92 KATO, K., GOTO, R., KATOH, K., SHIBAKAMI, M. Microtubule–cyclodextrin conjugate: functionalization of motile filament with molecular inclusion ability. *Biosci. Biotechnol. Biochem.* **2005**, *69*, 646–648.

93 AACHMANN, F. L., OTZEN, D. E., LARSEN, K. L., WIMMER, R., Structural background of cyclodextrin–protein interactions. *Protein Eng.* **2003**, *16*, 905–912.

94 CLARK, S. L., REMCHO, V. T. Aptamers as analytical reagents. *Electrophoresis* **2002**, *23*, 1335–1340.

95 KOJIMA, H., MUTO, E., HIGUCHI, H., YANAGIDA, T. Mechanics of single kinesin molecules measured by optical trapping nanometry. *Biophys. J.* **1997**, *73*, 2012–2022.

96 HANCOCK, W. O. Unpublished observations.

97 YOKOKAWA, R., YOSHIDA, Y., TAKEUCHI, S., KON, T., SUTOH, K., FUJITA, H. Evaluation of cryopreserved microtubules immobilized in microfluidic channels for bead-assay-based transportation system. *IEEE Adv. Packaging* **2005**, *28*, 577–583.

98 MCDONALD, J. C., DUFFY, D. C., ANDERSON, J. R., CHIU, D. T., WU, H., SCHUELLER, O. J., WHITESIDES, G. M. Fabrication of microfluidic systems in poly(dimethylsiloxane). *Electrophoresis* **2000**, *21*, 27–40.

99 KIM, T. S., NANJUNDASWAMY, H. K., LIN, C.-T., LAKAMPER, S., CHENG, L. J., HOFF, D., HASSELBRINK, E. F., GUO, L. J., KURABAYASHI, K., HUNT, A. J., MEYHÖFER, E. Biomolecular motors as novel prime movers for microTAS: microfabrication and materials issues. In: *Proceedings of the 7th International Conference on Micro Total Analysis Systems*, NORTHRUP, M. A., JENSEN, K. F., HARRISON, D. J. (Eds.), Transducers Research Foundation, Squaw Valley, CA, **2003**, vol. 2, pp. 33–36.

100 BRUNNER, C., ERNST, K. H., HESS, H., VOGEL, V. Lifetime of biomolecules in polymer-based hybrid nanodevices. *Nanotechnology* **2004**, *15*, S540–S548.

101 VERMA, V., HANCOCK, W. O., CATCHMARK, J. M. Micro- and nanofabrication processes for hybrid synthetic and biological system fabrication. *IEEE Adv. Packaging* **2005**, *28*, 584–593.

102 UPPALAPATI, M., HUANG, Y. M., JACKSON, T. N., HANCOCK, W. O. unpublished observations.

103 McCLAIN, M. A., CULBERTSON, C. T., JACOBSON, S. C., ALLBRITTON, N. L., SIMS, C. E., RAMSEY, J. M. Microfluidic devices for the high-throughput chemical analysis of cells. *Anal. Chem.* **2003**, *75*, 5646–5655.

104 TONER, M., IRIMIA, D. Blood-on-a-chip. *Annu. Rev. Biomed. Eng.* **2005**, *7*, 77–103.

105 HAN, F., WANG, Y., SIMS, C. E., BACHMAN, M., CHANG, R., LI, G. P., ALLBRITTON, N. L., Fast electrical lysis of cells for capillary electrophoresis. *Anal. Chem.* **2003**, *75*, 3688–3696.

106 IRIMIA, D., TOMPKINS, R. G., TONER, M. Single-cell chemical lysis in picoliter-scale closed volumes using a microfabricated device. *Anal. Chem.* **2004**, *76*, 6137–6143.

9
Self-assembly and Bio-directed Approaches for Carbon Nanotubes: Towards Device Fabrication

Arianna Filoramo

9.1
Introduction

Silicon-CMOS technology is the base of present hardware technology for information processing. Until now, its evolution has been governed by Moore's law (from the 1970s), stating that microprocessor performance (defined as the number of transistors on a chip) doubles every 18 months. However, the International Technology Roadmap for Semiconductors (ITRS) [1] predicts that the present CMOS technology will reach its fundamental limits in terms of miniaturization by 2010–2015, concurrently with a dramatic increase in the cost of the production units.

This prompts us to study and develop alternative nanofabrication technologies that will enable the production and manipulation of well-defined structures at the nanoscale level. Indeed, it is well accepted that conventional technologies based on the "top-down" approach are foreseen to experience experimental difficulties. This is due to the existence of various physical effects that do not down-scale properly and, most importantly, to the fabrication cost issues at the nanoscale dimension. At this scale, as also stressed in the IST Technology Roadmap for Nanoelectronics [2], self-assembly and, more generally, "bottom-up" approaches appear to be a more reasonable way to assemble nano-objects into circuits with a two- and/or three-dimensional layout. In particular, self-assembly is also identified as the most promising way to significantly reduce the fabrication costs compared to what is expected for standard top-down silicon-based devices. Indeed, the basic idea of self-assembly is to use a process involving the spontaneous self-ordering of substructures into superstructures. This spontaneous self-ordering is due to specific chemical or physical properties of matter and relies on the natural tendency of the system to search for a stable configuration. This chapter is not meant to be an exhaustive review of self-assembly techniques, but rather a focused discussion on the application of this concept to a particular nano-object – the carbon nanotube (CNT). At present there are several materials with potential for the fabrication of new-generation nanodevices. They vary from conventional semiconductors to conjugated molecules, but among them CNTs are one of the most interesting. This is

Nanotechnologies for the Life Sciences Vol. 4
Nanodevices for the Life Sciences. Edited by Challa S. S. R. Kumar
Copyright © 2006 WILEY-VCH Verlag GmbH & Co. KGaA, Weinheim
ISBN: 3-527-31384-2

due to the wide range of physical properties they can show as a function of their shape and geometry.

CNTs have already been used to fabricate nanodevices, like field-effect transistors (FETs), with very interesting performance characteristics. In this chapter we describe their fabrication techniques, as well as their principal characteristics. The chapter also contains a more general discussion about these unique materials and their nano-applications. In this sense, we provide a discussion on the relevance of using CNTs as active elements in future nanoelectronic devices. We also critically consider the existing techniques that, by self-assembly, enable us to selectively place the nanotubes at specific locations on a substrate in a circuit configuration.

The self-assembly methods we will consider in this chapter are: *in situ* chemical vapor deposition (CVD) growth, the self-assembled monolayer (SAM) technique and the DNA-directed approach. We present their advantages, drawbacks and perspectives. In particular, we discuss in detail the DNA-directed approach. This bio-directed method constitutes a genuine and complete molecular-scale bottom-up method, since it relies on recognition properties inherent to biological entities and can be employed without using any standard lithography technique (a finger-print of the top-down strategy). As well as the economically very appealing interest of a self-assembly technique is generated by the perspectives of important reductions in fabrication costs, DNA-directed self-assembly could in addition bring truly new technological perspectives by introducing a disruptive strategy and by stimulating the development of novel architecture paradigms, which could include concepts like self-repairing features, reconfigurability, three-dimensional design and massive parallel processing.

The chapter is organized as follows. Section 9.2 presents a brief review of the basic physical features of CNTs, their synthesis methods and the different families of CNT applications reported in the literature. Section 9.3 is devoted to the discussion of the techniques of fabrication of CNT transistors. Here, the interest in self-assembly is highlighted. The first self-assembly method to be detailed is *in situ* CVD in Section 9.4, whereas two post-growth approaches are extensively discussed in the following two sections: in Section 9.5 we review the SAM method and in Section 9.6 we discuss the different aspects of the DNA-directed self-assembly technique for nanotube electronic applications. Finally, we draw our conclusions in Section 9.7.

The scope of the chapter is to provide an introduction and an overview of the current state of the fabrication of devices made of CNTs by using self-assembly. Nonetheless the degree of maturity of the presented nanofabrication methods is variable; we provide a critical comparison of their perspectives and an analysis of the issues to be solved before they could be integrated to real-world applications. This aims to give to the reader the elements to think about the strategies to be conceived for the future developments of the different nanofabrication techniques. The focus of the chapter is more on the fabrication of devices by self-assembly methods than on the details of their functioning, which are left to the appropriate cited references. This chapter, with the help of the included references, will give the reader the necessary tools to acclimate, evaluate and enter into this interesting field of research.

9.2
CNTs: Basic Features, Synthesis and Device Applications

Since their discovery in the early 1990s by Sumio Ijima [3], CNTs have been a privileged subject of fundamental and applicative research due to their extraordinary physical properties [4]. For instance, their excellent resilience, tensile strength and thermal stability can be used for resistant car bodies, earthquake-resistant buildings or microscopic robots. More generally, the wide range of potential applications in vastly different domains is a result of the numerous and varied aspects of their physical, chemical, mechanical and electronics properties. Long-term exploitation of these unique properties is expected to allow the elaboration of a whole new family of valuable devices.

9.2.1
Basic Features

CNTs can be divided in two classes: single-wall CNTs (SWNTs) and multiwall CNTs (MWNTs). A SWNT is basically a rolled-up shell of graphene sheet made of hexagonal carbon rings with half-fullerenes capping the shell ends. MWNTs were discovered first and are a stack of SWNTs of different diameters disposed in a concentric geometry (like Russian dolls). These two kinds of CNTs are schematically presented in Fig. 9.1.

Figure 9.2(a) shows a small part of the graphene lattice, with carbon atoms labeled according to the Dresselhaus notation [4]. The nomenclature (n, m), with $n > m$, used to identify SWNT refers to integer indices of two graphene unit lattice vectors corresponding to a nanotube's wrapping index known as the *chiral vector*.

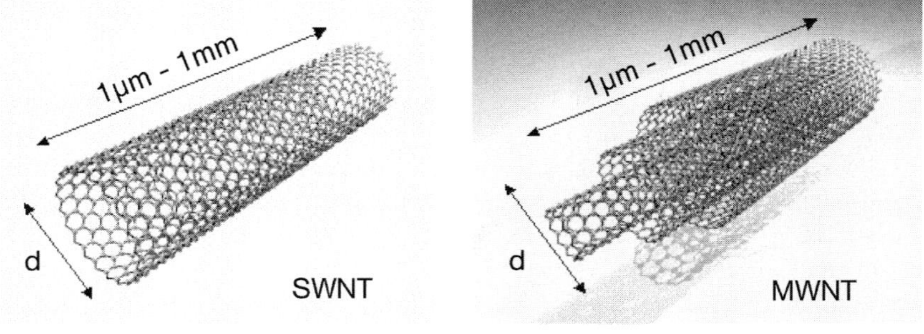

Figure 9.1. Schematic representations of a SWNT (a) and a MWNT (b).

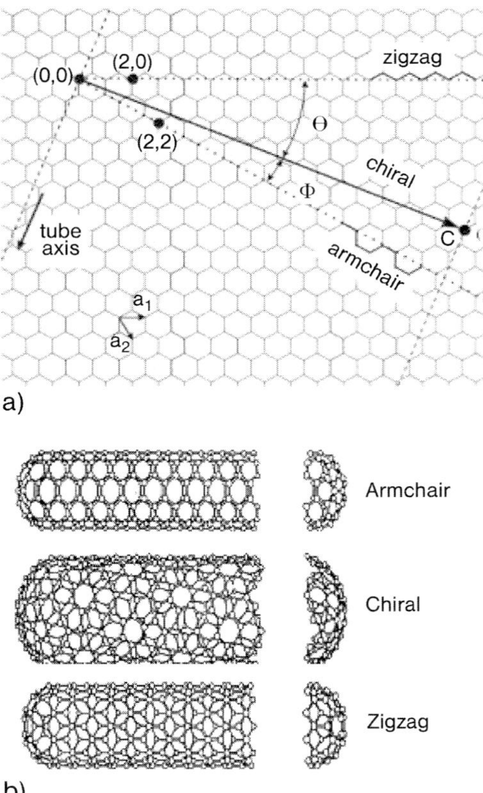

Figure 9.2. The graphene sheet is a planar hexagonal lattice of carbon atoms (a). We can recognize the unrolled honeycomb lattice of a nanotube. SWNTs with different chiralities are obtained by rolling-up parts of the graphene sheet along different axes: an arm-chair, a chiral and a zigzag (from top to bottom in b).

The chiral vector determines the direction along which the graphene sheet is rolled to form the nanotube. It is, by definition, perpendicular to the tube axis vector. The tubes of type (n, n) are commonly called armchair nanotubes because of their shape $(\backslash\_/\backslash\_/)$ perpendicular to the tube axis. They make a chiral angle of 30°. The tubes of type $(n, 0)$ are called zigzag nanotubes due to their shape $(\backslash\backslash\backslash/)$ perpendicular to the tube axis. They make a chiral angle of 0°. All the remaining tubes are called chiral nanotubes and have a chiral angle θ with $0 < \theta < 30°$ (see Fig. 9.2b).

CNTs thus possess both nanometric (related to their radius) and microscopic (related to their typical length) characteristic. Correspondingly, most of their original physical properties result from this hybrid nature. For electronic purposes we are principally concerned with the transport properties of CNTs.

In this context, single-wall nanotubes are particularly interesting because of the following singular physical property: a SWNT can show either metallic or semi-conducting behavior, depending on its chiral vector (n, m). The general rule is that if the difference $n - m$ is an integer multiple of 3 then the nanotube is metallic, whereas in all the other cases the nanotube is a semiconductor. As a consequence, if all chiralities are equiprobable, the metallic:semiconductor abundance ratio for SWNTs should be 1:3. Most importantly, one can envisage the realization of structures with specific transport properties by combining CNTs of different nature into the same device. For instance, nanotubes with different chiralities can be connected to obtain heterojunctions. Applications of SWNTs as electronic devices are discussed in a later section. Concerning MWNTs, the stacking of nanotubes with different chiralities can bring more unexpected results, as suggested by first-principles calculations. Indeed, the outcomes of such model calculations show that the interlayer coupling, which in most cases has little effect on the electronics properties of the individual tubes, can, under specific conditions of the relative positions of one tube shell with respect to the other, strongly modify their individual properties [5].

Finally, it is also important to recall that nanotubes have good elastomechanical properties and that they are thermally very stable. This is related to the fact that the two-dimensional arrangement of carbon atoms in a graphene lattice allows large out-of-plane distortions while the strength of carbon–carbon in-plane bonds keeps the graphene sheets exceptionally strong against distortions [6–8]. Experimental observations of distortions induced to a nanotube demonstrate this high elasticity and point towards its possible use as lightweight, highly elastic material [9–12]. Therefore, CNTs provide one-dimensional wires that enable the fabrication of different complete families of electronic devices.

9.2.2
Synthesis of Nanotubes

There are mainly three methods for the synthesis of nanotubes: arc-discharge [13–18], laser ablation [19–21] and CVD [22–26]. The first two are evaporation methods that employ solid-state carbon precursors as carbon sources for nanotube growth and involve carbon vaporization at high temperatures (in an oven at 1100–1200 °C) assisted, respectively, by an arc-discharge or laser ablation. In order to achieve nanotube growth some metal catalysts are added in the solid graphite source. The third method utilizes hydrocarbon gases as sources for carbon atoms. It also employs metal catalyst particles as "seeds" for nanotube growth, but the process takes place at relatively lower temperatures (500–1000 °C). More details about the CVD method are given in Section 9.4. These three methods have not been equally used in the literature.

Currently, the majority of the work published on the synthesis of CNTs (MWNTs and SWNTs) deals with different types of production by using the CVD route. A large variety of successful production approaches based on this method have been reported in recent years. Indeed, CVD processes are expected to be the solution for the mass production of CNTs.

The state-of-the-art synthesis achievements using CVD methods are: (a) production of centimeter-long CNTs [27, 28], (b) synthesis of CNTs along the direction of an applied electric field [29] and (c) very regular ordering of CNTs grown on templates [30]. The major problems of CVD-based approaches for SWNTs synthesis are: (a) the difficulty of producing CVD-SWNTs with a narrow diameter distribution and (b) the tubes produced generally have more defects. It is worth pointing out that although the best-quality SWNTs so far are produced by evaporation-related methods, the difference with CVD ones is becoming less and less significant.

It should be stressed that an important issue in the growth of SWNTs is the control of the tube diameter and of the chirality distributions. These two parameters are of prime importance in transport studies, since (a) as mentioned in Section 9.2.1, the chirality of a SWNT determines its intrinsic metallic or semiconducting conductive properties and (b) the current of a nanotube transistor in its off-state is strongly dependent upon the nanotube bandgap, which is inversely proportional to the tube diameter d (the bandgap of a semiconducting SWNT $\propto 1/d$). Despite the importance of having a good control of these distribution parameters, none of the three synthesis methods has yielded bulk materials with homogeneous diameters and chiralities. So far, evaporation techniques remain the best for the selective synthesis of SWNTs with a narrow diameter distribution [31, 32]. With respect to the chirality issue, two interesting reports should be mentioned: the first achievement concerns a preferential CVD growth of semiconducting SWNTs (with a yield of 90%) [33], while the second one concerns a postsynthesis method to separate metallic from semiconducting nanotubes [34].

9.2.3
Device Applications

A large number of studies on CNT applications have been reported in the literature [35]. They span from field emission electron sources [36–41], supercapacitors [42–44], artificial muscles [45–47], nanoelectromechanical systems [48–53], photoactuators [54], controlled drug delivery/release [55, 56], reinforcement of materials [12, 57–61], composite printable conductors [62], optical components [63, 64], nanoelectronic components [65], scanning probe tips [66–68], etc.

Here, we consider nanoelectronic device applications. In particular, we focus on transistors made of nanotubes and only briefly discuss their nanosensor applications. Indeed, it has been experimentally demonstrated that an individual semiconducting SWNT can be used as the channel of a CNTFET [69, 70]. In 2001, different achievements were accomplished: the demonstration of a room temperature single-electron transistor (SET) [71] and of nanotube transistors showing gain above unity [72], as well as the realization of logical gates mimicking CMOS ones (but with a lateral channel extension reduced to 1 nm) [72, 73]. Recent experimental [74–79] and theoretical [79–82] studies on CNTFETs showed that most of them work as Schottky barrier transistors. Their switching characteristics are limited by the Schottky barriers at the metal/nanotube junction that brings to the fore the crucial role of interfaces in such a transistor. By optimizing the nanotube/electrode interface [74, 78, 83] and the gate coupling, the device characteristics can be improved both in the "on" and "off" states. Some CNTFETs [84, 85] already exhibit a level of performance comparable to state-of-the-art silicon MOSFETs (at comparable geometry).

It is worth pointing out that SWNTs are also an ideal material for nanosensor fabrication – their one-dimensional structure and their high specific area make the conductance of semiconducting SWNTs highly sensitive to very small amounts of molecules [86].

More precisely, in the context of sensor applications, the exceptional properties of nanotubes are exploited mainly in two ways. The first approach is based on the changes in electrical characteristic of a device constituted by an individual (or very few) SWNT in a transistor configuration. In this case the current in the device is modified by the interaction of the molecules with the semiconducting SWNT channel and/or with the electrodes of the transistor. In some cases the SWNT can be appropriately functionalized in order to enhance its interaction with and its sensitivity to the molecules to be detected. We can quote sensors of gaseous molecules [52, 87], proteins [88, 89], pH/glucose [90], aromatic compounds [91], humidity [92] and DNA hybridization with high sensitivity (concentrations as low as 6.8 fM solution) [93].

In the second approach, the CNTs are used as electrode materials in electrochemical cells [94]. Since their first application for the oxidation of dopamine [95], an increasing number of studies have been devoted to the CNT electrocatalytic behavior towards the oxidation of biomolecules. We can quote studies on dopamine [96–98], ascorbic acid [96], epinerphrine [96, 98], 3,4-dihydroxyphenylacetic acid [99], homocysteine [100], thyroxine [101], glucose [102, 103], total cholesterol in blood [104] and DNA hybridization sensors [105–107]. The advantages presented by the CNT electrodes with respect to traditional carbon electrodes are better conducting characteristics, higher chemical stability, increased reaction rates, improved detection limit and reversibility. However, it should be noted that in most of these applications the nanotubes were utilized as a bulk material and not as an individual nano-objects. The nanotubes are generally dispersed in solution by chemical treatments, then incorporated in a matrix and deposited as (or onto) the electrodes. In some more recent reports the nanotubes are directly grown on the electrodes by CVD methods. A recent overview of this type of biosensors can be found in Ref. [108].

9.3
Fabrication of CNT Transistors and Self-assembly Approaches

The majority of the works reported on CNT transistors concern the study of their fundamental properties and performance. In this framework, the techniques that have been developed for the fabrication of such devices do not take into account yield/cost issue and are mainly based on random deposition methods.

Generally, the nanotubes are produced by an evaporation technique (laser ablation or arc-discharge) and, after a purification step, are dispersed in solution (typically in 1,2-dichloroethane). Then, they are deposited on an oxidized silicon substrate [69–78]. The result of this deposition is a completely spatially random distribution of the nanotubes on the substrate. Indeed, in this condition it is only possible to control the nanotube density on the surface by varying their concentration in the solution and not their positioning.

To connect some of these randomly dispersed nanotubes, they are first localized by atomic force microscopy (AFM) imaging with respect to some location marks fabricated on the substrate. Then, the best candidates (in terms of position) are chosen and a set of dedicated electrodes is fabricated. In some cases, a set of pre-patterned electrodes is already present on the substrate before the nanotube deposition and it is a matter of "luck" if the deposited nanotubes are correctly positioned on them. This is normally checked by AFM images of the sample. The same idea is followed if the nanotubes are randomly grown on the substrate by CVD. Indeed, after the growth, the nanotubes are imaged and the best candidates are electrically connected. Finally, it is also worth noting that each nanotube transistor does not always have its own gate electrode and that generally the conductive reverse side of the silicon substrate is used as a back-gate.

However, the possible use of CNTs as active elements in future nanoelectronics is closely related to the question of legacy/compatibility with present information technology. Indeed, it is quite unlikely that a system based on a new technology consisting of architecture with completely random disposition of such devices could be introduced and accepted. Therefore, to take full advantage of the unique electrical properties of SWNTs in device/circuit applications, it is desirable to be able to selectively place them at specific locations on a substrate with a low-cost, high-yield self-assembly-based technique.

Nowadays, the state-of-the-art of self-assembly of CNTs devices can be divided in two different classes of self-assembly methods: (a) *in situ* CVD growth, where the localization arises from the catalyst controlled positioning, and (b) post-growth localized deposition on a substrate. In the latter case, the nanotubes are first grown, handled in solution and only subsequently positioned on the substrate. Obviously, the technique chosen for selective placement of the nanotubes must not degrade the electrical characteristics of the devices and must leave open to all the possibilities of such interesting nano-objects.

The advantage of a post-growth deposition method is that, before deposition, CNTs can be purified and chemically treated in order to separate them by diameter [109–111], length [112] or chirality [34, 113, 114]. Moreover, in this predeposition step the nanotubes can also be chemically functionalized to add to their exceptional features other interesting chemical or physical properties [115, 116]. As discussed previously, the drawback to overcome in this case is mainly related to the deposition issue since it is generally random on the substrate. To solve this SWNT random deposition issue, two post-synthesis methods can be used.

The first one is to achieve a selective placement of SWNTs on regions of the substrate that are predefined by surface treatments. This post-growth selective placement approach is based on the use of SAMs to modify the surface properties of certain regions of a substrate. This, in turn, affects the interactions between the sidewalls of a CNT and the surface, and the CNTs are preferentially attracted there.

The second approach could solve the deposition challenge using biological scaffolds, as DNA molecules, to realize site-controlled implementation of nanocomponents. Indeed, the unique intra- and intermolecular recognition properties of DNA have already been used to build-up scaffold structures and position nanoparticles [117–121].

These two post-synthesis methods are discussed Sections 9.5 and 9.6, while the *in situ* CVD approach is discussed in the next section.

9.4
In situ CVD Growth

The *in situ* CVD method is based on the selective growth of CNTs on specified locations on the substrate. We now present the main aspects of this technique.

As previously mentioned, the first step of the CVD is the energy-activated decomposition of some hydrocarbon gas. The energy source can be either a plasma or a resistively heated coil and its function is to "crack" the gaseous molecules to provide reactive carbon atoms. Such carbon atoms diffuse towards the substrate, which is heated and coated with a transition metal catalyst. Then, the carbon atoms are fixed on the substrate and, if the appropriate conditions are fulfilled, CNT growth takes place.

The most commonly used gaseous carbon sources are methane, carbon monoxide and acetylene. Acetylene is widely used as a carbon precursor for the growth of MWNTs, which occurs at temperatures typically in the 600–800 °C range. Carbon monoxide or methane have proven to be more effective for the growth of SWNTs, since the temperature required is usually higher (800–1000 °C) and acetylene is not stable at these temperatures.

It should be noted that the CVD method has undergone dramatic and important developments over the past few years. The yield and average diameter of SWNTs were optimized by controlling the process parameters [122], and, although the diameter distribution is not as narrow as for laser ablation synthesis, impressive progress has been made in the optimization of this growth technique. For instance, as already quoted in Section 9.2.2, some reports have been found to move from the standard yield of 70% for semiconductor nanotube species [30] to a more interesting 90% [33].

Concerning the fabrication of CNT devices, the basic idea is to achieve the *in situ* localized growth of nanotubes by controlling the localization of the metal catalyst. Indeed, CVD CNT synthesis is essentially a two-step process consisting of an initial catalyst preparation step followed by the actual growth of the nanotubes, which starts at the places where the catalysts are present. Following this strategy, the first example of localized growth of SWNTs was realized in 1998 [86]. The authors produced individual SWNTs on silicon wafers patterned with micrometric islands of catalytic material. Their synthesis procedure begins with the patterning of catalytic islands on silicon substrates by electron beam lithography (EBL). More precisely, they first use EBL to define micrometric square islands in a poly(methyl methacrylate) (PMMA) resist. Then, they deposit a liquid phase of the metal catalysts [$Fe(NO_3)_3 \cdot 9H_2O, MoO_2(acac)_2$] mixed with alumina nanoparticles. Finally, they obtain the catalyst islands by lift-off of the nonexposed PMMA resist. For nanotube growth, they used methane as carbon precursors and a temperature of about 1000 °C. Moreover, they limited the formation of amorphous carbon by limiting the CVD synthesis to short times (maximum 10 min). They found that all the grown nanotubes were rooted in the islands and that some of them bridged two metallic islands. Following this pioneering work, much activity on *in situ* CVD for CNT devices fabrication has taken place and transistor arrays have been fabricated [30].

However, there is an important issue to be solved before integrating such a CVD method with current CMOS technology. Indeed, for the direct growth by CVD of CNTs on silicon, the temperature regime for the growth of SWNTs and MWNTs must be compatible with CMOS integration. In this sense, substantial progress has been recently achieved by the use of a plasma-enhanced CVD (PECVD) method [33]. In this work, nanotube growth was carried out at 600 °C on SiO_2/Si wafers on which some discrete ferritin particles were randomly adsorbed to act as catalysts. The idea is that the plasma-assisted dissociation of CH_4 into more reactive higher hydrocarbons and more reactive radicals must be favorable for efficient SWNT growth at lower temperatures. Moreover, the method uses a low-density plasma that propagates down and reaches the sample placed 40 cm from the plasma coil, thus preventing any local heating of the substrate. It should be noted that another advantage of lowering the CVD growth temperature is related to the diameter distribution and chirality issue. Indeed, it is likely that the size and shape of the catalytic nanoparticles should be more stable at lower temperatures, leading to a better control of the size and potential chirality of nanotubes.

In order to fully overcome the growth temperature issue and ensure compatibility with CMOS technology, a very interesting (but fundamentally different) solution can be envisaged. Indeed, these kinds of limitations can be completely avoided by preparing the nanotubes *ex situ*, functionalizing them and then selectively depositing the nanotubes into the CMOS circuit. This is the philosophy of the postgrowth strategies, as discussed in the following sections.

9.5
Selective Deposition of CNTs by SAM-assisted Techniques

The technique of localized deposition assisted by SAMs is nowadays the subject of large interest. It started with the pioneering works of the groups of Liu [123], Muster [124] and Choi [125]. It relies either on a local chemical functionalization of the surface [123] or on an electrostatic anchoring of surfactant-covered SWNTs on amino-silane functionalized surfaces [124, 125]. The basic idea beyond these processes is the same, but the use of amino-silane surfaces has allowed the control of both deposition density and selective placement in predefined areas of the substrate of isolated SWNTs. Due to its great importance in the selective deposition of SWNTs, the SAM-assisted technique is discussed in detail in the following subsection.

9.5.1
Methodology and Key Parameters

This process can be summarized as follows [126–128]: the patterns (regions for future selective placement) are defined by performing EBL on a PMMA resist deposited on the SiO_2 surface of a silicon substrate. After a cleaning step [reactive ion etching (RIE)], a monolayer of aminopropyltriethoxysilane (APTS) is deposited by CVD [129–131] to form a "sticky patch" in the regions opened in the resist (see schematic representation in Fig. 9.3).

Exposure to ethylenediamine (EDA) is used to increase both the surface concentration and the orientation of APTS [132], and consequently to improve interactions with SWNTs. It is likely that this EDA molecule plays the role of a catalyst during the chemical anchoring of APTS on Si-OH. Indeed, it blocks the hydrogen bond formation between the amino group and Si-OH [133]. As a consequence, at the end of the chemical reaction, each amino group is well oriented on the top of the monolayer and their density increases by around 50%. Gas deposition is chosen instead of silanization from an aqueous solution, since it yields a much better control of layer thickness [125]. Once the sticky patch has been formed, it is expected that the adhesion of SWNTs would be enhanced in the functionalized

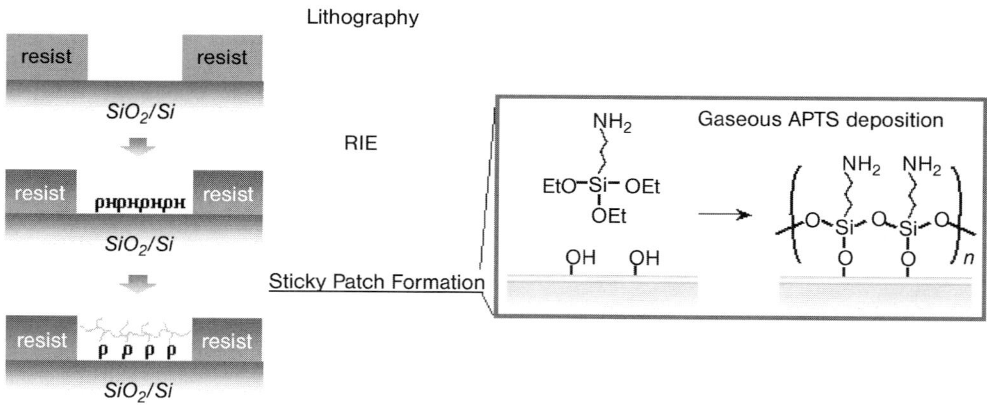

Figure 9.3. Schematic vision of the fabrication of the APTS patterns on the substrate surface. The patterns are defined in a PMMA resist by standard electronic lithography. Then, the motifs are developed and, after a cleaning step (RIE), the sticky patches of APTS are deposited in vapor phase.

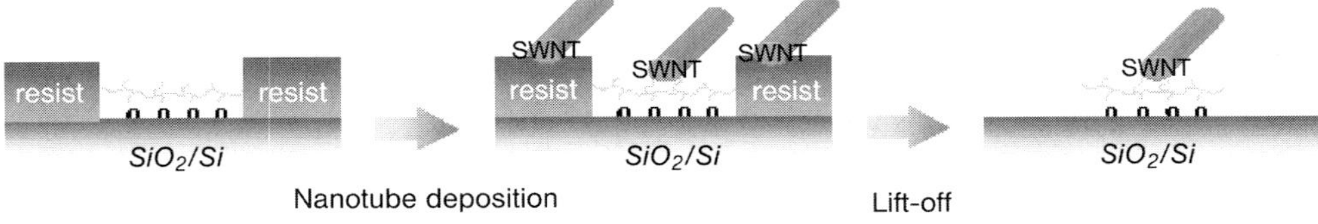

Figure 9.4. The sticky patches of APTS are exposed to the nanotube solution. In this case, to enhance placement selectivity, the residual PMMA resist is lifted off after nanotube deposition.

regions with the APTS "sticky patch". Essentially, two ways can be chosen to selectively place SWNT.

The first method consists of working with an aqueous solution, where sodium dodecylsulfate (SDS) surfactant is used to disperse SWNTs. In this case, the sample is exposed to the SWNT suspension and then the resist is lifted off to enhance selectivity by removing any nanotubes that would have been adsorbed on the PMMA (see schematic representation in Fig. 9.4).

However, in this case, the following bottleneck has to be faced: the density of adsorbed tubes on the surface is too low for the realization of integrated circuits. It turns out that increasing the concentration of SWNTs in the solution leads mainly to the deposition of bundles on the surface. The two problems to be solved within this approach are (a) to improve the dispersion of SWNTs in the solution, and (b) avoid the competition, in the electrostatic anchoring on the amino-silane surface, between the SDS micelles present in the solution and the SDS-covered tubes.

In order to avoid these two drawbacks, a second approach has been developed [126, 127]. The main point is that nanotubes are dispersed not in an aqueous solvent, but in *N*-methyl pyrrolidone (NMP). It has been observed [134] that this solvent allows dispersing nanotubes without any kind of surfactant. Therefore, it can solve point (a) and eliminates point (b). However, due to NMP interaction with the PMMA resist, it has been necessary to modify the process as shown in Fig. 9.5.

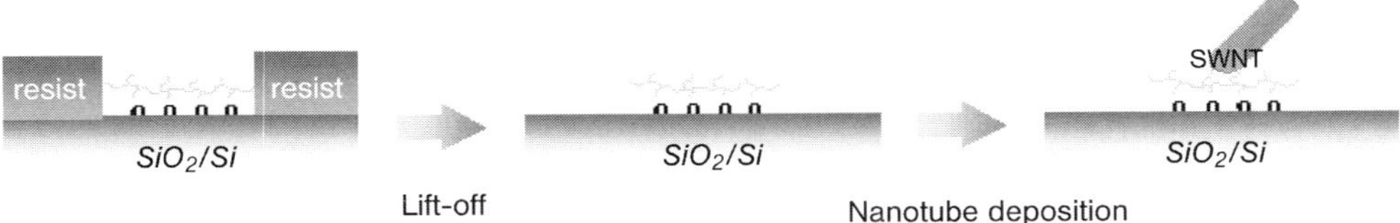

Figure 9.5. After the formation of the APTS patterns, the residual PMMA resist is lifted off. Then the sample is exposed to the nanotube solution. In this case, the selectivity is ensured only by the different affinity of nanotubes with different (APTS/SiO₂) surfaces.

The main difference with respect to the previous sequence (Fig. 9.4) is the reversed order of the last two steps. Thus, in the present case, the resist is removed before exposure to the nanotube solution. *A priori*, one would think that the selectivity of the placement would be partially lost; however, this is not the case, as shown in Refs. [126, 127].

The detailed studies reported in Refs. [126–128] have shown that each of the different steps in Figs. 9.3 and 9.5 play an important role in the selective placement, and that various parameters critically contribute to the success of the deposition process. Since these points are of particular importance in mastering the selective placement technique, we reconsider below the whole deposition process in much more detail. The discussion will follow Refs. [127, 128].

The nanotubes used in Refs. [127, 128] were obtained by the laser ablation technique [135]. The pristine SWNT samples first underwent a purification stage as described in Ref. [136]. Then, they were dispersed in NMP solvent. The solution was sonicated [137] and centrifugation was performed [126]. The SWNT concentration in NMP was varied from 0.1 to 0.005 mg mL^{-1} and the variation of the density of deposited SWNTs was recorded. The substrates for that study consisted of 200-nm thick thermally grown SiO_2 on silicon, covered with a monolayer of APTS. The SWNTs in NMP solution were deposited on the surface for 1 min. Finally, the density of deposited SWNTs was quantified by AFM experiments. The main results of this study are presented below.

Before discussing the results concerning the selective placement, it is worth stressing the excellent deposition yield of the NMP-based approach. For the sake of comparison, Fig. 9.6 presents AFM images of APTS-treated substrates after exposure to either an aqueous SDS solution at 1.2 CMC (Fig. 9.6a) or to a NMP solution (Fig. 9.6b) for the same SWNT concentration and exposure time. Contrary to aqueous solvents, for the NMP solution the adsorption process on APTS seems to be independent of any charge effect (thought they cannot be completely excluded [138]). Indeed, in the case of NMP, the NH_2 conversion of the silane group to NH_3^+ by exposure to HCl vapor does not seem to be relevant for the deposition yield. The adsorption is likely due to an interaction between the amine group of the APTS and the nanotube, as shown by Kong and Dai [139].

At a concentration of 0.1 mg mL^{-1}, the density is 150 SWNTs on a 4-μm^2 area. This result is comparable to that of Liu and coworkers [123], with 240 CNTs on a 6.25-μm^2 area. Between 60 and 70% of the nanotubes on the recorded AFM images are less than 1.6 nm high, indicating individual SWNTs or small bundles. Therefore, the bundling problem observed with surfactants in aqueous solvents [point (a) above] seems to be much less significant in NMP.

The influences of several parameters on the deposition have been analyzed [126, 128], such as the deposition time, centrifugation speed, sonication time and SWNT concentration. The distribution of the nanotube diameter appeared similar for all concentrations (below 0.1 mg mL^{-1}), indicating that there is no significant reduction of the bundles in diluted solutions. Concerning the centrifugation, the SWNT solutions were centrifuged for 10 min at different speeds up to 28 000 r.p.m. Unlike the case of SWNTs in aqueous solvents, the rotation speed seems to have no

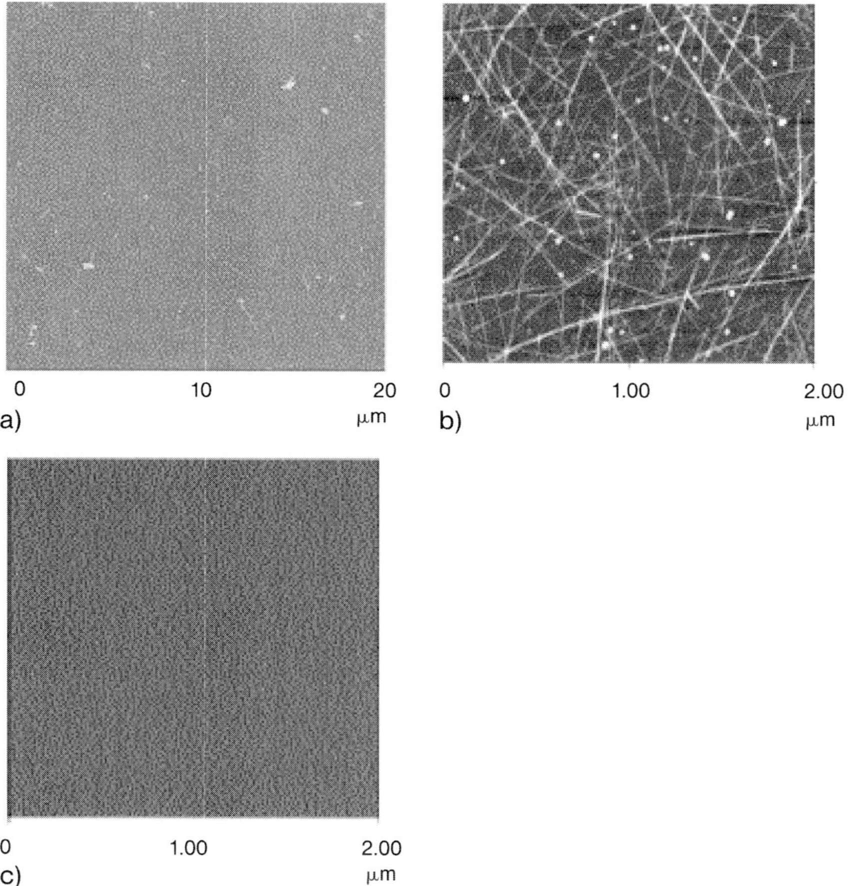

Figure 9.6. AFM images of APTS-treated substrates after exposure to two kinds of nanotube solutions. (a) The substrate was exposed to an aqueous solution with SDS surfactant, while in (b) it was exposed to a NMP solution. (c) A bare SiO_2 substrate was exposed to the same NMP solution as in (b), note that in this latter case no nanotubes deposition is observed.

significant effect on the dispersion of nanotubes. Concerning the sonication, after 24 h the tubes were up to 1–2 µm long, but they were severely shortened to less than 400 nm length if sonicated for 36 h. Finally, as expected, by increasing the deposition time we observed, accordingly, an increase of the density of SWNTs deposited.

The principal aim of the study reported in Refs. [127, 128] was to achieve selective placement of CNTs. For that, the experiments were repeated on patterned substrates. As already mentioned, the resist was removed before SWNT deposition (Fig. 9.5). In this case, the selectivity is ensured only by the different behavior of

Figure 9.7. AFM image of a sample with a series of continuous APTS stripes 100 nm wide and spaced 1 μm from each other. The stripes are defined by lithographic patterning of PMMA, then silanized and finally exposed to the CNTs in NMP solution.

the APTS-treated regions with respect to the nontreated regions. This point was checked by a preliminary experiment where a nonsilanized and a silanized substrate were exposed to the same nanotube solution. The results are reported in Fig. 9.6 [b (APTS-treated substrate) and c (nontreated substrate)]. It is clear from these figures that no deposition is observed for the nonsilanized sample.

An important parameter for the quality of the selective deposition is the geometry of the pattern. For the simple stripes geometry, the key parameter is the width of the stripes. Continuous stripes 500, 200 and 100 nm wide were patterned in PMMA, then silanized and exposed to the SWNTs in NMP solutions. Figure 9.7 shows the selective deposition obtained for the sample with 100-nm trenches. Suitable densities can be achieved for any width by varying the deposition time and/or the SWNT concentration in the solution within reasonable limits. For the same experimental conditions, the SWNT density increases roughly by a factor of 2 when the stripe width is increased from 100 to 200 nm. Moreover, was seen in this study that longer tubes (length $\gg$ 1 μm) are better aligned than shorter ones and that the quality of alignment is improved for narrower stripes, as already observed for aqueous solutions [125]. Finally, the use of 100-nm wide stripes represents the best way to limit the number of aligned SWNTs to one, which is crucial for the reliable study of electrical transport in individual SWNTs [127, 128].

Furthermore, in order to reliably control the fabrication of a large number of SWNT transistors on the same wafer, it is necessary not only to anticipate the statistics of deposition on a given pattern, but also the existence of any "proximity" effect, i.e. any effect on the deposition yield due to a possible combined interaction of patterned areas when they are in close vicinity to each other. SWNT deposition has been checked on groups of stripes 100 nm wide and 2 μm long spaced by 1, 3

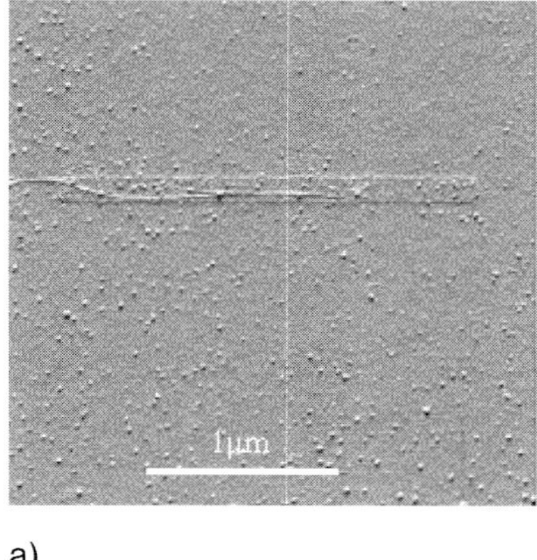
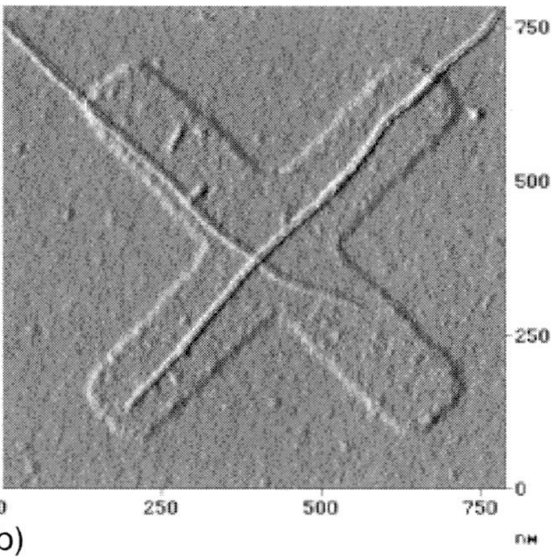

a)   b)

Figure 9.8. AFM images of two samples with APTS pre-patterned areas of different geometries, after exposure to a SWNT solution. (a) A single finite APTS stripe with one aligned nanotube. (b) A cross APTS pattern with two nanotubes aligned in each direction of the cross. No combing technique was applied during the nanotube deposition.

and 20 μm on the same substrate. AFM observation showed a placement yield superior to 85% in all groups. The constant density of deposition obtained for all groups clearly indicated that no "proximity" effects were involved in the placement process. An additional test was performed with the realization of a crossed SWNT configuration. Finite size crosses 100 nm wide and typically with 750-nm long arms were submitted to SWNT deposition. Typical results for the deposition of SWNT on striped and crossed APTS motifs are reported in Fig. 9.8.

It should be stressed that this kind of result (in terms of yield and selective deposition) critically depends on the quality of deposited APTS, and can be obtained only if the monolayer is perfectly uniform, homogeneous and well ordered.

In conclusion, for SAM-assisted selective placement, the NMP-based method provides an excellent deposition yield for both unpatterned surfaces and substrates patterned with different geometries (single and groups of stripes, and arrays of crosses), and the feasibility of deposition of a limited number of aligned nanotubes per stripe (ideally one) has already been demonstrated.

The last step in the fabrication of SWNT transistors is the realization of contacts. The choice of depositing SWNTs by a selective placement approach considerably simplifies the subsequent contacting process. Indeed, since we define the deposition areas by EBL, the patterning of electrodes on top of precisely localized SWNTs is simple. No specific and tedious AFM imaging is required to locate SWNTs, as is the case for randomly deposited SWNTs. The transistor is fabricated by first selectively depositing the tubes on a silane pattern prepared on a substrate fitted with position markers. After nanotube deposition, the contact electrodes are patterned and made using standard lithographic techniques (see Fig. 9.9 for electrodes in the crossed and striped geometries).

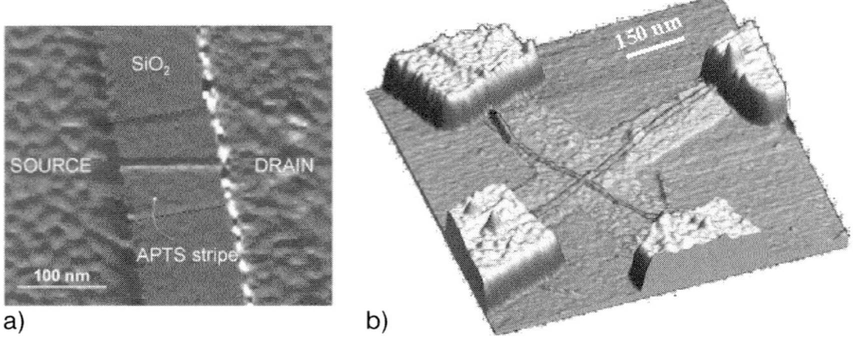

Figure 9.9. AFM images of two connected samples with APTS areas of different geometries. (a) A single finite APTS stripe with one aligned nanotube connected by two electrodes (source and drain). (b) A 3-D view of a connected crossed device.

Altogether, these results obviously open the way for the controlled fabrication of a large number of SWNT devices by SAM self-assembly.

9.5.2
Performance of CNTFETs Fabricated by the SAM Method

CNTFETs obtained by random deposition of nanotubes were introduced in Section 9.2.3. The present subsection is devoted to a comparison of the characteristics of transistors obtained by random and self-assembled depositions.

As already mentioned in Section 9.2.3, CNTFETs have been demonstrated to be Schottky barrier transistors. As a consequence, their transport characteristics can depend on the choice of the metal for electrode fabrication. One can check this point, e.g. by comparing the performance of devices obtained with the same tubes, but with different types of electrodes; this has actually been done for both randomly deposited [74] and self-assembled [83, 127] CNTFETs.

Figure 9.10 reports the characteristics of self-assembled CNTFETs in two configurations: (a) a "standard" configuration with 0.2-nm titanium/40-nm gold electrodes (Fig. 9.10a), and (b) an optimized configuration, hereafter called "TiC" (Fig. 9.10b), with deposition of 20-nm titanium/20-nm platinum electrodes and application of rapid thermal annealing (RTA). This RTA process took place at temperatures in the range 650–850 °C in inert ambient gas to convert the electrode contacts to titanium carbide (as in Ref. [74]). As the electrodes were fabricated on top of SWNTs deposited on 200-nm SiO_2 film grown on a silicon wafer, the wafer itself was used as the gate electrode ("back-gate" configuration). Electrical measurements were performed in vacuum.

"Standard" CNT field effect transistors behaved as p-type FETs, i.e. the dominant carriers are holes. As shown in Fig. 9.10(a), the transconductance (dI_D/dV_G) for

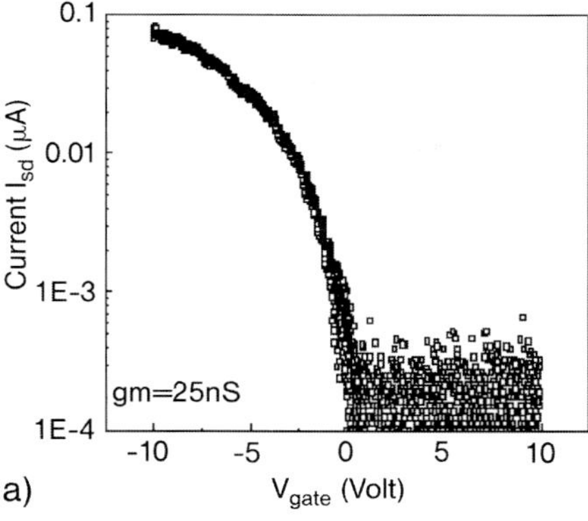
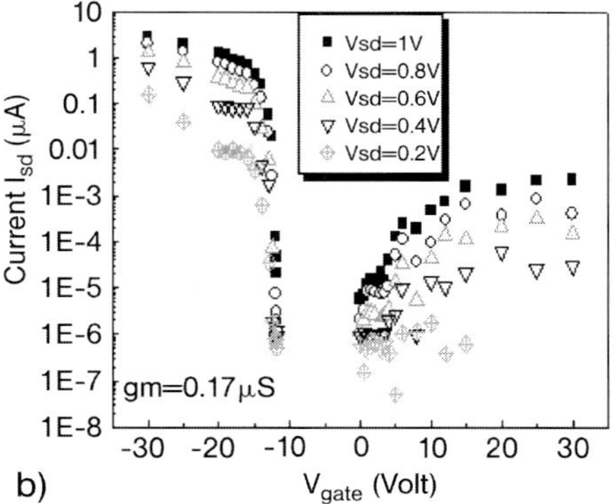

a)

b)

Figure 9.10. Transport characteristics of the "standard" (a) and "TiC" (b) devices. (a) Transfer characteristics at room temperature with source to drain voltage $V_{SD} = 200$ mV; $dI_D/dV_G = 25$ nA V^{-1}. The on:off current ratio is 10^4. (b) Transfer characteristics with source to drain voltage $V_{SD} = [0.2$ V:1 V]; $dI_D/dV_G = 0.17$ µA V^{-1}. The on:off current ratio is 10^7. (Adapted from Ref. [127].)

this kind of device is in the 10^{-9} A V^{-1} range and the current modulation occurs through 4 orders of magnitude. "TiC" CNTFETs exhibited a drastic improvement of performance. According to several authors, "TiC" decreases the contact resistance for the injection of both p- and n-type carriers [74]. Indeed, the "TiC" CNTFETs in Fig. 9.10 are ambipolar, i.e. they carry a strong current at both negative and positive values of V_G. dI_D/dV_G increases by 2 orders of magnitude and results in the 10^{-7} A V^{-1}, range while the current modulation occurs through 6–7 orders of magnitude, which is also 2 orders of magnitude better than "classical" devices.

The characteristic values of the self-assembled "TiC" devices are close to those obtained on similar back-gated devices fabricated by random deposition directly on SiO$_2$ (e.g. as evidenced by the comparison of the results in Fig. 9.10 with those reported in Ref. [74] for a p-type device). Moreover, it has been shown that the transport characteristics of "TiC" CNTFETs obtained with both random and self-assembly techniques are strongly dependent upon the temperature of the annealing process, with an optimum around $800 - 850$ K [74, 127].

The results clearly indicate that the electrical performance of self-assembled CNTFETs is mainly determined by the nanotube/metal contact interfaces and by their response to the applied electric fields. This obviously suggests that after RTA, the use of a SAM technique to direct the assembly of the nanotube does not perturb the transport characteristics of such fabricated CNTFETs. It is worth noticing the importance of this result, which validates the self-assembly approaches for large-scale production of nanotube-based electronics. Moreover, it should also be noted that the SAM-based process discussed above is fully compatible with the realization of top-gate devices [140] and/or low thickness and high effective dielectric constant oxide films [141]. In conclusion, this self-assembly technique allows con-

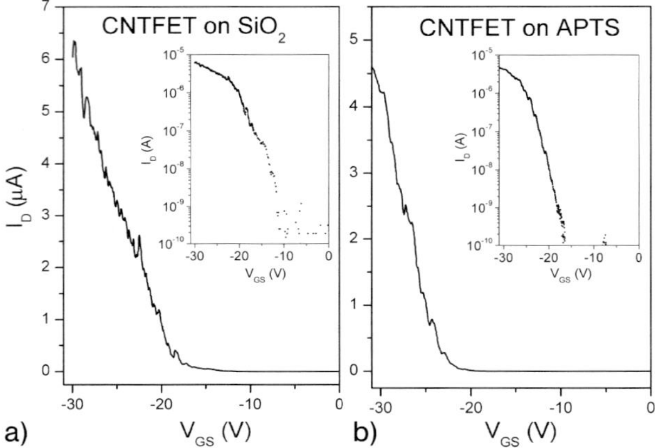

Figure 9.11. Transfer characteristics $I_D(V_{GS})$ at $V_{DS} = -1$ V of two CNTFETs with a gate oxide thickness of 200 nm. (a) CNTFET made by random deposition on SiO_2. (b) CNTFET made by the self-assembly technique. Insets show the same data in log-scale. (From Ref. [83].)

trolled and systematic fabrication of CNTFETs with performances reaching the actual state-of-the-art of CNTFETs as obtained by other techniques.

It should be also mentioned that the APTS monolayers are destroyed in the annealing process. Thus, it is still pertinent to question the role of the APTS monolayer on device characteristics when the RTA step is not performed, as it is the case of standard Ti/Au or Cr/Au electrodes. In order to elucidate this point, it is necessary to compare the performances of devices made by the APTS self-assembly technique (called CNTFETs on APTS) with those of CNTFETs made by random deposition (called CNTFETs on SiO_2). A comparison is presented in Fig. 9.11, which shows the $I_D(V_{GS})$ characteristics for the two kinds of devices. The measurements were performed at room temperature, in air, with the gate bias swept from the on- to the off-state [83]. We notice that their performances are very similar: on-state current up to around 5 µA, on:off ratio of 4 orders of magnitude, transconductance of $0.4 - 0.5$ µS and subthreshold slope $S = 2000 \pm 300$ mV dec^{-1}. This means that, under typical atmospheric conditions, the performances are set by the quality of the contacts and the gate efficiency (set by the oxide thickness), independently of the placement technique. Note, finally, that the performances of the devices used in the comparative study shown in Fig. 9.11 are similar to the best performances of CNTFETs of comparable geometry reported in the literature and thus representative of the state-of-the-art of CNTFETs.

In conclusion, the chemical functionalization of SiO_2 substrates by an APTS monolayer brings a relevant solution to the problems of (a) systematic placement of nanotubes in a transistor geometry and (ii) their subsequent connection to electrodes. Moreover, the use of the SAM deposition process not only does not deteriorate the device characteristics (as one may think), but it is also fully compatible with the production of high-quality, state-of-the-art CNTFETs. Finally, more recent works have shown that it is possible to take even further advantage of the APTS monolayer to perform chemical optimization of CNTFETs [142]. Indeed, when compared to a CNTFET on SiO_2, a CNTFET on APTS includes an additional, tunable, chemical interface (nanotube/APTS) to act onto.

In Sections 9.4 and 9.5 we have considered two approaches (CVD and SAMs) to perform selective placement of SWNTs. An important motivation was the development of a bottom-up technology for the implementation of nanodevices. However, standard top-down lithography techniques are still necessary in both two approaches. Indeed, even if the random deposition of nanotubes is avoided and large-scale fabrication can be envisaged, the patterns for the catalyst or for the APTS monolayer, as well as the electrodes, are realized by standard lithographic techniques. A real technological breakthrough in self-assembly would be to develop a complete molecular-scale bottom-up method. In this context, in the following section we discuss a promising technique based on the use of a DNA scaffold to realize nanoscale site-controlled implementation of nanocomponents.

9.6
DNA-directed Self-assembly

Among the new methodologies based on bottom-up approaches for future nanotechnology, the exploration of bio-directed assembly for organizing nano-objects is one of the most promising. Indeed, the nanoscale is the natural scale on which biological systems build up their structural elements, and biological molecules have already shown great potential in the fabrication and construction of nanostructures and devices.

In this context, the DNA molecule is of particular interest, as highlighted by the increasing number of recent works devoted to the study of its physical properties and implementation in nanoelectronics. Indeed, the DNA molecule has already been successfully used to build up nanostructures [117, 143] or scaffolds for nanoparticle assembly [118–121]. Moreover, one can envisage its use for the assembly of devices. The key advantage in using DNA as a scaffold for these constructions is that its intra- and inter-molecular interactions are the most readily known, engineered and reliably predicted. The information contained in DNA sequences can be envisioned to code:

- The assembly of the scaffold.
- Its selective attachment on the surface microscale electrodes.
- The positioning of nano-objects or nanodevices on the scaffold.
- The realization of electrical connections and circuitry.

This idealized pathway to assemble circuit in two (or three)-dimensional geometry is very appealing, and, as we discuss in the following, some of the necessary steps have already been realized and reported in the literature, as well as a first mono-device demonstrator [144]. In the following we discuss separately these four aspects of the use of DNA molecules for the realization of bio-assembled nanodevices and nanocircuits.

9.6.1
The Assembly of the Scaffold

In recent years, an increasing number of works have been devoted to the realization of nanostructured scaffolds with DNA molecules. The pioneering experiments concerned the assembly of sticky-ended linear DNA molecules [145]. Since then, a long way has been covered to evolve towards the vision, developed by Seeman and coworkers, of a construction with "smart bricks" made of DNA molecules. The main idea is based on the following simple scheme: the sticky ends of the envisaged DNA "smart bricks" would have the property to recognize each other and act as nanovelcro to realize the required assembly [146]. Obviously, such "smart bricks" must be more complex than linear DNA in order to realize a nanostructured two (or three)-dimensional scaffold. Therefore, in order to implement this scheme, synthetic molecules have been designed to produce branched motifs, taking inspiration from the natural phenomenon of reciprocal exchange crossover between DNA molecules [147]. Indeed, it is well known that natural DNA is not always in the linear configuration, but it passes, during its metabolism, through the configuration of an *unstable* branched molecule. The main idea of the method is to synthesize single-strand (ss) DNA molecules that self-assemble into *stable* macromolecular branched building blocks (called DNA tiles or "smart bricks"). Thus, the tip is to prepare synthetic oligonucleotides that break the homologous sequence symmetry of the natural branched molecules, avoiding in this way the instability due to the isomerization via branch migration [148]. Following this prin-

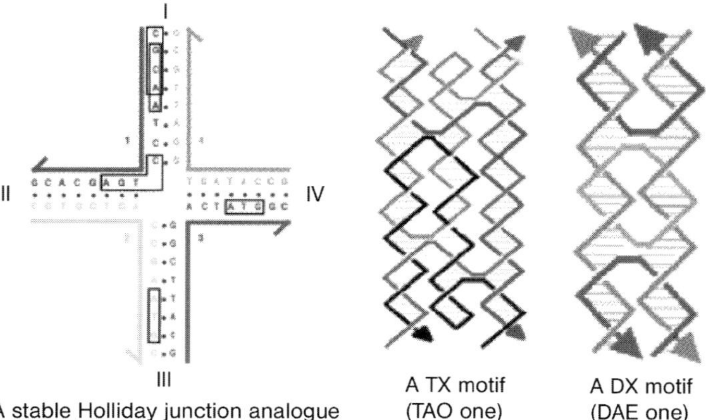

Figure 9.12. Schematic representations of various motifs constructed with DNA tiles: Holliday junction analogues, and DX and TX molecules. (From Ref. [148].)

ciple, various motifs have been fabricated, like Holliday junction analogues, double (DX) and triple (TX) crossover molecules, knots, and parallelograms (see Fig. 9.12).

The most interesting motifs in terms of scaffolds for nanotechnology purposes are those leading to a defined geometrical arrangement. In fact, simple branched junctions do not automatically lead to geometrical control [149, 150], essentially due to their lack of rigidity. The famous lattices reported by Seeman and coworkers rely on stiffer motifs, like the DX [151] or TX [152] tiles. More recent studies report the construction by biological recognition properties of a DNA-based nanostructure made of four four-arm junctions (4 × 4 tile, see Fig. 9.13) [153].

This programmed self-assembly gives rise to two distinct lattice morphologies: uniform-width nanoribbons and two-dimensional nanogrids that have been used to template protein and/or silver nanowires [153].

Figure 9.13. SEM and AFM images of different lattices constructed with DNA tiles. From left to right: (SEM image of) a DX, (SEM image of) TX and (AFM image of) 4 × 4 lattices. (Adapted from Ref. [146].)

Another interesting approach has been developed by Bergstrom and coworkers, who used rigid tetrahedral linkers with arylethynylaryl spacers to direct the assembly of attached oligonucleotide linker arms into novel DNA macrocycles [154]. The main originality of this method consists in the use of rigid tetrahedral organic vertices and in the fact that a variable number of oligonucleotide arms serve as connectors for the design of more complex architectures.

9.6.2
Selective Attachment of the DNA Scaffold on the Surface Microscale Electrodes

Another important step in the pathway for the use of DNA-directed assembly of nanocircuits is the fixation/linkage of the DNA scaffolds on the substrate and its connection to external microelectrodes. The issue is to have some anchoring sticky-end points on the electrodes (or substrate) to selectively deposit the scaffold. In this context, the biochip community has made very strong efforts and different methods have been developed for DNA probe technology. The goal to achieve is to fix a specific sequence on each electrode of the chip. In present-day technology, the DNA probes are nearly always attached to inorganic substrates (silicon and glass are the most widely used) [155–157], while the presence of the electrodes is not always required. Nevertheless, the majority of the most advanced techniques proposed in the literature can be transposed to our electrode-specific linkage problem. The solutions that have been found are essentially of two types. The first is to label the surface with defined oligonucleotides sequences, by means of either (a) addressing by micro-nanospotting or (b) addressing assisted by an applied electric field. The second type of specific labeling is more exactly an *in situ* localized synthesis of the desired oligonucleotide sequence. These different techniques are briefly described below.

Historically, the micro-spotting technique was the first method developed. In this case, the immobilization is achieved passively either by covalent bonding or adsorption [158–161]. Nowadays, it is possible to find different variations of the same principle, like the robotic deposition on a prepared substrate [162] or the use of an ink-jet printer [161]. The more recent evolution of this technique is "dip-pen" nanolithography (DPN), which allows a lateral resolution of the order of 50 nm [163]. This technique is based on scanning probe technology, as described in the following. First, an AFM tip is "inked" with a solution of the material to be transferred to the surface. Then the AFM tip "writes" the desired pattern on the surface. However, although advances have been made by the introduction of parallel multipen approaches (by multicantilever AFM), this technique is still fundamentally slow and its throughput cannot compete with standard printing process.

Electrical addressing has also been intensively developed. As one example, Nanogen recently developed an electric-field-assisted DNA immobilization process [164], designed to give pixel-by-pixel selectivity. The Nanogen DNA chip (NanoChip^{TM}) uses affinity-based immobilization (noncovalent bonding of the capture probes to the surface). This technology uses the electrophoresis principle of migration of

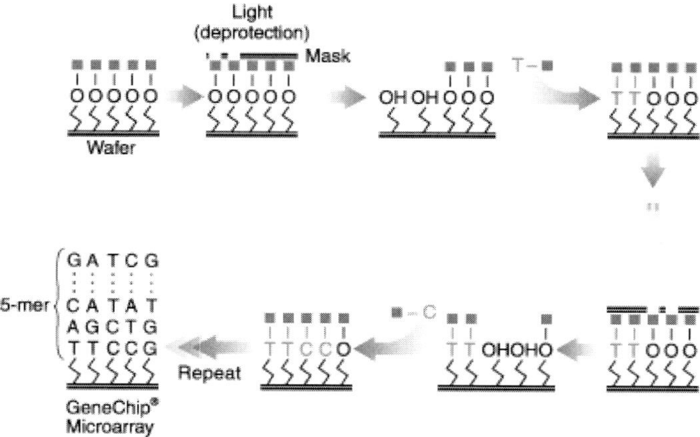

Figure 9.14. Schematic representation of the *in situ* oligonucleotides synthesis process using a light-directed method. Adapted from Ref. [167].

the negatively charged DNA molecule: "… when a biotinylated sample solution is introduced onto the array, the negatively charged sample moves to the selected positive electrode, where it is concentrated and bounds to the streptavidin in the permeation layer. The array is then washed and another sample can be added. In this way, site by site, an array of [oligonucleotides] samples is assembled on the [electrodes] array" [165]. A more advanced method of selective attachment of DNA strands to an electrode, also based on electrical addressing, consists of the successive (electrochemically addressed) copolymerization of 5′ pyrrole-labeled oligonucleotide and pyrrole. By this method, each electrode is covered by a conducting polymer (polypyrrole) grafted by an oligonucleotide [166].

The second solution concerns the *in situ* synthesis of the oligonucleotide sequence on the electrodes, using a light-directed method. Affymetrix commercializes this kind of array and the fabrication process can be schematized as follows (see Fig. 9.14) [167, 168]. A solid support is derivatized with a covalent linker molecule terminated with a photolabile protecting group. Light is then directed through a mask to de-protect and activate selected sites, and protected nucleotides couple to the activated sites. The process is repeated, activating different sets of sites and coupling different bases, allowing arbitrary DNA probes to be constructed at each site.

9.6.3
Positioning of Nano-objects or Nanodevices on the Scaffold

In order to position nano-objects on the scaffold, it is necessary to master their linkage to a DNA strand and then use the DNA recognition properties to insert them on the scaffold. The approach can differ depending on whether the nano-object is inserted during the construction of the scaffold or linked to it successively. In any case, the linkage of the nano-object to the DNA strand is a key step of the process. Many works have been reported aiming to develop methods for functionalizing small inorganic building blocks with DNA and then direct their assembly into extended structures by using the molecular recognition properties associated with DNA. So far, DNA has already been used to functionalize gold nanoparticles [120, 169–172], semiconductor quantum dots [173] and CNTs.

Concerning the chemical derivatization of CNTs, end chemistry of oxidatively etched nanotubes has been largely investigated [174–177]. This functionalization method was mostly used to improve nanotube solubility by reacting the nanotubes with hydrophilic dendra, alkyl chains or polymers. Nonetheless, it is still not clear if this strategy leads uniquely to functionalization of the defect sites of the oxidized ends of the nanotubes or also their side-wall surface. Indeed, the oxidation step performed during the usual purification process of the SWNTs produces such defects in relatively low amounts (estimated to about 2–3%) [178, 179], but they are not specifically localized only at the nanotube ends.

Currently, most of the work performed on nanotube functionalization is based on side-wall chemistry. Both covalent and noncovalent routes have been developed. A series of covalent side-wall functionalizations has been reported recently, including the reaction of nanotubes with nitrene, carbene and radical compounds. These methods should open routes to a wide variety of new nanotubes derivative [180–182]. In contrast, noncovalent routes have been mainly used to wrap nanotubes with polymer to enhance solubility and form composite materials [183–188]. An interesting study about solubilization of SWNTs by means of noncovalent chemistry (peptide wrapping) has been recently reported. In this work the authors first solubilize the nanotubes in aqueous solution by peptide wrapping [189] and then they increase the stability of such solution by crosslinking the wrapped peptides to each other [190].

Most results reported on CNT–DNA linkage also deal with covalent chemistry based on carboxylic acid defect groups present on SWNTs [191–195]. In 2002, a work reported the covalent coupling of peptide nucleic acid (PNA) rather than DNA [195]. The authors chose PNA as an intermediate for the covalent chemistry before hybridizing it with a DNA strand. The reason of this choice was the higher robustness of PNA to environmental conditions, as compared to other oligonucleotides. Nevertheless, in this report, as in each of the other described methods for covalent grafting of DNA onto nanotubes, an additional aggressive oxidation was performed in order to increase the defect density [192–195] and no indication was given on the yield of DNA–SWNT linkages. Moreover, the effect of introducing a large number of defects along the nanotubes is still not clearly known, but is believed to strongly affect their original (mainly sp^2) structure and electronic properties. Recently, the same type of covalent chemistry between ssDNA and nanotubes (MWNTs and SWNTs) has been exploited to form multicomponent structures including 150-nm gold nanoparticles [196]. Finally, there is an interesting report about the binding of DNA to nanotubes by photochemistry [197]. In this work, the covalent linkage is not performed on a carboxylic defect, but by using acid photochemistry and *in situ* DNA synthesis.

Different works have also been reported on the association of biological molecules (DNA and/or proteins) with nanotubes by means of noncovalent chemistry. A strong interaction of DNA with the nanotube surface has been suggested for MWNTs [198] and more recently a demonstration of wrapping of SWNTs by well-defined ssDNA molecules has been published [34]. The great advantage of a noncovalent method is its independence of the presence of the carboxylic groups, so that the DNA binding can also be successfully achieved for completely defect-free nanotubes.

In terms of noncovalent methods, the attachment of streptavidin protein [199] to the nanotube is of particular interest, as we will discuss below. Streptavidin is a relatively small protein (60 000 Da) composed of four identical subunits. The mechanism that binds streptavidin to a sp^2 nanotube surface is probably related to hydrophobic interactions [199]. Indeed, this molecule is known to bind to hydrophobic surfaces [200]. The streptavidin protein is particularly well studied for its various biochemical applications because of its high affinity to biotin. Indeed, each of the streptavidin subunits has an active binding site for biotin molecules and the streptavidin–biotin system has one of the largest free energies of association yet observed for noncovalent binding of a protein and small ligand in aqueous solution ($K_{assoc} = 10^{14}$ M^{-1}). Moreover, these complexes are also extremely stable over a wide range of temperature and pH. Correspondingly, the simple idea that has been followed for the DNA–nanotube attachment process is to react a biotinylated DNA strand with a streptavidin-coated nanotube (see scheme in Fig. 9.15). In other words, this technique uses noncovalent chemistry through a biological recognition complex (streptavidin–biotin) in order to link DNA and nanotubes [201, 202].

One great advantage of this method is related to the yield and the robustness of the reaction, while its main inconvenience is its lack of site specificity. However, it is worth pointing out that none of the methods reported above, particularly those based on covalent binding, is genuinely site specific. In fact, it would be extremely naïve to think that the carboxylic defects are present only at the ends of the nanotubes. On the contrary, it is more likely to think that the chemistry on a side-wall defect will be extremely favored with respect to a reaction at the ends, due to the nanotube's particular cylindrical shape and geometry factor (diameter versus length). In order to avoid this unspecific site chemistry and ensure the effective-

Figure 9.15. Schematic representation of the linking process between a biotinylated DNA strand with a streptavidin-coated nanotube.

ness of a covalent reaction on carboxylic defects only at the ends, Willner and co-workers [203] developed a method to mask the nanotube side-walls by a wrapping. In this work the authors succeeded in performing real end chemistry and they were able to link covalently the flavin adenine dinucleotide cofactor only to the nanotube ends. It is thought likely that such a method could be extended/adapted to perform covalent end chemistry between nanotubes and DNA strands in solution.

9.6.4
Realization of Electrical Connections and Circuitry

So far we have discussed methods to construct and position nanotubes on a DNA scaffold opportunely linked on the substrate microelectrodes. However, an essential step for electronic purposes is the electrical connection of such nanotubes. This latter relies on the transport properties of DNA molecules and has recently generated heated debate among scientists, as evidence both for and against the hypothesis of DNA as a conducting wire has piled up. While no full consensus has been reached, we feel that the extensive transport measurements carried out on single DNA molecules and DNA bundles strongly suggests that DNA in the dry state deposited on a substrate is a good insulator and thus not useful as a conducting element. We believe that in spite of its somewhat negative sense, such a conclusive statement is of great importance in defining strategies for implementing DNA-based technology. Indeed, it now becomes clear that to achieve an electrical connection using DNA strands, it is necessary to proceed to their metallization. During the past 10 years we have seen the development of numerous methods to metalize DNA scaffolds and a recent review of these metalization processes can be found in Ref. [204]. In the following we summarize the main aspects and results concerning this topic.

The feasibility of this biotemplating approach was first shown by Braun and coworkers [205]. The authors first immobilized a DNA strand between two electrodes. Then, they treated it with silver ions in order to perform an Ag^+/Na^+ ion exchange and replace the natural sodium counterions of the DNA backbone with silver ones. Successively, these silver ions were subjected to a chemical reduction process by the reducing agent hydroquinone to form small silver aggregates. Finally, the silver nanoclusters fixed on the DNA strand were autocatalytically grown (using an acidic solution of hydroquinone and silver ions) to give a granular (100 nm width) nanowire contacting the two electrodes. The majority of DNA metalization processes follow the same principle and can be decoupled in terms of successive steps, as schematically shown in Fig. 9.16 and discussed below.

The first step consists of biomolecule activation – the metal ions or metal complexes bind to DNA, creating reactive metal sites (Fig. 9.16a). The activation can take place by an ion-exchange mechanism (as discussed above for silver [205]) on the DNA backbone or by insertion of the metal complexes between the DNA bases (like platinum or palladium complexes [206]).

In the second step, the bound seeds are usually treated with a reducing agent

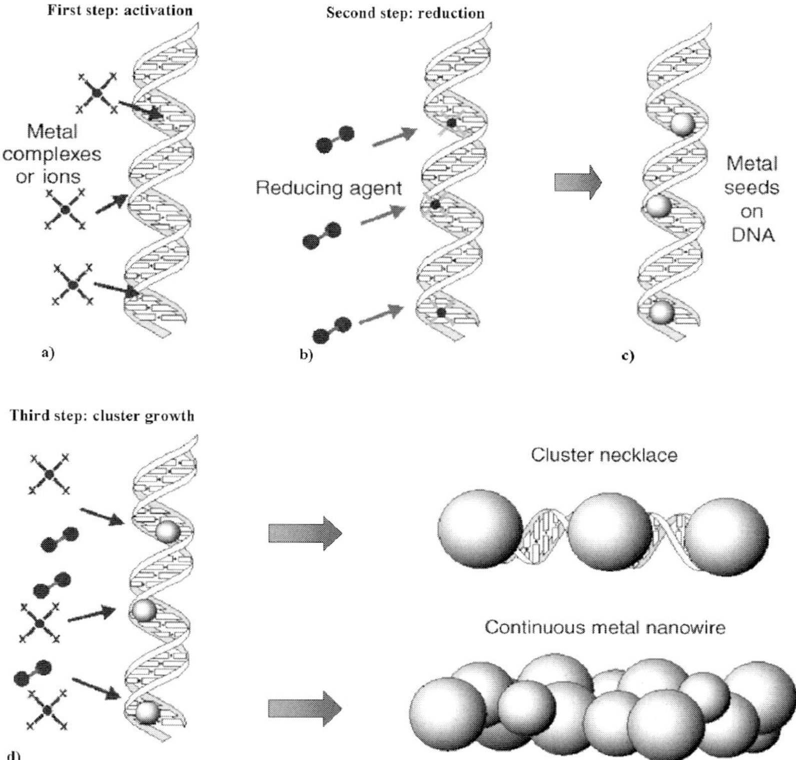

Figure 9.16. Schematic representation of the different steps involved in the DNA metallization process. (Adapted from Ref. [204].)

(Fig. 9.16b). This converts the metal ions or metal complexes in metal nanoclusters fixed on the DNA strand. The more often used reducing agents are dimethylaminoborane [207, 208], hydroquinone [205] and sodium borohydride [209]. An interesting variant has been proposed by Keren and coworkers [210, 211], who fixed the reducing agent (glutaraldehyde) directly to the DNA strand in order to enhance specificity and reduce parasitic unwanted background metalization. At the end of these two steps, the DNA strand has some small metal nanoclusters fixed on it (as represented in Fig. 9.16c), which will successively act as "seeds" for the metalization of the DNA molecules.

The third step of the metalization process consists of autocatalytic growth of the fixed metal seeds on the DNA strand (Fig. 9.16d) by the addition of new metal ions (or metal complexes solution) and new reducing agent solution. The idea of this autocatalytic process is that metal complexes or ions from solution are preferably reduced on already reduced metal nanoclusters (the seeds) fixed on the DNA strands. It should be noted that this autocatalytic cluster growth can be generalized to metal nanoclusters fixed on DNA strands by any other method. Indeed, effective metalization has been reported on *ex situ* prepared gold nanoclusters fixed on the DNA strand by appropriate chemical functionalization [212], by DNA construction [213] or simply by electrostatic interactions [214].

As mentioned above, different metals have been used in the metalization process. For the ion-exchange mechanisms on the DNA backbone we can quote silver [153, 205] or copper [215]. The results are quite convincing for the silver process, where a silver wire is formed consisting of a chain of contiguous 30- to 50-nm silver grains along the DNA backbone. However, the electrical measurements performed on the obtained silver necklace wire were not completely satisfactory. In a following work, the same group improved the process by replacing the silver clusters growth with an electroless gold coating of the silver-loaded DNA molecules [210]. In this way, using silver ions as catalysts, conductive gold DNA-templated wires with widths ranging from 50 to 100 nm were obtained. This procedure can be generalized and the final metal coating does not necessarily have to use the same metal as the seeding one [209, 210, 216].

Concerning the intercalation mechanism of metal complexes between the DNA bases, palladium or platinum complexes have been the more extensively studied. Indeed, the binding process of Pt(II) complexes to DNA is well investigated in the case of cisplatin (cis-[Pt(NH$_3$)$_2$Cl$_2$]), which is widely used as an anticancer drug [217]. It follows from these studies that when DNA is incubated with Pt(II) complexes such as cisplatin, the Pt(II) atom binds to one or two stacked DNA bases, forming monofunctional and bifunctional DNA–Pt(II) adducts, respectively. The most favorable binding site for cisplatin to the DNA is the N_7 position of guanine, followed by the N_7 position of adenine [217, 218]. Indeed, the bases A, G and C have exocyclic amine groups as well as ring amines, but it is the ring amines that act as Lewis bases. The Lewis base acidities differ from base to base, with the N_7 position of guanosine being the most basic. Other Lewis bases found in the nucleobases are N_7 of adenosine, N_3 of cytosine and the deprotonated N_3 of thymidine or uridine. The amines are all soft ligands and as such preferably complex to soft metals such as Pt(II), Pd(II) or Ru(II). When the DNA is in double-strand configuration, the arrangement of the basis is controlled by π stacking and then the Lewis base sites available for coordination to the metal (Pd, Pt, etc.) are limited to the exposed portion of the nucleobases found in the major groove (the N_7 position of guanosine and adenosine). It is commonly thought that, of these two sites, the N_7 of guanosine is the preferred one [219]. After these sites are occupied, the binding reaction proceeds more slowly and indiscriminately with other metal-binding sites of all bases [218–221]. Using this intercalation mechanism, palladium and platinum DNA-coated nanowires have been obtained. Generally speaking, these metalization processes of the DNA strands have been performed either (a) in solution and then the metalized DNA molecule is deposited on substrate for the characterization purpose or (b) on the DNA previously deposited on the substrate. The first case concerns the works reported by Ford and coworkers [209] and Mertig and coworkers [222], who showed the formation of tiny platinum nanocluster necklaces consisting of well-separated clusters of 3 – 5 nm diameter with a spacing from one to several nanometers. On the contrary, Richter and coworkers [223], Deng and coworkers [224], Dupraz and coworkers [216] and Ongaro and coworkers [214] have metalized DNA strands already deposited on the substrate. In more detail, Richter and coworkers fabricated continuous palladium nanowires with average diameters of 60 − 100 nm on DNA strands aligned on interdigitated gold electrodes and obtained interesting transport properties. On the contrary, Deng and coworkers reported very granular 30-nm palladium nanowires without any data about their conduction properties [224]. Actually, both Dupraz and coworkers [216] and Ongaro and coworkers [214] started the metalization process in solution (seeds fixation), and successively stretched the metal-loaded DNA on the substrate. Then they completed the metalization process. They used, respectively, platinum and gold nanoparticles as catalytic seeds, but they both finished by a gold electroless plating process. They both obtained DNA-templated gold nanowires (average diameters about 20 − 50 nm) with estimated resistivities of between 10^{-5} and 10^{-4} Ωm.

However, among all the methods discussed above, only a few are really promising in nanocircuit applications. First, the more interesting procedure is the one where the DNA metalization occurs as one of the last steps. Indeed, after the metalization process all the recognition properties of the DNA molecule are completely lost, and the circuit architecture must be necessarily fixed and deposited on the substrate. Moreover, another important point is that some parts of the DNA scaffold must not be metalized to avoid shorts and preserve the device characteristics. In this sense, three convincing studies have been reported [210, 211, 213]. They are all based on the RecA protein properties. *In vivo*, the RecA protein is a central component in recombinational DNA repair pathways and homologous genetic recombination (in *Escherichia coli*). *In vitro*, RecA protein promotes the pairing and exchange of complementary DNA strands in reactions. The mechanism is as follows: RecA catalyzes the pairing of ssDNA with complementary regions of double-stranded (ds) DNA. The RecA monomers first polymerize to form a helical filament around ssDNA (Fig. 9.17a). Duplex DNA is then bound to the polymer (see Fig. 9.17b).

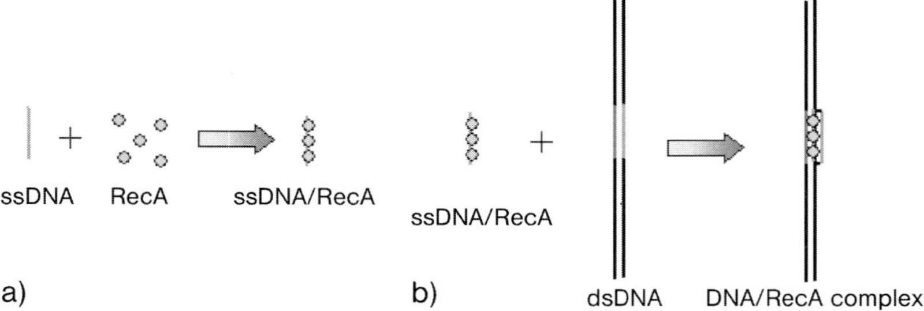

Figure 9.17. Schematic representation of the homologous recombination process that leads to binding of the ssDNA–RecA nucleoprotein filament at the complementary address on the dsDNA.

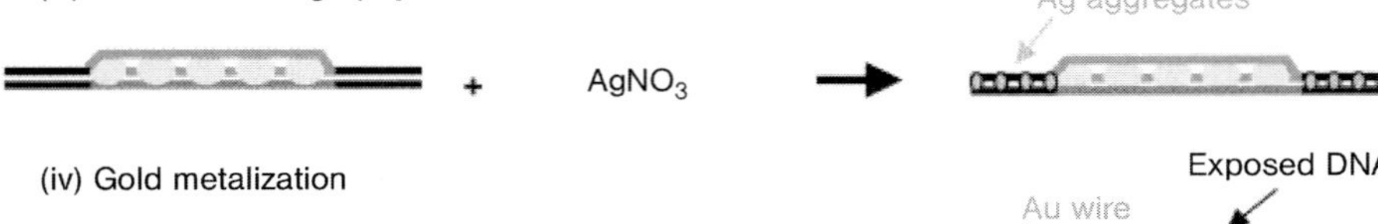

Figure 9.18. Mechanism of the sequence-selective metallization process. Thanks to the homologous recombination,
the RecA protein acts as a sequence-specific resists for the
creation of the silver seeds and successive gold metallization.
(From Ref. [210].)

This particular feature has been used to differentiate a part of the sequence of
the DNA strand in the metalization process. Indeed, it is clear that the targeted sequence is perfectly identified by the RecA polymerized ssDNA. The idea exploited
is that the creation of the complex between the RecA–ssDNA polymer and the
complementary regions of dsDNA acts as a mask for the metalization process. In
the first report on sequence-specific metalization [210] it is shown that this complex avoids the Ag^+/Na^+ ion exchange. This blocks the formation of the silver
seeds on the targeted sequence of the DNA molecule and, consequently, the successive gold metalization (see Fig. 9.18).

In a successive work [211], the same team showed that sequence-specific lithography can also be achieved by sequence-specific patterning of the local reducing
agent (glutaraldehyde). This patterning was performed both by hybridization between aldehyde-derivatized and underivatized DNA molecules and by sequence-
specific protection against aldehyde derivatization using homologous recombination processes by the RecA protein. Then, the sequence-specific patterning of the
reducing agent is reflected by the sequence-specific creation of silver metalization
seeds and successive gold metalization.

The more recent report on selective metalization also uses the homologous
recombination properties of the RecA protein and its originality is to employ a
modified RecA to act as a "linking factor" for sequence-specific fixation of gold
nanoparticles [213]. Then, these gold nanoparticles are used as "seeds" for the

metalization process. To be more specific, the authors use a genetically engineered cysteine derivative RecA protein (Cys-RecA) and, thanks to this derivatization, the gold nanoparticles can be fixed to the Cys-RecA–DNA filament. However, the gold nanoparticles are not fixed on unmodified RecA filament. Therefore, the strategy they developed to achieve selective metalization is to use separately cysteine derivatized RecA and unmodified wild-type RecA (RecA) to complex different sequences of the DNA strand. Then, the fixation of the gold nanoparticles respects the targeted sequences Cys-RecA–DNA and RecA–DNA, and the successive metalization presents a sequence specific gap corresponding to the unmodified RecA–DNA complex.

9.6.5
Fabrication of DNA-directed CNT Devices

Among the studies that use DNA to fabricate CNT devices [144, 225–227], the most impressive one is the report on the DNA templating of a CNTFET [144]. Indeed, in this work some of the crucial ingredients of the vision discussed in Section 9.6 were tackled and demonstrated, even if in a simple linear back-gate geometry. The authors employed (a) a selective placement of the nanotube on the DNA scaffold and (b) a sequence-selective metalization of the DNA strands. In this way, they were able to realize the electrical connection between the standard (lithographically defined) electrodes and the nanotube device. In both tasks (a) and (b) they exploited the sequence-specific homologous recombination of the RecA protein. They first anchored the SWNT in the desired part of the DNA scaffold and then, after deposition on a substrate, proceeded to the selective metalization process. The SWNT–DNA linkage was performed in buffered solution by the molecular recognition of a streptavidin-functionalized SWNT towards biotin and by the antibodies properties to link the biotin to the RecA–DNA filament, as schematized in Fig. 9.19.

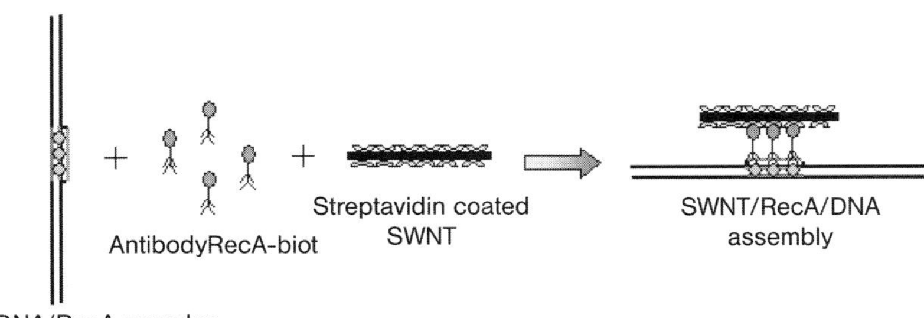

Figure 9.19. Schematic representation of the sequence-specific binding of a streptavidin-coated SWNT. The homologous recombination process of RecA is used to target the binding site. The streptavidin-coated nanotube is fixed to the DNA-bound RecA by using a complex antibodyRecA–biot (a primary antibody to RecA and a biotin-conjugated secondary antibody).

Then, to fabricate the nanotube device, they stretched the SWNT–RecA–DNA assembly on a silicon substrate and performed the selective Ag/Au metalization process as described in Ref. [210]. One of the interesting features of this experiment is that, even in this more complicated configuration (i.e. presence of the SWNT, streptavidin, biotin and antibodies species), the masking properties of the RecA are preserved and the "protected" segment of the DNA strand is not metalized. Finally, the DNA-templated gold wires were connected by fabricating standard lithography electrodes and the device characteristics were recorded.

However, even if all the steps for DNA-directed self-assembly of a CNT device have been demonstrated and reported in literature, it should be noted that it is still missing a study where all these steps are combined. Indeed, in the work of Keren and coworkers [144] there is no selective attachment to the microscale electrodes. This implies that in order to fabricate these electrodes, the deposited SWNT–RecA–DNA strand must be localized by imaging, as for nanotube devices obtained by random deposition. On the contrary, in the report of Hazani and coworkers [226, 227], the SWNT–DNA complex is fixed to the electrodes by DNA hybridization, but no selective metalization is performed. In this case, it is likely that the contact between the nanotube and the electrodes is ensured by the fact that the DNA strands used are very short and the nanotube touches the metal anyway.

In conclusion, the demonstration of a completely DNA-directed self-assembled CNT device is, at the time of writing, still to be shown, as well as the realization of a more structured scaffold hosting more than one nanotube device in a circuit configuration. The knowledge and mastery of the different steps needed for the implementation of such a demonstrator have already been reported in literature, and it is now a matter of multidisciplinary will and teams to accomplish the task. Finally, for such a still open and exploratory research domain, new findings are expected to further enlarge the present vision and generate novel strategies for the short- to medium-term development of nanoelectronics.

9.7
Conclusion

In this chapter we have presented a review on the self-assembly techniques for fabrication of CNT devices. First, we discussed their main physical features and sketched a few reasons why CNTs benefit from their present status as a serious potential candidate for future nanoelectronic applications. Then, we focused our attention on three methods to manipulate them by self-assembly. Indeed, this promising material could be envisaged to take part in the future nanoelectronic framework only if a cheap, massive parallel technology for fabricating nanotube devices is developed. The timing of this kind of research is particularly appropriate, as it results from an estimation of the ITRS roadmap [1]. From the analysis of this document it appears that, even if the scaling of CMOS device structures is a well-explored science, its limits will be reached in the near future due to various physical effects that do not scale properly, including quantum mechanical tunneling, the discreteness of dopants, voltage-related effects such as subthreshold swing, fabrication costs and reliability related to very small size variation, built-in voltage and minimum logic voltage swing, and application-dependent power-dissipation limits. Self-assembly of new and interesting materials could bring some answers to these limitations. Indeed, it is extremely naïve to think that a non-CMOS device (based on nanotubes, molecules or other) that simply mimics a CMOS device can solve this scaling limit only because of its reduced size. Among the various issues that CMOS technology is facing, the power dissipation effect is the most serious one and it can actually apply also for any non-CMOS device. Based on this assumption, the study and implementation of non-CMOS devices cannot be motivated only by the scaling issue, but must present other kind of advantages like (a) performing a particular functionality not well covered by standard CMOS technology, (b) be economically very appealing with important reductions of fabrication costs, (c) really introduce a disruptive technology by presenting new physical effects or (d) be prone to the development of new architecture paradigms, which could include concepts like three-dimensional design, defect/fault tolerances, reconfigurability, self-repairing features and massive parallel processing. This is a strong reason to increase the investments in time and resources in studying alternative strategies like self-assembly and, in particular, bio-inspired technologies.

References

1 The Internet Technology Roadmap for Semiconductors (ITRS) is available at http://public.itrs.net.http://www.itrs.net.

2 European Commission IST Programme Future and Emerging Technologies – Microelectronics Advanced Research Initiative – MELARI NANO. *Technology Roadmap for Nanoelectronics*, **1999**. Available at ftp://ftp.cordis.lu/pub/esprit/docs/melnarm.pdf.

3 IIJIMA S. Helical microtubules of graphitic carbon, *Nature* **1991**, *354*, 56–58.

4 SAITO R., DRESSELHAUS G., DRESSELHAUS M. S. *Physical Properties of Carbon Nanotubes*. Imperial College Press, London, **1998**.

5 SAITO R., DRESSELHAUS G., DRESSELHAUS M. S. Electronic structure of double-layer graphene tubules, *J. Appl. Phys.* **1993**, *73*, 494–500.

6 REICH S., THOMSEN C., ORDEJON P. Elastic properties of carbon nanotubes under hydrostatic pressure, *Phys. Rev. B* **2002**, *65*, 153407.

7 NATSUKI T., TANTRAKARN K., ENDO M. Prediction of elastic properties for single-walled carbon nanotubes, *Carbon* **2004**, *42*, 39–45.

8 LU J. P. Elastic properties of carbon nanotubes and nanoropes, *Phys. Rev. Lett.* **1997**, *79*, 1297–1300.

9 TREACY M. M. J., EBBESEN T. W., GIBSON J. M. Exceptionally high Young's modulus observed for individual carbon nanotubes, *Nature* **1996**, *381*, 678–680.

10 KRISHNAN A., DUJARDIN E., EBBESEN T. W., YIANILOS P. N., TREACY M. M. J. Young's modulus of single-walled nanotubes, *Phys. Rev. B* **1998**, *58*, 14013–14019.

11 DEMCZYK B. G., WANG Y. M., CUMINGS J., HETMAN M., HAN W., ZETTL A., RITCHIE R. O. Direct mechanical measurement of the tensile strength and elastic modulus of multiwalled carbon nanotubes, *Mater. Sci. Eng.* **2002**, *A334*, 173–178.

12 CADEK M., COLEMAN J. N., BARRON V., HEDICKE K., BLAU W. J. Morphological and mechanical properties of carbon-nanotube-reinforced semicrystalline and amorphous polymer composites, *Appl. Phys. Lett.* **2002**, *81*, 5123–5125.

13 EBBESEN T. W., AJAYAN P. M. Large-scale synthesis of carbon nanotubes, *Nature* **1992**, *358*, 220–222.

14 IIJIMA S., ICHIHASHI T. Single-shell carbon nanotubes of 1-nm diameter, *Nature* **1993**, *363*, 603–605.

15 BETHUNE D. S., KIANG C. H., DE VRIES M. S., GORMAN G., SAVOY R., VASQUEZ J., BEYERS R. Cobalt-catalysed growth of carbon nanotubes with single-atomic-layer walls, *Nature* **1993**, *363*, 605–607.

16 KIANG C. H., GODDARD III W. A., BEYERS R., BETHUNE D. S. Carbon nanotubes with single-layer walls, *Carbon* **1995**, *33*, 903–914.

17 MASER W. K., LAMBERT J. M., AJAYAN P. M., STEPHAN O., BERNIER P. Role of Y-Ni–B mixtures in the formation of carbon nanotubes and encapsulation into carbon clusters, *Synthetic Metals* **1996**, *77*, 243–247.

18 JOURNET C., BERNIER P. Production of carbon nanotubes, *Appl. Phys. A* **1998**, *67*, 1–9.

19 THESS A., LEE R., NIKOLAEV P., DAI H. J., PETIT P., ROBERT J., XU C. H., LEE Y. H., KIM S. G., RINZLER A. G., COLBERT D. T., SCUSERIA G. E., TOMANEK D., FISHER J. E., SMALLEY R. E. Crystalline ropes of metallic carbon nanotubes, *Science* **1996**, *273*, 483–487.

20 GUO T., NIKOLAEV P., THESS A., COLBERT D. T., SMALLEY R. E. Catalytic growth of single-walled manotubes by laser vaporization, *Chem. Phys. Lett.* **1995**, *243*, 49–54.

21 MASER W. K., BENITO A. M., MARTINEZ M. T., DE LA FUENTE G. F., MANIETTE Y., ANGLARET E., SAUVAJOL J. L. Production of high-density single-walled nanotube material by a simple laser-ablation method, *Chem. Phys. Lett.* **1998**, *292*, 587–593.

22 JOSÉ-YACAMAN M., YOSHIDA M. M., RENDON L., SANTIESTEBAN J. G. Catalytic growth of carbon microtubules with fullerene structure, *Appl. Phys. Lett.* **1993**, *62*, 657–659.

23 ENDO M., TAKEUCHI K., IGARASHI S., KOBORI K., SHIRAISHI M., KROTO H. W. The production and structure of pyrolytic carbon nanotubes (PCNTs), *J. Phys. Chem. Solids* **1993**, 1841–1848.

24 IVANOV V., NAGY J. B., LAMBIN P., ZHANG X. B., ZHANG X. F., BERNAERTS D., VANTENDELOO G., AMELINCKS S., VANLANDUYT J. The study of carbon nanotubules produced by catalytic method, *Chem. Phys. Lett.* **1994**, *223*, 329–335.

25 FONSECA A., HERNARDI K., PIEDIGROSSO P., COLOMER J. F., MUKHOPADHYAY K., DOOME R., LAZARESCU S., BIRO L. P., LAMBIN P., THIRY P. A., BERNAERTS D., NAGY J. B. Synthesis of single- and multi-wall carbon nanotubes over supported catalysts, *Appl. Phys. A* **1998**, *67*, 11–22.

26 KONG J., CASSEL A. M., DAI H. Chemical vapor deposition of methane for single-walled carbon nanotubes, *Chem. Phys. Lett.* **1998**, *292*, 567–574.

27 ZHU H. W., XU C. L., WU D. H., WEI B. Q., VAJTAI R., AJAYAN P. M. Direct synthesis of long single-walled carbon nanotube strands, *Science* **2002**, *296*, 884–886.

28 ZHENG L., O'CONNEL M., DOORN S., LIAO X., ZHAO Y., AKHADOV E., HOFFBAUER M., ROOP B., JIA Q., DYE R., PETERSON D., HUANG S., LIU J., ZHU Y. Ultralong single-wall carbon nanotubes, *Nat. Mater.* **2004**, *3*, 673–676.

29 URAL A., LI Y., DAI H. Electric-field-aligned growth of single-walled carbon nanotubes on surfaces, *Appl. Phys. Lett.* **2002**, *81*, 3464–3466.

30 JAVEY A., WANG Q., URAL A., LI Y., DAI H. Carbon nanotube transistor arrays for multistage complementary logic and ring oscillators, *Nano Lett.* **2002**, *2*, 929–932.

31 JOST O., GORBUNOV A. A., MÖLLER J., POMPE W., LIU X., GEORGI P., DUNSCH L., GOLDEN M. S., FINK J. Rate-limiting processes in the formation of single-wall carbon nanotubes: pointing the way to the nanotube formation mechanism, *J. Phys. Chem. B* **2002**, *106*, 2875–2883.

32 JOST O., GORBUNOV A., LIU X. J., POMPE W., FINK J. Single-walled carbon nanotube diameter, *J. Nanosci. Nanotechnol.* **2004**, *4*, 433–440.

33 LI Y., MANN D., ROLANDI M., KIM W., URAL A., HUNG S., JAVEY A., CAO J., WANG D., YENILMEZ E., WANG Q., GIBBONS J. F., NISHI Y., DAI H. Preferential growth of semiconducting single-walled carbon nanotubes by a plasma enhanced CVD method, *Nano Lett.* **2004**, *4*, 317–321.

34 ZHENG M., JAGOTA A., SEMKE E. D., DINER B. A., MCLEAN R. S., LUSTIG S. R., RICHARDSON R. E., TASSI N. G. DNA-assisted dispersion and separation of carbon nanotubes, *Nat. Mater.* **2003**, *2*, 338–342.

35 BAUGHMAN R. H., ZAKHIDOV A. A., DE HEER W. A. Carbon nanotubes – the route toward applications, *Science* **2002**, *297*, 787–792.

36 DEHEER W. A., CHATELAIN A., UGARTE D. A carbon nanotube field-emission electron source, *Science* **1995**, *270*, 1179–1180.

37 DE JONGE N., BONARD J. M. Carbon nanotube electron sources and applications, *Philos. Trans. Royal Soc. London A* **2004**, *362*, 2239–2266.

38 BONARD J. M., CROCI M., KLINKE C., KURT R., NOURY O., WEISS N. Carbon nanotube films as electron field emitters, *Carbon* **2002**, *40*, 1715–1728.

39 MILNE W. I., TEO K. B. K., AMARATUNGA G. A. J., LEGAGNEUX P., GANGLOFF L., SCHNELL J. P., SEMET V., BINH V. T., GROENING O. Carbon nanotubes as field emission sources, *J. Mater. Chem.* **2004**, *14*, 933–943.

40 BONARD J. M., STOCKLI T., NOURY O., CHATELAIN A. Field emission from cylindrical carbon nanotube cathodes: possibilities for luminescent tubes, *Appl. Phys. Lett.* **2001**, *78*, 2775–2777.

41 CHOI Y. S., CHO Y. S., KANG J. H., KIM Y. J., KIM I. H., PARK S. H., LEE H. W., HWANG S. Y., LEE S. J., LEE C. G., OH T. S., CHOI J. S., KANG S. K., KIM J. M. A field-emission display with a self-focus cathode electrode, *Appl. Phys. Lett.* **2003**, *82*, 3565–3567.

42 KIM I. H., KIM J. H., KIM K. B. Electrochemical characterization of electrochemically prepared ruthenium oxide/carbon nanotube electrode for supercapacitor application, *Electrochem. Solid State Lett.* **2005**, *8*, A369–A372.

43 LEE C. Y., TSAI H. M., CHUANG H. J., LI S. Y., LIN P., TSENG T. Y. Characteristics and electrochemical performances of supercapacitors with manganese oxide–carbon nanotube nanocomposites electrodes, *J. Electrochem. Soc.* **2005**, *152*, A716–A720.

44 SNOW E. S., PERKINS F. K., HOUSER E. J., BADESCU S. C., REINECKE T. L. Chemical detection with a single-walled carbon nanotube capacitor, *Science* **2005**, *307*, 1942–1945.

45 BAUGHMAN R. H., CUI C., ZAKHIDOV A. A., IQBAL Z., BARISCI J. N., SPINKS G. M., WALLACE G. G., MAZZOLDI A., DE ROSSI D., RINZLER A. G., JASCHINSKI O., ROTH S., KERTESZ M. Carbon nanotube actuators, *Science* **1999**, *284*, 1340–1344.

46 VOHRER U., KOLARIC I., HAQUE M. H., ROTH S., DETLAFF-WEGLIKOWSKA U. Carbon nanotube sheets for the use as artificial muscles, *Carbon* **2004**, *42*, 1159–1164.

47 LANDI B. J., RAFFAELLE R. P., HEBEN M. J., ALLEMAN J. L., VANDERVEER W., GENNETT T. Single wall carbon nanotube–Nafion composite actuators, *Nano Lett.* **2002**, *2*, 1329–1332.

48 CAO J., WANG Q., DAI H. Electromechanical properties of metallic, quasimetallic, and semiconducting carbon nanotubes under stretching, *Phys. Rev. Lett.* **2003**, *90*, 157601.

49 SAPMAZ S., BLANTER Y. M., GUREVICH L., VAN DER ZANT H. S. J. Carbon nanotubes as nanoelectromechanical systems, *Phys. Rev. B* **2003**, *67*, 235414.

50 BOURLON B., GLATTLI D. C., MIKO C., FORRO L., BACHTOLD A. Carbon nanotube based bearing for rotational motions, *Nano Lett.* **2004**, *4*, 709–712.

51 Li C., Chou T.-W. Single-walled carbon nanotubes as ultrahigh frequency nanomechanical resonators, *Phys. Rev. B* **2003**, *68*, 073405.

52 Lee S. W., Lee D. S., Morjan R. E., Jhang S. H., Sveningsson M., Nerushev O. A., Park Y. W., Campbell E. E. B. A three-terminal carbon nanorelay, *Nano Lett.* **2004**, *4*, 2027–2030.

53 Cha S. N., Jang J. E., Choi Y., Amaratunga G. A. J., Kang D.-J., Hasko D. G., Jung J. E., Kim J. M. Fabrication of a nanoelectromechanical switch using a suspended carbon nanotube, *Appl. Phys. Lett.* **2005**, *86*, 083105.

54 Ahir S. V., Terentjev E. M. Photomechanical actuation in polymer–nanotube composites, *Nat. Mater.* **2005**, *4*, 491–495.

55 Pantarotto D., Briand J. P., Prato M., Bianco A. Translocation of bioactive peptides across cell membranes by carbon nanotubes, *Chem. Commun.* **2004**, 16–17.

56 Kostarelos K., Lacerda L., Partidos C. D., Prato M., Bianco A. Carbon nanotube-mediated delivery of peptides and genes to the cells: translating nanobiotechnology to therapeutics, *J. Drug Deliv. Sci. Technol.* **2005**, *15*, 41–47.

57 Salvetat J. P., Briggs G. A. D., Bonard J. M., Bacsa R. R., Kulik A. J., Stockli T., Burnham N. A., Forro L. Elastic and shear moduli of single-walled carbon nanotube ropes, *Phys. Rev. Lett.* **1999**, *82*, 944–947.

58 Dalton A. B., Collins S., Munoz E., Razal J. M., Ebron V. H., Ferraris J. P., Coleman J. N., Kim B. G., Baughman R. H. Super-tough carbon-nanotube fibres – these extraordinary composite fibres can be woven into electronic textiles, *Nature* **2003**, *423*, 703–703.

59 Li X. D., Gao H. S., Scrivens W. A., Fei D. L., Xu X. Y., Sutton M. A., Reynolds A. P., Myrick M. L. Nanomechanical characterization of single-walled carbon nanotube reinforced epoxy composites, *Nanotechnology* **2004**, *15*, 1416–1423.

60 Cha S. I., Kim K. T., Arshad S. N., Mo C. B., Hong S. H. Extraordinary strengthening effect of carbon nanotubes in metal–matrix nanocomposites processed by molecular-level mixing, *Adv. Mater.* **2005**, *17*, 1377–1381.

61 Qian D., Dickey E. C., Andrews R., Rantell T. Load transfer and deformation mechanisms in carbon nanotube–polystyrene composites, *Appl. Phys. Lett.* **2000**, *76*, 2868–2870.

62 Blanchet Gr. B., Subramoney S., Bailey R. K., Jaycox G. D., Nuckolls C. Self-assembled three-dimensional conducting network of single-wall carbon nanotubes, *Appl. Phys. Lett.* **2004**, *85*, 828–830.

63 Wu Z. C., Chen Z. H., Du X., Logan J. M., Sippel J., Nikolou M., Kamaras K., Reynolds J. R., Tanner D. B., Hebard A. F., Rinzler A. G. Transparent, conductive carbon nanotube films, *Science* **2004**, *305*, 1273–1276.

64 Sakakibara Y., Rozhin A. G., Kataura H., Achiba Y., Tokumoto M. Carbon nanotube–poly(vinylalcohol) nanocomposite film devices: applications for femtosecond fiber laser mode lockers and optical amplifier noise suppressors, *Jpn. J. Appl. Phys. 1* **2005**, *44*, 1621–1625.

65 Avouris Ph., Appenzeller J., Martel R., Wind S. J. Carbon nanotube electronics, *Proc. IEEE* **2003**, *91*, 1772–1784.

66 Lee H. W., Kim S. H., Kwak Y. K., Han C. S. Nanoscale fabrication of a single multiwalled carbon nanotube attached atomic force microscope tip using an electric field, *Rev. Sci. Instrum.* **2005**, *76*, 046108.

67 Chen L. W., Cheung C. L., Ashby P. D., Lieber C. M. Single-walled carbon nanotube AFM probes: optimal imaging resolution of nanoclusters and biomolecules in ambient and fluid environments, *Nano Lett.* **2004**, *4*, 1725–1731.

68 Nguyen C. V., Chao K. J., Stevens R. M. D., Delzeit L., Cassell A., Han J., Meyyappan M. Carbon nanotube tip probes: stability and lateral resolution in scanning probe microscopy and application to surface science in semiconductors, *Nanotechnology* **2001**, *12*, 363–367.

69 TANS S. J., VERSCHUEREN A. R. M., DEKKER C. Room-temperature transistor based on a single carbon nanotube, *Nature* **1998**, *393*, 49–52.

70 MARTEL R., SCHMIDT T., SHEA H. R., HERTEL T., AVOURIS Ph. Single- and multi-wall carbon nanotube field-effect transistors, *Appl. Phys. Lett.* **1998**, *73*, 2447–2449.

71 POSTMA H. W. Ch., TEEPEN T., YAO Z., GRIFONI M., DEKKER C. Carbon nanotube single-electron transistors at room temperature, *Science* **2001**, *293*, 76–79.

72 BACHTOLD A., HADLEY P., NAKANISHI T., DEKKER C. Logic circuits with carbon nanotube transistors, *Science* **2001**, *294*, 1317–1320.

73 DERYCKE V., MARTEL R., APPENZELLER J., AVOURIS Ph. Carbon nanotube inter- and intramolecular logic gates, *Nano Lett.* **2001**, *1*, 453–456.

74 MARTEL R., DERYCKE V., LAVOIE C., APPENZELLER J., CHAN K. K., TERSOFF J., AVOURIS Ph. Ambipolar electrical transport in semiconducting single-wall carbon nanotubes, *Phys. Rev. Lett.* **2001**, *87*, 256805.

75 FREITAG M., RADOSAVLJEVIC M., ZHOU Y., JOHNSON A. T., SMITH W. F. Controlled creation of a carbon nanotube diode by a scanned gate, *Appl. Phys. Lett.* **2001**, *79*, 3326–3328.

76 DERYCKE V., MARTEL R., APPENZELLER J., AVOURIS Ph. Controlling doping and carrier injection in carbon nanotube transistors, *Appl. Phys. Lett.* **2002**, *80*, 2773–2775.

77 APPENZELLER J., KNOCK J., DERYCKE V., MARTEL R., WIND S. J., AVOURIS Ph. Field-modulated carrier transport in carbon nanotube transistors, *Phys. Rev. Lett.* **2002**, *89*, 126801.

78 CUI X., FREITAG M., MARTEL R., BRUS L., AVOURIS Ph. Controlling energy-level alignments at carbon nanotube/Au contacts, *Nano Lett.* **2003**, *3*, 783–787.

79 APPENZELLER J., RADOSAVLJEVIC M., KNOCH J., AVOURIS Ph. Tunneling versus thermionic emission in one-dimensional semiconductors, *Phys. Rev. Lett.* **2004**, *92*, 048301.

80 HEINZE S., TERSOFF J., MARTEL R., DERYCKE V., APPENZELLER J., AVOURIS Ph. Carbon nanotubes as schottky barrier transistors, *Phys. Rev. Lett.* **2002**, *89*, 106801.

81 NAKANISHI T., BACHTOLD A., DEKKER C. Transport through the interface between a semiconducting carbon nanotube and a metal electrode, *Phys. Rev. B* **2002**, *66*, 073307.

82 GUO J., DATTA S., LUNDSTROM M. A numerical study of scaling issues for Schottky-barrier carbon nanotube transistors, *IEEE Trans. Electon. Devices* **2004**, *51*, 172–177.

83 AUVRAY S., BORGHETTI J., GOFFMAN M. F., FILORAMO A., DERYCKE V., BOURGOIN J. P., JOST O. Carbon nanotube transistor optimization by chemical control of the nanotube–metal interface, *Appl. Phys. Lett.* **2004**, *84*, 5106–5108.

84 JAVEY A., GUO J., WANG Q., LUNDSTROM M., DAI H. J. Ballistic carbon nanotube field-effect transistors, *Nature* **2003**, *424*, 654–657.

85 JAVEY A., GUO J., FARMER D. B., WANG Q., WANG D. W., GORDON R. G., LUNDSTROM M., DAI H. J. Carbon nanotube field-effect transistors with integrated ohmic contacts and high-κ gate dielectrics, *Nano Lett.* **2004**, *4*, 447–450.

86 KONG J., FRANKLIN N. R., ZHOU C. W., CHAPLINE M. G., PENG S., CHO K. J., DAI H. J. Nanotube molecular wires as chemical sensors, *Science* **2000**, *287*, 622–625.

87 ADU C. K. W., SUMANASEKERA G. U., PRADHAN B. K., ROMERO H. E., EKLUND P. C. Carbon nanotubes: a thermoelectric nano-nose, *Chem. Phys. Lett.* **2001**, *337*, 31–35.

88 KOJIMA A., HYON C. K., KANIMURA T., MAEDA M., MATSUMOTO K. Protein sensor using carbon nanotube field effect transistor, *Jpn. J. Appl. Phys. Part 1* **2005**, *44*, 1596–1598.

89 STAR A., GABRIEL J. C. P., BRADLEY K., GRUNER G. Electronic detection of specific protein binding using nanotube FET devices, *Nano Lett.* 2003, *3*, 459–463.

90 BESTEMAN K., LEE J.-O., WIERTZ F. G. M., HEERING H. A., DEKKER C. Enzyme-coated carbon nanotubes as single-molecule biosensors, *Nano Lett.* 2003, *3*, 727–770.

91 STAR A., HAN T. R., GABRIEL J. C. P., BRADLEY K., GRUNNER G. Interaction of aromatic compounds with carbon nanotubes: correlation to the hammett parameter of the substituent and measured carbon nanotube FET response, *Nano Lett.* 2003, *3*, 1421–1423.

92 STAR A., HAN T. R., JOSHI V., STETTER J. R. Sensing with nafion coated carbon nanotube field effect transistors, *Electroanalysis* 2004, *16*, 108–112.

93 MAEHASHI K., MATSUMOTO K., KERMAN K., TAKAMURA Y., TAMIYA E. Ultrasensitive detection of DNA hybridization using carbon nanotube field-effect transistors, *Jpn. J. Appl. Phys.* 2004, *43*, L1558–L1560.

94 CAMBELL J. K., SUN L., CROOKS R. M. Electrochemistry using single wall nanotubes, *J. Am. Chem. Soc.* 1999, *121*, 3779–3780.

95 BRITTO P. J., SANTHANAM K. S. V., AJAYAN P. M. Carbon nanotube electrode for oxidation of dopamine, *Bioelectrchem. Bioenerg.* 1996, *41*, 121–125.

96 LUO H., SHI Z., LI N., GU Z., ZHUANG Q. Investigation of the electrochemical and electrocatalytic behavior of single-wall carbon nanotube film on a glassy carbon electrode, *Anal. Chem.* 2001, *73*, 915–920.

97 WANG J., DEO R. P., POULIN P., MANGEY M. Carbon nanotube fiber microelectrodes, *J. Am. Chem. Soc.* 2003, *125*, 14706–14707.

98 CHEN R.-S., HUANG W.-H., TONG H., WANG Z.-L., CHENG J.-K. Carbon fiber nanoelectrodes modified by single-walled carbon nanotubes, *Anal. Chem.* 2003, *75*, 6341–6345.

99 WANG J., LI M., SHI Z., LI N., GU Z. Electrocatalytic oxidation of 3,4-dihydroxyphenylacetic acid at a glassy carbon electrode modified with

100 GONG K. P., DONG Y., XIONG S. X., CHEN Y., MAO L. Q. Novel electrochemical method for sensitive determination of homocysteine with carbon nanotube-based electrodes, *Biosens. Bioelectron.* 2004, *20*, 253–259.

101 WU K., JI X., FEI J., HU S. The fabrication of a carbon nanotube film on a glassy carbon electrode and its application to determining thyroxine, *Nanotechnology* 2004, *15*, 287–291.

102 LIN Y., LU F., REN Z. Glucose biosensors based on carbon nanotube nanoelectrode ensembles, *Nano Lett.* 2004, *4*, 191–195.

103 WANG J., MUSAMEH M. Carbon-nanotubes doped polypyrrole glucose biosensor, *Anal. Chim. Acta* 2005, *539*, 209–213.

104 LI G., LIAO J. M., HU G. Q., MA N. Z., WU P. J. Study of carbon nanotube modified biosensor for monitoring total cholesterol in blood, *Biosens. Bioelectron.* 2005, *20*, 2140–2144.

105 CAI H., CAO X., JIANG Y., HE P., FANG Y. Carbon nanotube-enhanced electrochemical DNA biosensor for DNA hybridization detection, *Anal. Bioanal. Chem.* 2003, *375*, 287–293.

106 KERMAN K., MORITA Y., TAKAMURA Y., OZSOZ M., TAMIYA E. DNA-directed attachment of carbon nanotubes for enhanced label-free electrochemical detection of DNA hybridization, *Electroanalysis* 2004, *16*, 1667–1672.

107 WANG S. G., WANG R., SELLIN P. J., ZHANG Q. DNA biosensors based on self-assembled carbon nanotubes, *Biochem. Biophys. Res. Commun.* 2004, *325*, 1433–1437.

108 KATZ E., WILLNER I. Biomolecule-functionalized carbon nanotubes: applications in nanobioelectronics, *ChemPhysChem* 2004, *4*, 1084–1104.

109 MENNA E., DELLA NEGRA F., DALLA FONTANA M., MENEGHETTI M. Selectivity of chemical oxidation attack of single-wall carbon nanotubes in solution, *Phys. Rev. B* 2003, *68*, 193412.

110 KAVAN L., DUNSCH L. Diameter-selective electrochemical doping of hipco single-walled carbon nanotubes, *Nano Lett.* **2003**, *3*, 969–972.

111 WILTSHIRE J. G., KHLOBYSTOV A. N., LI L. J., LYAPIN S. G., BRIGGS G. A. D., NICHOLAS R. J. Comparative studies on acid and thermal based selective purification of HiPCO produced single-walled carbon nanotubes, *Chem. Phys. Lett.* **2004**, *386*, 239–243.

112 DUESBERG G. S., MUSTER J., BYRNE H. J., ROTH S., BURGHARD M. Towards processing of carbon nanotubes for technical application, *Appl. Phys. A* **1999**, *69*, 269–274.

113 STRANO M. S. Probing chiral selective reactions using a revised kataura plot for the interpretation of single-walled carbon nanotube spectroscopy, *J. Am. Chem. Soc.* **2003**, *125*, 16148–16153.

114 CHEN Z., DU X., DU M.-H., RANCKEN C. D., CHENG H.-P., RINZLER A. G. Bulk separative enrichment in metallic or semiconducting single-walled carbon nanotubes, *Nano Lett.* **2003**, *3*, 1245–1249.

115 BALASUBRAMANIAN K., BURGHARD M. Chemically functionalized carbon nanotubes, Small **2005**, *1*, 180–192.

116 SUN Y. P., FU K., LIN Y., HUANG W. Functionalized carbon nanotubes: properties and applications, *Acc. Chem. Res.* **2002**, *35*, 1096–1104.

117 SEEMAN N. C. DNA in a material world, *Nature* **2003**, *421*, 427–431.

118 ALIVISATOS A. P., JOHNSSON K. P., PENG X. G., WILSON T. E., LOWETH C. J., BRUCHEZ M. P., SCHULTZ P. G. Organization of "nanocrystal molecules" using DNA, *Nature* **1996**, *382*, 609–611.

119 NIEMEYER C., CEYHAN B. DNA-directed functionalization of colloidal gold with proteins, *Angew. Chem. Int. Ed. Engl.* **2001**, *40*, 3685–3688.

120 MIRKIN C. A. Programming the assembly of two- and three-dimensional architectures with DNA and nanoscale inorganic building blocks, *Inorg. Chem.* **2000**, *39*, 2258–2272.

121 LI H., PARK S. A., REIF J. H., LaBEAN T. H., YAN H. DNA-templated self-assembly of protein and nanoparticle linear arrays, *J. Am. Chem. Soc.* **2004**, *126*, 418–419.

122 NIKOLAEV P., BRONIKOSWSKI M. J., BRADELEY K., ROHMUND F., COLBERT D. T., SMITH K. A., SMALLEY R. E. Gas-phase catalytic growth of single-walled carbon nanotubes from carbon monoxide, *Chem. Phys. Lett.* **1999**, *313*, 91–97.

123 LIU J., CASAVANT M. J., COX M., WALTERS D. A., BOUL P., LU W., RIMBERG A. J., SMITH K. A., COLBERT D. T., SMALLEY R. E. Controlled deposition of individual single-walled carbon nanotubes on chemically functionalized templates, *Chem. Phys. Lett.* **1999**, *303*, 125–129.

124 MUSTER J., BURGHARD M., ROTH S., DUESBERG G. S., HERNANDEZ E., RUBIO A. Scanning force microscopy characterization of individual carbon nanotubes on electrode arrays, *J. Vac. Sci. Technol. B* **1998**, *16*, 2796–2801.

125 CHOI K. H., BOURGOIN J. P., AUVRAY S., ESTEVE D., DUESBERG G. S., ROTH S., BURGHARD M. Controlled deposition of carbon nanotubes on a patterned substrate, *Surf. Sci.* **2000**, *462*, 195–202.

126 VALENTIN E., AUVRAY S., GOETHALS J., LEWENSTEIN J., CAPES L., FILORAMO A., RIBAYROL A., TSUI R., BOURGOIN J. P., PATILLON J. N. High density selective placement methods for carbon nanotubes, *Microelecton. Eng.* **2002**, *61*, 491–496.

127 VALENTIN E., AUVRAY S., FILORAMO A., RIBAYROL A., GOFFMAN M. F., CAPES L., BOURGOIN J. P., PATILLON J. N. Self-assembly fabrication of high performance carbon nanotubes based FETs, *Mater. Res. Soc. Symp. Proc.* **2003**, *772*, 201–206.

128 VALENTIN E., AUVRAY S., FILORAMO A., RIBAYROL A., GOFFMAN M., GOETHALS J., CAPES L., BOURGOIN J. Ph., PATILLON J. N. Fabrication by self-assembly of carbon nanotubes field effect transistors, in: *NATO Science Series II: Molecular Nanowires and other Quantum Objects*, ALEXANDROV S., DEMSAR J., YANSON I. (Eds.). Kluwer, Dordrecht, **2004**, pp. 57–66.

129 HALLER I. Covalently attached organic

monolayers on semiconductor surfaces, *J. Am. Chem. Soc.* **1978**, *100*, 8050–8055.

130 PETRI D. F. S., WENZ G., SCHUNK P., SCHIMMEL T. An improved method for the assembly of amino-terminated monolayers on SiO_2 and the vapor deposition of gold layers, *Langmuir* **1999**, *15*, 4520–4523.

131 TSUKRUK V. V., BLIZNYUK V. N. Adhesive and friction forces between chemically modified silicon and silicon nitride surfaces, *Langmuir* **1998**, *14*, 446–455.

132 KANAN S. M., TZE W. T. Y., TRIPP C. P. Method to double the surface concentration and control the orientation of adsorbed (3-aminopropyl)dimethylethoxysilane on silica powders and glass slides, *Langmuir* **2002**, *18*, 6623–6627.

133 JEHOULET G. I. PhD Thesis, University of Grenoble, **1996**.

134 AUSMAN K. D., PINER R., LOURIE O., RUOFF R. S., KOROBOV M. Organic solvent dispersions of single-walled carbon nanotubes: toward solutions of pristine nanotubes, *J. Phys. Chem. B* **2000**, *104*, 8911–8915.

135 GORBUNOV A. A., FRIEDLEIN R., JOST O., GOLDEN M. S., FINK J., POMPE W. Gas-dynamic consideration of the laser evaporation synthesis of single-wall carbon nanotubes, *Appl. Phys. A* **1999**, *69*, S593–S596.

136 CAPES L., VALENTIN E., ESNOUF S., RIBAYROL A., JOST O., FILORAMO A., PATILLON J.-N. High yield non destructive purification of single wall carbon nanotubes monitored by EPR measurements, in: *Proceedings of the 2nd IEEE Conference on Nanotechnology*, Washington, DC, USA, **2002**, pp. 439–442.

137 Typically about 12 h at 24 W in a 40-kHz Fisherbrand US bath 2.8 l.

138 DIEHL M. R., YALIRAKI S. N., BECKMAN R. A., BARAHONA M., HEATH J. R. Self-assembled, deterministic carbon nanotube wiring networks, *Angew. Chem. Int. Ed. Engl.* **2002**, *41*, 353–356.

139 KONG J., DAI H. Full and modulated chemical gating of individual carbon nanotubes by organic amine compounds, *J. Phys. Chem. B* **2001**, *105*, 2890–2893.

140 WIND S. J., APPENZELLER J., MARTEL R., DERYCKE V., AVOURIS Ph. Vertical scaling of carbon nanotube field-effect transistors using top gate electrodes, *Appl. Phys. Lett.* **2002**, *80*, 3817–3819.

141 JAVEY A., KIM H., BRINK M., WANG Q., URAL A., GUO J., MCINTYRE P., MCEUEN P., LUNDSTROM M., DAI H. High-kappa dielectrics for advanced carbon-nanotube transistors and logic gates, *Nat. Mater.* **2002**, *1*, 241–246.

142 AUVRAY S., DERYCKE V., GOFFMAN M., FILORAMO A., JOST O., BOURGOIN J. P. Chemical optimization of self-assembled carbon nanotube transistors, *Nano Lett.* **2005**, *5*, 451–455.

143 SEEMAN N. C. The use of branched DNA for nanoscale fabrication, *Nanotechnology* **1991**, *2*, 149–159.

144 KEREN K., BERMAN R. S., BUCHSTAB E., SIVAN U., BRAUN E. DNA-templated carbon nanotube field-effect transistor, *Science* **2003**, *302*, 1380–1382.

145 COHEN S. N., CHANG A. C., BOYER H. W., HELLING R. B. Construction of biologically functional bacterial plasmids *in vitro*, *Proc. Natl Acad. Sci. USA* **1973**, *70*, 3240–3244.

146 http://scai.snu.ac.kr/cec2001/ selfassemble.talk.pdf.

147 SEEMAN N. DNA nicks and nodes and nanotechnology, *Nano Lett.* **2001**, *1*, 22–26.

148 SEEMAN N. Biochemistry and structural DNA nanotechnology: an evolving symbiotic relationship, *Biochemistry* **2003**, *42*, 7259–7269.

149 MA R.-I., KALLENBACH N. R., SHEARDY R. D., PETRILLO M. L., SEEMAN N. C. Three-arm nucleic acid junctions are flexible, *Nucleic Acids Res.* **1986**, *14*, 9745–9753.

150 PETRILLO M. L., NEWTON C. J., CUNNINGHAM R. P., MA R.-I., KALLENBACH N. R., SEEMAN N. C. The ligation and flexibility of four-arm DNA junctions, *Biopolymers* **1998**, *27*, 1337–1352.

151 WINFREE E., LIU F., WENZLER L. A., SEEMAN N. C. Design and self-assembly of two-dimensional DNA crystals, *Nature* **1998**, *394*, 539–544.

152 LaBean T. H., Yan H., Kopatsch J., Liu F., Winfree E., Reif J. H., Seeman N. C. Construction, analysis, ligation, and self-assembly of DNA triple crossover complexes, *J. Am. Chem. Soc.* **2000**, *122*, 1848–1860.

153 Yan H., Park S. H., Finkelstein G., Reif J. H., LaBean T. H. DNA-templated self-assembly of protein arrays and highly conductive nanowires, *Science* **2003**, *301*, 1882–1884.

154 Shi J., Bergstrom D. E. Assembly of novel DNA cycles with rigid tetrahedral linkers, *Angew. Chem. Int. Ed. Engl.* **1997**, *36*, 111–113.

155 Lenigk R., Carles M., Ip N. Y., Sucher N. J. Surface characterization of a silicon-chip-based DNA microarray, *Langmuir* **2000**, *17*, 2497–2501.

156 Joos B., Kuster H., Cone R. Covalent attachment of hybridizable oligonucleotides to glass supports, *Anal. Biochem.* **1997**, *247*, 96–101.

157 Rogers Y., Baucom P. J., Huang Z. J., Bogdanov V., Anderson S., Jacino M. T. Immobilization of oligonucleotides onto a glass support via disulfide bonds: a method for preparation of DNA microarrays, *Anal. Biochem.* **1999**, *266*, 23–30.

158 Strother T., Cai W., Zhao X., Hamers R. J., Smith L. M. Synthesis and characterization of DNA-modified silicon (111) surfaces, *J. Am. Chem. Soc.* **2000**, *122*, 1205–1209.

159 Chrisey L. A., O'Ferral C. E., Spargo B. J., Dulcey C. S., Calvert J. M. Fabrication of patterned DNA surfaces, *Nucleic Acids Res.* **1996**, *15*, 3040–3047.

160 Chrisey L. A., Lee G. U., O'Ferral C. E. Covalent attachment of synthetic DNA self-assembled monolayer films, *Nucleic Acids Res.* **1996**, *15*, 3031–3039.

161 Okamoto T., Suzuki T., Yamamoto N. Microarray fabrication with covalent attachment of DNA using bubble jet technology, *Nat. Biotechnol.* **2000**, *18*, 438–441.

162 Cheung V. G., Morley M., Aguilar F., Massimi A., Kucherlapati R., Childs G. Making and reading microarrays, *Nat. Genet.* **1999**, *21 (Suppl.)*, 15–19.

163 Demers L. M., Gingers D. S., Park S.-J., Li Z., Chung S.-W., Mirkin C. A. Direct patterning of modified oligonucleotides on metals and insulators by dip-pen nanolithography, *Science* **2002**, *296*, 1836–1838.

164 Edman C. F., Raymond D. E., Wu D. J., Tu E., Sosnowski R. G., Butler W. F., Nerenberg M., Heller M. J. Electric field directed nucleic acid hybridization on microchips, *Nucleic Acids Res.* **1997**, *25*, 4907–4914.

165 http://www.nanogen.com/ technologies/microarray/.

166 Livache T., Fouque B., Roget A., Marchand J., Bidan G., Teoule R., Mathis G. Polypyrrole DNA chip on a silicon device: example of hepatitis C virus genotyping, *Anal. Biochem.* **1998**, *255*, 188–194.

167 http://www.affymetrix.com/ technology/manufacturing/index.affx.

168 Lipshutz R. J., Fodor S. P. A., Gingeras T. R., Lockhart D. J. High density synthetic oligonucleotide arrays, *Nat. Genet.* **1999**, *21 (Suppl.)*, 20–24.

169 Storhoff J. J., Mirkin C. A. Programmed materials synthesis with DNA, *Chem. Rev.* **1999**, *99*, 1849–1862.

170 Niemeyer C. M., Ceyhan B., Noyong M., Simon U. Bifunctional DNA–gold nanoparticle conjugates as building blocks for the self-assembly of cross-linked particle layers, *Biochem. Biophys. Res. Commun.* **2003**, *311*, 995–999.

171 Taton T. A., Mucic R. M., Mirkin C. A., Letsinger R. L. The DNA-mediated formation of supramolecular mono- and multilayered nanoparticle structures, *J. Am. Chem. Soc.* 122, 6305–6306.

172 Csaki A., Maubach G., Born D., Reichert J., Fritzche W. DNA-based molecular nanotechnology, *Single Molecules* **2002**, *3*, 275–280.

173 Fu A., Micheel C. M., Cha J., Chang H., Yang H., Alivisatos A. P. Discrete nanostructures of quantum dots/Au with DNA, *J. Am. Chem. Soc.* **2004**, *126*, 10832–10833.

174 Chen J., Hamon M. A., Hu H., Chen Y., Rao A. M., Eklund P. C., Haddon R. C. Solution properties of single-walled carbon nanotubes, *Science* **1998**, *282*, 95–98.

175 Fu K., Huang W., Riddle L. A., Carroll D. L., Sun Y.-P. Defunction-alization of functionalized carbon nanotubes, *Nano Lett.* **2001**, *1*, 439–441.

176 Chen J., Rao A. M., Lyuksyutov S., Itkis M. E., Hamon M. A., Hu H., Cohn R. W., Eklund P. C., Colbert D. T., Smalley R. E., Haddon R. C. Dissolution of full-length single-walled carbon nanotubes, *J. Phys. Chem. B* **2001**, *105*, 2525–2528.

177 Czerw R., Guo Z., Ajayan P. M., Caroll D. L., Sun Y.-P. Identification of electron donor states in n-doped carbon nanotubes, *Nano Lett.* **2001**, *1*, 457–460.

178 Rinzler A. G., Liu J., Dai H., Nikolaev P., Huffman C. B., Rodríguez-Macías F. J., Boul P. J., Lu A. H., Heymann D., Colbert D. T., Lee R. S., Fisher J. E., Rao A. M., Eklund P. C., Smalley R. E. Large-scale purification of single-wall carbon nanotubes: process, product, and characterization, *Appl. Phys. A* **1996**, *67*, 29–37.

179 Hu H., Bhowmik P., Zhao B., Hamon M. A., Itkis M. E., Haddon R. C. Determination of the acidic sites of purified single wall carbon nanotubes by acid–base titration, *Chem. Phys. Lett.* **2001**, *345*, 25–28.

180 Banerjee S., Wong S. S. Synthesis and characterization of carbon nanotube-nanocrystal heterostructures, *Nano Lett.* **2002**, *2*, 49–54.

181 Bahr J. L., Yang J., Kosynkin D. V., Bronikowski M. J., Smalley R. E., Tour J. M. Functionalization of carbon nanotubes by electrochemical reduction of aryl diazonium salts: a bucky paper electrode, *J. Am. Chem. Soc.* **2001**, *123*, 6536–6542.

182 Bahr J. L., Tour J. M. Highly functionalized carbon nanotubes using in situ generated diazonium compounds, *Chem. Mater.* **2001**, *13*, 3823–3824.

183 Tang B. Z., Xu H. Preparation, alignment, and optical properties of soluble poly(phenylacetylene)-wrapped carbon nanotubes, *Macromolecules* **1999**, *32*, 2569–2576.

184 Chen G. Z., Shaffer M. S. P., Coleby D., Dixon G., Zhou W., Fray J., Windle A. H. Carbon nanotube and polypyrrole composites: coating and doping, *Adv. Mater.* **2000**, *12*, 522–526.

185 O'Connell M. J., Boul P., Ericson L. M., Huffman C., Wang Y., Haroz E., Kuper C., Tour J., Ausman K. D., Smalley R. E. Reversible water-solubilization of single-walled carbon nanotubes by polymer wrapping, *Chem. Phys. Lett.* **2001**, *342*, 265–271.

186 Star A., Stoddart J. F., Steuerman D., Diehl M., Boukai A., Wong E. W., Yang X., Choi S. W., Heath J. R. Preparation and properties of polymer-wrapped single-walled carbon nano-tubes, *Angew. Chem. Int. Ed. Engl.* **2001**, *113*, 1721–1725.

187 Mc Carthy B., Coleman J. N., Czerw R., Dalton A. B., Caroll D. L., Blau W. J. Microscopy studies of nanotube-conjugated polymer interactions, *Synthetic Metals* **2001**, *121*, 1225–1226.

188 Dalton A. B., Blau W. J., Chambers G., Coleman J. N., Henderson K, Lefrant S., Mc Carthy B., Stephan C., Byrne H. J. A functional conjugated polymer to process, purify and selectively interact with single wall carbon nanotubes, *Synthetic Metals* **2001**, *121*, 1217–1218.

189 Zorbas V., Ortiz-Acevedo A., Dalton A. B., Yoshida M. M., Dieckmann G. R., Draper R. K., Baughman R. H., Jose-Yacaman M., Musselman I. H. Preparation and characterization of individual peptide-wrapped single-walled carbon nanotubes, *J. Am. Chem. Soc.* **2004**, *126*, 7222–7227.

190 Xie H., Ortiz-Acevedo A., Zorbas V., Baughman R., Draper R. K., Musselman I. H., Dalton A. B., Dieckmann G. R. Peptide cross-linking modulated stability and assembly of peptide-wrapped single-walled carbon nanotubes, *J. Mater. Chem.* **2005**, *15*, 1734–1741.

191 HAZANI M., NAAMAN R., HENNRICH F., KAPPES M. M. Confocal fluorescence imaging of DNA-functionalized carbon nanotubes, *Nano Lett.* **2003**, *3*, 153–155.

192 DWYER C., GUTHOLD M., FALVO M., WASHBURN S., SUPERFINE R., ERIE D. DNA-functionalized single-walled carbon nanotubes, *Nanotechnology* **2002**, *13*, 601–604.

193 BAKER S. E., CAI W., LASSETER T. L., WEIDKAMP K. P., HAMERS R. J. Covalently bonded adducts of deoxyribonucleic acid (DNA) oligonucleotides with single-wall carbon nanotubes: synthesis and hybridization, *Nano Lett.* **2002**, *2*, 1413–1417.

194 NGUYEN V. C., DELZEIT L., CASSELL A. M., LI J., HAN J., MEYYAPPAN M. Preparation of nucleic acid functionalized carbon nanotube arrays, *Nano Lett.* **2002**, *2*, 1079–1081.

195 WILLIAMS K. A., VEENHUIZEN P. T. M., DE LA TORRE B. G., ERITJA R., DEKKER C. Nanotechnology – carbon nanotubes with DNA recognition, *Nature* **2002**, *420*, 761–761.

196 LI S., HE P., DONG J., GUO Z., DAI L. DNA-directed self-assembling of carbon nanotubes, *J. Am. Chem. Soc.* **2005**, *127*, 14–15.

197 MOGHADDAM M. J., TAYLOR S., GAO M., HUANG S., DAI L., McCALL M. J. Highly efficient binding of DNA on the sidewalls and tips of carbon nanotubes using photochemistry, *Nano Lett.* **2004**, *4*, 89–93.

198 GUO Z., SADLER P. J., TSANG S. C. Immobilization and visualization of DNA and proteins on carbon nanotubes, *Adv. Mater.* **1998**, *10*, 701–703.

199 BALAVOINE F., SCHULTZ P., RICHARD C., MALLOUH V., EBBESEN T. W., MIOSKOWSKI C. Helical crystallization of proteins on carbon nanotubes: a first step towards the development of new biosensors, *Angew. Chem. Int. Ed. Engl.* **1999**, *38*, 1912–1915.

200 FURUNO T., SASABE H. Two-dimensional crystallization of streptavidin by nonspecific binding to a surface film: study with a scanning electron microscope, *Biophys. J.* **1993**, *65*, 1714–1717.

201 FILORAMO A., DEKKER C., SIVAN U., SCHONENBEGER C., MICHEL-BEYERLE M. E. Highligth from DNA-based electronic project, *Phantoms Newslett.* **2003**, *10/11*, 4–6.

202 GOUX-CAPES L., FILORAMO A., COTE D., VALENTIN E., BOURGOIN J. P., PATILLON J. N. Non-covalent binding of DNA to carbon nanotubes controlled by biological recognition complex, *AIP Conf. Proc.* **2004**, *725 (1)*, 17–24.

203 PATOLSKY F., WEIZMANN Y., WILLNER I. Long-range electrical contacting of redox enzymes by SWCNT connectors, *Angew. Chem. Int. Ed. Engl.* **2004**, *43*, 2113–2117.

204 RICHTER J. Metallization of DNA, *Physica E* **2003**, *16*, 157–173.

205 BRAUN E., EICHEN Y., SIVAN U., BEN-JOSEPH G. DNA-templated assembly and electrode attachment of a conducting silver wire, *Nature* **1998**, *391*, 775–778.

206 SEIDEL R., COLOMBI CIACCHI L., WEIGEL M., POMPE W., MERTIG M. Synthesis of platinum cluster chains on DNA templates: conditions for a template-controlled cluster growth, *J. Phys. Chem. B* **2004**, *108*, 10801–10811.

207 RICHTER J., SEIDEL R., KIRSCH R., MERTIG M., POMPE W., PLASCHKE J., SCHACKERT H. K. Nanoscale palladium metallization of DNA, *Adv. Mater.* **2000**, *12*, 507–510.

208 SEIDEL R., MERTIG M., POMPE W. Scanning force microscopy of DNA metallization, *Surf. Interface Anal.* **2002**, *33*, 151–154.

209 FORD W. E., HARNACK O., YASUDA A., WESSELS J. M. Platinated DNA as precursors to templated chains of metal nanoparticles, *Adv. Mater.* **2001**, *13*, 1793–1797.

210 KEREN K., KRUEGER M., GILAD R., BEN-JOSEPH G., SIVAN U., BRAUN E. Sequence-specific molecular lithography on single DNA molecules, *Science* **2002**, *297*, 72–75.

211 Keren K., Berman R., Braun E. Patterned DNA metallization by sequence-specific localization of a reducing agent, *Nano Lett.* **2004**, *4*, 323–326.

212 Harnack O., Ford W. E., Yasuda A., Wessels J. Tris(hydroxymethyl)-phosphine-capped gold particles templated by DNA as nanowire precursors, *Nano Lett.* **2002**, *2*, 919–923.

213 Nishinaka T., Takano A., Doi Y., Hashimoto M., Nakamura A., Matsushita Y., Kumaki J., Yashima E. Conductive metal nanowires templated by the nucleoprotein filaments, complex of DNA and RecA protein, *J. Am. Chem. Soc.* **2005**, *127*, 8120–8125.

214 Ongaro A., Griffin F., Beecher P., Nagle L., Iacopino D., Quinn A., Redmond G., Fitzmaurice D. DNA-templated assembly of conducting gold nanowires between gold electrodes on a silicon oxide substrate, *Chem. Mater.* **2005**, *17*, 1959–1964.

215 Monsoon C. F., Woolley A. T. DNA-templated construction of copper nanowires, *Nano Lett.* **2003**, *3*, 359–363.

216 Dupraz C. J.-F., Nickels P., Beierlein U., Huynh W. U., Simmel F. C. Towards molecular-scale electronics and biomolecular self-assembly, *Superlattices and Microstructures* **2003**, *33*, 369–379.

217 Lippert B. (Ed.). *Cisplatin: Chemistry and Biochemistry of a Leading Anti-cancer Drug*. Wiley-VCH, Weinheim, **1999**.

218 Macquet J. P., Theophanides T. Specificity of the interaction of DNA–platinum, amount of platinum, and pH measurement, *Biopolymers* **1975**, *14*, 781–799.

219 Colombi Ciacchi L., Mertig M., Seidel R., Pompe W., de Vita A. Nucleation of platinum clusters on biopolymers: a first principles study of the molecular mechanisms, *Nanotechnology* **2003**, *14*, 840–848.

220 Macquet J. P., Theophanides T. DNA–platinum interactions. Characterization of solid DNA–$K_2[PtCl_4]$ complexes, *Inorg. Chim. Acta* **1976**, *18*, 189–194.

221 Macquet J. P., Butour J. L. A circular dichroism study of DNA–platinum complexes. Differentiation between monofunctional, *cis*-bidentate and *trans*-bidentate platinum fixation on a series of DNAs, *Eur. J. Biochem.* **1978**, *83*, 375–385.

222 Mertig M., Colombi Ciacchi L., Seidel R., Pompe W., De Vita A. DNA as a selective metallization template, *Nano Lett.* **2002**, *2*, 841–844.

223 Richter J., Mertig M., Pompe W., Monch I., Schackert H. K. Construction of highly conductive nanowires on a DNA template, *Appl. Phys. Lett.* **2001**, *78*, 536–538.

224 Deng Z., Mao C. DNA-templated fabrication of 1D parallel and 2D crossed metallic nanowire arrays, *Nano Lett.* **2003**, *3*, 1545–1548.

225 Braun E., Keren K. From DNA to transistors, *Adv. Phys.* **2004**, *53*, 441–496.

226 Hazani M., Shvarts D., Peled D., Sidorov V., Naaman R. Self-assembled carbon-nanotube-based field-effect transistors, *Appl. Phys. Lett.* **2004**, *85*, 5025–5027.

227 Hazani M., Hennrich F., Kappes M., Naaman R., Peled D., Sidorov V., Shvarts D. DNA-mediated self-assembly of carbon nanotube-based electronic devices, *Chem. Phys. Lett.* **2004**, *391*, 389–392.

10
Nanodevices for Biosensing: Design, Fabrication and Applications

Laura M. Lechuga, Kirill Zinoviev, Laura G. Carrascosa, and Miguel Moreno

10.1
Introduction

There is increasing interest in obtaining biosensor devices based on nanotechnology developments which can detect, in a fast and selective way, any type of substance in air and liquid samples at very low concentrations – ideally at the single-molecule level [1]. Clinical diagnosis, genomics and proteomics are some of the fields where new laboratory analysis methods (faster, direct, more accurate, more selective, having a high throughput and cheaper than conventional methods) are in high demand. Due to their small size, ultra-sensitive transduction and the possibility of integration in "lab-on-a-chip" microsystems, biosensing devices fabricated with nanotechnologies are potential candidates for fulfilling all the above requirements.

In recent years several interesting nano-developments have been proposed as highly sensitive transducers for biosensing [as nanoparticles, carbon nanotubes (CNTs), photonic crystals, micro- and nanocatilevers, etc.], but few biomolecular interactions using such developments have been demonstrated. Many of those developments are still in their infancy, and further research and technological development is needed before real functional biosensing devices will be available. One of the main problems is the implementation. The path to connect such nano-developments to operations in real-world environments has not be paved, and large and complicated laboratory setups are still needed for signal acquisition and processing.

However, micro/nanobiosensor devices based on microelectromechanical systems (MEMS) and related (BIO)MEMS and (bio)nanoelectromechanical systems [BIO(NEMS)] technologies could provide a technological solution to achieve label-free devices which could be operated in stand-alone fashion outside from laboratory environment. For that reason, in this chapter we focus mainly on two important branches of nanodevices for biosensing: (a) nanodevices based on nanophotonics/optoelectronics and (b) nanodevices based on nanomechanics. Nanobiosensors based on optoelectronics and nanomechanics platforms are excellent examples of devices developed with microelectronics technologies, and constitute the platforms with more possibilities for being used in real applications in the near future.

Nanotechnologies for the Life Sciences Vol. 4
Nanodevices for the Life Sciences. Edited by Challa S. S. R. Kumar
Copyright © 2006 WILEY-VCH Verlag GmbH & Co. KGaA, Weinheim
ISBN: 3-527-31384-2

The chapter covers the design, fabrication and testing of both types of biosensor nanodevices. Further integration of nanosensors, microfluidics, optical and electronic functions on a single sensing circuit could lead to a complete "lab-on-a-chip" technological solution which could be used in field applications and *in situ* analysis. Examples of fabrication, characterization and real applications of the devices will be discussed as well as the way of their integration. Although there are a number of reviews covering some aspects of the devices described here, this is the first summary and critical discussion of the developments based on the fields of photonics and nanomechanics, and the development of a new (nano)device combining both fields.

The chapter is organized as follows:

- An overview of the reasons for using nanobiosensors instead of the classical biosensor approaches.
- A brief overview of the immobilization techniques which could be employed for immobilization of the biological receptors in the transducers.
- A complete description of nanophotonic biosensors, starting from a general overview and showing one example, based in integrated optics, of the latest developments in this field.
- A complete description of nanomechanical biosensors, with an extensive overview of these devices, and showing some examples of design, fabrication and testing. One of the last developments in this field, combining nanomechanics and integrated optics, is presented for the first time.
- Finally, the future trends of this exciting nanobiosensor field are discussed.

10.2
From Biosensor to Nanobiosensor Devices

10.2.1
Overview

In addition to the excellent results obtained with existing biosensor technologies [2], there is still a need for devices able to detect, in a direct way, very low levels (picomolar to femtomolar and ideally at the single-molecule level) of a great number of chemical and biochemical substances in the areas of environmental monitoring, industrial and food processes, health care, biomedical technology, clinical analysis, etc. In addition to an extreme sensitivity, if we want to apply biosensor technology to real situations, we would need a high selectivity, short time analysis, and must be reversible, stable, simple to operate, robust, low cost and capable of multianalyte determination [2].

To achieve the above characteristics, the application of recent progress in micro- and nanotechnologies seem to be the most appealing alternative [3]. These technologies are already improving both the miniaturization and the sensivity of the biosensor devices by using nanomaterials for their construction, allowing the intro-

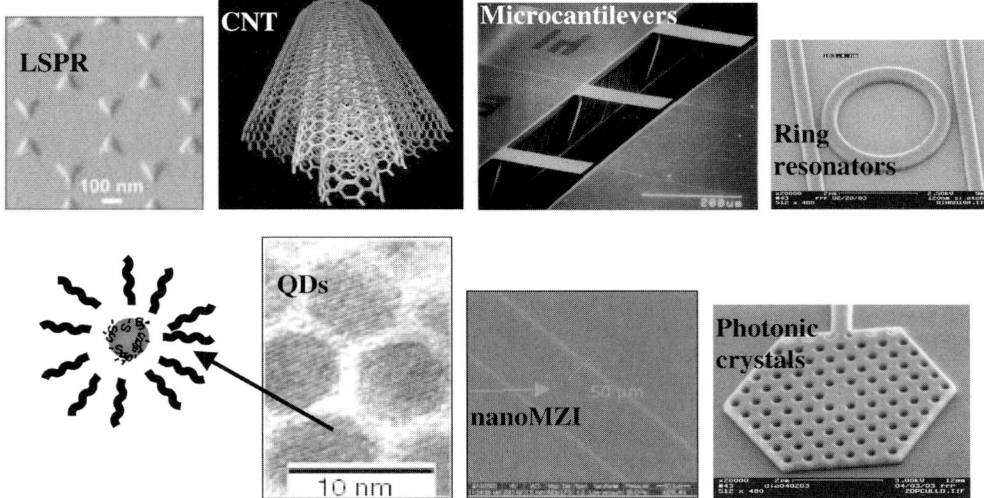

Figure 10.1. Schemes and photographs of some nanodevices proposed for biosensing. (Reprinted with permission.)

duction of new signal transduction technologies. Due to their submicron dimensions, nanosensors, nanoprobes and other nanosystems allow simple and rapid analyses *in vivo*. We are assisting to the birth of new biosensor devices at the nanoscale which could be easily integrated in portable "lab-on-chip" platforms to perform "point-of-care" analysis and which in the future could even work inside the human body to detect, at the very early stages, the presence of cancer cells or infectious agents.

Several interesting biosensing developments based on nanotechnology have appeared in the literature during recent years, such as the use of semiconductor, metal or magnetic nanoparticles [4], localized surface plasmon resonance sensors (LSPR) [5], different types of CNT biosensors [6], probes encapsulated by biologically localized embedding (PEBBLE) [7] and many others [8] (Fig. 10.1). Although all of these developments are interesting from a scientific point of view, the real implementation of much of them will be hampered by several factors. Firstly, most of them still require using labels to defect the biomolecular interaction, which is undesirable for real applications (direct reading is much more precise). Secondly, the path to connect such nano-developments to the operation in a realworld environment has not been paved, and large and complicated laboratory setups are needed for signal acquisition and processing.

In contrast, micro/nanobiosensor devices based on microelectronics and related (BIO)MEMS/NEMS technologies could provide a technological solution for achieving label-free devices which could be operated in a stand-alone fashion outside of a laboratory environment [9]. This fabrication approach allows the flexible development of miniaturized compact sensing devices, microfluidics delivery systems and the possibility of fabricating multiple sensors on one chip, opening the way for high-throughput screening. Additional advantages are the robustness, reliability, potential for mass production with consequent reduction of production costs, low energy consumption and simplicity in the alignment of the individual elements [9].

Such nanobiosensors based on optoelectronics and nanomechanics platforms are excellent examples of devices developed with such technologies, and these platforms offer more possibilities for being used in real applications in the near future.

10.2.2
Biological Functionalization of Nanobiosensors

For biosensing purposes, a layer of receptor molecules (proteins, DNA, etc.) capable of selectively binding the substances to be analyzed has to be previously immobilized on the biosensor surface. We must not forget that the immobilization of the receptor molecule on the nanosensor surface is a key step towards the final performance of any biosensor device as it affects to the reproducibility, selectivity and resolution of the device.

The immobilization procedure employed must be stable and reproducible, and must retain the stability and activity of the receptors. Even though the nanodevice that we develop would be the most sensitive one, if the immobilization fails, then the performance of the device will be poor and the theoretical extreme sensitivity of the nanodevice will never be achieved. For handling such diminutive areas, immobilization can be performed with *ex situ* techniques as ink-jet, dip-pen or micro/nanospotting [3] or by *in-situ* techniques through dedicate microfluidics and nanofluidics.

Generally, direct adsorption is not adequate, giving significant losses in biological activity and random orientation of the receptors. Two immobilization strategies are the most employed at the biosensor field: (a) covalent coupling and (b) affinity noncovalent interactions. Covalent coupling gives a stable immobilization as the receptors do not dissociate from the surface or exchange with other receptors in solution. In affinity bonding, a high-affinity capture ligand is nonreversibly immobilized on the sensor surface. The most employed method is the immobilization on gold-coated surfaces using thiol self-assembled monolayers (SAMs) [10]. For example, a widespread method is functionalization of single-stranded (ss) DNA (or proteins) with an alkane chain terminating in a thiol (-SH) or disulfide group (-SS) as sulfurs form a strong bond with gold. This can also be applied for silicon surfaces, using silane monolayers covalently attach to silicon, SiO_2 or Si_3N_4 sensor surfaces [11]. Several aspects must be taken into account in the development of the immobilization procedures, such as how to avoid nonspecific interactions, getting an optimized surface density of the receptor in order to prevent steric hindrance phenomenon or regenerating the receptor for continuous measurements.

Further details about this subject can be found in the specific literature [12, 13];

it is beyond the scope of this chapter to discuss in detail the immobilization procedures which can be used for (nano)biosensing.

10.3
Nanophotonic Biosensors

10.3.1
Overview

Photonic biosensors are providing an increasingly important analytical technology for the detection of biological and chemical species [2, 14]. Most optical biosensors make use of optical waveguides as the basic element of their structure for light propagation and are based on the same operation principle – evanescent field sensing (EFS). With the emergence of nanotechnology, new photonic structures have been suggested as possible highly sensitive transducers for biosensing, such as photonic crystals [15], ring resonators [16] or hollow waveguides [17], but almost no biosensing demonstrations have appeared in the literature using such structures.

Evanescent wave detection combined with nanophotonics structures is proving to be one of the most highly sensitive biosensors. In evanescent wave detection, a receptor layer is immobilized onto the waveguide and the exposure of such a surface to the partner analyte molecules produces a biochemical interaction, which induces a change in its optical properties. This change is detected by the evanescent wave. The extent of the optical change will depend on the concentration of the analyte and on the affinity constant of the interaction, in this way obtaining a quantitative sensor of the interaction. The evanescent wave decays exponentially as it penetrates the outer medium and, therefore, only detects changes taking place on the surface of the waveguide since the intensity of the evanescent field is much higher in this particular region. For that reason, it is not necessary to carry out *a priori* separation of nonspecific components (as in conventional analysis) because any change in the bulk solution will hardly affect the sensor response. In this way, evanescent wave sensors are selective and sensitive devices for the detection of very low levels of chemicals and biological substances, and for the measurement of molecular interactions *in situ* and in real-time [18].

The advantages of optical sensing are significantly improved when the above approach is used within an integrated optics context [19]. Integrated optics technology allows the integration of passive and active optical components (including fibers, emitters, detectors, waveguides and related devices) onto the same substrate, permitting the flexible development of miniaturized compact sensing devices, with the additional possibility to fabricate multiple sensors on a single chip. The integration offers some additional advantages to the optical sensing systems, such as miniaturization, robustness, reliability, potential for mass production with consequent reduction of production costs, low energy consumption and simplicity in the alignment of the individual optical elements [19].

10.3.2
Integrated Mach–Zehnder Interferometer (MZI) Nanodevice

One of the most sensitive direct biosensor is the MZI. This device fabricated at the micro/nanoscale has shown sensitivity levels close to 10^{-3} nm in adsorbed molecular layers, which means a sensitivity in the picomolar range for biomolecular interactions in a direct assay (without labels) [20].

In a MZI device [18] the light from a laser beam is divided into two identical beams that travel through the MZI arms (sensor and reference areas) and are recombined again into a monomode channel waveguide, giving a signal which is dependent on the phase difference between the sensing and the reference branches. Any change in the sensor area (in the region of the evanescent field) produces a phase difference (and therein a change of the effective refractive index of the waveguide) between the reference and the sensor beam, and thus in the intensity of the outcoupled light. A schematic diagram of this sensor is shown in Fig. 10.2.

When a chemical or biochemical reaction takes place in the sensor area, only the light that travels through this arm will experience a change in its effective refractive index. At the sensor output, the intensity (I) of the light coming from both arms will interfere, showing a sinusoidal variation that depends on the difference of the effective refractive indexes of the sensor ($N_{\text{eff},S}$) and reference arms ($N_{\text{eff},R}$) and on the interaction length (L):

$$I = \frac{1}{2} I_\text{o} \left[1 + \cos\left(\frac{2\pi}{\lambda} (N_{\text{eff},S} - N_{\text{eff},R}) L \right) \right] \tag{1}$$

where λ is the wavelength. This sinusoidal variation can be directly related to the concentration of the analyte to be measured.

For evaluation of specific biosensing interactions, the receptor is covalently attached to the sensor arm surface, while the complementary molecule binds to the receptor from free solution. The recognition of the complementary molecule by the receptor causes a change in the refractive index and the sensor monitors that change. After the molecular interaction, the surface can be regenerated using a

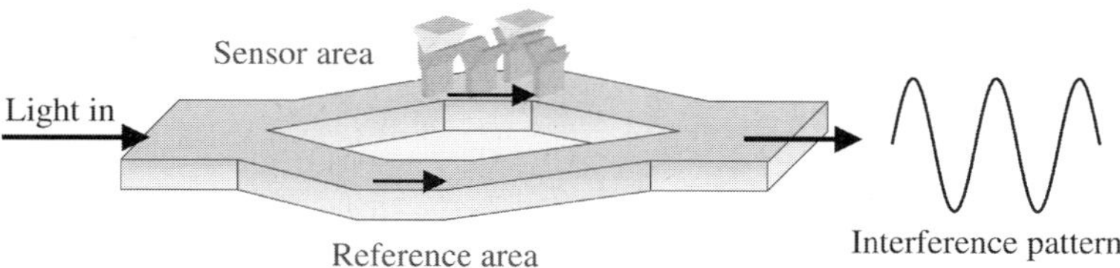

Figure 10.2. Scheme of the MZI nanodevice configuration and its working principle. The biomolecular interaction in the sensing area, where the receptors are attached, induces a phase change of the light traveling through that area as compared to the light traveling through the reference area.

suitable reagent in order to remove the bound analyte without denaturing the immobilized receptor.

The interferometric sensor platform is highly sensitive, and is the only one that provides an internal reference for compensation of refractive index fluctuations and unspecific adsorption. Interferometric sensors have a broader dynamic range than most other types of sensors and show higher sensitivity as compared with other integrated optical biosensors [18, 21]. Due to the high sensitivity of the interferometer sensor, direct detection of small molecules (e.g. environmental pollutants where concentrations down to 0.1 ng mL^{-1} must be detected) would be possible with this device.

The detection limit is generally limited by electronic, chemical and mechanical noise, thermal drift, and light source instabilities. However, the intrinsic reference channel of the interferometric devices offers the possibility of reducing common mode effects like temperature drifts and nonspecific adsorptions. A detection limit of 10^{-7} (or better) in the refractive index can be achieved with this device [20], which opens the possibility of the development of highly sensitive devices for *in situ* chemical and biologically harmful agent detection, for example.

10.3.2.1 Design and Fabrication

For biosensing applications, the waveguides of the MZI device must be designed to work in the monomode regime and to have a very high surface sensitivity at the sensor arm towards biochemical interactions. If several modes were propagated through the structure, each of them would detect the variations in the characteristics of the outer medium and the information carried by all the modes would interfere between them. The design of the optical waveguide satisfying the above requirements and the dimensions of the Mach–Zehnder structure is achieved by using modeling programs such as the finite difference methods in a nonuniform mesh, the effective index method and the beam-propagation method. Parameters such as propagation constants, attenuation and radiation losses, evanescent field profile, modal properties, and field evolution must been calculated [22]. In order to quantify and optimize the surface sensitivity, the variation of the effective refractive index of the guided modes must be calculated when the thickness of a homogeneous biological layer (d_1) changes:

$$n_{\text{sup}} = \partial N / \partial d_1 \tag{2}$$

In Fig. 10.3(b) the surface sensitivity is represented as a function of the core thickness, assuming that the refractive index of this layer is $n_b = 1.45$, the external medium is water $(n_e = 1.33)$ and the light wavelength is 632 nm.

If we want to use total internal reflection (TIR) waveguides for the sensors, we must come to an agreement between single-mode behavior, low attenuation losses for the fundamental mode and high surface sensitivity. For those reasons, the structure that has been finally chosen [20], for an operating wavelength of 0.633 μm, has the configuration shown in Fig. 10.3(b). In this configuration, the monomode behavior is obtained for core thickness below 200 nm and rib depths

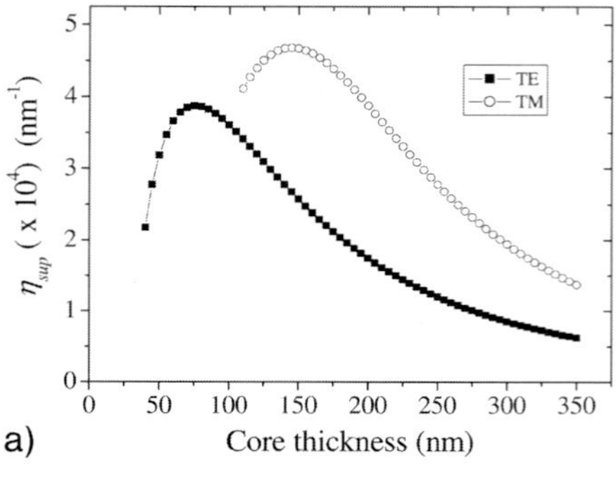

a)

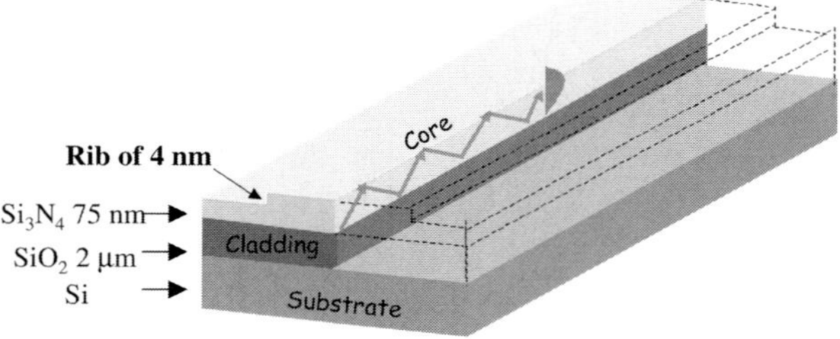

b)

Figure 10.3. Cross-section of the optical monomode waveguides used in the Mach–Zehnder device. Note that a rib of only 4 nm is needed for monomode and high-biomolecular sensitivity characteristics.

below 4 nm when the rib width is 4 μm. The small dimensions of this device imply some drawbacks: the reduced core dimensions for monomode behavior (thickness of less than hundreds nanometers and rib depths of a few nanometers) introduces a technological disadvantage for mass production and large insertion losses when coupling light with single-mode optical fibers (with a core thickness of several micrometers). However, the high surface sensitivity for biosensing applications justifies the development of these devices.

The fabrication is done through the following geometry: (a) a conducting silicon wafer of 500 μm thickness, (b) a 2-μm thick thermal SiO_2 layer on top with a refractive index of 1.46, and (iii) a low-pressure chemical vapor deposition (LPCVD) Si_3N_4 layer of 75 nm thickness and a refractive index of 2.00, which is used as a guiding layer. To achieve monomode behavior we needed to define a rib structure, with a depth of only 4 nm, on the Si_3N_4 layer by a lithographic step. This rib structure is performed by reactive ion etching (RIE) and is the most critical step in the microfabrication of the device. Finally, a SiO_2 protective layer is deposited by LPCVD over the structure with a 2 μm thickness and a refractive index of 1.46,

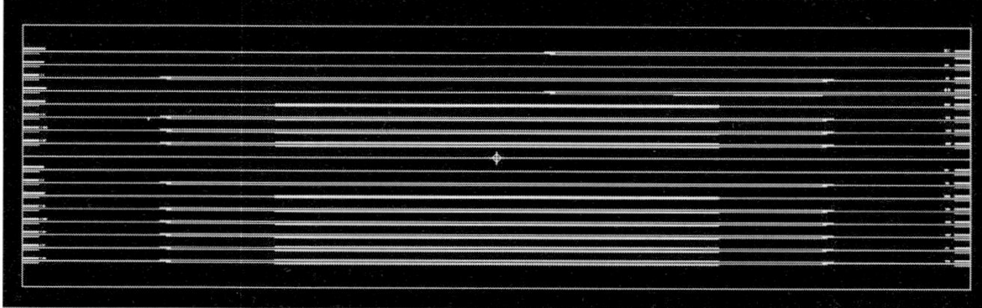

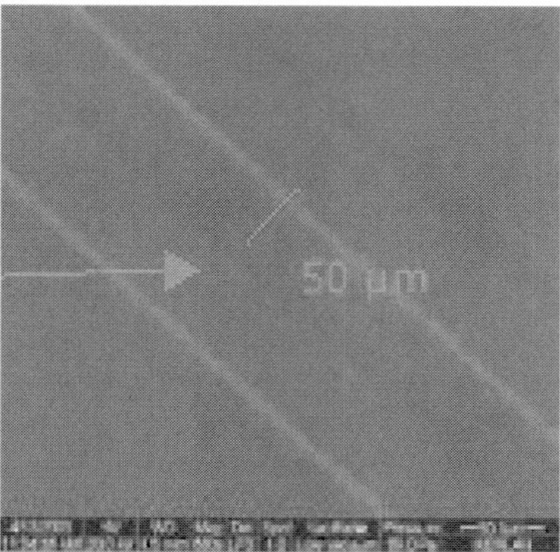

Figure 10.4. Photographs of the mask used for MZI fabrication and some of the fabricated devices. The details of the marks alignments and the reference/sensor areas of a device can be observed. Note that the MZI waveguide cannot be observed due to its dimensions, but is underneath the reference and sensor area.

which is patterned and etched by RIE to define the sensing and reference arms of the interferometer. The final devices (within all the fabrication processes) are CMOS compatible. The MZI configuration is designed to be symmetric with a circular Y-junction (radii of 80 mm). Separation between the sensor and reference arms is of 50 µm to avoid coupling between modes traveling through both branches. Finally, the sensors must be cut in individual pieces and polished for light coupling by the end faces.

In Fig. 10.4, the mask designed for the fabrication of such devices is shown as well some details of the alignment marks and the reference/sensor areas of one fabricated device. Due to the nanometric rib dimension, the device can only be observed by AFM as shown in Fig. 10.5.

10.3.2.2 Characterization and Applications

The devices must be implemented with a microfluidics unit, electronics, data acquisition and software for optical and biochemical testing. For the experimental

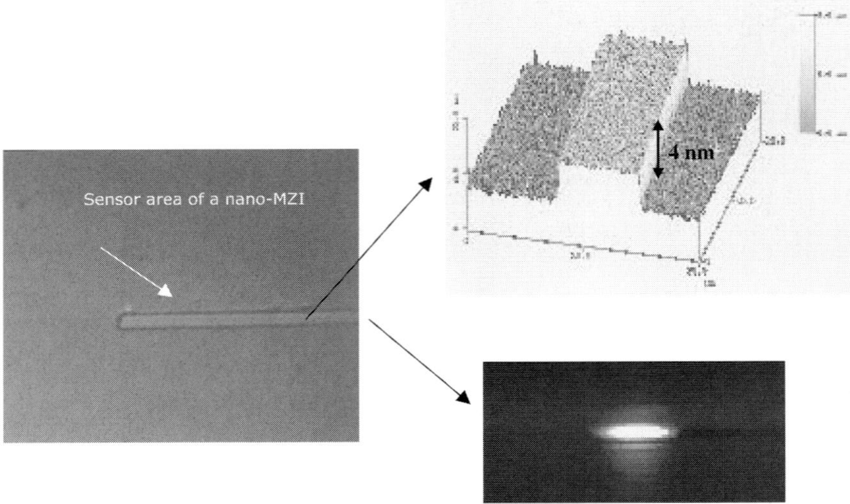

Figure 10.5. Image of the sensing area of a nano-MZI device. The AFM photograph of the rib of the waveguide clearly shows the 4-nm step achieved during fabrication. The monomode behavior can be observed at the output light collected from the device.

characterization, the light coming from a connected laser diode at a wavelength of 632 nm is end-fire coupled into the interferometers by a single-mode optical fiber. The light coming out of the interferometer is collected by another single-mode fiber, which is connected to a photodiode. The photodiode signal is amplified, digitalized and processed. The different dissolutions used in the characterizations are controlled by a peristaltic pump and a polymeric flow cell.

The evaluation of the sensitivity is done by flowing dissolutions of varying refractive index and measuring the output signal of the MZI in real-time. With these measurements, a calibration curve is constructed where the phase response of the sensor is plotted versus the variation in the refractive index as depicted in Fig. 10.6.

The lower detection limit measured is $\Delta n_{o,\mathrm{min}} = 2.5 \times 10^{-6}$, corresponding to an effective refractive index change of $\Delta N = 1.4 \times 10^{-7}$. It can be estimated that the lowest phase shift measurable would be around $0.03 \times 2\pi$. The detection limit value corresponds to a very high surface sensitivity around 2×10^{-4} nm^{-1}, which means that picomolar detection of a biomolecular interaction in a direct way is feasible using this nanodevice.

As a proof of the utility of MZI technology towards biosensing detection, the application of MZI nanobiosensors for the direct detection of DNA is described. The first step is the immobilization of the biomolecular receptors in the sensor area. This immobilization must be strong and stable to perform the sensitivity measure-

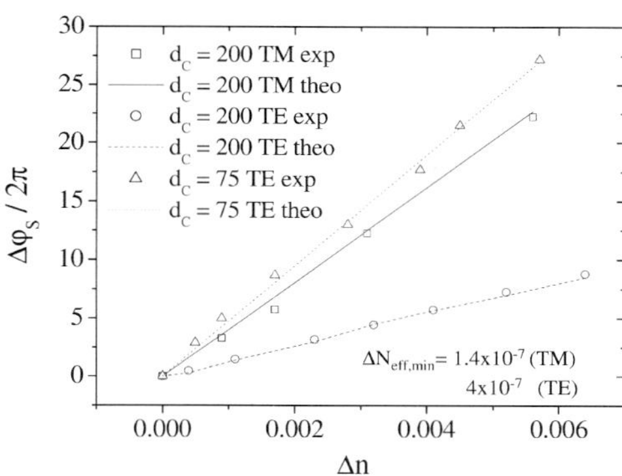

Figure 10.6. Experimental and theoretical evaluation of the sensitivity to changes of refractive index for MZI sensors with Si_3N_4 core layers of 75 and 200 nm, and for the TE and TM polarizations. In the case of the sensor with a 75-nm core layer, the TM mode is not guided.

ments of the device and for reusability. For that reason, a covalent immobilization protocol through SAMs using silane chemistry is employed.

First, the Si_3N_4 surface is cleaned with oxygen plasma and 10% nitric acid in distilled water to oxidize the surface. Second, the Si_3N_4 layer is immersed in 10% 3-mercaptopropyltrimethoxysilane (MPTMS) in toluene at room temperature for 12 h. The MPTMS functionalizes the sensor surface with a thiol group, allowing the thioled DNA to be covalently immobilized on the silanized Si_3N_4 surface by a disulfide bond. A ssDNA probe (28 nucleotides) with the thiol linker group [SH-$(CH_2)_6$] at the 5′ is used. The 15-T tail is employed as a vertical spacer chain to increase the accessibility to the complementary DNA. For the immobilization, a 45-nM solution of the ssDNA probe in phosphate-buffered (PB) solution (pH 7) is used. Figure 10.7(a) shows the real-time detection of the covalent immobilization of the DNA probes by means of a phase change of $\Delta\varphi_S = 7.75\ 2\pi$ rad.

The hybridization with the complementary ssDNA strand is detected using a 58-nucleotide strand and flowing a 100-nM solution in the same PB buffer. Figure 10.7(b) shows the real-time detection of the hybridization between the complementary sequences inducing a total phase change $\Delta\varphi_S = 2.5\ 2\pi$ rad. In order to test the specificity of the DNA hybridization, 100 nM dissolutions of a noncomplementary DNA sequence flow after the regeneration of the surface with 10 mM NaOH. This measurement shows a null response of the sensors, ensuring the specificity of the DNA binding. A calibration curve has been obtained by using different DNA concentrations. The lower experimental limit of detection is 10 pM, as can be observed in Fig. 10.7(c). This result clearly demonstrated the high sensitivity which can be obtained for the direct detection of biomolecular interactions by using this nanophotonic concept for a biosensor.

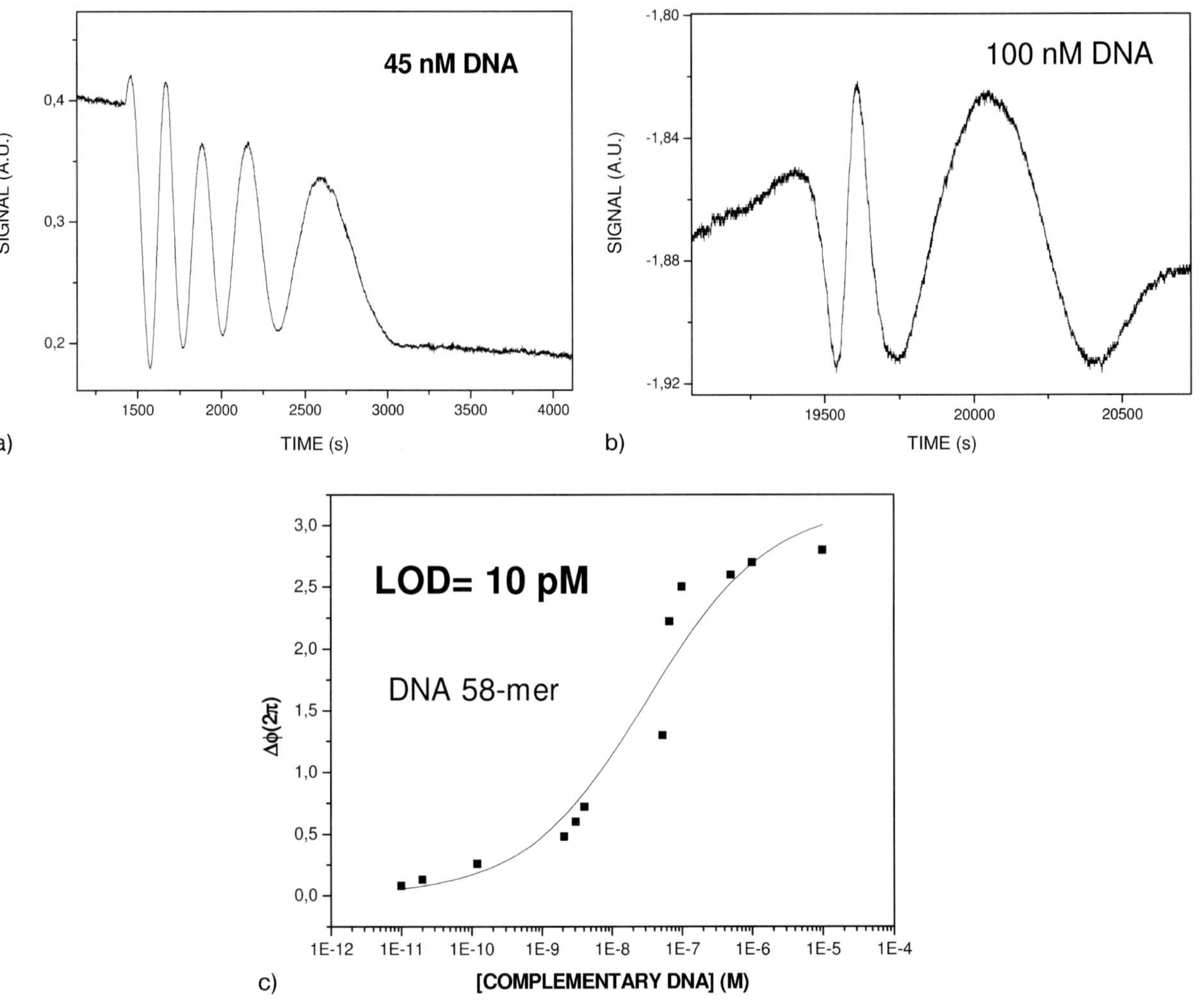
45 nM DNA
SIGNAL (A.U.)
TIME (s)
a)
100 nM DNA
SIGNAL (A.U.)
TIME (s)
b)
LOD= 10 pM
DNA 58-mer
Δφ(2π)
[COMPLEMENTARY DNA] (M)
c)

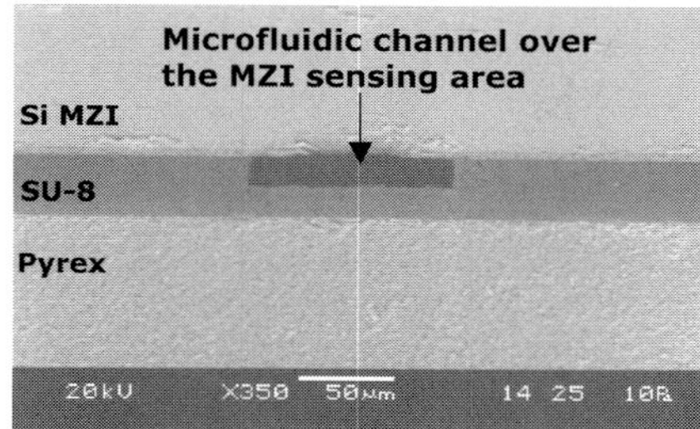

Figure 10.8. Scanning electron microscopy image of the cross-section of the polymer microfluidic channels over the sensor area of the photonic MZI device.

10.3.3
Integration in Microsystems

The integration of optical transducers is a key issue in the further development of "lab-on-a-chip" microsystems [23]. The main advantage of the Mach–Zehnder devices fabricated with standard microelectronics technology comes from the possibility to develop a complete "lab-on-a-chip" by optoelectronic integration of the light source, photodetectors and sensor waveguides on a single semiconductor package together with the flow system and the CMOS electronics [19]. There are many advantages to shrinking down these devices and integrating them for use in high-throughput microsystem applications such as single-molecule detection and DNA sequencing. A complete system fabricated with integrated optics will offer low complexity, robustness, a standardized device and, what is more important, portability. Devices for on-site analysis or point-of-care operations for biological and chemical detection are geared for portability, ease of use and low cost. In this sense, integrated optical devices have a compact structure and could allow fabricating optical sensor arrays on a single substrate for simultaneous detection of multiple analytes. Mass production of sensors will be also possible with the fabrication of miniaturized devices by using standard microelectronics technology.

For the development of a complete MZI microsystem, several units must be incorporated on the same platform: (a) the micro/nanodevices, (b) the flow cells and the flow delivery system, (c) a modulation or compensation system for translating the interferometric signals into direct ones, (d) integration of the light sources and the photodetectors, and (e) CMOS processing electronics. As an example, Fig. 10.8 shows the integration of a nano-MZI device with a microfluidic sys-

---

Figure 10.7. (a) Detection of the covalent immobilization of ssDNA receptor probes from a 45-nM solution by a MZI sensor. (b) Detection of the hybridization with the complementary ssDNA strand (100 nM). (c) Calibration curve for DNA hybridization. A lower detection limit of 10 pM can be achieved.

tem. The microflow cells are specifically designed and fabricated using a novel fabrication method of three-dimensionally embedded microchannels using the polymer SU-8 as structural material [24]. Integration of sources will be achieved by connection with optical fibers or using embedded diffraction gratings.

10.4
Nanomechanical Biosensors

10.4.1
Overview

Over recent years, biosensors based on microcantilevers have arisen as interesting devices for measuring biomolecular interactions in a direct way with very high sensitivity [25]. These sensors derive from the microfabricated cantilevers used in AFM and are based on the bending induced in the cantilever when a biomolecular interaction takes place on one of its surfaces. Microcantilevers transduce the molecular recognition of biomolecules into a nanomechanical motion [26] (from a few to hundreds of nanometers), which is commonly detected by an optical or piezoresistive readout system [27–29]. Research in this new type of sensors grew exponentially after the landmark paper of Fritz and coworkers in 2000, where the ability of microcantilever sensors for discerning single-base variations in DNA strands without using fluorescent labels was demonstrated [26]. This paper made a deep impact on the biotechnology area and marked the beginning of a major research effort on this field. Shortly after, microcantilever sensors were used in other works like DNA hybridization [30, 31], detection of proteins involved in cancer [32] and other diseases [33, 34] with increased accuracy, as well as in applications in environmental sciences [35]. Cantilever sensors have also been used for the detection of such chemical molecules as volatile compounds, warfare pathogens, explosives, glucose and even ionic species [25].

Microcantilevers are fabricated by using standard microelectronics technology in arrays of tens to thousands of microcantilevers. For that reason they are a promising alternative to current DNA and protein chips because they could permit the parallel, fast and real-time monitoring of thousands of analytes (proteins, pathogens, DNA strands, etc.) without the need for labeling. When fabricated at the nanoscale (*nanocantilevers*) the sensitivity increases and expected limits of detection are in the femto–atto regime with the astonish possibility of detection at the single-molecule level in real-time [29].

10.4.2
Working Principle

The physical working principle is based on the bending of the cantilever when a biomolecular interaction takes place. The bending arises as consequence of a sur-

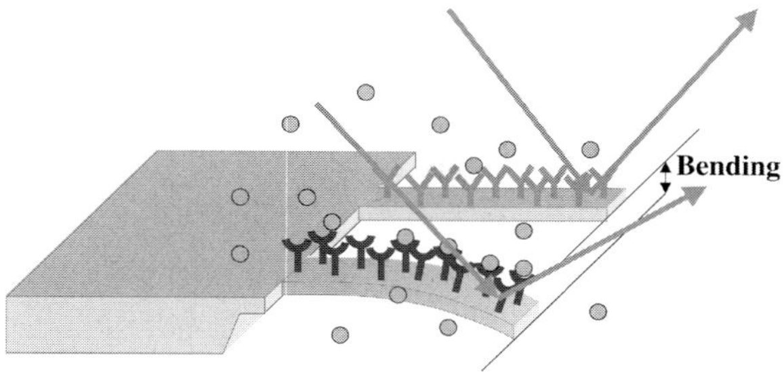

Figure 10.9. Scheme of cantilever bending due to a biomolecular interaction between an immobilized receptor and its complementary target. Only the specific recognition causes a change on the surface stress driving the bending of the cantilever.

face stress change induced by the molecular recognition when this phenomena happens just on one of its sides (with regard to the other). Hence, immobilization has to be selectively performed only on one side of the cantilever, allowing the target molecule to react onto the functionalized side (see Fig. 10.9 for details). Detection based on cantilever bending is known as *static mode* detection.

At the same time, the cantilever resonance frequency also varies due to mass loading. This type of detection is known as *dynamic mode* detection. Resonance frequency changes can be detected by measuring the thermal cantilever noise. However, to achieve high-sensitivity resolution, especially when working in liquids, it is necessary to produce a previous excitation of the cantilevers by using alternated electric, magnetic or acoustic fields.

Both static and dynamic modes have proven to be very sensitive when working in air. However, when operated in liquids, the resonance peak and the quality value shift toward much lower values than in air due to the damping effect of the liquid. This factor dramatically affects measurements based on the dynamic mode, making this method less suitable to monitor biochemical process in aqueous environments than when using the static mode. For this reason, as biological reactions take place in liquids, microcantilever sensors operating in the static mode are especially suitable as a platform for performing nanomechanical biomolecular assays. In fact, there are only a few demonstrations of biomolecular interaction detection by using the resonant frequency method [36].

Factors and phenomena responsible for the surface stress response during molecular recognition remain unclear. Several factors are considered to be involved, but a great controversy already exists in the scientific community [30, 37, 38]. Electrostatic interaction between neighboring adsorbates, changes in surface hydrophobicity and conformational changes of the adsorbed molecules can all induce stresses which may contrast with each other and make the change in stress not directly related to the receptor–ligand binding energy. This is particularly the case for biological adsorption due to the complexity of the interactions involved.

10.4.3
Detection Systems

The readout signal is critical to the real-time measurement, accuracy and possibility of integration of microcantilever biosensors. Therefore, a crucial area is the implementation a readout system capable of monitoring changes with subnanometer accuracy. The bending or/and the resonant frequency changes can be monitored by several techniques including optical beam deflection, piezoresistivity, piezoelectricity, interferometry, capacitance and electron tunneling amongst the most important [27]. In addition, under real conditions, sensors have to be stable long-term, selective and sensitive to the target molecule with no crosstalk reactions. Nonspecific binding of molecules and noise sources such as vibrations and temperature changes have to be avoided. These problems can be overcome by using differential measurements using cantilever array platforms in which a passivated cantilever is used as a reference. Optical and piezoresistive readouts are the most popular, and are compatible with array formats.

For the optical readout, the displacement of the free end of the cantilever is measured using the optical deflection of an incident laser beam on a position-sensitive photodetector which allows us to calculate the absolute value of the cantilever displacement [Fig. 10.10(a)]. This method provides sub-angstrom resolution and can be easily implemented for one cantilever; however, implementation for readout of arrays is technologically challenging, as it requires an array of laser sources with the same number of elements as the cantilever array. This technique is employed

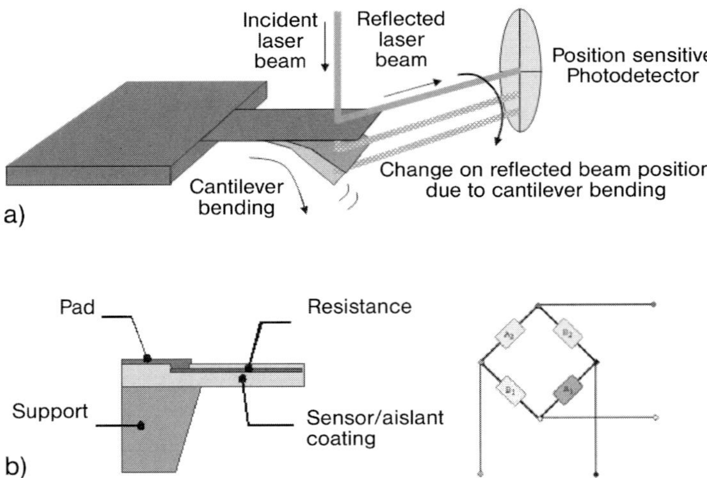

Figure 10.10. (a) Scheme of the optical readout method for cantilever bending evaluation. (b) Scheme of the piezoresistive readout and the Wheatstone bridge configuration.

in optically based commercialized array platforms, but sequential switching on and off of each laser source is necessary to avoid overlapping of the reflected beams on the photodetector. This problem can be elegantly solved by using a scanning laser source, where the laser beam is scanned along the array in order to sequentially illuminate the free ends of each microcantilever [39].

Recently, Zinoviev and coworkers introduced a new type of optical waveguide cantilever [40, 41], where the cantilever act as a waveguide for the light. Light going out from the cantilever can be collected by other waveguides or by a photodetector. This new device has shown good performance and offers an interesting approach for further integration in "lab-on-a-chip" microsystems. The design and fabrication of this device is covered in Section 10.4.3.2.

Piezoresistive readout is based on the changes observed in the resistivity of the material of the cantilever as a consequence of a surface stress change [42, 43]. To measure the change on the resistance, silicon cantilevers must be included into a DC-biased Wheatstone bridge (Fig. 10.10b). This configuration is very suitable for further integration using arrays of cantilevers [44]. However, the main disadvantage is the intrinsic high noise level that directly affects to the resolution and the sensitivity when compared to optically detected cantilevers [45], although the reduction of the thickness of piezoresistive cantilevers could increase the sensitivity. However, the cross-sectional structure of piezoresisitive cantilevers is complex, with the consequent technological limits in fabricating thin and highly sensitive cantilevers. Moreover, the piezoelectric readout requires electrical connections to the cantilever and their isolation from the solution. For all those reasons, the optical method is the one most employed. In addition, the best detection limits found in the literature are achieved with the optical method [30, 32].

10.4.4
Design of a Standard Microcantilever Sensor

Microcantilevers are typically made on silicon/Si_3N_4 or polymer materials, displaying dimensions ranging from tens to hundred of micrometers long, some tens of micrometers wide and hundreds of nanometers thick. Silicon, Si_3N_4 and SiO_2 cantilevers are available commercially with different shapes and sizes in analogy to AFM cantilevers, with typical lengths between 10 and 500 μm, and ultra-thin cantilevers up to 12 nm thick. However, for specific applications (as in highly sensitive biosensors) cantilevers must be designed and fabricated to satisfy such requirements.

Previous modeling is needed in order to know the ranges of thickness, length and width which could give the highest sensitivity. Several factors must be taken into account. Reducing the thickness and increasing the length results in an increase of sensitivity of the device, but also leads to complex fabrication technology. The width of cantilevers is rather important when cantilevers are used in dynamic mode. A reduction in the width of the beam, in a certain range, results in an increase of the damping and subsequently in a decrease in the quality factor [27]. The effect of the frame and the material must be also taken into account. Cantilever sensitivity depends critically on their spring constant. The lower the constant, the higher the sensitivity for measurements in liquids based on the static method. A key factor that dramatically affects the spring constant of a cantilever is the Young modulus, which is directly related to the characteristics of the cantilever material. Cantilevers are normally made of silicon or related materials that have a high Young's modulus. A cantilever made of a softer material would be more sensitive for static deflection measurements. For that reason, polymers with a much lower Young's modulus than that of silicon have been used as a substitute material for fabricating cantilevers [46]. Among polymers, SU-8 has been shown to be very sensitive, exhibiting a Young's modulus about 40 times lower than for silicon. In addition, the cantilever fabrication process is relatively inexpensive, fast and reliable. It also provides a convenient way to realize arrays of multiple sensors and to integrate them into a miniaturized biochemical analysis system. However, there is still no proof of biosensing testing using polymer cantilever sensors, mainly due to the difficulty in achieving a stable immobilization of the receptor layer.

Modifications of cantilever shape and dimensions could also improve the cantilever spring constant – longer and thinner [47, 48] cantilevers can address very small spring constants. Microfabrication technologies allow fabricating micrometer-sized cantilevers with a high length:thickness ratio in a reproducible and inexpensive way. However, thermal motion of the cantilever severely limits the extent to which the spring constant of the cantilever can be reduced [29]. Modeling can be done by using, finite element programs (ANSYS), for example.

10.4.4.1 Fabrication of a Standard Microcantilever Sensor

Cantilevers are batch fabricated using well-established thin-film-processing technologies which provide low cost, high yield and good reproducibility. Such fabrication techniques include thin-layer deposition, photolithographic patterning and etching, and surface and bulk micromachining. Usually, a sacrificial layer is first deposited on a pre-patterned substrate before the deposition of the cantilever structural material. This structural layer must be free of stress gradients, otherwise problems with the initial bending of cantilevers will appear. The thickness of the layer must be uniform enough around the wafer to make sure that all the beams will be identical. It is possible to fabricate arrays of thousands of identical cantilevers on one wafer.

The cantilevers might be fabricated extended over the border of the chip or they might be located in individual cavities inside the chip. This depends on the type of flow cells to be employed – a common one or a discrete one with independent inlets and outlets for each cantilever. For the chip with cantilevers in a common window, the immobilization of the receptor can be conducted on each cantilever individually using, for example, ink-jet and nanojet printing. The individual cavity design is more complicated from a technological point of view, but allows the immobilization of different receptors in each cantilever *in situ* by using the discrete flow cell and also allows parallel screening of different substances.

As an example, the technology for the fabrication of arrays composed of 20 silicon cantilevers is described. Both types of chips with discrete and common win-

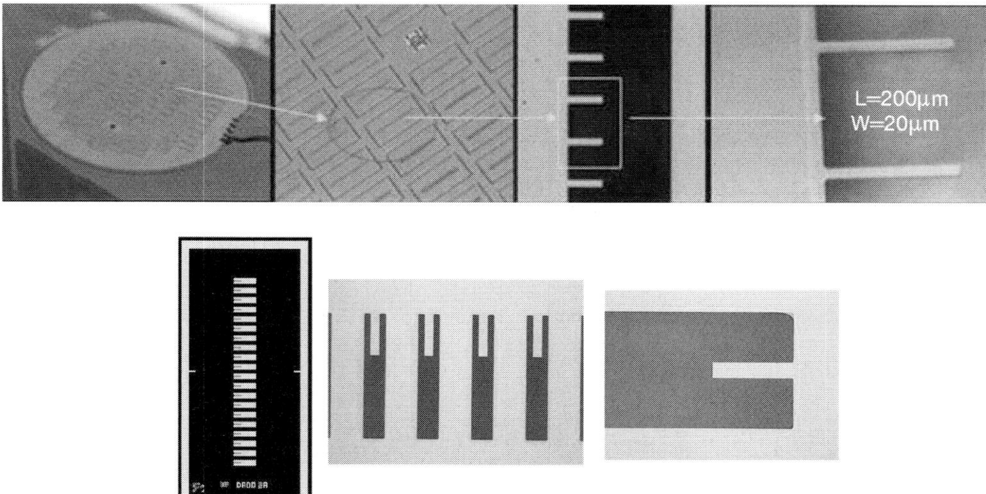

Figure 10.11. Photographs of nanomechanical sensors based on microcantilever arrays fabricated at the author's Clean Room facilities. Two different arrays of microcantilevers can be observed, with common and discrete windows (see text for explanation).

dows were fabricated. The cantilevers in the array were separated by a distance of 250 μm. Photographs of the fabricated devices with common and discrete windows are shown in the Fig. 10.11. The chips were 3×7 mm^2 in size. The dimensions were chosen small enough to fabricate as many as possible devices on one wafer and large enough to be conveniently handled for measurements.

SOI wafers were chosen as a starting material. The structural silicon layer was free of intrinsic stress gradients and any superficial defects. In this way, it is possible to fabricate arrays of thousands of identical cantilevers on one wafer. To fabricate the cantilevers, the most simple approach is to use bulk micromachining using anisotropic etching of silicon, but this method does not allow us to form a gap between the cavities due to lateral etching. For that reason, deep RIE (DRIE) must be employed to obtain windows with vertical walls.

The sequence of the technological steps is shown in Fig. 10.12. The front side contains a structural silicon layer (1) and a sacrificial SiO$_2$ layer (2). The reverse side of the silicon substrate (3) has a SiO$_2$ layer (4). As a first step, the reverse side was covered with an aluminum layer (5), the most adequate mask material for the following DRIE. The initial multilayer structure used for the fabrication of the cantilevers is shown in Fig. 10.12(a). Cantilevers on the front side were defined by dry etching of silicon through the pattern obtained by previous photolithography. The aluminum mask was deposited on the reverse side and the SiO$_2$ was removed from the areas where the silicon substrate must be etched. Before the next step, the components side was covered with a photoresist layer (6) (Fig. 10.12b).

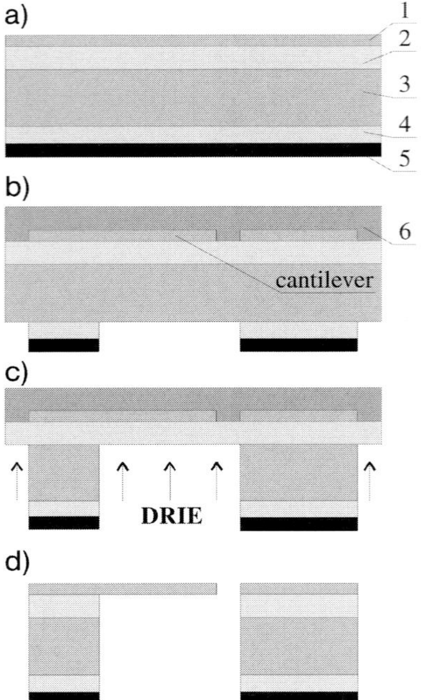

Figure 10.12. Fabrication steps for microcantilever arrays (see text for details).

The etching with DRIE of the silicon substrate resulted in almost vertical walls in the cavities with a small (about 20 μm) lateral undercut at the top. At this stage, clean SiO_2 membranes with cantilevers and a photoresist layer on top were obtained (Fig. 10.12c). The photoresist layer (6) prevented the membranes cracking. To release the cantilevers, the membranes were etched in vapors of HF (49%). Afterwards, the wafers were briefly rinsed with deionized water. The photoresist film was removed by oxygen plasma etching.

The next step was dicing the wafers. Common sawing could not be employed as it would break the cantilevers by the cooling water flow and might leave residuals on the sensor surface. As a solution, a DRIE step was applied to have wafers "preliminary diced". As the cavities under the cantilevers were etched, a groove around every chip was etched to make the array of cantilevers joined to the wafer by two thin hinges, which could be broken manually. With this technology a 100% yield for the cantilever fabrication (2500 cantilevers per wafer) was obtained [49]. All cantilevers were identical and the initial bending was almost negligible. Typical profiles of the fabricated cantilevers are shown in Fig. 10.13. The dispersion did not exceed 0.6 μm, which corresponded to 0.005 rad dispersion of angular deflection. Even cantilevers with an extreme low spring constant ($k = 0.000061$ N m^{-1}) can be obtained following this fabrication procedure (cantilever dimensions $800 \times 20 \times 0.334$ μm^3).

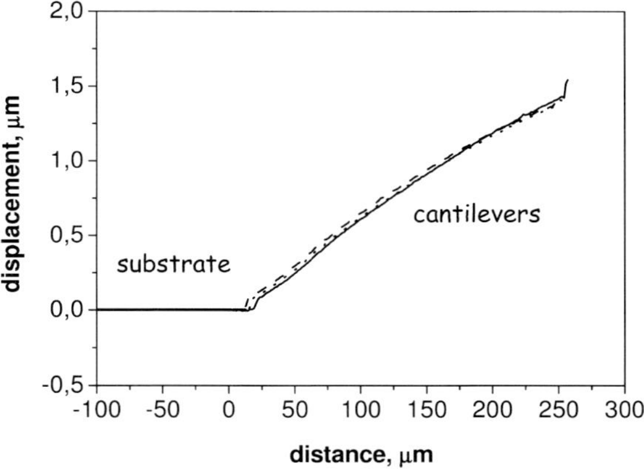

Figure 10.13. Typical cantilever profiles after fabrication. Cantilevers are 200 µm long, 40 µm wide and 0.334 µm thick.

Table 10.1 shows the main parameters experimentally evaluated for the fabricated devices and the comparison with the values for the commercial ones, demonstrating the feasibility of the designed and fabricated devices for higher sensitivity than the commercial devices.

10.4.4.2 Optical Waveguide Microcantilever: Design and Fabrication

The optical method is normally employed for microcantilever sensors readout, but has several disadvantages – the major one being the difficulties experienced while performing parallel monitoring of several cantilevers at the same time. In order to achieve further integration, a new optical cantilever sensor has been recently proposed [40, 41]. The detection method is based on monitoring the light exiting a

Table 10.1. Main mechanical parameters of fabricated microcantilevers, and comparison with commercial and polymer microcantilevers.

Length × width × thickness	K (N m^{-1})	Frequency (kHz)	Q
500 × 20 × 0.3	9.1×10^{-3}	1.5	1.8
200 × 20 × 0.3	7.4×10^{-3}	9.9	6.3
100 × 20 × 0.3	6.1×10^{-3}	34.9	13.6
50 × 40 × 0.3	7.9×10^{-3}	86.5	20.7
200 × 20 × 1.5 (polymer)	2.5×10^{-3}	17	15
550 × 40 × 0.8 (silicon commercial)	11×10^{-3}	23.9	34.7

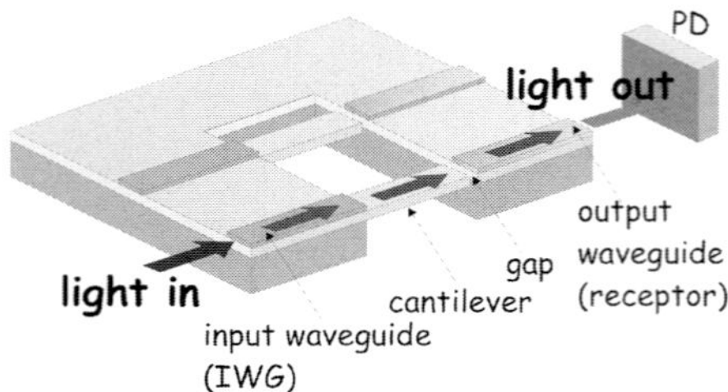

Figure 10.14. Scheme of the novel optical waveguide microcantilever device.

waveguide microcantilever (see Fig. 10.14 for details). This integrated waveguide cantilever sensor can be fabricated as an array of independent waveguide channels designed for monitoring bio-specific reactions. The sensor can work in static or dynamic modes, either by monitoring the deflection or by monitoring the changes in the resonance frequency of the cantilever. The advantage of the device is that the transducer is integrated with the receptor on one chip and the external photodetector is only used for optical power readout. No preliminary alignment or adjustment is needed, except for light coupling into the chip, which does not seriously affect the performance of the device if the coupler is well designed. The sensitivity of the device is comparable to standard microcantilever sensors discussed above.

10.4.4.2.1 Principle of Operation and Theoretical Analysis The principle of operation is based on monitoring the coupling efficiency between two butt-coupled waveguides. The energy transfer between the waveguides is very sensitive to their misalignment with respect to each other. In this device the transducer is an optically transparent cantilever beam of submicron thickness. It is located in a cavity and acts as a symmetrical optical waveguide. Light from the cantilever is injected through a short gap into an output waveguide, called a receptor. After exiting the cantilever, light diverges very quickly in the transversal direction and after a few microns its intensity distribution is much larger compared to the distribution of the receptor waveguide modes. Thus, the near field of the cantilever is probed by the receptor, which is a single-mode asymmetrical waveguide. The changes in the power of light exiting the output waveguide, which are attributed to the cantilever bending caused either by the surface stress and/or by vibration of the cantilever, are monitored by a conventional photodetector [41].

The simulations of the coupling efficiency and the sensitivity of the waveguides to their misalignment with respect to each other can be done using overlap integrals [50]. The simulations are performed separately for the fundamental and for the first propagating modes of the cantilever. Two electric field distributions have been overlapped. The first one was the distribution of light exiting the cantilever after propagation through the gap. It was obtained using finite difference beam propagation method (FDBPM) [51]. The second distribution was the fundamental mode of the output waveguide. It was built using the solution of Maxwell equations with appropriate boundary conditions [50]. Waveguide parameters close to

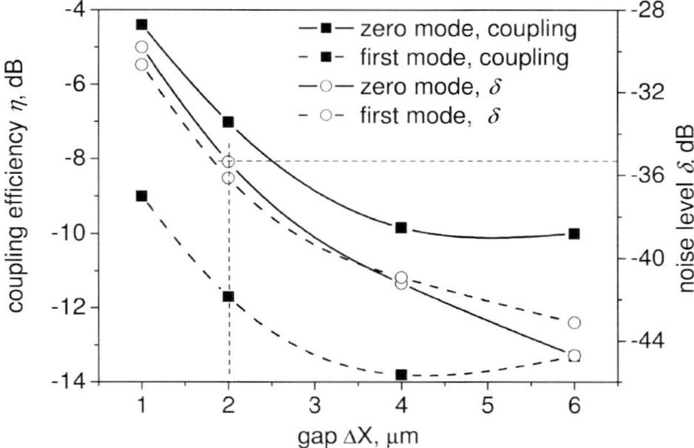

Figure 10.15. Modeling of the coupling efficiency and the noise level permitted in the acquisition system for the detection of a 1-nm optical cantilever displacement versus the gap width.

those of the fabricated device were used for the simulations. The cantilever and the receptor were made of SiO_2 and Si_3N_4 materials, respectively.

The sensitivity of the device is defined as the relative change in the output signal per unit cantilever free end displacement. It was calculated as the change in the output power required for the detection of 1-nm cantilever displacement with respect to the power of light exiting the cantilever. This accuracy is equivalent to the noise level allowed in the system. This parameter, further called the noise level, is expressed in relative units. The results of simulations are shown in Fig. 10.15. The noise level demonstrates similar behavior for both modes. The coupling efficiency for the zero mode is higher than that for the first mode. The curves of Fig. 10.15 were produced assuming the cantilever is biased to the most sensitive point which depends on the gap width as well. In general, the width of the gap is a trade-off since a small gap allows for high sensitivity and efficiency, whereas a wide gap makes fabrication tolerances less strict.

10.4.4.2.2 Fabrication and Characterization

The most difficult step in fabrication of the device is the fabrication of the cantilevers aligned with the output waveguides, implying that the cantilevers should be very flat. A thermally grown SiO_2 layer was used for the fabrication. The film demonstrated no stress gradient if the bottom layer of a few hundred nanometers was previously eliminated. This allows us to fabricate straight cantilever beams 200 μm long, 40 μm wide and 600 nm thick. The gap between the cantilever and the receptor waveguide was fixed to 3 μm. As the refractive index of SiO_2 is low, it is not possible to conform a total internal reflection waveguide over the silicon substrate, unlike a SiO_2 cantilever in air. Therefore, light was launched over the substrate to the cantilever by using a Si_3N_4 waveguide, called an input waveguide (IWG). At the cantilever anchoring area, the IWG deposited over the silica buffer forms a junction with the cantilever beam, which is

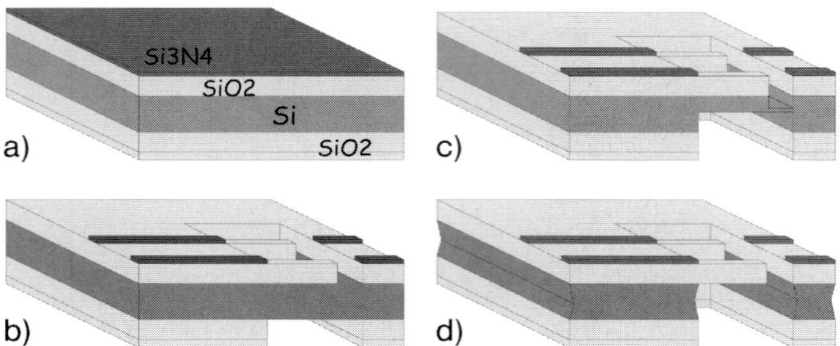

Figure 10.16. Technology for the fabrication of the optical waveguide cantilevers.

an extension of the buffer (Fig. 10.14). Light coupling into the cantilever was carried out by means of the evanescent field of the fundamental mode of the IWG. The efficiency of coupling is inversely proportional to the square root of the thickness of the IWG. It can reach 75% if the thickness is close to the value defined by the cut-off condition for the fundamental mode. The silica buffer thickness was 1.0 µm – enough to avoid leakage of energy into the substrate assuming the thickness of the input and output waveguides was 140 nm.

The device was fabricated using standard silicon technologies (Fig. 10.16). First, a SiO_2 layer was thermally grown by wet oxidation on both sides of a silicon wafer. Then, a high-temperature LPCVD Si_3N_4 layer was deposited on the front side and a PECVD SiO_2 on the reverse side. Standard photolithography and RIE were applied to define the waveguides on the Si_3N_4 layer and the cantilevers on the SiO_2 layer. The same processes were used to obtain the mask for DRIE on the reverse side of the wafer. DRIE was applied to both sides in order to define the cavities under and around the cantilevers. Finally, the cantilevers were released by etching the rests of silicon using TMAH solution. A yield close to 100% was achieved, which means more than 2500 cantilevers per wafer. Figure 10.17 shows some photographs of a fabricated array of waveguide cantilevers. The cantilevers on the chip are located in a common cavity, which is a reach-through chip hole located in the center.

For characterization, the chip was located on a piezoelectric actuator connected to a sine waveform synthesizer. Light from a He–Ne laser (632.8 nm, 7.5 mW) was coupled into the chip using direct focusing through an objective lens at the exiting light collected by another objective and then directed to a silicon photodetector. The coupling efficiency into the IWG was about 5%. Near 40% of light was transmitted from the IWG into the cantilever. The power of light exiting the output waveguide was 0.015 mW. Total losses were −27 dB with respect to the laser output power. The mechanical resonance of the cantilevers was close to of 13.1 kHz. The spectrum of the output signal measured with AC (11.1 kHz) voltage of 50 mV supplied to the piezoactuator is shown in Fig. 10.18. At this frequency, a small modulation of the output signal was observed, attributed to the modulation of coupling efficiency at the input. AC voltage (50 mV) applied to the piezoactuator

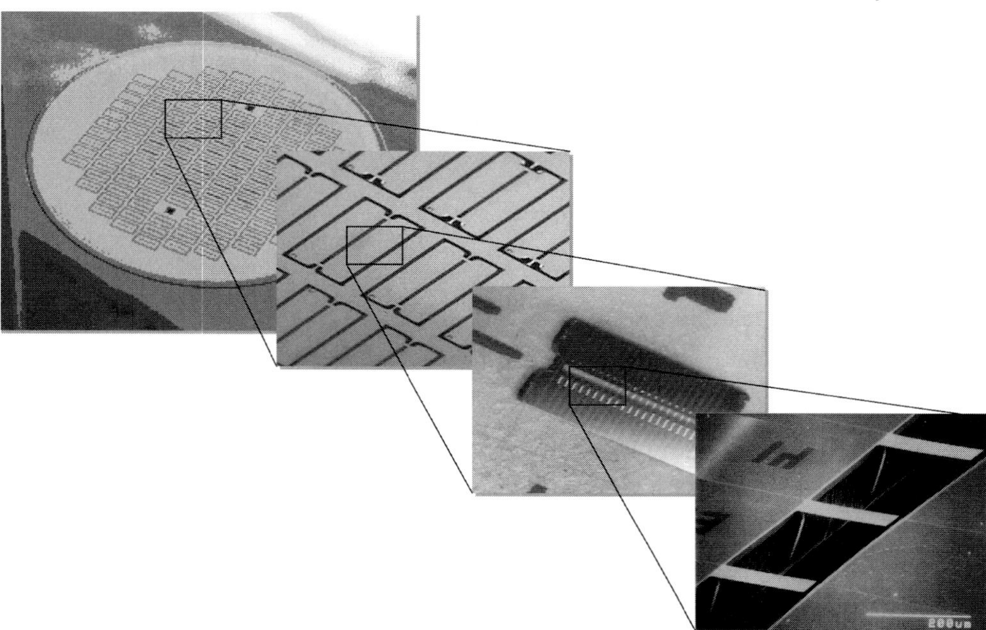

Figure 10.17. Photographs of the fabricated waveguide cantilevers.

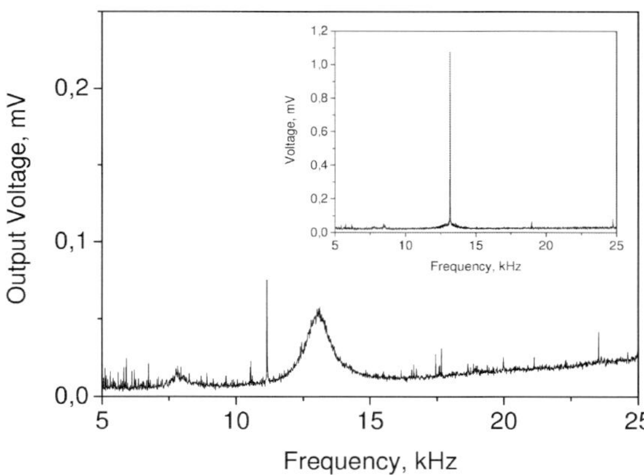

Figure 10.18. Spectrum of the output signal of a waveguide cantilever. The peak at 13.1 kHz corresponds to the cantilever vibration at resonance frequency induced by a piezoelectric actuator. The excitation voltage of the piezoactuator was 50 mV. The cantilever oscillation amplitude was about 1.7 nm at this frequency.

produced a periodic change in the output signal. At a frequency of 13.1 kHz, the cantilever was subject to resonance vibration with an amplitude of about 1.7 nm. The sensitivity, calculated as fractional change ($\Delta U_{out} = 1.0$ mV after amplification by factor 100 and subtracting the modulation at the input, see the inset in Fig. 10.18) in the output voltage ($U_{out} = 60$ mV) per unit cantilever displacement was 10^{-4} nm^{-1}. The signal to noise ratio, which was about 40 at these frequencies, would allow us to register a 0.05-nm cantilever free end displacement. This shows the potential for using this novel structure as a nanomechanical sensor for biomolecular interaction detection with high sensitivity and with a much more integrated approach that that for standard microcantilever sensors.

10.4.5
Biosensing Applications of Nanomechanical Sensors

One of the first applications of nanomechanical sensors was in the field of genomics. Fritz and coworkers [26] demonstrated the detection of a single-base mismatch with a detection limit of 10 nM using an array of two cantilevers. One of the cantilevers was functionalized with a control (noncomplementary) oligonucleotide and the DNA probe (complementary) was immobilized in the other. They achieved hybridization deflection signals as small as 10 and 16 nm for 12 and 16mer DNA targets, respectively, within a deflection noise of 0.5 nm. More recently, McKendry and coworker reached a detection of 75 nM for target oligonucleotides in an array of eight microcantilevers [30]. Both results include the specific immobilization by microcapillarity with a 40-µM solution of the thiolated DNA probe. Figure 10.19 shows the cantilever response for a DNA immobilization step. The discrimination of single-nucleotide polymorphisms has been also reported by Thundat and coworkers [31], although with a single cantilever. Other DNA detection schemes have been reported, e.g. the one which used a capture oligonucleotide combined with a DNA probe attached to a gold nanoparticle. This method can detect at least 0.05 nM and is able to discriminate single mismatch measured by resonance [52].

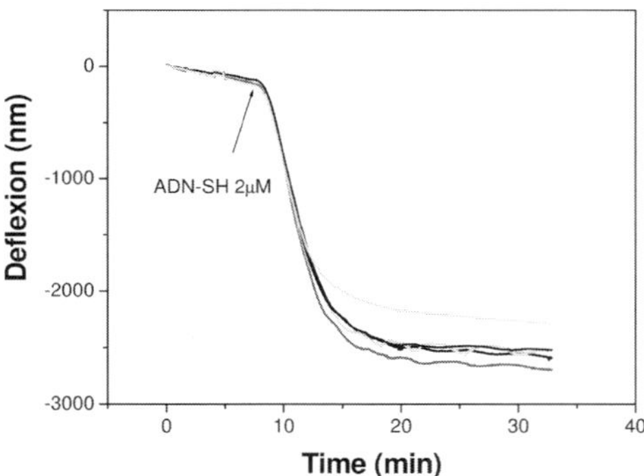

Figure 10.19. Simultaneous 2 µM thiolated DNA (27mer) immobilization detection using the cantilever array platform showed in Fig. 10.11.

Proteins have also caught the attention of nanomechanics applications, mainly motivated by the possibility of achieving protein microarrays based on arrays of cantilevers with lower-cost fabrication methods, no labeling of the target protein and improved sensitivity. Recently, Wee and coworkers [53] reported the detection of prostate-specific antigen (PSA), a useful marker for earlier detection of prostate cancer, and C-reactive protein (CRP), a specific marker of cardiac disease, by an electromechanical biosensor using self-sensing piezoresistive microcantilevers. Majumdar and coworkers [32] reported the detection of two isoforms of PSA with an excellent range of discrimination and a detection limit of 6 ng mL^{-1} (deflection signal of 20 nm) in a background of 1 mg mL^{-1} of BSA protein. In addition, a novel development for early osteosarcoma detection has been described, sensing the interactions between vimentin antibodies and antigens with a single-cantilever-based biosensor [54]. Other clinical applications included the detection of different pathogens like *Salmonella enterica* by Weeks and coworkers [55], vaccinia virus by Gunter and coworkers [56] or fungal spores from *Aspergillus niger* by Nugaeva and coworkers [57]. The use of aptamers as the bioreceptor element has also been widely probed on a number of microcantilever biosensing platforms [58–61].

Biosensing with microcantilevers also extends to applications in environmental sciences. Alvarez and coworkers [35] used this nanodevice for the detection of the organochlorine insecticide compound dichlorodiphenyltrichloroethane (DDT). A competitive assay was performed, in which the cantilever was exposed to a mixed solution of the monoclonal antibody and DDT, and direct detection was proven. With this detection strategy DTT concentrations as low as 10 nM were detected involving deflection signals in the range of 50 nm (Fig. 10.20). Many other applica-

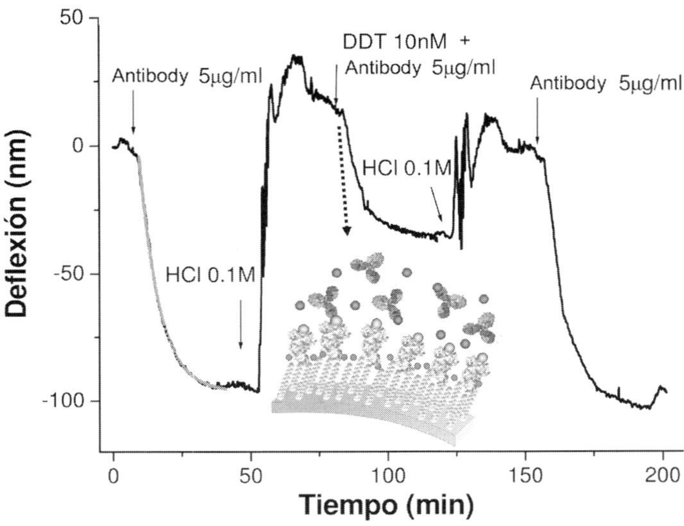

Figure 10.20. DTT pesticide detection using a single microcantilever sensor in a real-time competitive immunoassay. The cantilever surface was regenerated with 100 mM HCl.

tions have been described for the detection of pesticides and avidin–streptavidin [28].

Nowadays, there are several commercial platforms based on cantilever array sensors available that demonstrate the potential of nanomechanical biosensors as a reliable sensing tool for biochemical applications [62–65].

10.5
Conclusions and Future Goals

This chapter has provided an overview of most of the technical aspects related to two important branches of nanodevices for biosensing (nanomechanical and nanophotonic biosensors), including design, fabrication and applications of some specific devices. Nanomechanical and nanophotonic biosensors constitute a promising technology as a suitable solution for an important number of problems in the biosensor field. The improvement of the reproducibility and sensitivity along with the integration of microfluidics and detections systems is the main aim of the present research.

Nanobiosensors are still undergoing considerable diversification with respect to technologies, but those fabricated by using standard microelectronics and related MEMS/NEMS approaches could allow the development of portable microsystems platforms which could be employed outside the laboratory. However, limitations in the technology, problems in the integration of all the components in one microsystem and the connection of such tiny devices with the "real" world must be appropriately solved. There is no doubt that such limitations will be overcome in the near future, opening up immense possibilities for early, personalized diagnosis and high-throughput screening.

Acknowledgments

The authors would like to acknowledge financial support from European and national projects, and thank their colleagues working within the Biosensors Group for their contribution. The authors thank F. J. Blanco and K. Mayora (Ikerlan Corp., Spain) for the microfluidics image.

References

1 JAIN, K. K. Nanotechnology in clinical laboratory diagnostics. *Clin. Chim. Acta* **2005**, *358*, 37–54.

2 LECHUGA, L. M. Optical biosensors. In: *Biosensors and Modern Biospecific Analytical Techniques, Comprehensive Analytical Chemistry Series XLIV*, GORTON, L. (Ed.). Elsevier, Amsterdam, **2005**.

3 NIEMEYER, C. M., MIRKIN, C. A. (Eds.). *Nanobiotechnology. Concepts, Applications and Perspectives*. Wiley-VCH, Weinheim, **2004**.

4 GAO, X., CHAN, W. C. W., NIE, S.,

Quantum-dot nanocrystals for ultrasensitive biological labeling and multicolor optical encoding. *J. Biomed. Opt.* **2002**, *7*, 532–537.

5 HAES, A. J., ZOU, S., SCHATZ, G. C., VAN DUYNE, R. P. A Nanoscale optical biosensor: the long range distance dependence of the localized surface plasmon resonance of noble metal nanoparticles. *J. Phys. Chem. B* **2004**, *108*, 109–116.

6 WANG, J., LIU, G. D., JAN, M. R., Ultrasensitive electrical biosensing of proteins and DNA: carbon-nanotube derived amplification of the recognition and transduction events. *J. Am. Chem. Soc.* **2004**, *126*, 3010–3011.

7 SUMNER, J. P., AYLOTT, J. W., MONSON, E., KOPELMAN, R. A fluorescent PEBBLE nanosensor for intracellular free zinc. *Analyst* **2002**, *127*, 11–16.

8 FORTINA, P., KRICKA, L. J., SURREY, S., GRODZINSKI, P. Nanobiotechnology: the promise and reality of new approaches to molecular recognition. *Trends Biotechnol.* **2005**, *23*, 168–173.

9 MEDEA+ Program (http://www.medea.org).

10 KASEMO, B. Biological surface science. *Surf. Sci.* **2002**, *500*, 656–677.

11 MANNELLI, I., MINUNNI, I., TOMBELLI, S., WANG, R., SPIRITI, M. M., MASCINI, M., Direct immobilization of DNA probes for the development of affinity biosensors. *Bioelectrochemistry* **2005**, *66*, 129–138.

12 FERRETTI, S., PAYNTER, S., RUSSELL, D. A., SAPSFORD, K. E., RICHARDSON, D. J. Self-assembled monolayers: a versatile tool for the formulation of bio-surfaces. *Trends Anal. Chem.* **2000**, *19*, 530–540.

13 GOODING, J. J., HIBBERT, D. B. The application of alkanethiol self-assembled monolayers to enzyme electrodes. *Trends Anal. Chem.* **1999**, *18*, 525–533.

14 LIEGLER, F. S., TAITT, C. R. (Eds.). *Optical Biosensor: Present and Future.* Elsevier, Amsterdam, **2002**.

15 BURR, G. W., CHOW, E., MIRKARIMI, L. W., SIGALAS, M., GROT, A. Photonic crystals microcavities as ultracompact film-thickness monitors for bio-sensing. Nanophotonics for Information Systems, paper NPIS-ThC4.

16 BOYD, R. W., HEEBNER, J. E. Sensitive disk resonator photonic biosensor. *Appl. Optics* **2001**, *40*, 5742–5747.

17 YIN, D., BARBER, J. P., HAWKINS, A. R., SCHMIDT, H. Low-loss integrated optical sensors based on hollow-core ARROW waveguide. *Proc. SPIE* **2005**, *5730*, 218–225.

18 LECHUGA, L. M., PRIETO, F., SEPÚLVEDA, B. Interferometric biosensors for environmental pollution detection. In: *Optical Sensors for Industrial E (Springer Series on Chemical Sensors and Biosensors)*, NARAYANASWAMY, R., WOLFBEIS, O. S. (Eds.). Springer, Berlin, **2003**, pp. 227–250.

19 DOMÍNGUEZ, C., LECHUGA, L. M., RODRÍGUEZ, J. A. Integrated optical chemo- and Biosensors. In: *Integrated Analytical Systems*, ALEGRET, S. (Ed.). Elsevier, Amsterdam, **2003**, pp. 541–586.

20 PRIETO, F., SEPÚLVEDA, B., CALLE, A., LLOBERA, A., DOMÍNGUEZ, C., ABAD, A., MONTOYA, A., LECHUGA, L. M. An integrated optical interferometric nanodevice based on silicon technology for biosensor applications. *Nanotechnology* **2003**, *14*, 907–912.

21 CAMPBELL, D. P., MCCLOSKEY, J. Interferometric biosensor. In: *Optical Biosensors: Present and Future.* LIEGLER, F., ROWE, C. (Eds.). Elsevier, Amsterdam, **2002**, pp. 277–304.

22 PRIETO, F., LLOBERA, A., CALLE, A., DOMÍNGUEZ, C., LECHUGA, L. M. Design and analysis of silicon antiresonant reflecting optical waveguides for highly sensitive sensors. *J. Lightwave Technol.* **2000**, *18*, 966–972.

23 THRUSH, E., LEVI, O., COOK, L. J., DEICH, J., KURTZ, A., SMITH, S. J., MOERNER, W. E., HARRIS, J. S. Mono-lithically integrated semiconductor fluorescence sensor for microfluidic applications. *Sens. Actuators B* **2005**, *105*, 393–399.

24 BLANCO, F. J., AGIRREGABIRIA, M., GARCIA, J., BERGANZO, J., TIJERO, M.,

Arroyo, M. T., Ruano, J. M., Aramburu, I., Mayora, K. Novel three-dimensional embedded SU-8 microchannels fabricated using a low temperature full wafer adhesive bonding. *J. Micromech. Microeng.* **2004**, *14*, 1047–1056.

25 Carrascosa, L. G., Moreno, M., Alvarez, M., Lechuga, L. M. Nano-mechanical biosensors: a new sensing tool. *Trends Anal. Chem.* **2006**, *25*(3), 196–206.

26 Fritz, J., Baller, M. K., Lang, H. P., Rothuizen, H., Vettiger, P., Meyer, E., Güntherodt, H. J., Gerber, Ch., Gimzewski, J. W. Translating bio-molecular interactions into nano-mechanics. *Science* **2000**, *88*, 316–318.

27 Lavrik, N., Sepaniak, M., Datskos, P. Cantilever transducers as a platform for chemical and biological sensors. *Rev. Sci. Instrum.* **2004**, *75*, 2229–2253.

28 Raiteri, R., Grattarola, M., Butt, H.-J., Skládal, P. Micromechanical cantilever-based biosensors. *Sens. Actuators B* **2001**, *79*, 115–126.

29 Datskos, P. G., Thundat, T., Lavrik, N. V. Micro and nanocantilever sensors. In: *Encyclopedia of Nanoscience and Nanotechnology*, Nalwa, H. S. (Eds.), American Scientific Publishers, Stevenson Ranch, CA, **2004**, vol. X, pp. 1–10.

30 Mckendry, R., Zhang, J., Arntz, Y., Strunz, T., Hegner, M., Lang, H. P., Baller, M. K., Certa, U., Meyer, E., Güntherodt, H.-J., Gerber, Ch. Multiple label-free biodetection and quantitative DNA-binding on a nano-mechanical cantilever array. *Proc. Natl Acad. Sci. USA* **2002**, *99*, 9783–9788.

31 Hansen, K. M., Ji, H.-F., Wu, G., Datar, R., Cote, R., Majumdar, A. Thundat, T., Cantilever-based optical deflection assay for discrimination of DNA single-nucleotide mismatches. *Anal. Chem.* **2001**, *73*, 1567–1571.

32 Wu, G., Datar, H., Hansen, K. M., Thundat, T., Cote, R. J., Majumdar, A. Bioassay of prostate-specific antigen (PSA) using microcantilevers. *Nat. Biotechnol.* **2001**, *19*, 856–860.

33 Arntz, Y., Seelig, J., Lang, H., Zhang, J., Hunziker, P., Ramseyer, J., Meyer, E., Hegner, M., Gerber, C., Label-free protein assay based on a nanomechanical cantilever array. *Nanotechnology* **2003**, *14*, 86–90.

34 Savran, C., Knudsen, S., Ellington, A., Manalis, S., Micromechanical detection of proteins using aptamer-based receptor molecules. *Anal. Chem.* **2004**, *76*, 3194–3198.

35 Alvarez, M., Calle, A., Tamayo, J., Abad, A., Montoya, A., Lechuga, L. M. Development of nanomechanical biosensors for detection of the pesticide DDT. *Biosens. Bioelectron.* **2003**, *18*, 649–653.

36 Ghatkesar, M. K., Barwich, V., Braun, T., Bredekamp, A. H., Drechsler, U., Despont, M., Lang, H. P., Hegner, M., Gerber, C., Real-time mass sensing by nanomechanical resonators in fluid. *Proc. IEEE Sensors* **2004**, *2004*, 1060–1063.

37 Alvarez, M., Carracosa, L. G., Moreno, M., Calle, A., Zaballos, A., Lechuga, L. M., Martínez-A, C., Tamayo, J. Nanomechanics on the formation of DNA self-assembled monolayers and hybridization on microcantilevers. *Langmuir* **2004**, *20*, 9663–9676.

38 Wu, G., Ji, H., Hansen, K., Thundat, T., Datar, R., Cote, R., Hagan, M. F., Chakraborty, A. K., Majumdar, A. Origin of nanomechanical cantilever motion generated from biomolecular interactions. *Proc. Natl Acad. Sci. USA* **2001**, *98*, 1560–1564.

39 Tamayo, J., Álvarez, M., Lechuga, L. M. System and method for detecting the displacement of a plurality of micro- and nanomechanical elements, such as microcantilevers. *European Patent PCT/EP-05/002356*, **2004**.

40 Zinoviev, K., Domínguez, C., Lechuga, L. M., Plaza, J. A., Cadarso, V. Cantilever-based detector device and method of manufacturing such a device. *European Patent PCT/EP-05/380137*, **2005**.

41 Zinoviev, K., Dominguez, C., Plaza, J. A., Lechuga, L. M. Light coupling into an optical microcantilever by an embedded diffraction grating. *Appl. Optics* **2005**, *45*(2), 229–234.

42 Linnemann, R., Gotszalk, T., Hadjiiski, L., Rangelow, I. W., Characterization of a cantilever with an integrated deflection sensor. *Thin Solid Films* **1995**, *264*, 159–164.

43 Minne, S. C., Manalis, S. R., Quate, C. F. Parallel atomic force microscopy using cantilevers with integrated piezoresistive sensors and integrated piezoelectric actuators. *Appl. Phys. Lett.* **1995**, *67*, 3918–3920.

44 Vettiger, P., Despont, M., Drechsler, U., Durig, U. T., Haberle, W., Lutwyche, M., Rothuizen, H., Stutz, R., Widmer, R., Binnig, G. K. The "Millipede" – more than one thousand tips for future AFM data storage. *IBM J. Res. Dev.* **2000**, *44*, 323–340.

45 Yu, X., Thaysen, J., Hansen, O., Boisen, A., Optimization of sensitivity and noise in piezoresistive cantilevers. *J. Appl. Phys.* **2002**, *92*, 6296–6307.

46 Thaysen, J., Yalvinkaya, A. D., Vettiger, P., Menon, A., Polymer-based stress sensor with integrated readout. *J. Phys. D* **2002**, *35*, 2698–2703.

47 Harley, J. A., Kenny, T. W. High-sensitivity piezoresistive cantilevers under 1000 Å thick. *Appl. Phys. Lett.* **1999**, *75*, 289–291.

48 Stowe, T. D., Yasumura, K., Kenny, T. W. Attonewton force detection using ultrathin silicon cantilevers. *Appl. Phys. Lett.* **1997**, *71*, 288–230.

49 Optonanogen project, http://eprints.ecs.soton.ac.uk/view/projects/optonanogen.html.

50 Kim, C. M., Ramaswamy, R. V. Modeling of graded-index channel waveguides using non-uniform finite difference method. *J. Lightwave Technol.* **1989**, *7*, 1581–1589.

51 Anemogiannis, E., Glytsis, E. N. Multilayer waveguides: efficient numerical analysis of general structures. *J. Lightwave Technol.* **1992**, *10*, 1344–1351.

52 Su, M., Li, S., Dravid, V. P. Microcantilever resonance-based DNA detection with nanoparticle probes. *Appl. Phys. Lett.* **2003**, *82*, 3562–3564.

53 Wee, K. W., Kang, G. Y., Park, J., Kang, J. Y., Yoon, D. S., Park, J. H., Kim, T. S., Novel electrical detection of label-free disease marker proteins using piezoresistive self-sensing micro-cantilevers. *Biosen. Bioelectron.* **2005**, *20*, 1932–1938.

54 Milburn, C., Zhou, J., Bravo, O., Kumar, C., Soboyejo, W. O., Sensing interactions between vimentin antibodies and antigens for early cancer detection. *J. Biomed. Nanotechnol.* **2005**, *1*, 30–38.

55 Weeks, B. L., Camarero, J., Noy, A., Miller, A. E., Stanker L., De Yoreo, J. J. A microcantilever-based pathogen detector. *Scanning* **2003**, *25*, 297–299.

56 Gunter, R. L., Delinger, W. G., Manygoats, K., Kooser, A., Porter, T. L., Viral detection using an embedded piezoresistive microcantilever sensor. *Sens. Actuators A* **2003**, *107*, 219–224.

57 Nugaeva, N., Gfeller, K., Backmann, N., Güntherodt, H.-J., Hegner, M., Micromechanical cantilever array sensors for selective fungal immobilization and fast growth detection. *Biosens. & Bioelect.* **2005**, *21*(6), 849–856.

58 Gokulrangan, G., Unruh, J. R., Holub, D. F., Ingram, B., Johnson, C. K., Wilson, Q. S., DNA aptamer-based bioanalysis of IgE by fluorescence anisotropy. *Anal. Chem.* **2005**, *77*, 1963–1970.

59 Ikebukuro, K., Kiyohara, C., Sode, K. Novel electrochemical sensor system for protein using the aptamers in sandwich manner. *Biosens. Bioelectron.* **2005**, *20*, 2168–2172.

60 Potyrailo, R., Conrad, R., Ellington, A., Hieftje, G., Adapting selected nucleic acid ligands (aptamers) to biosensors. *Anal. Chem.* **1998**, *703*, 419–3425.

61 Ziegler, C. Cantilever-based biosensors. *Anal. Bioanal. Chem.* **2004**, *379*, 946–959.

62 http://www.concentris.ch.

63 http://www.cantion.com.

64 http://www.veeco.com.

65 http://www.protiveris.com.

11

Fullerene-based Devices for Biological Applications

Ginka H. Sarova, Tatiana Da Ros, and Dirk M. Guldi

11.1
Introduction

The great excitement commencing with the advent of fullerenes and their large-scale production was the inception of a wealth of scientific projects focused on two major disciplines – material chemistry and biomedical applications. Research and development of biomedical applications was, however, penalized by the necessity to work with soluble and biocompatible materials. Still, the outstanding physicochemical features of fullerenes, in general, and [60]fullerene, in particular, together with their unique shape, renders this class of carbon nanostructures as a candidate *par excellence* for biological applications. The ability, for example, to uptake electrons causes fullerenes to act as a very appropriate radical sponge, which is unequivocally reflected in their neuroprotective action. Photoexcitation of fullerenes, however, produces selectively reactive oxygen species (ROS) that are known to be cytotoxic. In this chapter we report on the solubility, toxicity and major biological applications of C_{60}, considering both milestone achievements and the most recent advances in these intriguing areas.

11.2
Solubility

Considering the unique and hydrophobic structure of fullerenes, it is not surprising that C_{60} renders absolutely insoluble in polar solvent. This aspect is, however, detrimental for developing biological and medical applications, and it consequently slowed down research activities in this intriguing field. To overcome this limitation, different approaches have been used – suspension in co-solvents, encapsulation into water-soluble hosts and chemical functionalization were mainly pursued as potent alternatives. In a number of instances the last two methods have been combined through the preparation, for example, of covalently linked cyclodextrins (CDs), which play a double role, i.e. providing solubilizing features and also physically entrapping the carbon cage.

Nanotechnologies for the Life Sciences Vol. 4
Nanodevices for the Life Sciences. Edited by Challa S. S. R. Kumar
Copyright © 2006 WILEY-VCH Verlag GmbH & Co. KGaA, Weinheim
ISBN: 3-527-31384-2

Suspensions of fullerenes are achieved by dissolving them first in a nonpolar organic solvent, such as benzene, adding medium polar tetrahydrofuran (THF) and then acetone. In the last steps, the addition of water to the mixture and the subsequent evaporation of the organic solvents are performed to produce yellow stable suspensions [1, 2]. Notably, such suspensions are composed of fullerene aggregates of variable sizes.

The encapsulation and/or the association of fullerenes into CDs or calixarenes eventually led to the solubilization of fullerenes in aqueous media [3–12]. In this context, the complex formation, for example, between sulfonated thiacalix[4]arene and calix[6]arene and fullerene has been reported [13]. In this case, the stoichiometry between C_{60} and complexing molecules depends mainly on the dimension of the latter, and it is 1:1 in the case of calix[6]arene and 1:2 for calix[4]arene.

Poly(vinyl pyrrolidone) (PVP), dimethyldioctadecylammonium bromide, Triton X-100, dihexadecyl hydrogen phosphate and lecithin also gave interesting results, leading to rather concentrated solutions of pristine fullerenes with concentrations as high as 10^{-5} M [14].

The covalent functionalization of fullerenes with hydrophilic groups is, however, by far the most powerful strategy to obtain water-soluble derivatives. Different functionalities, including ethylene glycol chains [15, 16], hydroxyl groups [17], carboxylic acids [18], ammonium groups [19] and CDs (Fig. 11.1) [7, 20–22], have been introduced onto the carbon cage in recent years with good results.

In the same manner, preparation of monosaccharide and oligosaccharide fullerene derivatives has been recently re-proposed [23–26]. Similarly, fullerene amino acids and fulleropeptides emerged as other important classes of soluble fullerene compounds that are abundantly reported [27]. Functionalization of the carbon cage has been performed by either direct attachment of amino acids [28] or by tra-

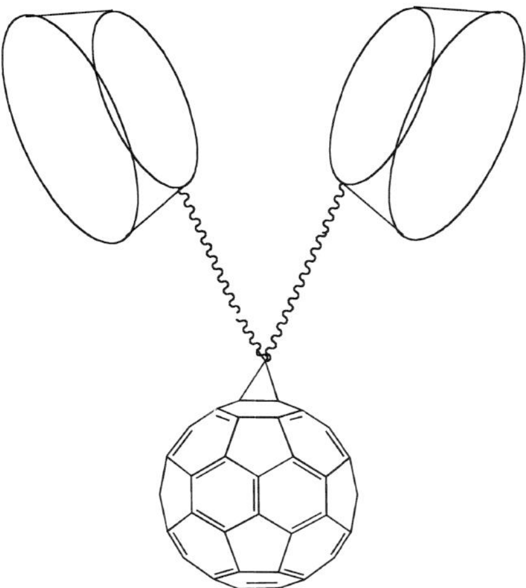

Figure 11.1. Example of covalent functionalization of fullerene with CDs.

ditional cycloaddition reactions that are followed by subsequent modification of the appendage [29–31].

A parallel development involves the synthesis of polyadducts to increase fullerene solubility. Nevertheless, the best results are currently achieved through the monofunctionalization of fullerenes – in the form of dendrofullerene **1** – that reaches water solubilities of the order of 34 or 254 mg mL^{-1} at pH 7.4 or 10, respectively [32].

1

11.3
Toxicity

The wide-ranging appeal of fullerenes has triggered concern about their potential toxicity. If they are employed widely, it is important that the toxicity of fullerenes be measured and counteracted, if possible. Of key consideration is fullerene solubility in water. Although they are inherently hydrophobic, their water solubility is essential for many emerging biomedical technologies.

Early work focused on studying the effects of unfunctionalized C_{60}, raw soot and fullerene black on bovine alveolar macrophage cells and macrophage-like cells. Raw soot and fullerene black are often the intermediaries and byproducts created during the bulk production of fullerenes in the laboratory. Enzyme tests for lysosomal damage and generation of ROS were undertaken as indicators for toxicity. Such tests demonstrate that C_{60} and the raw soot did induce a slight cytotoxic effect on alveolar and HL-60 cells after 48 h of incubation [33].

Functionalization of normally hydrophobic fullerenes is primarily done to render them hydrophilic. Cells, tissues and living systems all operate in a hydrophilic environment, and functionalization aids in the interaction between the fullerene and the biological system of interest. Their toxicity, both in tissue culture and *in vivo*, is an important characteristic for defining and constraining these applications. In fact, there have been a few papers on the influence of C_{60} and its derivatives on cell growth as well [1]. Surprisingly, no toxic effects of unfunctionalized C_{60} on biological samples were noted in these early assays.

Some water-soluble fullerene derivatives were found to cross the cellular membrane and localize in the mitochondria of HS 68 human fibroblast and COS-7 monkey kidney cells, but no toxicity studies of this derivative have been performed [34]. Carboxylic acid C_{60} derivatives inhibited excitotoxic cell death of mouse cortical neurons *in vitro*, and delayed the death and functional deterioration of a transgenic mouse containing the human superoxide dismutase (SOD) gene responsible for the neurodegenerative disease called familial amyotrophic lateral sclerosis (ALS; also known as Lou Gehrig's disease) [35].

When C_{60} was solubilized with PVP in water and the aqueous solution was applied to a mouse midbrain cell differentiation system, both cell differentiation and proliferation were potently inhibited [36, 37]. However, water-soluble fullerene carboxylic acids derived from C_{60} and C_{70}, which were examined for photocytotoxicity toward Raji cells (B lymphocytes), did not show any photocytotoxic effect even at 50 μM [38].

In some cases, the phototoxicity of fullerene molecules has been identified as a feature useful for therapeutics [39, 40]. Other work has sought to minimize the toxicity of water-soluble fullerenes so as to permit their use in drug-delivery applications.

Recent attention has been drawn to the environmental effects of nanoscale aggregates of C_{60}. This form results when pristine C_{60}, from either the solid state or organic solution, is placed into contact with neutral water [41]. Rather than completely precipitating, some C_{60} will form suspended and water-stable aggregates up to 100-p.p.m. concentrations. The environmental and biological significance of fullerenes in water was examined by comparing the cytotoxicity of several important types of water-soluble fullerenes using human liver carcinoma cells and dermal fibroblasts. It has been shown that nanoscale aggregates of C_{60} are cytotoxic to HDF and HepG2 cells at the 20-p.p.b. level [42].

In conclusion, fullerenes are toxic to human cells as well, but surface modifications can significantly reduce the effect. Fullerenes are toxic because they have a high affinity for electrons and in the presence of oxygen they can generate radical species that damage cell walls. Some studies even postulate that blockage of the cell cycle might be a mechanism of this activity [43]. The low-energy, unoccupied molecular orbitals that make it easy for fullerenes to accept electrons also make them electrically conductive. Such results could help pave the way for the general use of fullerenes.

11.4
DNA Photocleavage

DNA cleaving and lipid peroxidation activities of fullerenes have attracted considerable attention [44–46]. Importantly, C_{60}, if exposed to light, can either make singlet oxygen (type II energy transfer pathway) or be an electron donor to make superoxide radicals. For example, photoirradiation of C_{60} results in the formation of the singlet excited state $^{1}C_{60}{}^{*}$, which undergoes efficient intersystem crossing to give the triplet excited state $^{3}C_{60}{}^{*}$. $^{3}C_{60}{}^{*}$ reacts with nearly diffusion-controlled rate constants with molecular oxygen to yield singlet oxygen [47]. In principle, $^{1}O_{2}$ would then cleave DNA [48].

Possible involvement of $^{1}O_{2}$ was examined in some detail by comparing the reactivity of a fullerene–oligonucleotide linked system with a similarly linked eosin–oligonucleotide [49]. Fullerene–oligonucleotides that either bind single- or double-stranded DNA cleave the strand(s) proximal to the fullerene moiety upon exposure to light and oxygen. DNA damage occurs predominantly at guanine bases, without revealing a significant specificity between the various G sites [44]. Only when C_{60} was conjugated to an oligonucleotide was a good selectivity observed [50]. A key observation was that addition of a singlet oxygen quencher, sodium azide, largely inhibits the eosin–oligonucleotide cleavage, while no discernable effect was noted for the fullerene–oligonucleotide cleavage. A likely rationale suggests that the fullerene–oligonucleotide cleavage does not involve the singlet oxygen mechanism, but rather an alternative mechanism must be involved.

The alternative mechanism implies an electron transfer scenario – the type I electron transfer pathway. More precisely, the triplet excited state $^{3}C_{60}{}^{*}$ is subject to reactions with reductants to afford the radical anion of C_{60}. In this case, the rate constant for electron transfer quenching as a function of donor oxidation potential typically follows the Weller relationship [51]. Evidence for this pathway comes from early work by Foote and coworkers, who reported that $^{3}C_{60}{}^{*}$ directly oxidizes guanine in a DNA stack, because the oxidation potential of a guanosine derivative is located at 1.26 V, which is close to the reduction potential of $^{3}C_{60}{}^{*}$ (1.14 V). Conceptually similar is the sequence of reacting the radical anion of C_{60} with molecular oxygen to generate superoxide radicals and then indirectly cleaving DNA through reductive electron transfer. Such a reductive activation of molecular oxygen by photoexcited fullerenes was shown to be highly feasible under physiological conditions.

However, the poor solubility of fullerenes in water precluded in most instances the detailed mechanistic studies of C_{60} photosensitized DNA damage. However, of course, the binding to and recognition of DNA by synthetic organic compounds – fullerenes – is important.

Photoinduced DNA cleavage occurs efficiently by C_{60} in O_{2} saturated solutions containing NADH – the most important redox coenzyme acting as the source of electrons in the living system – through superoxide radicals. In fact, spectroscopic and kinetic studies using a laser flash photolysis technique enabled the detection of superoxide ($O_{2}{}^{-}$) through the use of a radical scavenger – 5-diethoxyphosphoryl-5-methyl-1-pyrroline *N*-oxide (DEPMPO) [52, 53].

Highly efficient DNA cleavage by small amounts of water-soluble poly(fullerocyclodextrin)s under visible light conditions and the photocleavage of DNA by pristine fullerene in mixed organic solvent system have also been observed [54]. The cleaving is followed by a strong interaction between cleaved DNA fragments and C_{60} leading to DNA–C_{60} conjugates in high yields. Similarly, water soluble C_{60}–

homooxacalix[3]arene complexes and lipid-membrane-incorporated C_{60} acted as efficient DNA photocleavage reagents [11, 55].

Several concepts were followed to enhance the fullerene DNA interactions. A promising strategy involves fullerene derivatives that bear DNA minor groove binders, such as triple-helix-forming oligonucleotides. The rational design of this synthesis is based on a reinforced effect due to the simultaneous presence of two different agents able to confer sequence selectivity [56]. This, for example, is expected to assist in targeting the fullerene moiety to a desired DNA sequence. Although the triplex formation was demonstrated, the presence of fullerene moieties gives rise to a high degree of instability. In this context, the approach to introduce nucleic acid-specific agents such as acridine [57] or netropsin [45] to understand the mechanism of action of these classes of conjugates and to increase both cytotoxicity and sequence selectivity was unsuccessful.

An alternative concept implies that DNA can be used as a framework for the assembly of fullerene materials – bearing cationic functionalities – through electrostatic interactions with the phosphate groups along the DNA backbone [58, 59]. This concept was driven by the finding that a simple carboxylic acid derivative does not bind to DNA and hence it is unlikely that DNA undergoes direct chemical reaction with the fullerene derivative. However, some mediating function – most likely due to the involvement of superoxide radical anions – helped in cleaving DNA fragments at guanine residues upon exposure to light [60].

11.4.1
Photodynamic Therapy (PDT)

Photosensitizers that are typically employed in PDT are aromatic molecules, able to efficiently form long-lived triplet excited states. The latter display the potential to generate ROS with high quantum yields. In addition, these compounds should possess low energy absorptions, low toxicity *in vivo* and high selectivity towards biological target in order to avoid side-effects.

Due to the small singlet-triplet energy gap and to the forbidden nature of the radiative $S_1 \rightarrow S_0$ transition, the dominant deactivation pathway of the fullerene photoexcited singlet state is $S_1 \rightarrow T_1$ intersystem crossing. The quantum yield of triplet formation is close to unity for both C_{60} and C_{70} [61, 62]. The triplet excited state with lifetimes in the range of hundreds of microseconds is photoactive and gives rise to the formation of cytotoxic species such as $O_2^{\cdot -}$, $\cdot OH$ and $H^{\cdot}$. For instance, quenching of the fullerene triplet by oxygen through triplet–triplet annihilation is close to diffusion control:

$$^3C_{60}{}^* + {}^3O_2 \rightarrow {}^1C_{60} + {}^1O_2{}^*$$

The fact that the reverse process, i.e. quenching of the singlet oxygen by ground state fullerene, proceeds at a rate much lower than diffusion control, together with the low fluorescence quantum yield of fullerenes, makes this class of compounds good photosensitizers with a great potential in PDT [61].

The main drawback for the application of fullerenes as photosensitizers *in vivo* is the lack of significant absorption at longer wavelengths. However, this disadvantage can be overcome when functional groups are appended that act as a "light harvesting" antenna. Indeed, this pathway was considered and reported first by Cheng and coworkers [63] and later by other authors [64].

To characterize fullerenes C_{60} and C_{70} as photosensitizers in biological systems, the generation of active oxygen species through energy and electron transfer was studied *in vitro* [60]. It was found that $^1O_2{}^*$ is generated effectively in nonpolar solvents, such as benzene, whereas in water, $O_2{}^{\cdot-}$ and $\cdot OH$ were produced instead, especially in the presence of a physiological concentration of reductants including NADH. These redox active species were shown to contribute to the photoinduced DNA cleavage under physiological conditions.

The effect of the fullerene core substitution on singlet oxygen formation has revealed interesting trends [65]. The efficiency of singlet oxygen production from a series of fullerene derivatives – epoxides and diethylmalonate derivatives – was evaluated by measuring the near-IR emission at 1268 nm, which corresponds to the O_2 $(^1\Delta_g) \rightarrow O_2$ $(^3\Sigma_g{}^-)$ transition. Overall, it was shown that functionalization of the fullerene core reduced the photodynamic activity. Fascinatingly, the effect was independent of the nature of the addend, but dependent on the number of addends. Higher degrees of functionalization, for example, lead to decreasing efficiencies of singlet oxygen production with greater effects imposed by adjacent addends compare to more remotely placed addends. Complementary studies showed that even multifunctionalized fullerene compounds, such as fullerol $C_{60}(OH)_n$ ($n > 18$) and fullerene-core star-like polymer, $C_{60}[>N\text{-}CH_2CH_2(OCH_2CH_2OCH_3]_6$, are efficient $^1\Delta_g$ generators *in vitro* [66]. Although the photophysical capabilities of fullerols have been shown to be reduced by hydroxylation, when compared to pristine C_{60}, sufficient photoactivity remains to trigger sizeable effects on aqueous systems [67]. Indeed, fullerol $C_{60}(OH)_{24}$ has been shown to produce a mixture of singlet oxygen and superoxide under both visible and UV irradiation.

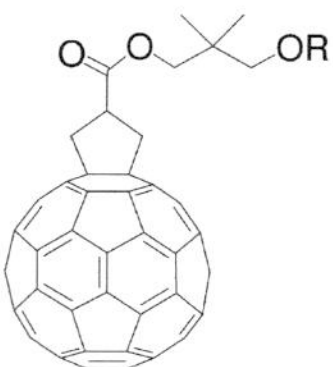

2

R = CO(CH$_2$)$_2$CO$_2$H (a)

R = H (b)

R = CO(CH$_2$)$_2$CO$_2$H.NEt$_3$ (c)

Fullerenes **2a** and **2c**, incubated with HeLa S3 cells, were shown to be cytotoxic upon irradiation [44]. Cells in the presence of fullerenes were not affected when not exposed to light. Compound **2b** was found to be noncytotoxic even when irradiated, possibly due to its lower water solubility compared to **2a** and **2c**. A reference compound lacking the fullerene moiety was inactive when incubated with cells and exposed to light. These interesting results were explained by the formation of singlet oxygen through fullerenes.

Li and coworkers reported a 2-fold proliferation of human cervix cancer cells incubated with liposome-encapsulated C_{60} and irradiated for a short period of time relative to unirradiated cells [68]. The authors concluded that the cancer cells were likely to be killed only if the laser power was above a certain threshold during the irradiation time, otherwise it would have promoted cell growth.

C_{60} derivatives bearing poly(ethylene glycol) (PEG) chains were found to be cytotoxic to L929 cells, but only when photoirradiated with visible light – measurements in the dark disclosed no appreciable impact [69]. The authors considered the generation of singlet oxygen as the critical step. Moreover, cytochrome *c* was reduced when irradiated in the presence of the fullerene derivative and addition of SOD suppressed the cytochrome *c* reduction. However, when SOD, cytochrome *c* and the C_{60} derivative were administered all together to the cells, no change in cytotoxicity was noted that would have been caused by adding SOD.

$[NH-CH_2-CH_2-O-(CH_2-CH_2-O)_n-CH_2-CH_2-OH]_m$

3

PEG-modified C_{60} fullerene **3** was shown to accumulate in tumor tissues to a greater extent than in normal tissues upon intravenous injection into tumor-bearing mice, exhibiting prolonged C_{60} retention at the tumor tissue [70, 71]. Light irradiation that was applied after the injection of **3** significantly suppressed the volume increase of the tumor. The compound was basically released from the mice within 1 week.

Derivative **4** was not active at all up to concentrations of 50 µM, whereas at higher concentrations the cell growth was inhibited – even in the absence of light [38]. The light-triggered oxidative properties of singlet oxygen were considered responsible for a series of biological activities exhibited by fullerene solutions.

4

In the search of new photosensitizers for PDT, the attachment of sugar moieties to the fullerene chromophores attracted considerable attention, since carbohydrates play, in general, an important role in cell–cell interactions [72, 73]. Sugar-pendant monofunctionalized and bisfunctionalized fullerene derivatives were recently synthesized starting with carbohydrate-linked azides, and their photocytotoxicity against HeLa cells was studied [74]. The photosensitizing ability of the sugar-pendant derivatives to produce singlet oxygen in dimethylsulfoxide solution was demonstrated by the direct observation of the emission due to the $O_2\,(^1\Delta_g) \rightarrow O_2$ $(^3\Sigma_g^-)$ transition at 1268 nm. In agreement with the established trend that functionalization of fullerenes reduces the efficiency of singlet oxygen production [65], mono-adducts were shown to produce singlet oxygen with higher efficiencies than the corresponding bis-adducts. The cells were incubated for 12 h prior to the addition of photosensitizer and further incubated for an additional 6 h. After washing and irradiating with 10 J cm^{-2} of laser energy, cells were further incubated for 24 h before the numbers of living cells were counted by the MTT [3-(4,5-dimethylthiazol-2-yl)-2,5-diphenyltetrazolium bromide] assay. Overall, the dark cytotoxicity of all investigated compounds was found to be quite small, while significant phototoxicities were observed for D-glucose, D-mannose and D-galactose derivatives bearing a single carbohydrate unit. However, D-xylose and maltose derivatives were less toxic despite the fact that comparable amounts of singlet oxygen were produced *in vitro*. Also, under the chosen experimental conditions the bis-adducts failed to display any meaningful phototoxicity, since they gave rise to poorer performance in terms of generating singlet oxygen than the analogous mono-adducts.

Recently, two new fullerene-bis-pyropheophorbide *a* derivatives, i.e. a mono-adduct **5** and a hexakis-adduct **6**, were synthesized and tested *in vitro* with regard to their intracellular uptake and photosensitizing activity towards human leukemia T lymphocytes (Jurkat cells) [75, 76].

5

The uptake of **5**, **6** and the reference compounds **7–9** by Jurkat cells was investigated with confocal laser scanning microscopy and by measuring the fluorescence intensity of cell extracts at the emission wavelength of pyropheophorbide *a*. The intracellular concentrations of fullerene complexes and **9** were 27 times lower than that of the free sensitizers after 24 h of incubation. They also showed slower accumulation in cells. The authors rationalized these results on the basis of different uptake mechanisms – lipophilic molecules with molecular weights lower than 100 Da normally diffuse through the membranes, while bigger molecules, such as fullerene–sensitizer complexes **5** and **6**, and **9** can be taken up only by endocytosis or pinocytosis, which have slower kinetics than passive diffusion through cell membranes.

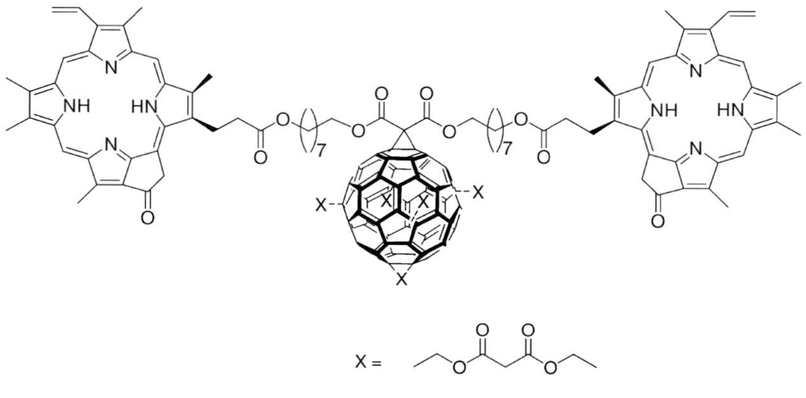

6

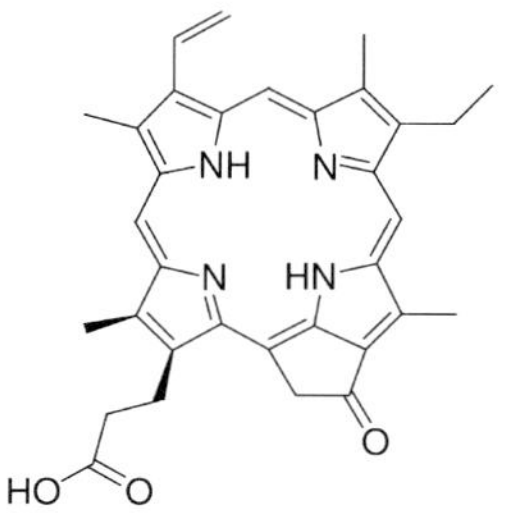

7

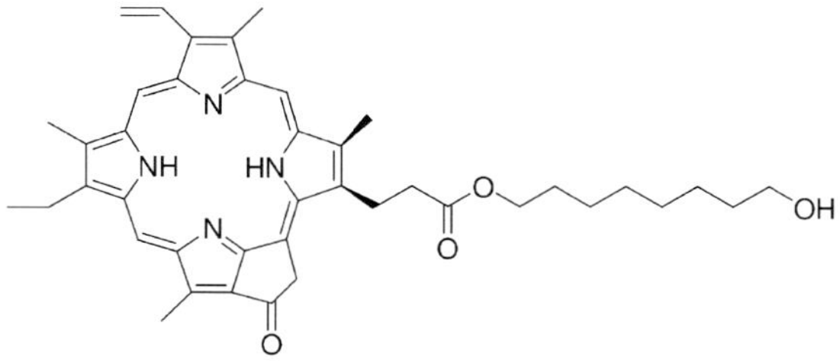

8

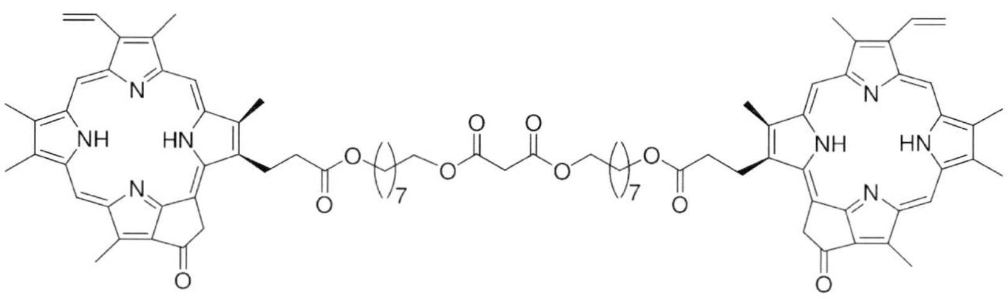

9

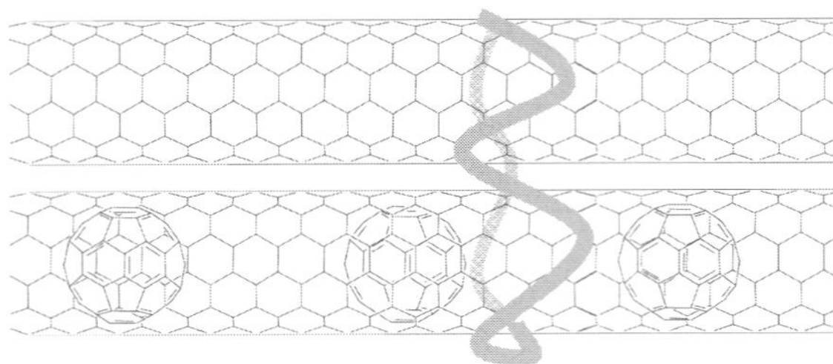

Figure 11.2. Hybrid nanotube rope controlling protein activity *in vivo* by near-IR radiation.

The phototoxicity of fullerene–sensitizer complexes was tested in comparison with the fullerene-free sensitizers. The rates of necrotic and apoptotic cells were determined 24 h after irradiation with a laser diode (688 nm, 2.12 mW cm^{-2}). The fullerene complexes were found to be less phototoxic compared to the fullerene-free sensitizers **7** and **8**. This is mainly due to the high molecular weight of the fullerene complexes, leading to lower intracellular concentration.

The hexakis-adduct **6** showed significant phototoxic activity (58% dead cells after a dose of 400 mJ cm^{-2}, 688 nm), while the mono-adduct **5** showed very low toxicity even at higher doses of irradiation. The latter was attributed to a low intracellular uptake for **5** and to an efficient electron transfer process from the pyropheophorbide singlet state to the fullerene moiety, resulting in low yield of the intersystem crossing (ISC) and low quantum yield of singlet oxygen formation.

No dark cytotoxicity was found towards Jurkat cells after 24 and 48 h of incubation with all studied sensitizers.

Recently, it has been demonstrated that chemical reactions of importance for PDT could be directly activated by a hybrid carbon nanotube (CNT) rope consisting of two adjacent metallic CNTs, where one is filled, i.e. the peapod, and the other is empty [77]. Under electric bias, a substantial charge transfer from the entrapped metal ion to the fullerene cage in metallofullerenes occurs. In the CNT rope this process also involves the peapod and the twin (empty) CNT, thus the two CNTs become oppositely charged. This process can be partly inverted at elevated temperatures, since the metallofullerenes have levels close to the Fermi levels. In this way, by heating with near-IR excitation, recharging of the CNTs is expected that would result in a change of the local electric field and further to a deformation of proteins selectively attached to the CNT (Fig. 11.2).

11.4.2
Fullerene-mediated Electron Transfer Across Membranes

Both C$_{60}$ and C$_{70}$ embedded within a lipid bilayer act as efficient electron acceptors at interfaces [78, 79]. Moreover, fullerenes can efficiently transport negative charges across membranes [80]. Mauzerall and coworkers have observed that transmembrane electron transfer takes place via electronic conduction mediated by fullerene aggregates [81]. The possible role of small fullerene aggregates, which act as an antenna for collecting photons, in the presence of monomers was also discussed by Seta and coworkers [82]. In a study of the photoconductivity of ultra-thin bilayer lipid membranes doped with C_{60}, the authors suggested that since the extinction coefficient of the aggregated form is higher than that of the monomeric form, the absorption spectrum is dominated by the contribution of the aggregates. Thus, the energy absorbed by the aggregates is transferred to the monomers via singlet–singlet energy transfer and the excited monomers are then reduced by the donor species at the membrane interface. The reduction can take place either in the singlet excited state or in the triplet excited state of the fullerene. An alternative mechanism could be exciton-induced charge injection at the interface of the C_{60} aggregates with the bilayer membrane.

Photoinduced charge transfer across membranes has also been achieved in the composite assembly of C_{60} and CdS nanoparticles prepared by the Langmuir–Blodgett technique [83]. The CdS nanoparticles exciton emission was quenched due to efficient electron transfer from CdS nanoparticles to C_{60} across a lipid layer (2–3 nm). The fatty layer between C_{60} and CdS played an important role in preventing charge recombination in the composite assembly.

Amino acid derivatives of C_{60} were shown to penetrate through the lipid bilayer of liposomes without destroying the membrane integrity [84]. The L-isomer was able to diffuse through the phospholipid membrane into the liposome interior, whereas the D-isomer localized in the region of the outer membrane surface. These derivatives were also able to carry bivalent metal ions through phospholipid bilayers as a result of the formation of complexes. The study of the effects of the two stereoisomers on lipid peroxidation in mitochondria of rat cortex brain showed that C_{60}-L-Arg exhibits a prolonged inhibition in the malonaldehyde accumulation, whereas C_{60}-D-Arg led to no discernable effects on malonaldehyde accumulation. This data showed that amino acid derivatives of C_{60} affect membrane-bound enzymes. In particular, the L-isomer stimulates the catalytic activity of monoaminoxidase A (MAO) A, while the main action of the D-isomer is an increase in the catalytic activity of MAO B. It is worth pointing out that MAO is a known redox enzyme. This led to the postulate that the observed effect of the amino acid derivatives of C_{60} on the catalytic activity of MAO A and MAO B is due to an intra-protein electron transfer process.

10   **11**

Recently, fullerene derivatives **10** and **11** were studied as counteranions in the activation of oligo/polyarginine–anion complexes in living cells, and, thus, to modulate cellular uptake and anion carrier activity of oligo/polyarginines [85, 86]. The fact that cell-penetrating peptides (CPPs) are anion carriers and that bilayer penetration is regulated by amphiphilic anions suggested that the carrier activity of oligo/polyarginines occurs via counterion exchange, which facilitates the reversible adaptation of their solubility to a changing environment. An efficient counteranion activator should, therefore, have one or more negative charges for ion pairing with guanidinium cations of the oligo/polyarginine. In addition, an amphiphilic character with large hydrophobic domains and an aromatic surface is expected to favor interactions with bilayer membranes. R-rich CPPs usually exist as complexes $R_n(X)_m$ (X = hydrophilic counteranion) to minimize intramolecular charge repulsion and the exchange of the scavenged X^- with an activator Y^- yields the active complex $R_n(X)_{m'}(Y)_{m''}$ at the membrane water interface. This active complex can then shuttle across the bilayer. In this respect, the activator Y should have also the ability to form stable, but labile, guanidinium anion complexes.

The key finding in the aforementioned study was that the efficiency of counteranion activators significantly depends on activator membrane and activator carrier interactions. Specifically, the activator efficiency was found to increase with increasing aromatic surface of the activator, decreasing size of the transported anion, increasing carrier concentration as well as increasing membrane fluidity. Fullerenes **10** and **11** were the most efficient counteranions in egg yolk phosphatidylcholine, where compound **10** showed an extraordinary EC_{50} value. Bellow the binding of one activator per carrier, fullerene activators acted at catalytic concentrations [86]. At high concentrations, fullerenes **10** and **11** became inhibitors of CPP uptake. However, in sol-phase dipalmitoyl phosphatidylcholine (DPPC) and in gel-phase DPPC, as well as in HeLa cells, the efficiency of the planar activators (pyrene and coronene) exceeded that of spherical activators (fullerenes and calixarenes). These results showed that spherical activators, particularly fullerene **10**, may specifically mediate CPP uptake.

The transmembrane extraction of fullerenes C_{60} and C_{70} across a membrane having γ-CDs as molecular recognition sites attached to a poly(vinyl alcohol) matrix was achieved [87]. C_{60} molecules were transported from a feed aqueous phase, in which fullerene was rendered water soluble in the form of γ-CD:C_{60} complexes, to a stripping organic phase such as toluene, xylene or tetralin. The high affinity of the fullerene C_{60} towards γ-CD allows their interfacial exchange and penetration into the membrane core. The loading of the membrane was confirmed by the observed color change of the membrane to magenta. During the first 20 h of the experiment the flux of permanent C_{60} rapidly reached a constant value. However, the membrane was far from saturation as the majority of the γ-CD cavities were unoccupied. It was suggested that steric hindrance due to crosslinking prevents the formation of the inclusion complex and the occurrence of a complex involving more than two γ-CDs due to the high local concentration of these moieties is possible. Moreover, small aggregates of fullerenes, which are surrounded by several CD cavities, were observed in the membrane. In this way the membrane acted as a reservoir of C_{60} molecules that were afterwards slowly released into the receiving solution. The rate-limiting step of the membrane transport was demonstrated to be governed by the dissociation of the inclusion complex at the stripping interface. This led to a low overall membrane transport flux.

The good electron acceptor properties of fullerenes stimulated studies on their possible applications as biosensors. A novel biosensor for the amperometric detection of glutathione was obtained by co-immobilizing a redox enzyme with a redox mediator (fullerene derivative **12**) realized at the glassy carbon electrode by the use of an amphiphilic pyrrole derivative **13** ($n = 2$) [88]. The reversible reduction of **12**, which is entrapped within the poly-**13** film, was deduced from the coefficient of variation curves and the electron transfer was likely to occur via a vectorial charge-hopping mechanism involving neighboring fullerene molecules that are coupled to the diffusion of the counterions. The poly-**13**/**12**/GR film was investigated for its electrocatalytic properties. The biosensor showed a fast and reproducible response to glutathione.

$$H_3C\overset{\oplus}{N}CH_2CH_2OCH_2CH_2OCH_2CH_2OCH_3$$

$$2I^-$$

$$N{-}(CH_2)_{12}{-}N(C_nH_{2n+1})_3{}^+BF_4{}^-$$

13

$$H_2C\overset{\oplus}{N}CH_2CH_2OCH_2CH_2OCH_2CH_2OCH_3$$

12

A C_{60}-containing lipid bilayer membrane was shown to function as a light-sensitive diode and to be useful for electrochemical biosensor electronic device developments [89].

The potential of glycosylated fullerene layers as biosensors for glucoproteins was reported [90]. The amphiphilic C_{60} dendrimer conjugates **14** and **15** with one or two glucodendron headgroups form stable, ordered monomolecular Langmuir layers at the air/water interface, which were further transferred onto quartz slides as X-type Langmuir–Blodgett films. The bulky glucodendron headgroups were very effective in suppressing fullerene aggregation.

14

15

11.4.3
Neuroprotective Activity via Radical Scavenging

Excess production of superoxide and/or nitric oxide radicals, as well as nonradical H_2O_2 and hypochloric acid, is considered as one of the primary initiators of neurodegenerative diseases. Endogenous cellular antioxidants, vitamin E analogs and SOD enzymes, have evolved in organisms – from bacteria to mammals – to inactivate low levels of free radicals produced under regular metabolic conditions. However, their antioxidant defense may be overwhelmed by ROS in the cells that are generated in the pathogenesis of a number of neuronal injuries and, as a result, they are no longer able to achieve meaningful therapeutic results.

In this respect, compounds that act as free radical scavengers were shown to reduce neuronal death. Fullerenes and related derivatives reveal promising antioxidant properties, i.e. their ability to react with multiple radical species at diffusion-controlled rates [91]. To this end, several C_{60} derivatives, including carboxyfullerenes [18, 35, 92–112], polyhydroxyfullerenes [113–116] and a limited number of other derivatives [19, 107, 117–120] have been investigated as neuroprotective agents.

In particular, two different tris-adducts, C_3 (**16**) and D_3 (**17**), of $C_{60} \cdot [C(COOH)_2]_3$, have been confirmed to be the most promising compounds for preventing neuronal damage [35, 92]. The presence of six carboxylic functional groups improves the solubility in biological media and avoids the fullerene aggregation that is commonly encountered in polar solvents. As a result of an improved intercalation into biological membranes, the C_3 isomer has been studied in more detail. These studies demonstrated that the C_3 isomer is an excellent radical scavenger, which reacts both with hydroxyl radicals, $\cdot$OH, and H_2O_2 at micromolar concentrations [35, 93].

Dugan and coworkers also evaluated the ability of C_3 to eliminate superoxide generation in intact cells [94]. Using confocal microscopy and a superoxide-sensitive fluorescent compound, dihydroethidium, they determined that this isomer can reduce basal mitochondria production of superoxide in cortical astrocytes and neurons. In addition, C_3 blocks iron-induced lipid peroxidation *in vitro* [92] and *in vivo* [95]. The free radical nitric oxide, produced by three different synthase isoforms and responsible for the generation of citrulline, is also an important effector molecule in immune and cardiovascular system.

Although Dugan and coworkers have found that C_3 is unreactive with NO˙ itself, in another study [96] carboxyfullerenes **16** and **17** were shown to exert some inhibition on the three synthases. This work also substantiated a reduction in the level of nitric oxide production and the rate of citrulline formation. However, this effect was attributed to interactions of the fullerene molecules with the enzyme and not to their properties as radical scavengers. The same authors also reported that these compounds inhibit the Arg-independent nicotinamide adenine dinucleotide phosphate (NADPH) oxidase activity of one of the synthase isoforms without affecting its catalyzed cytochrome *c* reductive activity.

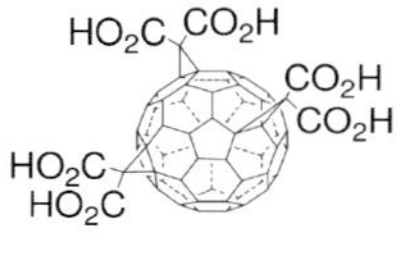

16 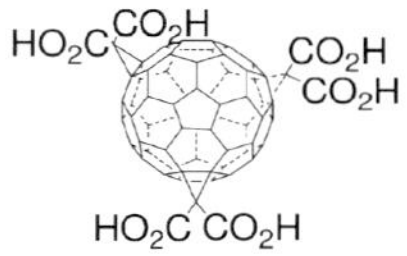 **17**

To explore the neuroprotective and cytoprotective properties of the malonic acid tris-adducts, Dugan and coworkers prepared neocortical cell cultures from fetal (E15) Swiss-Webster mice (Simonson). These were subjected to tests with different antioxidants after carrying out a brief exposure to *N*-methyl-D-aspartate (NMDA) [97]. Both the C_3 and D_3 isomers produced a dose-dependent decrease of the excitotoxic death of cultured cortical neurons. C_3 fully blocked NMDA receptor-mediated toxicity at 100 μM concentration, showing greater protection and efficacy against acute toxicity initiated by brief (10 min) exposure to high-dose NMDA. This isomer also provided protection against AMPA (α-amino-3-hydroxy-5-methyl-4-isoxazolepropionic acid) receptor-mediated injury in neurons [35] and oligodendroglia [98], the cell type responsible for central myelin formation. The authors stressed that the observed behavior not only reflects the antioxidant properties of the compound, but other features of the molecule, e.g. its amphiphilic nature.

Moreover, the carboxyfullerene **16** limited apoptotic neuronal death produced by several insults, including serum deprivation and exposure to the Alzheimer's peptide $A\beta_{1-42}$ [18], and has shown robust neuroprotection in a number of other cell culture models of neurological disease including Parkinson's disease [35].

In a study of neurotoxin-mediated death of dopamine neurons, the same authors have pursued the concept that carboxyfullerene derivatives could block the toxic effect of the neurotoxins 6-OHDA (6-hydroxydopamine) and MPP^+ (1-methyl-4-phenylpyridinium). 6-OHDA and MPP^+ are widely used to generate animal models of Parkinson's disease, since after administering *in vivo* they cause Parkinsonian conditions. Parkinsonian conditions are marked by decreased dopamine levels and tyrosine hydroxylase activity, impaired dopamine uptake, and an ensuing loss of dopaminergic neurons. With regard to oxidative injuries, the C_3 isomer dramatically rescued dopaminergic neurons from 6-OHDA-induced cell death in a dose-dependent manner (92% recovery at the highest dose), whereas it quickly plateaued in the case of MPP$^+$ (37.5% recovery). Concentrations of C_3 in excess of 100 μM were found to be cytotoxic [99]. The D_3 isomer was also found to be protective against 6-OHDA injury and MPP$^+$-mediated death, but to a lesser extent than the C_3 isomer. The C_3 isomer proved more effective than the dopaminergic neuroprotectant glial cell line-derived neurotrophic factor (GDNF) in rescuing dopamine neurons from 6-OHDA- or MPP$^+$-mediated cell death.

A parallel assay that focused on utilizing the two isomers, C_3 and D_3, of tris-adducts, among other antioxidant agents, showed that these fullerene derivatives inhibited apoptosis of human hepatoma HEP-3B cells, induced by transforming growth factor (TGF)-β [100]. The mechanism of action is supposed to be indirect, because all other actions induced by TGF-β remained, surprisingly, unaltered.

The same carboxyfullerenes have been demonstrated to protect human keratinocytes from apoptosis induced both by UV-B irradiation and exposure to deoxy-D-ribose [101]. However, **16** and **17** failed in preventing the downregulation by UV-B light of the Bcl-2 levels. This suggested that these compounds protect human keratinocytes from UV-B damage via generation of ROS from depolarized mitochondria without the possible involvement of Bcl-2.

The same group of authors reported the protective activity of carboxyfullerenes against oxidative stress-induced apoptosis in human peripheral blood mononuclear cells [102]. Two models of apoptosis were used – one, induced by 2-deoxy-D-ribose, a reducing sugar capable to induce oxidative stress and glutathione depletion, and second, induced by tumor necrosis factor (TNF)-α plus cycloheximide. In both models, the studied fullerene derivatives were able to inhibit apoptosis, likely acting as an antioxidizing drug and partially preventing the depolarization of mitochondrial membrane potential as in the case of keratinocyte cells [101].

In another study, the C_3 isomer was shown to inhibit lymphotoxin-β receptor-mediated apoptosis signal-regulating kinase 1 activation in Hep3BT2 cells treated with LIGHT-R228E [103].

In recent work, Prato and coworkers synthesized and investigated a water-soluble fullerene derivative **18** that bears three ethylene glycol chains and three ammonium groups for increasing the water solubility as a potential neuroprotecting agent [19]. This compound was designed as a monofunctionalized derivative in order to resemble the free radical scavenging ability of pristine C_{60}, as it is known that higher numbers of addends on the fullerene moiety decreases its radical scavenging features [91]. As observed by UV-vis spectroscopy, this mono-adduct showed aggregation at concentrations higher than 10^{-5} M not only in water, but also in water/ethanol and water/dimethylformamide mixtures. In a series of spectrophotometric assays employing the xanthine/xanthine oxidase system for the generation of superoxide radicals and ferricytochrome c reduction for measuring superoxide concentration, this new fulleropyrrolidine derivative did not show a significant re-

18

action with $O_2^{\cdot-}$. In a neuroprotection model, in cerebral cortical cell death induced by glutamate, the compound was not only found to be ineffective, but also showed a significant concentration-dependent toxicity. The authors explained this undesirable toxicity by the lipophilic character of the derivative, which coupled with its hydrophilic part confers surfactant properties favoring its interaction with cell membranes, with their possible disruption and subsequent cell death. An important conclusion from this work is that the neuroprotective activity of C_{60} derivatives correlates more with the number of substitutions than with their nature. Increased number of polar groups on C_{60} probably prevents its interaction with the cell membrane and, consequently, effects its disruption.

Although *in vivo* studies bear great importance for understanding how well the neuroprotective activity of novel antioxidant agents – observed in cell cultures – translates to an intact organism, only a limited number of *in vivo* studies have been published in this field so far [35, 92, 94–95, 109]. Continuous intraperitoneal infusion of **16** into a transgenic mouse model of ALS, carrying the human mutant (*G93A*) SOD gene responsible for a form of familial ALS, delayed both death and functional deterioration [35]. The initial *in vivo* studies of two models of Parkinson's disease have shown that **16** when co-injected with iron, intrastriatal injected to produce striatal injury, reduced dopamine depletion [95]. A first proposed rationale prompts the ability of **16** – once injected – to reduce injury, reflecting its interaction either directly with iron or direct involvement in lipid peroxidation. A second model, in which **16** was delivered systemically for 1 month to rats intrastriatally injected with 6-OHDA to provoke Parkinsonian conditions, indicated that dopaminergic terminals and behavior were significantly improved by C_3 treatment [94]. Due to their antioxidizing action, C_3 tris-adducts also prevent iron-induced stress in rat brain, and lipid peroxidation induced by superoxide and hydroxyl radicals [95, 109].

Towards the goal of reducing the oxidative stress in brain damage, the neuroprotective effect of carboxyfullerenes on transient focal ischemia reperfusion injury was recently studied in rat brains [95, 104]. Carboxyfullerene was administered either intravenously or intracerebroventricularly to chloral hydrate-anesthetized Sprague-Dawley rats 30 min prior to transient ischemia reperfusion. The data showed that intravenous injection did not inhibit the reperfusion injury. This is likely due to the limited permeability of this compound to the blood–brain barrier. In contrast, pre-treatment with local carboxyfullerene-attenuated cortical infarction prevented both elevated lipid peroxidation and depleted glutathione (GHS) levels in the infarcted cortex induced by transient ischemia reperfusion. Adverse behavior changes were simultaneously observed in rats receiving intracerebroventricular infusion of carboxyfullerene, including writhing accompanied by trunk stretching and even death.

Intravenous administration of hexasulfobutylated C_{60} (FC4S) has been found to reduce the total volume of infarction produced by transient ischemia reperfusion in rat brain [117].

The redox properties of carboxyfullerenes correlate with another interesting activity exhibited by this class of compounds – they were shown to inhibit bacterial meningitis [105, 106]. The effect of the C_3 isomer on the regulation of brain inflammatory responses after intracerebral injection of *Escherichia coli* in B6 mice that induced TNF-α and interleukin (IL)-1β production and recruited neutrophil infiltration at 6–9 h post-injection was studied [106]. Pre-treatment of each mouse with C_3 protected 40% of mice in a dose-dependent manner. Fullerene-treated showed less TNF-α and IL-1β production compared with the levels of production for non-treated mice. TNF-α and IL-1β production is typically detected in the cerebrospinal fluid of patients infected with bacterial meningitis and in experimental animals. The authors conclude their work by ascribing the inhibition of *E. coli*-induced meningitis not due to direct antimicrobial activity, since C_3 did not exhibit the growth of *E. coli* in LB broth culture. Moreover, the *E. coli* cells were cleared from the brain after 24 h in fullerene-treated mice. At the same time they were found to replicate significantly in untreated mice. This leads to the suggestion that C_3 might have enhanced the natural antibacterial defense in the brain. A series of similar fullerene mono- and bis-adducts was also able to suppress *E. coli* growth [105, 107].

Some cationic fullerene derivatives were found to exhibit bacteriostatic effects [118, 107]. C_{60}-bis(N,N-dimethylpyrrolidinium iodide), for example, in low concentrations inhibits *E. coli* growth and dioxygen uptake caused by *E. coli* and glucose [119], whereas at high concentrations dioxygen was consumed and converted to H_2O_2.

Derivative **16** was tested *in vitro* against apoptotic neuronal death in rat cerebellar granule cells [108]. Cerebellar granule cells represent one of the best *in vitro* models of neuronal apoptosis, both for the mitochondria and the nucleus, which is strictly related to the generation of ROS.

Dugan and coworkers studied the effect of polyhydroxy fullerols $C_{60}(OH)_n$ ($n = 6$–15, 24) on apoptotic death of culture neurons and found that these compounds reduced excitotoxic and apoptotic death of cultured neurons by 80% following NMDA treatment [113].

Polyhydroxylated fullerols also display excellent efficiency in quenching superoxide radicals ($O_2{}^{\cdot-}$) generated by xanthine/xanthine oxidase [114]. The same fullerols $C_{60}(OH)_n$ ($n = 6$–15) were shown to attenuate bronchoconstriction induced by the xanthine/xanthine oxidase system in guinea pigs [115]. The intratracheal instillation of xanthine/xanthine oxidase caused a marked decrease in dynamic respiratory compliance, which was significantly reduced by these fullerols [116].

Whereas carboxyfullerenes were indicated as effective neuroprotection against excitotoxic cell death, apoptosis initiated by several different types of triggers and metabolic insults in cultured cells, the molecular mechanism behind the antioxidant reactions of C_{60} compounds, particularly the mechanism of $O_2^{\cdot-}$ removal by malonic acid C_{60} derivative, remains controversial. Krusic and coworkers first suggested that direct reactions between radical species and the highly conjugated double-bond system of C_{60} were responsible for the antioxidant action of carboxyfullerenes and attributed this property to their delocalized π system [91]. The spherical structure of C_{60} was also believed to increase the reactivity towards free radicals compare to that of the planar aromatic and polyene compounds. The study by Bensasson and coworkers [110], however, failed to detect a C_{60} radical intermediate in the reaction of C_3 and several other malonic acid derivatives with $O_2^{\cdot-}$. In addition, no correlation was found between the IC_{50} values and the reduction potentials of various carboxyfullerenes [111]. A recent study of SOD mimetic properties of **16** has made a step forward showing that the reaction between **16** and the superoxide is not via stoichiometric scavenging, where $O_2^{\cdot-}$ donates an electron to C_{60}, but through catalytic dismutation of superoxide [112]. The catalytic mechanism was indicated by the time invariability of the concentration and structure of **16** during the exposure to a continuous flux of $O_2^{\cdot-}$, produced by xanthine/ hypoxanthine metabolism, regeneration of oxygen, production of hydrogen peroxide, as well as absence of paramagnetic intermediate products, detectable by electron paramagnetic resonance (EPR) spectroscopy. In particular, a mechanism has been proposed in which electron-deficient areas on the C_{60} sphere work in concert with the malonyl addends to electrostatically guide and stabilize superoxide, promoting dismutation. The letter process occurs at a rate of 2×10^6 M^{-1} s^{-1}, approximately 100-fold slower than what is known for SOD, but within the range of several biologically effective, metal-containing SOD mimetics. To determine whether **16** is capable of acting as a SOD mimetic *in vivo*, the authors tested its efficacy in $Sod2^{+/-}$ mice, which lack expression of mitochondrial manganese SOD (MnSOD) and found that treatment with this compound increased the lifespan of mice by 300%. These data, coupled with the evidence that **16** localizes to mitochondria [34], suggested that the C_3 isomer functionally replaces MnSOD, acting as a biologically effective SOD mimetic.

11.4.4
Enzyme Inhibition and Antiviral Activity

The inhibition of enzymes by fullerene derivatives became one of the most important fields of biological application of fullerene since the publication of milestone works in 1993 [44, 121]. Tokuyama and coworkers reported the inhibition of papain, cathepsin, trypsin, plasmin and trombine by derivative **3**, while Wudl's group described the inhibition of HIV protease (HIV-P) by compound **19**.

19

The HIV-P is one of the key enzymes in HIV replication. This protease bears a hydrophobic cavity, whose diameter is about 10 Å, in which two catalytic aspartic acid residues are present. Considering the dimensions of C_{60}, this fullerene is a perfect match for the cavity size of HIV-P. Furthermore, the presence of C_{60} might prevent interactions between the catalytic portions of HIV-P and any virus substrates. The inhibition constant (K_i) of compound **19** is 5.3 μM *in vitro*, but fullerene derivatives with positive charges, able to interact with the aspartic negative residues, are expected to raise the binding constant up to 1000 times and consequently dramatically decrease the K_i [121].

Although current therapy against HIV infection utilizes HIV-P inhibitors active at nanomolar and subnanomolar concentrations, the potential use of fullerene derivatives can be taken into account considering the common problem of resistance versus the used drugs – a phenomenon partially caused by the tendency of the virus to present frequent genetic mutation.

A selection of fullerene derivatives was studied by Schuster and coworkers as potential inhibitors of HIV on human peripheral blood mononuclear cells infected with HIV-1 and three of these compounds presented an EC_{50} in the range of 0.9–2.9 μM [122]. In the same way, a peptidic fullerene bearing the C-terminal sequence of peptide T, which is able to activate chemotaxis of human monocytes through CD4/T4 antigen and to inhibit HIV-P activity, was prepared by Toniolo and coworkers [123].

Better results were achieved with compound **1** [124] and the *trans-2* isomer of **20** [125]. In both cases the EC_{50} was 0.2 μM. Is it interesting to note that, in the cases of bis-adducts, the regioisomers are not equally active, with the *trans-2* regioisomers presenting the highest activity among the studied compounds [125, 126].

Recently, the same salts (**20**) have been used, together with fulleropyrrolidine **21**, to study the inhibition of HIV reverse transcriptase – yet another enzyme that plays a determinant role in the HIV replication. The results obtained are outstanding. In fact, the IC_{50} of derivative **21**, bearing three carboxylic functions, is 100 times lower than nevirapine. Interesting results have also been obtained on hepatitis C virus RNA-dependent RNA polymerase when assaying the same derivative [127].

20

21

Other important biological enzymatic targets blocked by fullerene derivatives are glutathione-S-transferase, P450-cytochrome-dependent monooxygenases, plasmatic reticulum enzymes of hepatic cells and mitochondrial ATPase in the process of oxidative phosphorylation [128, 129].

The inhibition of all the three forms of nitric oxide synthase (NOS), i.e. neuronal, epithelial and inducible, has been found by trimalonic tris-adducts of C_{60}, mainly C_3 (**16**) and D_3 (**17**) [96]. The same compounds have been reported as potential inhibitors of β-lactamase [130].

11.4.5
Antibacterial Activity

One of the first experiments developed to study the effect of fullerene derivatives on bacteria was performed in 1996 [16]. Derivative **22** was tested on *Candida albicans* (eukaryote), *Bacillus subtilis* (Gram-positive bacterium), *E. coli* (Gram-negative enteric bacterium) and a clinical isolate strain 261/6 of *Mycobacterium avium* (acid-fast, emerging pathogen resistant to most antimicrobial drugs) with promising results. No particular reaction mechanism was demonstrated for this action, but the disruption of the cell wall was hypothesized.

22

Further studies on the antibacterial activity of fulleropyrrolidinium salts have been reported [118]. Lately, however, attention has been re-focused on bis-adduct mixtures (Fig. 11.3), which demonstrate bacteriostatic effects on *E. coli* and the inhibition of oxygen uptake. In this context, blocking the energy metabolism seems to be the mechanism of action [107]. Additional studies on antimicrobial PDT have

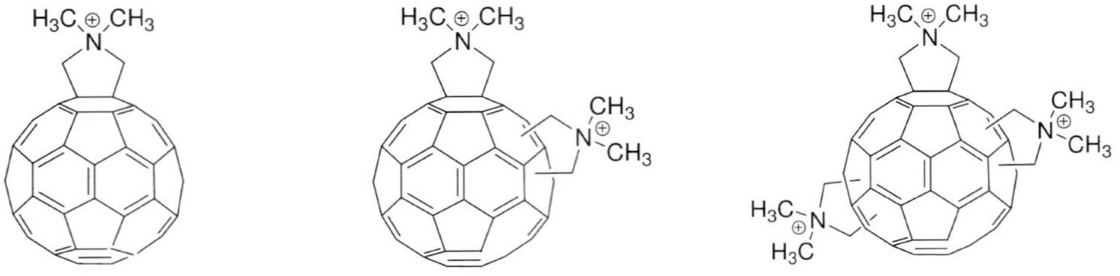

Figure 11.3. Mono-, bis- and tris-*N,N*-dimethylfulleropyrrolidinium salts.

been performed, testing the same mixture of cationic bis-adducts, cationic fullerene and a combination of tris-adducts/parent mono-adduct (Fig. 11.3). In this particular case a very interesting selectivity for microbes (*Staphylococcus aureus* and *E. coli*) over mammalian cells was demonstrated [131]. At present, the question remains if the reaction mechanism involves the formation of singlet oxygen or the generation of hydroxyl and superoxide radicals.

The *trans-2* and *trans-4* *N,N*-dimethyl-bisfulleropyrrolidinium salts exhibit their antibacterial action by inhibition of the respiratory chain, with a biphasic effect on oxygen uptake [132]. A fullerene derivative concentration up to 5 μM decreases the oxygen uptake, while at higher concentrations of C_{60} salts, an increase of oxygen uptake is found and oxygen is converted to H_2O_2 correspondingly. Fascinatingly, irradiation with light did not affect the results at all. In terms of mechanistic aspects this observation indicates that the electron transfer from the respiratory chain to C_{60} is independent of light.

Another water-soluble derivative, i.e. one that bears six sulfo-butyl chains, inhibits environmental bacteria. In contrast to the preceding studies, the noted effects have been ascribed to singlet oxygen – produced by photoexcitation. Apparently, in this case, the multifunctionalization does not seem to alter the yield of singlet oxygen formation with respect to comparable mono-adducts. Its incorporation into coated polymers and their subsequent irradiation with fluorescent visible light leads to germicidal effects on *E. coli* [133].

Very recently, the utilization of C_{60} water suspension, also called nano-C_{60}, was reported in studies on cell association and toxicity versus *E. coli* and *B. subtilis* [2]. For the first bacterium, the minimum inhibitory concentration (MIC) was 0.5–1 mg L^{-1}, while for the bacillus it is slightly higher (1.5–3.0 mg L^{-1}). The presence of this nano-C_{60} inhibits bacterial growth independent of the presence of light, both in aerobic and anaerobic conditions [134]. Notably, higher concentrations of fullerenes were detrimental to the antimicrobial action because of aggregating nanoparticles, followed by subsequent precipitation that consequently causes the reduction or loss of activity [2].

A different group of fullerene derivatives with good potential antibacterial activity is represented by fulleropeptides. Amino acid C_{60} derivative **23** has been utilized in solid-phase synthesis for the preparation of derivative **24**. It was tested on *S. aureus* and *E. coli*, with MICs of 8 and 64 mM, respectively [135, 136].

FmocHN...—OH

(Gly-Orn)$_6$-Gly-NH$_2$

H$_3$N$^+$...

23

24

11.4.6
Fullerenes as Nanodevices in Monoclonal Immunology

Recent studies with antibodies of the IgG isotype focused specifically on C_{60} derivatives conjugated to bovine thyroglobulin [137–139]. Such work pointed towards the potential of fullerenes in the development of nanodevices applicable in monoclonal immunology. However, only a limited number of reports on the use of fullerenes in this field have appeared so far.

To produce an antibody with specificity to fullerenes, Chen and coworkers [137] successfully linked water-soluble fullerene derivatives to large foreign proteins – including bovine thyroglobulin, and bovine and rabbit serum albumin. Immunization of mice with these C_{60}–protein conjugates produced a polyclonal immune response comprised of IgG antibodies specific for C_{60} fullerene and a subpopulation that crossreacted with a C_{70} derivative. This study showed for the first time that the immune system could recognize and process fullerenes as protein conjugates, and display the processed peptides for recognition by T cells to yield IgG antibodies. Later [138], the sequences of light and heavy chains of this IgG antibody were determined. Using X-ray crystallography of its Fab' fragment, it was found that the binding cavity was formed by the clustering of hydrophobic amino acids, several of which participate in stacking interactions with the fullerene core. Alternatively, weak hydrogen bond interactions between the fullerenes and the antibodies are considered to contribute to the C_{60}–antibody complex stability. Molecular dynamics simulation studies revealed that the fullerene molecule in the corresponding C_{60}–antibody complex is readily accommodated in the suggested binding site of the antibody [139]. In particular, an eminent trend towards predominant surroundings by hydrophobic amino acid side-chains, which are involved in π-stacking interactions with the fullerene molecule, evolves. Moreover, fullerenes inside the binding site undergo a small relative translational motion and a much larger rotation motion. However, no favored axis of rotation was found to exist. About 17% of the surface is exposed to the solvent that might potentially be used for functional derivatization.

The discovery that the mouse immune repertoire is diverse enough to recognize and produce antibodies specific for fullerenes has allowed the indirect determination of the distribution of fullerenes within a cellular environment. For the first time, Foley and coworkers reported that the water-soluble fullerene derivative $C_{61}(CO_2H)_2$ is able to cross the cell membrane and to bind preferentially to the mitochondria [34]. The finding that fullerenes locate close to this organelle indicates the protective effects of fullerenes, in general, with respect to ROS [35].

Furthermore, the selective binding of proteins to carbon nanostructures has been extended from fullerenes to CNTs [140–142]. For example, particular antigens have been attached to the CNT walls, while retaining their conformation and thereby inducing an antibody response with the right specificity [143].

Erlanger, Chen and coworkers have shown that the monoclonal IgG C_{60}-specific antibody recognizes and specifically binds to aqueous suspensions of single-wall CNT (SWNT) bundles [144]. The authors considered these findings to have practical application. For example, the antibody-coated SWNTs can be used as probes of cell or membrane function. The antifullerene antibody on the surfaces of CNTs can be covalently decorated with probes of cell function, e.g. redox or luminescent probes, and after insertion, the probe molecule can be optically excited or electrically addressed via the conducting SWNT wire.

Using a polymer coating that is receptive to selected proteins, while rejecting others, is an elegant way to render nanotubes capable of interacting selectively with biomolecules. A variety of polymer coatings and self-assembled monolayers (SAMs) have been used to prevent nonspecific binding of proteins on surfaces for biosensor and biomedical device applications [145]. Among the coating materials, PEG is one of the most effective and widely used polymer materials that irreversibly adsorbs onto SWNTs. In this context, it has been shown that the protein streptavidin adsorbs spontaneously onto multiwalled CNTs (MWNTs) via hydrophobic interactions and thereby forms close packs. On the contrary, in the case of SWNTs that are coated with a surfactant and PEG, the protein nonspecific binding is prevented. A feasible way to circumvent this problem and to achieve selective binding involves the co-functionalization of SWNTs with PEG and biotin [146]. Surfactants, such as Triton X-100 or Triton X-405, that are prior adsorbed onto the surface of SWNTs as a wetting layer, significantly enhance the PEG adsorption and, thus, lead to high resistance of CNTs to the nonspecific binding of streptavidin. In the final step biotin was added through the covalently linkage of amine-terminated PEG chains and amine-reactive biotin reagents – biotinamidocaproic acid 3-sulfo-*N*-hydroxysuccinimde ester. For CNTs that are coated with Triton/PEG–biotin, selective binding of streptavidin was revealed by the high density of adsorbed proteins along the CNTs after exposure to a streptavidin solution. No appreciable adsorption was seen when CNTs functionalized in the same manner were exposed to streptavidin plugged with four equivalents of free biotin. These results demonstrate that functionalization of CNTs is one way to achieve specific protein recognition, while eliminating or minimizing nonspecific protein binding, and, thus, this has important implications for the biocompatibility of CNTs and specifically for the development of potential bioelectronic devices that integrate CNTs [147].

The design of selective piezoelectric crystal immunosensors that assist in the detection of IgG and hemoglobin in aqueous media utilizing water-soluble immobilized C_{60}-antibodies as coating material has recently been demonstrated successfully [148]. The piezoelectric biosensors were prepared by coating C_{60} fullerene from a toluene solution of poly(vinyl chloride) and C_{60}, using spin coating at 10 cycles s^{-1} (Hz) and subsequent adsorption of the antibodies.

The prepared biosensors exhibited linear frequency responses to the concentration of human IgG and hemoglobin with sensitivities of 1.25×10^2 and 1.56×10^4 Hz $(mg\ mL)^{-1}$ and reveal detection limits of 10^{-4} mg mL^{-1}. The optimal conditions for the IgG sensor were found at pH 6.7 and 30 °C. Important implications from the findings are that species such as urea, uric acid, cysteine, tyrosine and ascorbic acid imposed no interference on the biosensor activities. This corroborates the high selectivity.

11.4.7
Fullerenes as Radiotracers

Encapsulating metal atoms inside the fullerene interior bears promising prospects for biomedical applications that might lead ultimately to nanodevices for diagnostic and therapeutic nuclear medicine. The rigid structure of the fullerene cage protects the encapsulated metal ion from external chemical attack and toxic metal ion release *in vivo*. Wilson and coworkers have, for example, demonstrated the feasibility of metallofullerenes as *in vivo* radiotracers [149]. Biodistribution studies of $^{166}Ho_x@C_{82}(OH)_y$ in mice over a 48-h period showed selective localization of the tracer in the liver, but with continued slow excretion as well as retainment in the bone.

The same group has explored the concept of utilizing the nontoxic scaffold of C_{60} for the development of new X-ray contrast agents [150, 151]. Like other contemporary contrast agents (**25**), the C_{60}-based agents use iodine active X-ray attenuating vehicles. However, these compounds take advantage of the unique structure of fullerene. For instance, in contrast to the disk-like shape of contemporary contrast agents, the inherited globular shape of the fullerene-based contrast agent reduces viscosity in clinical formulations, which allows rapid intravenous injection of the agent. The fullerene core may also block one side of the tri-iodinated phenyl rings in **26** and **27** from having hydrophobic interactions with blood plasma proteins, and, thus, would lead to decreased protein binding and increased *in vivo* tolerability. Additionally, the existence of more than three iodine atoms will allow lower concentrations of contrast agents to be used.

Gadolinium-containing metallofullerenes have recently emerged as a new generation of magnetic resonance imaging (MRI) contrast agents. This development is driven, largely, by the high proton relaxivities of these metallofullerenes and the complete lack of Gd^{3+} release under metabolic conditions. The most extensive study on the potential of gadofullerenes has been carried out by Wilson's group. Recently, they reported the synthesis and *in vivo* biodistribution study of the first water-soluble $Gd@C_{60}$ derivative [152–154]. The respective malonate derivative

$Gd@C_{60}.[C(COOH)_2]_{10}$ was found to possess a relaxivity (4.6 mM^{-1} s^{-1} at 20 MHz and 40 °C comparable to that of commercially available Gd(III) chelate-based MRI agents. Moreover, an *in vivo* MRI biodistribution study revealed that $Gd@C_{60}[C(COOH)_2]_{10}$ is the first-water soluble endohedral metallofullerene with decreased uptake by the reticuloendothelial system (RES) and facile excretion via the urinary tract, consistent with its lack of intermolecular aggregation in solution.

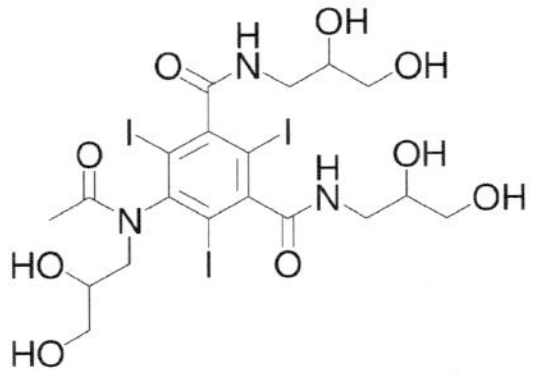

25

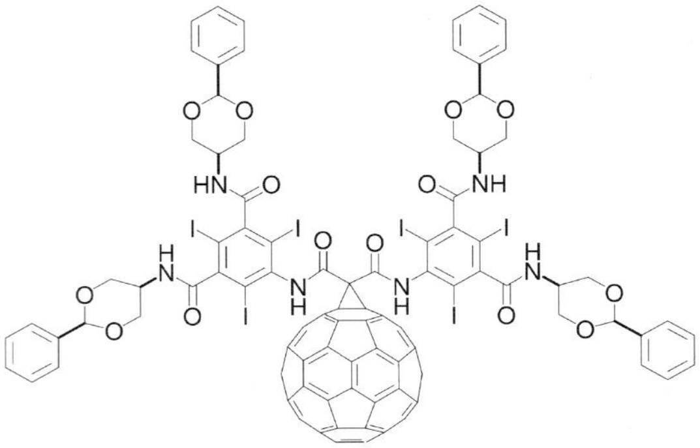

26

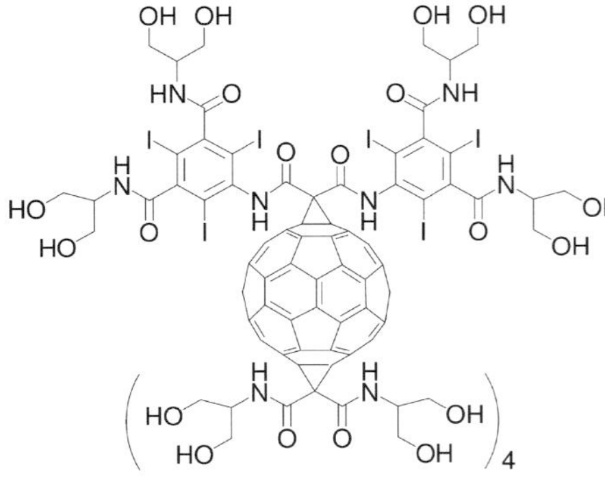

27

Other polyhydroxyl derivatives of gadolinium-containing metallofullerenes and endohedral fullerenes with a series of lanthanide metal ions for $M@C_{82}(OH)_n$ (M = La, Ce, Dy, Er) [155] have also been proposed as potential MRI agents, with $Gd@C_{82}(OH)_2$ being the most efficient relaxing agent [156, 157].

Studies by Dorn and coworkers are currently focusing on the development of tri-metallic nitride templated endohedral metallofullerenes as a powerful, new generation of imaging contrast agents and radiotracers. These endohedral fullerenes are designed at the nanoscale level with improved contrast features as well as multimodal imaging potential (X-ray, MRI). Recent milestones in the development of tri-metallic nitride templated endohedral metallofullerenes involve the synthesis of a lutenium-based series of mixed metal species of gadolinium/lutenium and holmium/lutenium $Lu_{3-x}A_xN@C_{80}$ ($x = 0$–2) endohedral metallofullerenes [158], and the subsequent functionalization of both diamagnetic $Sc_3N@C_{80}$ and para-magnetic $Er_3N@C_{80}$ [159].

11.4.8
Fullerenes as Vectors

Finally, biological use of fullerenes should be discussed that finds frequent mention over the years, but has only been scarcely pursued, i.e. the application of C_{60} as a drug vector. In particular, the hydrophobic nature of fullerenes should expedite membrane crossing. The paclitaxel fullerene derivative **28** has been prepared by Wilson and coworkers, who design a C_{60}-based slow release system of paclitaxel for a liposome aerosol delivery system in lung cancer treatment [160].

28

This initial work shows the first attempts to use fullerenes in pro-drug design and synthesis, while derivative **29** was the first compound – and up to now the only one – utilized to transfect DNA materials into cells [161]. The presence of four positive charges is particularly promising in light of facilitating DNA binding through electrostatic interactions with the phosphate groups of the oligonucleotide, but without specific recognition of the bases. The cells take up the complex through phagocytosis, triggering the release of duplex DNA in the cells.

29

Acknowledgments

This work was carried out with partial support from the EU (RTN network "CASSIUS CLAYS"), SFB 583, DFG (GU 517/4-1), FCI and the Office of Basic Energy Sciences of the US Department of Energy.

References

1 W. A. Scrivens, J. M. Tour, K. E. Creek, L. Pirisi. Synthesis of ^{14}C-labeled C_{60}, its suspension in water, and its uptake by human keratinocytes. *J. Am. Chem. Soc.* **1994**, *116*, 4517–4518.

2 D. Y. Lyon, J. D. Fortner, C. M. Sayes, V. L. Colvin, J. B. Hughes. Bacterial cell association and antimicrobial activity of a C_{60} water suspension. *Environ. Toxicol. Chem.* **2005**, *24*, 2757–2762.

3 Y. Liu, H. Wang, P. Liang, H.-Y. Zhang. Water-soluble supramolecular fullerene assembly mediated by metallobridged beta-cyclodextrins. *Angew. Chem. Int. Ed. Engl.* **2004**, *43*, 2690–2694.

4 S. Samal, K. E. Geckeler. Cyclodextrin–fullerenes: a new class of water-soluble fullerenes. *J. Chem. Soc. Chem. Commun.* **2000**, 1101–1102.

5 T. Braun. Water soluble fullerene-cyclodextrin supramolecular assemblies – preparation, structure, properties. *Fullerene Sci. Technol.* **1997**, *5*, 615–626.

6 T. Andersson, K. Nilsson, M. Sundahl, G. Westman, O. Wennerström. C_{60} embedded in gamma-cyclodextrin: a water-soluble fullerene. *J. Chem. Soc. Chem. Commun.* **1992**, 604–606.

7 S. D. M. Islam, M. Fujitsuka, O. Ito, A. Ikeda, T. Hatano, S. Shinkai. Photoexcited state properties of C_{60} encapsulated in a water-soluble calixarene. *Chem. Lett.* **2000**, *1*, 78–79.

8 A. Ikeda, Y. Suzuki, M. Yoshimura, S. Shinkai. On the prerequisites for the formation of solution complexes from [60]fullerene and calix[n]arenes: a novel allosteric effect between [60]fullerene and metal cations in calix[n]aryl ester complexes. *Tetrahedron* **1998**, *54*, 2497–2508.

9 A. Ikeda, S. Nobukuni, H. Udzu, Z. Zhong, S. Shinkai. A novel [60]fullerene-calixarene conjugate which facilitates self-inclusion of the [60]fullerene moiety into the homooxacalix[3]arene cavity. *Eur. J. Org. Chem.* **2000**, 3287–3293.

10 S. Shinkai, A. Ikeda. Calixarene–fullerene conjugates: marriage of the third generations of inclusion compounds and carbon clusters. *Gazz. Chim. It.* **1997**, *127*, 657–662.

11 A. Ikeda, T. Hatano, M. Kawaguchi, H. Suenaga, S. Shinkai. Water-soluble [60]fullerene–cationic homooxacalix[3]arene complex which is applicable to the photocleavage of DNA. *J. Chem. Soc. Chem. Commun.* **1999**, 1403–1404.

12 J. L. Atwood, G. A. Koutsantonis, C. L. Raston. Purification of C_{60} and C_{70} by selective complexation with calixarenes. *Nature* **1994**, *368*, 229–231.

13 S. Kunsági-Máté, K. Szabó, I. Bitter, G. Nagy, L. Kollar. Complex formation between water-soluble sulfonated calixarenes and C_{60} fullerene. *Tetrahedron Lett.* **2004**, *45*, 1387–1390.

14 Y. N. Yamakoshi, T. Yagami, K. Fukuhara, S. Sueyoshi, N. Miyata. Solubilization of fullerenes into water with polyvinylpyrrolidone applicable to biological tests. *J. Chem. Soc. Chem. Commun.* **1994**, 517–518.

15 Y. Tabata, Y. Murakami, Y. Ikada. Antitumor effect of poly(ethylene glycol)modified fullerene. *Fullerene Sci. Technol.* **1997**, *5*, 989–1007.

16 T. Da Ros, M. Prato, F. Novello, M. Maggini, E. Banfi. Easy access to water-soluble fullerene derivatives via 1,3-dipolar cycloadditions of azomethine ylides to C_{60}. *J. Org. Chem.* **1996**, *61*, 9070–9072.

17 L. Y. Chiang, J. B. Bhonsle, L. Wang, S. F. Shu, T. M. Chang, J. R. Hwu. Efficient one-flask synthesis of water-soluble [60]fullerenols. *Tetrahedron* **1996**, *52*, 4963–4972.

18 I. Lamparth, A. Hirsch. Water-soluble malonic acid derivatives of C_{60} with a defined three-dimensional structure. *J. Chem. Soc. Chem. Commun.* **1994**, 1727–1728.

19 C. Cusan, T. Da Ros, G. Spalluto, S. Foley, J.-M. Janot, P. Seta, C. Larroque, M. C. Tomasini, T. Antonelli, L. Ferraro, M. Prato. A new multi-charged C_{60} derivative: synthesis and biological properties. *Eur. J. Org. Chem.* **2002**, 2928–2934.

20 S. Filippone, F. Heimann, A. Rassat. A highly water-soluble 2:1 beta-cyclodextrin–fullerene conjugate. *J. Chem. Soc. Chem. Commun.* **2002**, 1508–1509.

21 J. Yang, Y. Wang, A. Rassat, Y. Zhang, P. Sinaÿ. Synthesis of novel highly water-soluble 2:1 cyclodextrin/fullerene conjugates involving the secondary rim of beta-cyclodextrin. *Tetrahedron* **2004**, *60*, 12163–12168.

22 Y. Liu, Y.-L. Zhao, Y. Chen, P. Liang, L. Li. A water-soluble beta-cyclodextrin derivative possessing a fullerene tether as an efficient photodriven DNA-cleavage reagent. *Tetrahedron Lett.* **2005**, *46*, 2507–2511.

23 A. Vasella, F. Uhlmann, C. A. Waldraff, F. Diederich, C. Thilgen. Fullerene sugars: preparation of enantiomerically pure, spiro-linked C-glycosides of C_{60}. *Angew. Chem. Int. Ed. Engl.* **1992**, *31*, 1388–1390.

24 H. Kato, A. Yashiro, A. Mizuno, Y. Nishida, K. Kobayashi, H. Shinohara. Syntheses and biological evaluations of alpha-d-mannosyl [60]fullerenols. *Bioorg. Med. Chem. Lett.* **2001**, *11*, 2935–2939.

25 Y. Nishida, A. Mizuno, H. Kato, A. Yashiro, T. Ohtake, K. Kobayashi. Stereo- and biochemical profiles of the 5–6- and 6–6-junction isomers of alpha-d-mannopyranosyl [60]fullerenes. *Chem. Biodivers.* **2004**, *1*, 1452–1564.

26 S. Abe, H. Moriyama, K. Niikura, F. Feng, K. Monde, S.-I. Nishimura. Versatile synthesis of oligosaccharide-containing fullerenes. *Tetrahedron Asymm.* **2005**, *16*, 15–19.

27 D. Pantarotto, N. Tagmatarchis, A. Bianco, M. Prato. Synthesis and biological properties of fullerene-containing amino acids and peptides. *Mini-Rev. Med. Chem.* **2004**, *4*, 805–814.

28 G. N. Bogdanov, R. A. Kotelnikova, E. S. Frog, V. N. Shtolko, V. S. Romanova, Yu. N. Bubnov. Enantiomers of the amino acid derivatives of fullerene C_{60} possess stereospecific membranotropic properties. *Doklady Biochem. Biophys.* **2004**, *396*, 165–167.

29 J. Yang, A. R. Barron. A new route to fullerene substituted phenylalanine derivatives. *J. Chem. Soc. Chem. Commun.* **2004**, 2884–2885.

30 L. Watanabe, M. Bhuiyan, B. Jose, T. Kato, N. Nishino. Synthesis of novel fullerene amino acids and their multifullerene peptides. *Tetrahedron Lett.* **2004**, *45*, 7137–7140.

31 A. Bianco. Efficient solid-phase synthesis of fullero-peptides using Merrifield strategy. *J. Chem. Soc. Chem. Commun.* **2005**, 3174–3175.

32 M. Brettreich, A. Hirsch. A highly water-soluble dendro[60]fullerene. *Tetrahedron Lett.* **1998**, *39*, 2731–2740.

33 T. Baierl, E. Drosselmeyer, A. Seidel, S. Hippeli. Comparison of immunological effects of fullerene C_{60} and raw soot from fullerene production on alveolar macrophages and macrophage like cells *in vitro*. *Exp. Toxicol. Pathol.* **1996**, *48*, 508–511.

34 S. Foley, C. Crowley, M. Smaihi, C. Bonfils, B. F. Erlanger, P. Seta, C. Larroque. Cellular localization of a water-soluble fullerene derivative. *Biochem. Biophys. Res. Commun.* **2002**, *294*, 116–119.

35 L. L. Dugan, D. M. Turetsky, C. Du, D. Lobner, M. Wheeler, C. R. Almli, C. K. F. Shen, T. Y. Luh, D. W. Choi, T. S. Lin. Carboxyfullerenes as neuroprotective agents. *Proc. Natl. Acad. Sci. USA* **1997**, *94*, 9434–9439.

36 T. Tsuchiya, I. Oguri, Y. N. Yamakoshi, N. Miyata. Novel harmful effects of [60]fullerene on mouse embryos *in vitro* and *in vivo*. *FEBS Lett.* **1996**, *393*, 139–145.

37 T. Tsuchiya, Y. N. Yamakoshi, N. Miyata. A novel promoting action of fullerene C_{60} on the chondrogenesis in rat embryonic limb bud cell-culture system. *Biochem. Biophys. Res. Commun.* **1995**, *206*, 885–894.

38 K. Irie, Y. Nakamura, H. Ohigashi, H. Tokuyama, S. Yamago, E. Nakamura. Photocytotoxicity of water-soluble fullerene derivatives *Biosci. Biotechnol. Biochem.* **1996**, *60*, 1359–1361.

39 S. Yamago, H. Tokuyama, E. Nakamura, K. Kikuchi, S. Kananishi, K. Sueki, H. Nakahara, S. Enomoto, F. Ambe. In-vivo behavior of a water-soluble fullerene – C14 labeling, absorption, distribution, excretion and acute toxicity. *Chem. Biol.* **1995**, *2*, 385–389.

40 A. Yang, D. L. Cardona, F. A. Barile. *In vitro* cytotoxicity testing with fluorescence-based assays in cultured human lung and dermal cells. *Cell Biol. Toxicol.* **2002**, *18*, 97–108.

41 J. P. Simonin. Solvent effects on osmotic second virial coefficient studied using analytic molecular models. Application to solutions of C_{60} fullerene. *J. Phys. Chem. B* **2001**, *105*, 5262–5270.

42 C. M. Sayes, J. D. Fortner, W. Guo, D. Lyon, A. M. Boyd, K. D. Ausman, Y. J. Tao, B. Sitharaman, L. J. Wilson, J. B. Hughes, J. L. West, V. L. Colvin. The differential cytotoxicity of water-soluble fullerenes. *Nano Lett.* **2004**, *4*, 1881–1887.

43 X. Yang, C. Fan, H. S. Zhu. Photo-induced cytotoxicity of malonic acid [C_{60}]fullerene derivatives and its mechanism. *Toxicol. In Vitro* **2002**, *16*, 41–46.

44 H. Tokuyama, S. Yamago, E. Nakamura, T. Shiraki, Y. Sugiura. Photoinduced biochemical activity of fullerene carboxylic acid. *J. Am. Chem. Soc.* **1993**, *115*, 7918–7919.

45 E. Nakamura, H. Tokuyama, S. Yamago, T. Shiraki, Y. Sugiura. Biological activity of water-soluble fullerenes. Structural dependence of DNA cleavage, cytotoxicity, and enzyme inhibitory activities including HIV-protease inhibition. *Bull. Chem. Soc. Jpn.* **1996**, *69*, 2143–2151.

46 A. S. Boutorine, H. Tokuyama, M. Takasugi, H. Isobe, E. Nakamura, C. Hélène. Fullerene–oligonucleotide conjugates – photoinduced sequence-specific DNA cleavage. *Angew. Chem. Int. Ed. Engl.* **1995**, *33*, 2462–2465.

47 D. M. Guldi, M. Prato. Excited-state properties of C_{60} fullerene derivatives. *Acc. Chem. Res.* **2000**, *33*, 695–703.

48 J. L. Anderson, Y.-Z. An, Y. Rubin, C. S. Foote. Photophysical characterization and singlet oxygen yield of a dihydrofullerene. *J. Am. Chem. Soc.* **1994**, *116*, 9763.

49 R. Bernstein, F. Prat, C. S. Foote. On the mechanism of DNA cleavage by fullerenes investigated in model systems: electron transfer from guanosine and 8-oxo-guanosine derivatives to C_{60}. *J. Am. Chem. Soc.* **1999**, *121*, 464–465.

50 E. Nakamura, H. Isobe. Functionalized fullerenes in water. The first 10 years of their chemistry, biology, and nanoscience. *Acc. Chem. Res.* **2003**, *36*, 807–815.

51 J. W. Arbogast, C. S. Foote, M. Kao. Electron transfer to triplet C_{60}. *J. Am. Chem. Soc.* **1992**, *114*, 2277–2278.

52 Y. Yamakoshi, S. Sueyoshi, K. Fukuhara, N. Miyata, $^{\bullet}OH$ and $O_2^{\bullet-}$ generation in aqueous C_{60} and C_{70} solutions by photoirradiation: an EPR study. *J. Am. Chem. Soc.* **1998**, *120*, 12363–12364.

53 I. Nakanishi, S. Fukuzumi, T. Konishi, K. Ohkubo, M. Fujitsuka, O. Ito, N. Miyata. DNA cleavage via superoxide anion formed in photoinduced electron transfer from

NADH to gamma-cyclodextrin-bicapped C_{60} in an oxygen-saturated aqueous solution. *J. Phys. Chem. B.* **2002**, *106*, 2372–2380.

54 S. Samal, K. E. Geckeler. DNA-cleavage by fullerene-based synzymes. *Macromol. Biosci.* **2001**, *1*, 329–331.

55 A. Ikeda, T. Sato, K. Kitamura, K. Nishiguchi, Y. Sasaki, J. Kikuchi, T. Ogawa, K. Yogo, T. Takey. Efficient photocleavage of DNA utilizing water-soluble lipid membrane-incorporated [60]fullerenes prepared using a [60]fullerene exchange method. *Org. Biomol. Chem.* **2005**, *3*, 2907–2909.

56 T. Da Ros, M. Bergamin, E. Vázquez, G. Spalluto, B. Baiti, S. Moro, A. Boutorine, M. Prato. Synthesis and molecular modeling studies of fullerene-5,6,7-trimethoxyindole-oligonucleotide conjugates as possible probes for study of photochemical reactions in DNA triple helices. *Eur. J. Org. Chem.* **2002**, 405–413.

57 Y. N. Yamakoshi, T. Yagami, S. Sueyoshi, N. Miyata. Acridine adduct of [60]fullerene with enhanced DNA-cleaving activity. *J. Org. Chem.* **1996**, *61*, 7236–7237.

58 A. M. Cassell, W. A. Scrivens, J. M. Tour. Assembly of DNA/fullerene hybrid materials. *Angew. Chem. Int. Ed. Engl.* **1998**, *37*, 1528–1531.

59 Q. Ying, J. Zhang, D. Liang, W. Nakanishi, H. Isobe, E. Nakamura, B. Chu. Fractal behavior of functionalized fullerene aggregates. I. Aggregation of two-handed tetraaminofullerene with DNA. *Langmuir* **2005**, *21*, 9824–9831.

60 Y. Yamakoshi, N. Umezawa, A. Ryu, K. Arakane, N. Miyata, Y. Goda, T. Masumizu, T. Nagano. Active oxygen species generated from photoexcited fullerene (C_{60}) as potential medicines: $O_2^{\cdot -}$ versus $^1O_2$. *J. Am. Chem Soc.* **2003**, *125*, 12803–12809.

61 J. W. Arbogast, A. O. Darmannyan, C. S. Foote, F. N. Diederich, R. L. Whetten, Y. Rubin, M. M. Alvarez, S. J. Anz. Photophysical properties of sixty atom carbon molecule (C_{60}). *J. Phys. Chem.* **1991**, *95*, 11–12.

62 J. W. Arbogast, C. S. Foote. Photophysical properties of C_{70}. *J. Am. Chem. Soc.* **1991**, *113*, 8886–8889.

63 P. Cheng, S. R. Wilson, D. I. Schuster. A novel parachute-shaped C_{60}–porphyrin dyad. *Chem. Commun.* **1999**, 89–90.

64 M. D. Meijer, G. P. M. van Klink, G. van Koten. Metal-chelating capacities attached to fullerenes. *Coord. Chem. Rev.* **2002**, *230*, 141–163.

65 T. Hamano, K. Okuda, T. Mashino, M. Hirobe, K. Arakane, A. Ryu, S. Mashiko, T. Nagano. Singlet oxygen production from fullerene derivatives: effect of sequential functionalization of the fullerene core. *J. Chem. Soc. Chem. Commun.* **1997**, 21–22.

66 B. Vileno, A. Sienkiewicz, M. Lekka, A. J. Kulik, L. Forró. *In vitro* assay of singlet oxygen generation in the presence of water-soluble derivatives of C_{60}. *Carbon* **2004**, *42*, 1195–1198.

67 K. D. Pickering, M. R. Wiesner. Fullerol-sensitized production of reactive oxygen species in aqueous solution. *Environ. Sci. Technol.* **2005**, *39*, 1359–1365.

68 W. Li, K. Qian, W. Huang, X. Zhang, W. Chen. Water-soluble C_{60}-liposome and the biological effect of C_{60} to human cervix cancer cells. *Chin. Phys. Lett.* **1994**, *11*, 207–210.

69 N. Nakajima, C. Nishi, F.-M. Li, Y. Ikada. Photo-induced cytotoxity of water-soluble fullerene. *Fullerene Sci. Technol.* **1996**, *4*, 1–19.

70 Y. Tabata, Y. Murakami, Y. Ikada, Antitumor effect of poly(ethylene glycol)modified fullerene. *Fullerene Sci. Technol.* **1997**, *5*, 989–1007.

71 Y. Tabata, Y. Murakami, Y. Ikada. Photodynamic effect of polyethylene glycol-modified fullerene on tumor. *Jpn. Cancer Res.* **1997**, *88*, 1108–1116.

72 A. Hamazawa, I. Kinoshita, B. Breedlove, K. Isobe, M. Shibata, Y. Baba, T. Kakuchi, S. Hirohara, M. Obata, Y. Mikata, S. Yano. Meso-tetraphenylporphyrin having hexa-maltosyl and decyl chain as an amphiphilic photosensitizer toward photodynamic therapy. *Chem. Lett.* **2002**, *3*, 388–389.

73 G. Zheng, A. Graham, M. Shibata, J. R. Missert, A. R. Oseroff, T. J. Dougherty, R. K. Pandey. Synthesis of beta-galactose-conjugated chlorins derived by enyne metathesis as galectin-specific photosensitizers for photodynamic therapy. *J. Org. Chem.* **2001**, *66*, 8709–8716.

74 Y. Mikata, S. Takagi, M. Tanahashi, S. Ishii, M. Obata, Y. Miyamoto, K. Wakita, T. Nishisaka, T. Hirano, T. Ito, M. Hoshino, C. Ohtsuki, M. Tanihara, S. Yano. Detection of 1270 nm emission from singlet oxygen and photocytotoxic property of sugar-pendant [60]fullerenes. *Bioorg. Medicinal Chem. Lett.* **2003**, *13*, 3289–3292.

75 S. A. Omari, E. A. Ermilov, M. Helmreich, N. Jux, A. Hirsch, B. Roder. Transient absorption spectroscopy of a monofullerene C-60–bis-(pyropheophorbide *a*) molecular system in polar and nonpolar environments. *Appl. Phys. B* **2004**, *79*, 617–622.

76 F. Rancan, M. Helmreich, A. Moelich, N. Jux, A. Hirsch, B. Roeder, C. Witt, F. Boehm. Fullerene–pyropheophorbide *a* complexes as sensitizer for photodynamic therapy: uptake and photo-induced cytotoxicity on Jurkat cells. *J. Photochem. Photobiol. B* **2005**, *80*, 1–7.

77 P. Král. Control of catalytic activity of proteins *in vivo* by nanotube ropes excited with infrared light. *Chem. Phys. Lett.* **2003**, *382*, 399–403.

78 K. C. Hwang, D. Mauzerall. Vectorial electron transfer from an interfacial photoexcited porphyrin to ground state fullerene C_{60} and C_{70} and from ascorbate to triplet C_{60} and C_{70} in a lipid bilayer. *J. Am. Chem. Soc.* **1992**, *114*, 9705–9706.

79 R. V. Bensasson, J. L. Geraud, S. Leach, G. Miquel, P. Seta. Transmembrane electron transport mediated by photoexcited fullerenes. *Chem. Phys. Lett.* **1993**, *210*, 141.

80 K. C. Hwang, D. Mauzerall. Photoinduced electron transport across a lipid bilayer mediated by C_{70}. *Nature* **1993**, *361*, 138–140.

81 S. Niu, D. Mauzerall. Fast and efficient charge transport across a lipid bilayer is electronically mediated by C_{70} fullerene aggregates. *J. Am. Chem. Soc.* **1996**, *118*, 5791–5795.

82 J. M. Janot, P. Seta, R. V. Bensasson, S. Leach. Involvement of C-60 fullerene monomers and aggregates in the photoconductivity of ultrathin bilayer lipid membranes. *Synthetic Metals* **1996**, *77*, 103–106.

83 L. Wang, X. Zhang, Z. Du, Y. Bai, T. Li. The photo-electronic transfer across membrane between C_{60} and CdS nanoparticles. *Chem. Phys. Lett.* **2003**, *372*, 331–335.

84 R. A. Kotelnikova, G. N. Bogdanov, E. C. Frog, A. I. Kotelnikov, V. N. Shtolko, V. S. Romanova, S. M. Andreev, A. A. Kushch, N. E. Fedorova, A. A. Medzhidova, G. G. Miller. Nanobionics of pharmacologically active derivatives of fullerene C_{60}. *J. Nanoparticle Res.* **2003**, *5*, 561–566.

85 M. Nishihara, F. Perret, T. Takeuchi, S. Futaki, A. N. Lazar, A. W. Coleman, N. Sakai, S. Matile. Arginine magic with new counterions up the sleeve. *Org. Biomol. Chem.* **2005**, *3*, 1659–1669.

86 F. Perret, M. Nishihara, T. Takeuchi, S. Futaki, A. N. Lazar, A. W. Coleman, N. Sakai, S. Matile. Anionic fullerenes, calixarenes, coronenes, and pyrenes as activators of oligo/polyarginines in model membranes and live cells. *J. Am. Chem. Soc.* **2005**, *127*, 1114–1115.

87 H. Eddaoudi, A. Deratani, S. Tingry, F. Sinan, P. Seta. Fullerene membrane transport mediated by γ-cyclodextrin immobilized in poly(vinyl alcohol) films. *Polym. Int.* **2003**, *52*, 1390–1395.

88 M. Carano, S. Cosnier, K. Kordatos, M. Marcaccio, M. Margotti, F. Paolucci, M. Prato, S. Roffia. A glutathione amperometric biosensor based on an amphiphilic fullerene redox mediator immobilized within an amphiphilic polypyrrole film. *J. Mater. Chem.* **2002**, *12*, 1996–2000.

89 H. T. Tien, L.-G. Wang, X. Wang, A. L. Ottova. Electronic processes in supported bilayer lipid membranes (s-BLMs) containing a geodesic form of carbon (fullerene C_{60}). *Bioelectrochem. Bioenerg.* **1997**, *42*, 161–167.

90 F. Cardullo, F. Diederich, L. Echegoyen, T. Habicher, N. Jayaraman, R. M. Leblanc, J. F. Stoddart, S. Wang. Stable Langmuir and Langmuir–Blodgett films of fullerene–glycodendron conjugates. *Langmuir* **1998**, *14*, 1955–1959.

91 P. J. Krusic, E. Wasserman, P. N. Keizer, J. R. Morton, K. F. Preston. Radical reactions of C_{60}. *Science* **1991**, *254*, 1183–1185.

92 L. L. Dugan, E. Lovett, S. Cuddihy, B.-W. Ma, T.-S. Lin, D. W. Choi. Carboxyfullerenes as neuroprotective antioxidants. In: *Handbook of Fullerenes*, K. Kadish, R. Ruoff, (Eds.). Wiley, New York, **2000**, pp. 467–479.

93 K. L. Quick, J. I. Hardt, L. L. Dugan. Rapid microplate assay for superoxide scavenging efficiency. *J. Neurosci. Methods* **2000**, *97*, 139–144.

94 L. L. Dugan, E. G. Lovett, K. L. Quick, J. Lotharius, T. T. Lin, K. L. O'Malley. Fullerene-based antioxidants and neurodegenerative disorders. *Parkinsonism Rel. Disord.* **2001**, *7*, 243–246.

95 A. M. Lin, B. Y. Chyi, S. D. Wang, H. H. Yu, P. P. Kanakamma, T. Y. Luh, C. K. Choi, L. T. Ho. Carboxyfullerene prevents iron-induced oxidative stress in rat brain. *J. Neurochem.* **1999**, *72*, 1634–1640.

96 D. J. Wolf, A. D. Papoiu, K. Mialkowski, C. F. Richardson, D. I. Schuster, S. R. Wilson. Inhibition of nitric oxide synthase isoforms by tris-malonyl-C_{60}-fullerene adducts. *Arch. Biochem. Biophys.* **2000**, *378*, 216–223.

97 L. L. Dugan, V. M. G. Bruno, S. M. Amagasu, R. G. Giffard. GLIA modulate the response of murine cortical-neurons to excitotoxicity – GLIA exacerbate AMPA neurotoxicity. *J. Neurosci.* **1995**, *15*, 4545–4555.

98 M. P. Goldberg, S. P. Althomsons, T. Chapman, D. W. Choi, L. L. Dugan. Carboxyfullerene free radical scavengers reduce hypoxic and excitotoxic oligodendrocyte death *in vitro. Neurology* **1998**, *50*, A370 (S59002 Suppl. 4).

99 J. Lotharius, L. L. Dugan, K. L. O'Malley. Distinct mechanisms underlie neurotoxin-mediated cell death in cultured dopaminergic. *J. Neurosci.* **1999**, *19*, 1284–1293.

100 Y. L. Huang, C. K. Shen, T. Y. Luh, H. C. Yang, K. C. Hwang, C. K. Chou. Blockage of apoptotic signaling of transforming growth factor-beta in human hepatoma cells by carboxyfullerene. *Eur. J. Biochem.* **1998**, *254*, 38–43.

101 C. Fumelli, A. Marconi, S. Salvioli, E. Straface, W. Malorni, A. M. Offidani, R. Pelicciari, G. Schettini, A. Giannetti, D. Monti, C. Franceschi, C. Pincelli. Carboxyfullerenes protect human keratinocytes from ultraviolet-B-induced apoptosis. *J. Invest. Dermatol.* **2000**, *115*, 835–841.

102 D. Monti, L. Moretti, S. Salvioli, E. Straface, W. Malorni, R. Pelicciari, G. Schettini, M. Bisaglia, C. Pincelli, C. Fumelli, M. Bonafé, C. Franceschi. C_{60} carboxyfullerene exerts a protective activity against oxidative stress-induced apoptosis in human peripheral blood mononuclear cells. *Biochem. Biophys. Res. Commun.* **2000**, *277*, 711–717.

103 M.-C. Chen, M.-J. Hwang, Y.-C. Chou, W.-H. Chen, G. Cheng, H. Nakano, T.-Y. Luh, S.-C. Mai, S.-L. Hsieh. The role of apoptosis signal-regulating kinase 1 in lymphotoxin-β-receptor-mediated cell death. *J. Biol. Chem.* **2003**, *278*, 16073–16081.

104 A. M.-Y. Lin, S.-F. Fang, S.-Z. Lin, C.-K. Chou, T.-Y. Luh, L.-T. Ho. Local carboxyfullerene protects cortical infarction in rat brain. *Neurosci. Res.* **2002**, *43*, 317–321.

105 K. Okuda, M. Hirobe, M. Mochizuki, T. Mashino. Effects of fullerene derivatives on active oxygen toxicity in *E. coli. Proc. Electrochem. Soc.* **1997**, *97*, 337–338.

106 N. Tsao, P. P. Kanakamma, T.-Y. Luh, C.-K. Chou, H.-Y. Lei. Inhibition of *Escherichia coli*-induced meningitis by carboxyfullerene. *Antimicrob. Agents Chemother.* **1999**, *43*, 2273–2277.

107 T. Mashino, K. Okuda, T. Hirota, M. Hirobe, T. Nagano, M. Mochizuchi. Inhibition of *E. coli* growth by fullerene derivatives and inhibition mechanism. *Bioorg. Med. Chem. Lett.* **1999**, *9*, 2959–2962.

108 M. Bisaglia, B. Natalini, R. Pellicciari, E. Straface, W. Malorni, D. Monti, C. Franceschi, G. Schettini. C_3-fullero-tris-methanodicarboxylic acid protects cerebellar granule cells from apoptosis. *J. Neurochem.* **2000**, *74*, 1197–1204.

109 I. C. Wang, L. Tai, D. Lee, P. Kanakamma, C. F. Shen, T. Y. Luh, C. Cheng, K. Hwang. C_{60} and water-soluble fullerene derivatives as antioxidants against radical-initiated lipid peroxidation. *J. Med. Chem.* **1999**, *42*, 4614–4620.

110 R. V. Bensasson, M. Brettreich, J. Frederiksen, H. Gottinger, A. Hirsch, E. J. Land, S. Leach, D. J. McGarvey, H. Schonberger. Reactions of e^-_{aq}, $CO_2^{\cdot-}$, $HO^{\cdot-}$, $O_2^{\cdot-}$ and $O_2(^1D_g)$ with a dendro[60]fullerene and $C_{60}[C(COOH)_2]_n$ ($n = 2–6$). *Free Radic. Biol. Med.* **2000**, *20*, 26–33.

111 K. Okuda, T. Hirota, M. Hirobe, T. Nagano, M. Mochizuki, T. Mashino. Synthesis of various water-soluble C_{60} derivatives and their superoxide-quenching activity. *Fullerene Sci. Technol.* **2000**, *8*, 89–104.

112 S. S. Ali, J. I. Hardt, K. L. Quick, J. S. Kim-Han, B. F. Erlanger, T.-T. Huang, C. J. Epstein, L. L. Dugan. A biologically effective fullerene (C_{60}) derivative with superoxide dismutase mimetic properties. *Free Radic. Biol. Med.* **2004**, *37*, 1191–1202.

113 L. L. Dugan, J. K. Gabrielson, S. P. Yu, T.-S. Lin, D. W. Choi. Buckminsterfullerenol free radical scavengers reduce excitotoxic and apoptotic death of cultured cortical neurons. *Neurobiol. Dis.* **1996**, *3*, 129–135.

114 L. Y. Chiang, F.-J. Lu, J.-T. Lin. Free-radical scavenging activity of water-soluble fullerenols. *J. Chem. Soc. Chem. Commun.* **1995**, 1283–1284.

115 Y.-L. Lai, L. Y. Chiang. Water-soluble fullerene derivatives attenuate exsanguination-induced bronchoconstriction of guinea-pigs. *J. Auton. Pharmacol.* **1997**, *17*, 229–235.

116 Y.-L. Lai, W.-Y. Chiou, L. Y. Chiang. Fullerene derivatives attenuate bronchoconstriction induced by xanthine–xanthine oxidase. *Fullerene Sci. Technol.* **1997**, *5*, 1337–1345.

117 H. C. C. Chen, Y. T. Huang, V. F. Paug, S. C. Liang, L. Y. Chiang. Water-soluble C_{60} and macrophages: morphologic features of FC4S-treated peritoneal macrophages *in vitro* and *in vivo* – a preliminary report. *Fullerene Sci Technol.* **1999**, *7*, 505–517.

118 S. Bosi, T. Da Ros, S. Castellano, E. Banfi, M. Prato. Antimycobacterial activity of ionic fullerene derivatives. *Bioorg. Med. Chem. Lett.* **2000**, *10*, 1043–1045.

119 T. Mashino, N. Usui, K. Okuda, T. Hirota, M. Mochizuki. Respiratory chain inhibition by fullerene derivatives: Hydrogen peroxide production caused by fullerene derivatives and a respiratory chain system. *Bioorg. Med. Chem.* **2003**, *11*, 1433–1438.

120 Y. I. Pukhova, G. N. Churilov, V. G. Isakova, A. Ya. Korets, Y. N. Titarenko. Biological activity of water-soluble fullerene complexes. *Doklady Akad. Nauk.* **1997**, *355*, 269.

121 R. Sijbesma, G. Srdanov, F. Wudl, J. A. Castoro, C. Wilkins, S. H. Friedman, D. L. DeCamp, G. L. Kenyon. Synthesis of a fullerene derivative for the inhibition of HIV enzymes. *J. Am. Chem. Soc.* **1993**, *115*, 6510–6512.

122 D. I. Schuster, S. R. Wilson, R. F. Schinazi. Anti-human immuno-deficiency virus activity and cytotoxicity of derivatized buckminsterfullerenes. *Bioorg. Med. Chem. Lett.* **1996**, *6*, 1253–1256.

123 C. Toniolo, A. Bianco, M. Maggini, G. Scorrano, M. Prato, M. Marastoni, R. Tomatis, S. Spisani, G. Palù, E. D. Blair. A bioactive fullerene peptide. *J. Med. Chem.* **1994**, *37*, 4558–4562.

124 D. I. SCHUSTER, L. WILSON, A. N. KIRSCHNER, R. F. SCHINAZI, S. SCHLUETER-WIRTZ, P. THARNISH, T. BARNETT, J. ERMOLIEFF, J. TANG, M. BRETTREICH, A. HIRSCH. Evaluation of the anti-HIV potency of a water-soluble dendrimeric fullerene. In: *Fullerene 2000 – Functionalized Fullerenes. Proceedings of the 197th Meeting of the Electrochemical Society.* The Electrochemical Society, Pennington, NJ, **2000**, vol. 9, p. 267.

125 S. MARCHESAN, T. DA ROS, G. SPALLUTO, J. BALZARINI, M. PRATO. Anti-HIV properties of cationic fullerene derivatives. *Bioorg. Med. Chem. Lett.* **2005**, *15*, 3615–3618.

126 S. BOSI, T. DA ROS, G. SPALLUTO, J. BALZARINI, M. PRATO. Synthesis and anti-HIV properties of new water-soluble bis-functionalized [60]fullerene derivatives. *Bioorg. Med. Chem. Lett.* **2003**, *13*, 4437–4440.

127 T. MASHINO, K. SHIMOTOHNO, N. IKEGAMI, D. NISHIKAWA, K. OKUDA, K. TAKAHASHI, S. NAKAMURA, M. MOCHIZUKI. Human immunodeficiency virus-reverse transcriptase inhibition and hepatitis C virus RNA-dependent RNA polymerase inhibition activities of fullerene derivatives. *Bioorg. Med. Chem. Lett.* **2005**, *15*, 1107–1109.

128 N. IWATA, T. MUKAI, Y. YAMAKOSHI, S. HARA, T. YANASE, M. SHOJI, T. ENDO, N. MIYATA. Effects of C_{60}, a fullerene, on the activities of glutathione S-transferase and glutathione-related enzymes in rodent and human livers. *Fullerene Sci. Technol.* **1998**, *6*, 213–226.

129 T. H. UENG, J. J. KANG, H. W. WANG, Y. W. CHENG, L. Y. CHIANG. Suppression of microsomal cytochrome P450-dependent monooxygenases and mitochondrial oxidative phosphorylation by fullerenol, a polyhydroxylated fullerene C_{60}. *Toxicol. Lett.* **1997**, *93*, 29–37.

130 S. L. McGOVERN, E. CASELLI, N. GRIGORIEFF, B. K. SHOICHET. A common mechanism underlying promiscuous inhibitors from virtual and high-throughput screening. *J. Med. Chem.* **2002**, *45*, 1712–1722.

131 G. P. TEGOS, T. N. DEMIDOVA, D. ARCILA-LOPEZ, H. LEE, T. WHARTON, H. GALI, M. R. HAMBLIN. Cationic fullerenes are effective and selective antimicrobial photosensitizers. *Chem. Biol.* **2005**, *12*, 1127–1135.

132 T. MASHINO, D. NISHIKAWA, K. TAKAHASHI, N. USUI, T. YAMORI, M. SEKI, T. ENDO, M. MOCHIZUKIA. Antibacterial and antiproliferative activity of cationic fullerene derivatives. *Bioorgan. Med. Chem. Lett.* **2003**, *13*, 4395–4397.

133 C. YU, T. CANTEENWALA, L. Y. CHIANG, B. WILSON, K. PRITZKER. Photodynamic effect of hydrophilic C_{60}-derived nanostructures for catalytic antitumoral antibacterial applications. *Synthetic Metals* **2005**, *153*, 37–40.

134 J. D. FORTNER, D. Y. LYON, C. M. SAYES, A. M. BOYD, J. C. FALKNER, E. M. HOTZE, L. B. ALEMANY, Y. J. TAO, W. GUO, K. D. ASMUS, V. L. COLVIN, J. B. HUGHES. C_{60} in water: nanocrystal formation and microbial response. *Environ. Sci Technol.* **2005**, *39*, 4307–4316.

135 F. PELLARINI, D. PANTAROTTO, T. DA ROS, A. GIANGASPERO, A. TOSSI, M. PRATO. A novel [60]fullerene amino acid for use in solid-phase peptide synthesis. *Org. Lett.* **2001**, *3*, 1845–1848.

136 D. PANTAROTTO, A. BIANCO, F. PELLARINI, A. TOSSI, A. GIANGASPERO, I. ZELEZETSKY, J.-P. BRIAND, M. PRATO. Solid-phase synthesis of fullerene–peptides. *J. Am. Chem. Soc.* **2002**, *124*, 12543–12549.

137 B.-X. CHEN, R. S. WILSON, M. DAS, D. J. COUGHLIN, B. F. ERLANGER. Antigenicity of fullerenes: antibodies specific for fullerenes and their characteristics. *Proc. Natl Acad. Sci. USA* **1998**, *95*, 10809–10813.

138 B. C. BRADEN, F. A. GOLDBAUM, B.-X. CHEN, A. N. KIRSCHNER, S. R. WILSON, B. F. ERLANGER. X-ray crystal structure of an anti-Buckminsterfullerene antibody Fab fragment: biomolecular recognition of

C_{60}. *Proc. Natl Acad. Sci. USA* **2000**, *97*, 12193–12197.

139 W. H. Noon, Y. Kong, J. Ma. Molecular dynamic analysis of a buckyball-antibody complex. *Proc. Natl Acad. Sci. USA* **2002**, *99*, 6466–6470.

140 D. Pantarotto, C. D. Partidos, R. Graff, J. Hoebeke, J.-P. Briand, M. Prato, A. Bianco. Synthesis, structural characterization, and immunological properties of carbon nanotubes functionalized with peptides. *J. Am. Chem. Soc.* **2003**, *125*, 6160–6164.

141 G. R. Dieckmann, A. B. Dalton, P. A. Johnson, J. M. Razal, J. Chen, G. M. Giordano, E. Munoz, I. H. Musselman, R. H. Baughman, R. Draper. Controlled assembly of carbon nanotubes by designed amphiphilic peptide helices. *J. Am. Chem. Soc.* **2003**, *125*, 1770–1777.

142 S. Q. Wang, E. S. Humphreys, S.-Y. Chung, D. F. Delduco, S. R. Lustig, H. Wang, K. Parker, N. W. Rizzo, S. Subramoney, Y.-M. Chiang. Peptides with selective affinity for carbon nanotubes. *Nat. Mater.* **2003**, *2*, 196–200.

143 D. Pantarotto, C. D. Partidos, J. Hoebeke, F. Brown, E. Kramer, J.-P. Briand, S. Muller, M. Prato, A. Bianco. Immunization with peptide-functionalized carbon nanotubes enhances virus-specific neutralizing antibody responses. *Chem. Biol.*, **2003**, *10*, 961–966.

144 B. F. Erlanger, B.-X. Chen, M. Zhu, L. Brus. Binding of an anti-fullerene IgG monoclonal antibody to single wall carbon nanotubes. *Nano Lett.* **2001**, *1*, 465–467.

145 E. Ostuni, R. G. Chapman, R. E. Holmlin, S. Takayama, G. M. Whitesides. A survey of structure–property relationships of surfaces that resist the adsorption of protein. *Langmuir* **2001**, *17*, 5605–5620 and references therein.

146 M. Shim, N. W. S. Kam, R. J. Chen, Y. Li, H. Dai. Functionalization of carbon nanotubes for biocompatibility and biomolecular recognition. *Nano Lett.* **2002**, *2*, 285–288.

147 M. P. Mattson, R. C. Haddon, A. M. Rao. Molecular functionalization of carbon nanotubes and use as substrates for neuronal growth. *J. Mol. Neurosci.* **2000**, *14*, 175–182.

148 N.-Y. Pan, J.-S. Shih. Piezoelectric crystal immunosensors based on immobilized fullerene C_{60}-antibodies. *Sens. Actuators B* **2004**, *98*, 180–187.

149 D. W. Cagle, S. J. Kennel, S. Mirzadeh, J. M. Alford, L. J. Wilson. *In vivo* studies of fullerene-based materials using endohedral metallofullerene radiotracers. *Proc. Natl Acad. Sci. USA* **1999**, *96*, 5182–5187.

150 T. Wharton, L. J. Wilson. Highly-iodinated fullerenes as a contrast agent for X-ray imaging. *Bioorg. Med. Chem.* **2002**, *10*, 3545–3554.

151 T. Wharton, L. J. Wilson. Towards fullerene-based X-ray contrast agents: design and synthesis of non-ionic, highly-iodinated derivatives of C_{60}. *Tetrahedron Lett.* **2002**, *43*, 651–564.

152 R. D. Bolskar, A. F. Benedetto, L. O. Husebo, R. E. Price, E. F. Jackson, S. Wallace, L. J. Wilson, J. M. Alford. First soluble M@C_{60} derivatives provide enhanced access to metallofullerenes and permit *in vivo* evaluation of Gd@$C_{60}[C(COOH)_2]_{10}$. *J. Am. Chem. Soc.* **2003**, *125*, 5471–5478.

153 S. Laus, B. Sitharaman, É. Tóth, R. D. Bolsklar, L. Helm, S. Asokan, M. S. Wong, L. J. Wilson, A. E. Merbach. Destroying gadofullerene aggregates by salt addition in aqueous solution of Gd@$C_{60}(OH)_x$ and Gd@$C_{60}[C(COOH)_2]_{10}$. *J. Am. Chem. Soc.* **2005**, *127*, 9368–9369.

154 É. Tóth, R. D. Bolsklar, A. Borel, G. González, L. Helm, A. E. Merbach, B. Sitharaman, L. J. Wilson. Water-soluble gadofullerenes: towards high-relaxivity, pH-responsive MRI contrast agents. *J. Am. Chem. Soc.* **2005**, *127*, 799–805.

155 H. Kato, Y. Kanazawa, M. Okumura, A. Taninaka, T. Yokawa, H. Shinohara. Lanthanoid endohedral metallofullerenols for MRI contrast agents. *J. Am. Chem. Soc.* **2003**, *125*, 4391–4397.

156 M. Mikawa, H. Kato, M. Okumura, M. Narazaki, Y. Kanazawa, N. Miwa, H. Shinohara. Paramagnetic water-soluble metallofullerenes having the highest relaxivity for MRI contrast agents. *Bioconjugate Chem.* **2001**, *12*, 510–514.

157 H. Shinohara, K. Yagi, J. Nakamura. *Japanese Patent 94-285395.*

158 E. B. Iezzi, J. C. Duchamp, K. R. Fletcher, T. E. Glass, H. C. Dorn. Lutetium-based trimetallic nitride endohedral metallofullerenes: new contrast agents. *Nano Lett.* **2002**, *2*, 1187–1190.

159 T. Cai, Z. Ge, E. B. Iezzi, T. E. Glass, K. Harich, H. W. Gibson, H. C. Dorn. Synthesis and characterization of the first trimetallic nitride templated pyrrolidino endohedral metallofullerenes. *J. Chem. Soc. Chem. Commun.* **2005**, 3594–3596.

160 T. Y. Zakharian, A. Seryshev, B. Sitharaman, B. E. Gilbert, V. Knight, L. J. Wilson. A fullerene–paclitaxel chemotherapeutic: synthesis, characterization, and study of biological activity in tissue culture. *J. Am. Chem. Soc.* **2005**, *127*, 12508–12509.

161 E. Nakamura, H. Isobe, N. Tomita, M. Sawamura, S. Jinno, H. Okayama. Functionalized fullerene as an artificial vector for transfection. *Angew. Chem. Int. Ed. Engl.* **2000**, *39*, 4254–4257.

12
Nanotechnology for Biomedical Devices

Lars Montelius

12.1
Introduction

The pace of employment of advanced microfabrication technology for realizing biomedical devices has been accelerated with the development of a seemingly never-ending increased packing density of functionalities in semiconductor chips [1] as a consequence of the downscaling of device dimensions in such chips [2]. This area of microelectronic development has been one of the major drivers in the development of our modern, technological society. However, this success has not only been brought to us by simple dimensional downscaling, but also by impressive developments in the fields of material sciences that have led to several new basic material discoveries [3–5]. Such inventions have, in turn, been brought into the engineering sciences, realizing these devices into everyday life. One can especially think about the invention of the concept of hetero-structures, making it possible to fabricate semiconductor lasers and optical detectors – the key to modern fiber optic-based communication [4–7]. In the area of biomedical devices, the development in material sciences has brought diagnostic tools such as computer-based imaging systems relying on the interplay between tissue and X-rays, magnetic radiation and ultrasound, etc. [8–10]. We have also witnessed various electronic devices being developed for enhancing impaired senses such as hearing aids, etc. Several devices have made life easier for those suffering from glucose-related diseases, for example [11, 12]. Handheld devices have been developed allowing individuals to monitor the sugar content of their blood in their own environment and, hence, also permitting the use of insulin delivery when needed.

With the introduction of nanotechnology for fabrication of significantly downscaled materials, we are now facing another even more rapid development. These materials do not only occupy an immensely small volume, but they also possess certain functions being created just due to the reduction in size [4–7, 13–15]. A conservative and often heard definition of nanodevices is that at least one of the scalar dimensions is below 100 nm and that the functionality of the device itself is changed due to its nanometer size.

Nanotechnologies for the Life Sciences Vol. 4
Nanodevices for the Life Sciences. Edited by Challa S. S. R. Kumar
Copyright © 2006 WILEY-VCH Verlag GmbH & Co. KGaA, Weinheim
ISBN: 3-527-31384-2

The development of microfabrication methods and technologies in the nanometer domain is considered by many people not only to be a simple evolutionary step, but also to have a more revolutionary aspect [13, 15]. Basically, this is achieved through the downscaling that led not only to an added packing density of functions, which means higher modality, but also to objects with certain unique properties. Such evolution has not yet been experienced by mankind on any large scale and right now we are in the middle of it. Although we have developed a lot of new tools for fabrication, characterization, imaging and manipulation of matter at the nanoscale [15–25], we must still consider ourselves to be in a time rather similar to the "hammer and sledge" stage in the stone-age. Nevertheless, we have already experienced an enormous impact in many of today's consumer products, e.g. recently a new cellular phone was announced having silver particles on its surface, killing bacteria and germs [26]. During the 5 last years we have also seen the introduction of various kinds of garments making the fabric more or less impossible to stain [27]. Recently, nanotechnology-based methods have been implemented to spin fibers, creating possibilities for enhanced fabrics having built-in intelligent functions [28]. Today we can buy self-cleaning windows that use a catalytic process to break down inorganic dirt that will eventually be washed off the window during next rain fall [29]. We can tailor-make surfaces with certain hydrophilic properties so that they can serve either as a surface that a liquid just wets or that a liquid will form droplets on. Such droplets may encapsulate drugs or reagents for future fluidic "factory-on-a-chip" applications, and can be manipulated by their surface tension [28] and/or by electro-wetting, making it possible to move droplets of liquids at will on a two-dimensional (2-D) surface [30]. In the area of biomedical devices, there are many devices that rely on absolute modern microfabrication standards; however, when talking about real nanotechnology-based devices (in the concept of the definition given above), such devises are much less numerous.

In this chapter, I introduce the reader to various forms of nanotechnologies and try to explain why some of these methods are more likely, and more suitable, than others to be employed in the biomedical field. I try to describe how modern nanotechnology may impact the fields of biomedical engineering by giving some examples of nanotechnologies applied in the domain of life sciences. Of course, it is impossible to cover fully the fields of nanobiological and nanomedical research and development, and there are various journals and other publications that cover these areas of nanotechnology. What I present here is solely an attempt to give some representative and, hopefully, interesting examples that may bring the interest and excitement in this field to the reader.

I have divided this chapter into various main parts, each part being divided in subsections. In the first part, I describe various forms of nanotechnologies and nanotools, mainly coming from the field of inorganic nanotechnology. After an overall introductory discussion, some technologies are described in more detail in various subsections. In the second part, I give an overview of various forms of applications having a relation to biomedical devices and technology. In further subsections, I describe a couple of such applications in more depth. The final part presents a discussion and a short speculation of what the future may bring.

12.2
Nanotechnologies

12.2.1
Overview of Nanotechnologies and Nanotools

In order to create nanostructures, one is bound to follow two approaches – either top-down or bottom-up. The former is considered to be the driving force for nanotechnology development, while the latter can often be described as a molecular nanotechnology approach [13]. However, these two approaches are more and more blending together, and one is often obliged to make use of both methods in order to fabricate functional devices. Hence, both technologies are just as important for the development of the field. In the top-down approach, one usually relies on a lithographic technology where one can pattern a surface into a relief structure by using some kind of a polymer layer that is put on top of a hard material, e.g. a piece of silicon. The lithographic tools necessary for making nanostructures are based on either electromagnetic [31–37] interactions with the polymer chains or mechanical deformation of it [20, 38–41]. Table 12.1 presents a summarized compilation showing some of the important characteristics. For the electromagnetic lithographies, one often uses optical illumination [31–34]. This is the method used by most semiconductor companies. However, the smallest dimensions using these methods are around 50 nm and if one needs smaller size definition, one is bound to utilize other technologies. Here, electron beam lithography (EBL) dominates [34–36], either using a vectorial or a circular exposure strategy depending on the application [42]. In all cases, the energized electrons either break or make molecular junctions in the polymer layer, making it either harder or softer for the subsequent solvent to develop the latent pattern, i.e. one can obtain both a positive and negative contrast [34–36]. With this method, one can obtain resolutions down to a few nanometers. After the exposure and development of the resist, one usually employs an etching method to induce a surface relief in the underlying substrate material [34]. One can also utilize lift-off methods [43] in order to make a metal pattern on the surface or a combination of both, possibly repeated time after time, thereby creating a complex pattern on a surface. Alternate technologies, also based on radiation, rely on the use of X-rays [35]. Most often the X-rays are produced as a consequence of Brehmsstrahlung in synchrotron rings [44] or due to laser-induced emission [45].

Another more recent approach is to employ various forms of mechanical deformations of the polymer layer, e.g. nanoimprint lithography (NIL). This technology was pioneered by Chou, presently at Princeton [19, 37–41]. The method resembles, in some sense, the way compact discs (CDs and DVDs) are fabricated. The difference is that for CDs one uses an embossing tool to make an all-plastic structure, while in NIL (Fig. 12.1) one makes a mechanical indentation into a polymer layer, having been spun onto a solid support, usually a silicon wafer [46]. It is also fair, in this connection, to mention the close resemblance of NIL to microcontact printing [41, 47–49]. Classically that denotes a technology that transfers a molecular layer

Table 12.1. Compilation of top-down lithographies. In the table is shown some key data of the major top-down lithographies. The numbers are not absolute but more given as indicative numbers to define in which "ball-park" the various lithographies presently belong. Interesting to note is that the various printing technologies show an overall good performance making them to be the possible preferred choices in a future perspective. For more discussions and references, see text.

Technologies	Generation of pattern	Exposure field per "exposure"	Possible feature resolution	Type of interaction	Throughput 10–12 inch diameter Wafer/hour
UV-lithographies	Mask	Chip (Dize)	≥ 50 nm	Electro magnetic	>50
X-ray	Mask	Chip	≥ 100 nm	Electro magnetic	≈ 10–30
Focussed Ion beam lithography	Serial	Chip	≥ 50 nm	Electro magnetic	1
E-beam lith	Serial	Chip	≥ 5 nm	Electro magnetic	0.1
SPM lithography	Serial	Chip	≥ 10 nm	Electro magnetic	<0.1
SPM manipulation	Serial	Chip	Individual atoms	Mechanical	<0.01
Micro-contact printing	Stamp	Wafers, ≈ 6 inch	≥ 20 nm	Mechanical	>20
Nanoimprint lithographies	Stamp	Wafer scale ≈ 8–10 inch	≥ 5 nm	Mechanical	>50
Injection moulding	Stamp	Wafer scale 130 mm	≥ 50 nm	Mechanical	$\geq$ several 100

by soft contacting of a stamp, made of a soft polymer inked with a layer of molecules, to the substrate. Thereby transferring the inked layer from the protruding features of the stamp. After this "fabrication" step, one may continue with etching or metal deposition, or by building another molecular layer on top of the first layer, sometimes utilizing covalent binding chemistry. It is also fair to mention the so-called step and flash technology [39, 41, 50–53], being heavily pushed as a NIL, although the process is a bit different. In this concept (Fig. 12.2) one utilizes a floating resist and a stamp that deforms the floating resist so that a relief structure is obtained (still in a liquid phase). By applying a UV exposure through the transparent mask/stamp, one can UV-harden the resist so that it is possible to remove the stamp leaving a UV-cured, structured polymer on the surface. These different "mechanical deformation" technologies can be denoted as a family of technologies, but one could also place the step and flash technology, for example, in the category of

NIL (Chou)

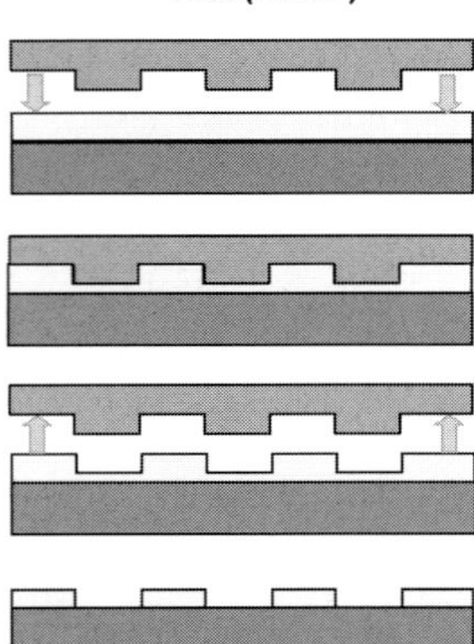

Figure 12.1. The process flow of NIL. (From Ref. [41].)

optical lithography. Nowadays, there is often a blending of technologies and one has come up with a hybrid technology, based on ideas borrowed from both areas. Here, one uses a hard stamp and spun on resist, but the actual hardening of the resist is not by temperature as in classical NIL, but by illumination. This technology, often denoted UV-NIL [41, 53–58], is maybe one of the most promising technologies for future high-volume manufacturing (HVM) of nanostructures. I will continue a bit longer on nanoimprint-related methods, since they offer an opportunity for fabrication of biomedical devices [59]. Using this technology, one can fabricate many samples having nanostructures in a very economical way [60]. This has previously been the limiting factor for not introducing nanotechnology in the biomedical field. There is a need in biomedical research to be able to make many structures in order to be able to build significance in the observations. This is in-

S-FIL (Willson)

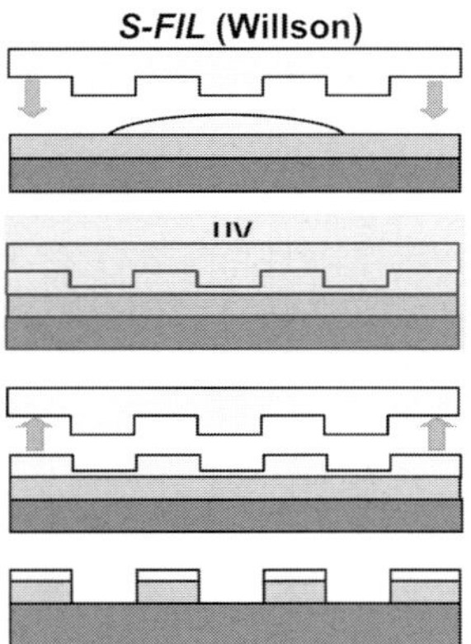

Figure 12.2. The process flow of step and flash NIL. (From Ref. [41].)

deed different from in physics, where one observable "always" will behave in a similar fashion from measurement to measurement. In biology, there is a natural spread due to the biological variability between species, etc., which necessitates that many samples need to be analyzed in order to obtain statistically significant results. Due to the importance that NIL may have in the future [61–63], I spend more time describing it in more depth in a separate subsection below.

The invention of scanning tunneling microscopy [64] and related atomic force microscopy (AFM) [18] has brought tools to nanoscience with which it is possible to image, characterize and manipulate individual atoms, molecules and structures [16–19, 21–25]. It is even possible to image a few magnetic spins of a few atoms below a sharp magnetic tip [66]. Also, as we see later, it has been instrumental for another class of sensor structures [65] as well as inspiring some of the nanotechnologies described above. Hence, it is fair to conclude that this technology has probably been the single most important one for the development of the nanotechnology field as we know it today. Scanning probe technology is discussed further in a separate subsection.

Lately, there has been considerable interest for making and exploring nanowires, e.g. carbon nanotubes (CNTs) and epitaxial wires [67–74], being made by a bottom-up approach (Figs. 12.3 and 12.4). This is a new class of materials with large pos-

Figure 12.3. Epitaxial nanowires of, in this case, InP as defined by a bottom-up approach. The lateral position is governed by catalytic metal particles (Au) deposited by a lift-off procedure before growth has taken place. The shape and length can be controlled in the process, while the diameter is determined by the size of the catalytic metal particle. The wires can be branched by putting additional metal particles on the sides, making an additional growth run and thereby creating a 3-D network. (From Ref. [68].)

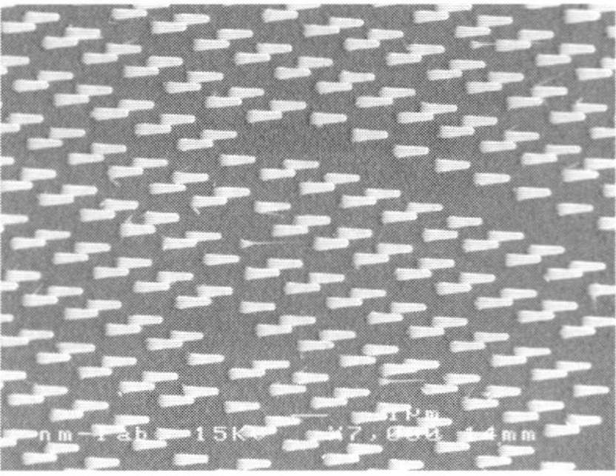

Figure 12.4. Epitaxial nanowires patterned by using NIL
allowing exact control of the wire positions over a large surface
area. (From Ref. [70].)

sibilities for future applications in many different areas of science. They offer
extremely small sizes with extremely large aspect ratios. They are not made by
top-down lithography; hence, their electronic properties have not been damaged
or affected by the processing. Their smallness offers mechanical properties that
may be explored for various nanomechanical applications. Some of the foreseeable
biomedical applications of nanowires are described below in more detail. The
nanomechanics of small objects [74–77] also offer several opportunities for applications as sensors in the biomedical area. Coupled with the increased efforts in
nanotechnology is the possibility to utilize nanoparticles for various applications.
These particles can be solid, semipermeable, hollow or core–shell particles. They
have immediate medical applications, such as magnetic nanoparticles [78–79]
that may be used as mobile reporters in magnetic imaging, allowing 3-D capabilities.

There are a lot of other research areas of great importance for handling molecular materials and cells at the nanoscale, e.g. the employment of biomimetic ideas,
which is a field on its own. Modifying surfaces using biomaterials can be very important for the development of methods allowing construction of model biological
structures for studies of various bio-reactions and for sensor applications, etc. For
instance, lipid bilayer membranes supported on surfaces can be very valuable since
they may offer possibilities to investigate membranes and membrane proteins in
an artificial environment [80–86].

Presently, there is an increased interest for using optical techniques based
on surface enhanced Raman scattering (SERS) and surface plasmon resonance
(SPR) as mediators for biomolecular interactions [87]. There is also an increased
interest in the use of light-emitting point sensors [88] as passive reporters for various chemical and biological reactions/processes. However, due to space limitations, these techniques will not be further explored in this chapter. The interested
reader is instead referred to other articles/reports (e.g. Ref. [88]).

12.2.1.1 **NIL**

The process of NIL (Fig. 12.1) is essentially a rather simple process based on pressing a pre-patterned substrate into a layer of polymer spun onto a hard support, the substrate, usually a silicon wafer. The stamp must be pre-processed containing the relief structure that is going to be printed into the polymer layer. For stamp fabrication, one usually employs EBL in combination with etching [89]. In order to be able to press the tool into the polymer layer one usually heats up the sandwiched tool–substrate to a process temperature, where the polymer layer has become fluid, i.e. one heats everything above the glass transition temperature (T_g) of the polymer. We have investigated several polymers [90] and found, as a rule of thumb, that the process temperature should be slightly above the T_g [91]. After applying a pressure high enough for the stamp to be pressed into the polymer layer, one cools down the sandwich, whilst relieving the pressure. Then one can separate the stamp and substrate from each other. This last separation is easier if one has put an anti-sticking layer onto the stamp [92–95]. There are several problem areas associated with such a process. If the stamp and substrate are made of different materials, heating may cause large problems due to the different thermal expansion coefficients, making the materials move laterally with respect to each other. Hence, one has been trying to come up with new polymers enabling patterning at lower temperatures [96]. Also, for HVM, one considers possibilities such as keeping the process temperature constant and pressing/removing at process temperature, thereby also minimizing the problem of any lateral mismatch between expansions of the stamp and substrate, respectively. An outstanding problem with NIL is the necessity of being able to align subsequent stamps to previous structures on a substrate [97]. A lot of activity is presently being pursued in order to solve this problem, [97–100]. Nevertheless, this technology is promising and it can fabricate nanostructures on a wafer scale [40, 46, 101, 102], and the imprint time can be less than seconds [103]. Hence, it is a very powerful technology for future nanoscale materials that may be employed in a variety of application fields.

12.2.1.2 **Other Lithography Techniques**

Just as NIL is a very good alternative as one of the next-generation technologies (NGL) [1, 13, 61, 62] of importance for the information technology sector, other printing techniques, such as microcontact printing [41, 51–53, 104], offer similar advantages. The method (to be more precise, the family of microcontact printing technologies) offers additional benefits as compared to NIL, since the process is simpler, it is a room temperature process, does not require a high pressure (although that problem can be circumvented by using soft NIL [41, 55–57]), has limited stamp–substrate sticking problems, etc. The major drawback is the flexible stamp commonly used that limits the resolution to the order of 100 nm and its use for wafer-scale single printing with high resolution. However, there is presently a tremendous development in nanoimprinting, microcontact and other printing technologies, so we will certainly soon experience an even larger variety of such technologies [105]. For instance, a method based on ''printing'' was reported recently by Yuand coworkers where they used DNA as a template in a replication "printing" process, making it possible to replicate DNA assemblies in a hybridization–contact–dehybridization process, which they denoted as "supramolecular nanostamping" [106]. In a similar way, high-density metal lateral nanowires were reported to be transferred from a template to a silicon surface through a contact mechanism. The authors employed "stamps", fabricated from a superlattice of III–V materials, combined with cleaving and selective etching and metallization. Using such an approach, it is possible to make very high-density and ordered lines that would be very difficult to make using other top-down approaches [107, 108].

A candidate for NGL has for many years been X-ray lithography [1, 35, 44, 45] and with the continuous push from the microelectronic industry for using UV lithographies there has been a merge of the two into the concept of extreme (E)-UV lithography where one utilizes light with wavelengths in the 10-nm regime (for a compilation of papers on E-UV lithography, see Ref. [109]). Usually, such light is generated from plasma sources [2, 45, 109]. The positive aspect of this technology is that the main processing infrastructure technology is rather similar that already employed in the big fabrication facilities. The drawback of E-UV, however, is the enormous costs required in order to put it into a working technology. The costs will not stop it, but it will effectively limit the number of facilities that can afford the investment. Hence, its importance for the biomedical field as a technology for making small series of chips is limited, since the initial cost for such a small series will be high. Another type of lithography technology is based on scanning probe microscopy (SPM) [110]. It will probably have a limited impact on wafer-scale processing. However, it is still one of the most important since it will allow small series of samples to be made in a rather inexpensive way. The needed development, which is making steady progress, is the increase of scanning speed, allowing both imaging and lithography to be made fast enough. However, the limitation of the method, even when it has gained real-time speeds of scanning, is the same as for EBL, i.e. the serial nature of pixel-by-pixel exposure. However, in contrast to several attempts made over the years for making an array of small electron-beam-based exposure sources that could expose a wafer in parallel, the use of an array of scanning probes might just overcome this problem (see also Sections 12.2.3 and 12.3.8). Dip-pen lithography [111, 112] has also been introduced, presenting us with another possibility to fabricate "molecular" lines or patterns without the need for going through the cumbersome pattern transferring method that is needed for other lithographies. Here, one has a tool that makes it possible to write only on those parts of a surface that one needs in a direct way, using the fountain pen principle. It is, to some extent at least, principally a scanning probe lithography method, based on the fact that there is always a water layer on a surface in air. If two surfaces are in contact with each other, the content of the water layer may, depending on the respective hydrophobicity of the two surfaces in contact with each other, be transferred from the tip to the surface and the water layer on the tip may not be consumed if it is connected to a reservoir of liquid or if it is dipped (inked) into a reservoir repeatedly. This is a method that naturally has a large potential for writing various forms of important molecular patterns for use in protein chip applications, etc. Some examples are given in the next section. In a further extension of the technique, one can fabricate a hollow cantilever tip with a small opening at its end allowing molecules to be transferred from the hollow tips to surfaces [113]. Using concepts from the old days, when one used shadow mask evaporation masks for making electrical (metal) contacts to semiconductors, one has refined the shadow mask technology into stencil mask nanolithography, enabling us to draw metal lines arbitrarily on surfaces in a vacuum [114, 115]. The technique allows a fine pattern to be made by moving the stencil (the shadow mask) while metal (or molecules or something else that can be evaporated) is flowing though the apertures of the stencil. This is a nice method that may have its specific areas application. It can easily be parallelized by having an array of such nanostencil masks moving in unison, allowing large areas to be covered in a short time. The drawback is the clogging of the nanoscale aperture that has to be overcome in order to find a real industrial use for it. The posibilities to develop the tools being described above have been enabled by the development of nanotechnologies in a pull–push fashion. One nanofabrication method that has been instrumental in this aspect, and actually is just in its initial phase of development at the moment, is the use of focused ion beams [116], allowing extremely nice structures to be fabricated in a true 3-D space. The method is often combined with scanning electron microscopy (SEM), allowing observation of the structures being made (although it is also possible to use the ions for scanning and imaging purposes). The principle is to make local and directed deposition or etching of materials due to the energetic ions impinging onto the surface. The method has the same limitations as EBL systems – it is a serial method. Still, it is not at all a mature technology at this stage and I guess that we will experience a tremendous development in other technologies, since it is such a generic technology.

12.2.1.3 Scanning Probes

Scanning probes offer many other possibilities in addition to imaging capabilities. For instance they have been utilized to make a new kind of memory device, the Millipede [117], which in principle is an array of cantilevers (Fig. 12.5) that can be controlled individually, allowing the possibility to melt a plastic material and press the tip into it, thereby cause cooling. Hence, it is possible to write information on the nanoscale. Such information, consisting of local deformations, can easily be read out with the cantilevers in imaging mode. Using the technology behind AFM [65], basically a cantilever beam that can be moved in the xyz directions with a sharp tip mounted at the end, one can make an analysis tool that will be able to detect interactions on the cantilever surface being observed through induced bending as a function of, for example, heat, or stress as a function of molecular adhesion to one of the surfaces, thereby creating a bimetal-like switching behavior [118] (Fig. 12.6). Such cantilever-based sensing, transforming molecular recognition into a mechanical displacement [119], has been used for many aspects, e.g. detection of antigen–antibody reactions [118]. These results are discussed further below. AFM

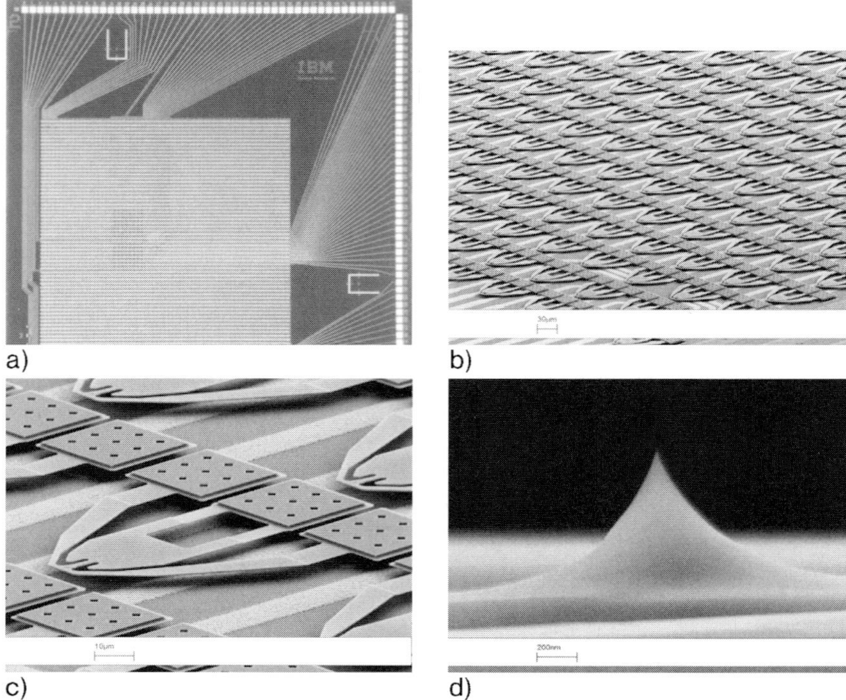

Figure 12.5. The Millipede chip: optical micrograph of a 64 × 64 cantilever array chip (A), and SEM images of the center of the array (B), one cantilever (C) and a tip apex (D). Data can be written at extremely high densities (greater than 1 TB in^{-1}). The cantilevers have separate heaters for reading and writing, and utilize electrostatic actuation for the z-direction. The cantilevers are around 70 μm long, with a 500–700 nm tip integrated directly above the write heater. (Reprinted with permission from IBM.)

has been utilized for force detection, allowing a detailed understanding of binding forces between molecules. One could denote this as "molecular fishing". It is based on coating the tip with a certain type of molecule and having another molecule that will bind to it on the substrate. When the tip is lowered into the molecules at the surface, a mechanical bond may be formed and, by subsequent retraction of the AFM tip from the surface, one can follow the sequence when breaking the chemical bonds. This provides valuable information about bond strengths and other molecular properties. This technology has been advanced to the level where it is possible to identify single molecular binding events [120]. Such detailed knowledge will be of great value in the biological area, e.g. in order to increase the understanding of cell signaling mechanisms and protein folding [121]. To fully use SPM for enhancing our understanding in biology, we must continue the development of probes so that they can obtain information from both time and space simultaneously, i.e. allow the possibility to capture images revealing important biological processes in real-time.

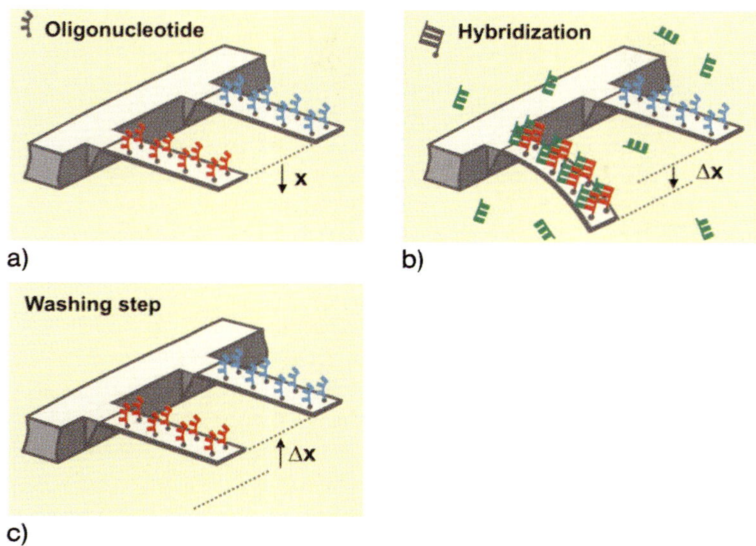

Figure 12.6. Operating modes of cantilevers employed as sensors: (A) static deflection mode, (B) dynamic resonance mode and (C) bimetallic heat mode. (From Ref. [118].)

12.3
Applications

12.3.1
Introduction

Here, I explore various directions of nanotechnology in the life sciences sector paving its way into the biomedical sector. However, so far, very few examples of real employment of nanotechnology in the biomedical sector can be found. Nanotechnology offers many chances for applications, but hitherto the development has been mainly focused on being used as sensors. In the following section, I show how passive and active nanostructures surfaces can be explored, how nanotechnology can affect the area of proteomics, and how nanowires and nanomechanics may be employed as sensors for molecular detection.

12.3.2
Biomedical Applications based on Nanostructured Passive Surfaces

For many life sciences applications, one might need to detect only a few substances dispersed into a solution and blended with other substances. One needs to have a functionality of the substances, giving them a kind of a handle. This can be solved by utilizing different kinds of chemistry, often in conjunction with attaching certain dyes in order to be able to detect various reactions. However, what one really would like to have is a system providing identification without the use of a marker, i.e. a label-free diagnostic method. The reason is 2-fold – a marker may change the biological function of the substance and there is a probability that markers may not attach with 100% fidelity. In order to find a species at low concentration, it would be advantageous to increase the concentration. Many of these aspects can be dealt with using nanotechnology and passive surfaces. By passive we mean a surface that may or should have a function, but we cannot influence that function from the outside during a measurement, observation or reaction.

12.3.2.1 Separation, Concentration and Enriching Structures

Separation can be achieved by employing nanotechnology in various ways. One form is to use a system of nanoscaled posts (functioning as bumpers) in an array format, together with time-alternating electric fields, making it possible to separate various molecules according to the entropic force they experience in the confined region, which in turn limits the natural conformation possibilities that they have [122, 123] (Figs. 12.7 and 12.8). Since the entropy of a molecule depends on its length, these confined space structures act as a sorter of molecules with various lengths, e.g. lengths of sequences of DNA. Such a molecular sorter at the nanoscale has been fabricated by EBL. Similar confinement-related devices, but instead of posts, using nanochannels in an array format have been proposed and realized utilizing NIL [41, 124, 125]. Here, NIL was employed (Figs. 12.9 and 12.10) creating ultra-narrow channels, forcing fluorescently labeled DNA to stretch out while passing through the narrow channel [124]. The vision for these kinds of structures is to be able to integrate light emitters along the channel and to determine the DNA code down to individual bases by reading them as they pass the emitters using near-field excitation/detection mechanisms. The first steps have been realized (Fig. 12.11), allowing the detection of a marker attached to the DNA at a certain position, using excitation through narrow slits positioned perpendicular to the channels [125, 126].

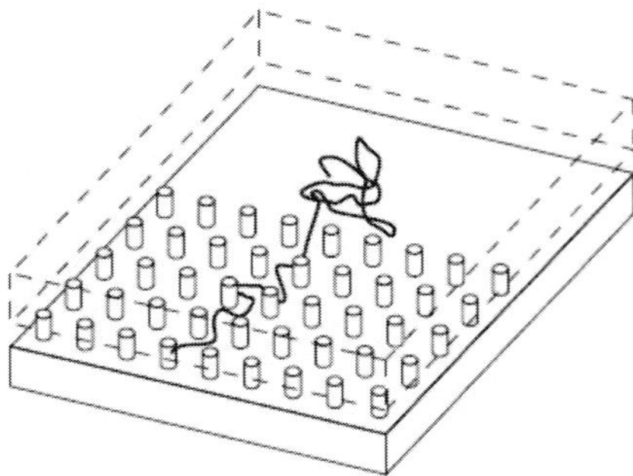

Figure 12.7. A fluidic device consisting of an array of nanopillars in between a roof and a floor. (From Ref. [122].)

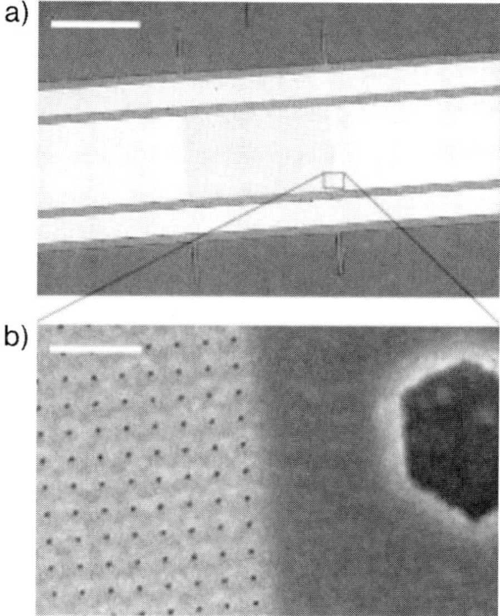

Figure 12.8. (A) An optical micrograph of the fabricated device. (B) A close-up showing the 35-nm pillars with a center-to-center distance of 160 nm. (From Ref. [122].)

It is especially interesting to note the use of another polymer, Topaz, with important benefits when it comes to biocompatibility and nanofluidicity in combination with optical spectroscopy [127]. The properties are such that it is an ideal candidate for integration of a waveguide directly in the polymer (Figs. 12.12 and 12.13). Furthermore, it would be possible to insert dye molecules acting as local emitters for locally addressing, for example, a DNA chain being swept in a channel close to the emitter. The full utilization of these concepts has not yet been reported. Similar kinds of nanofluidic structures have been employed to create a cell-based assay

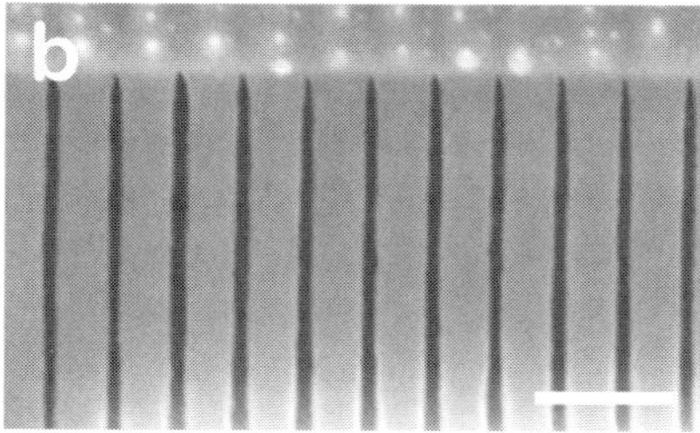

Figure 12.9. SEM images showing (a) the cross-section of the NIL-fabricated nanochannels and (b) the top view displaying the channels. (From Ref. [124].)

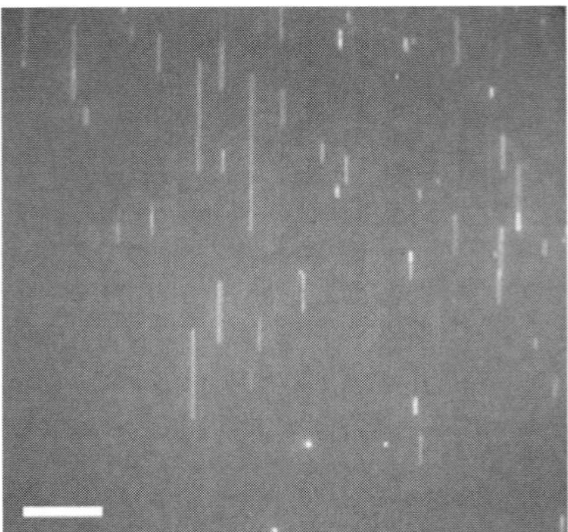

Figure 12.10. CCD image showing DNA stretched out in the nanochannels. Scale bar is 30 μm. (From Ref. [124].)

[128], and in a recent paper a nanofluidic transistor (Fig. 12.14) was reported enabling control of ion and molecular flow in a channel [129].

12.3.2.2 Molecular Motors Transported in Nanometer Channels

Other kinds of nanoscaled channel structures have been reported in the field of molecular motors (Figs. 12.15 and 12.16). This work, using nanoscaled structures, has been pioneered in a collaborative effort by the author and Mansson at Kalmar University [130–135]. Several other groups have reported micrometer-sized channels [136, 137], showing a performance similar to that obtained by us. However, we have, in contrast to the others, a precision in the motor guidance at the nanometer level [132, 135]. We have further refined the concepts and have built a com-

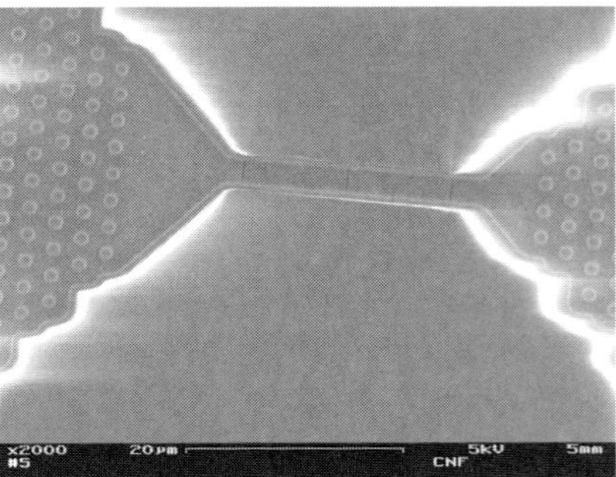

Figure 12.11. SEM image of a device with 100-nm wide slits for optical detection of DNA molecules as they pass by in the 5-μm wide channel. At the outer ends are nanopillars for stretching the DNA filaments before entry into the channel region. (From Ref. [125].)

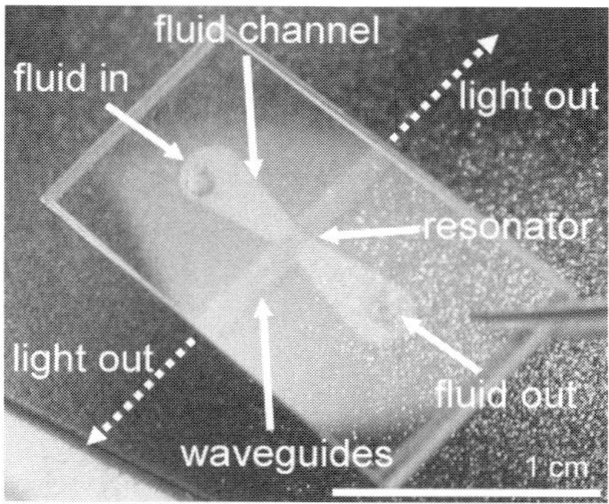

Figure 12.12. A photo of the microfluidic device made from Topaz with incorporated laser resonator. (From Ref. [127].)

plete toolbox system (Fig. 12.17) around this activity, where we can direct motion by fabrication of rectifiers, roundabouts, injectors, etc. [135]. Basically, we have combined nanolithography and surface functionalization (Figs. 12.18 and 12.19) using silane procedures, resulting in a very robust structure permitting detailed studies of the actomyosin system [134–136]. Such a system may be explored as a "lab-on-a-chip" system for use in drug development. For instance, one can imagine using a circular pattern into which individual motors are directed and, by counting the rotation speed of the motors in these circles as a function of the added drug, for instance, one may be able to investigate the effect of various formulas or concentrations of added drugs. If so, maybe such a system can work favorable as a pre-screening system for drug development.

12.3.2.3 Topographical Structures, Cells and Guidance of Neurons

Another class of passive structures is topographical patterns in conjunction with cell survival, growth and adhesion, with important implications for medical im-

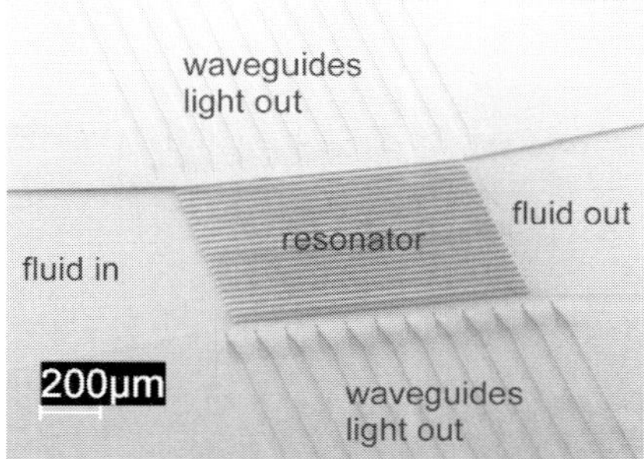

Figure 12.13. A SEM image showing the laser resonator structure located in the flow region. (From Ref. [127].)

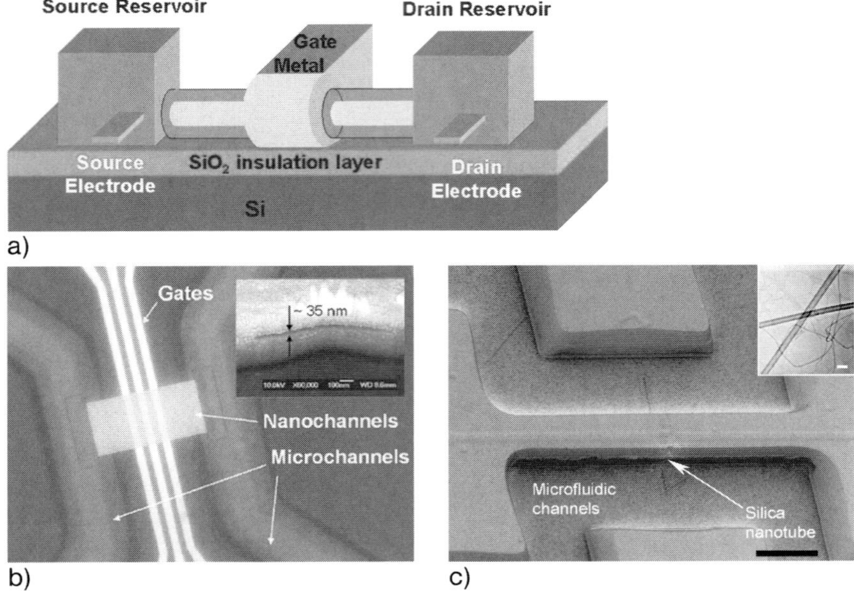

Figure 12.14. A nanofluidic transistor. (A) Schematic representation of the device. (B) Micrograph showing the 35 nanochannels running from left to right and the three electrode lines perpendicular to the nanochannels. (From Ref. [129].)

plants and prosthetic devices. Several investigations have been performed showing the positive effects of micrometer structures with respect to individual cell attachments, but results employing real nanometer-sized structures are sparse [138]. For the single-cell case and microstructures, grooved topography especially has been found to interact [139] and cells have been reported to change their shape in order to follow the grooves (and ridges) [140]. It has also been observed that cell regulation and transcription may be affected [141]. For nanostructures, macrophages, endothelial and fibroblast cells have been reported to interact in one way or another way [142–144]. We have recently performed a pilot study in which we have utilized nerve cells and studied the possibility of aligning nerve cells during growth [145]. Using NIL we made a large number of nanostructured surfaces containing grooves in the commonly employed polymer poly(methyl methacrylate) (PMMA). We made both positive and negative structures in 200 µm × 200 µm squares, containing lines with dimensions varying between 100 and 400 nm, while the distance in between the lines varied from 200 to 1200 nm. Ganglia, both sympathetic and sensory, were dissected from an adult female NMRI mouse and mounted on the chip surfaces, about 1 mm from 17 nanostructured squares, as described above. After 5–7 days in culture media, we found that most axons had regenerated along all of the lines with an apparent preference to grow along the line peripheries (Figs. 12.20–12.22) [145]. The aim of the study was to investigate if we could separate

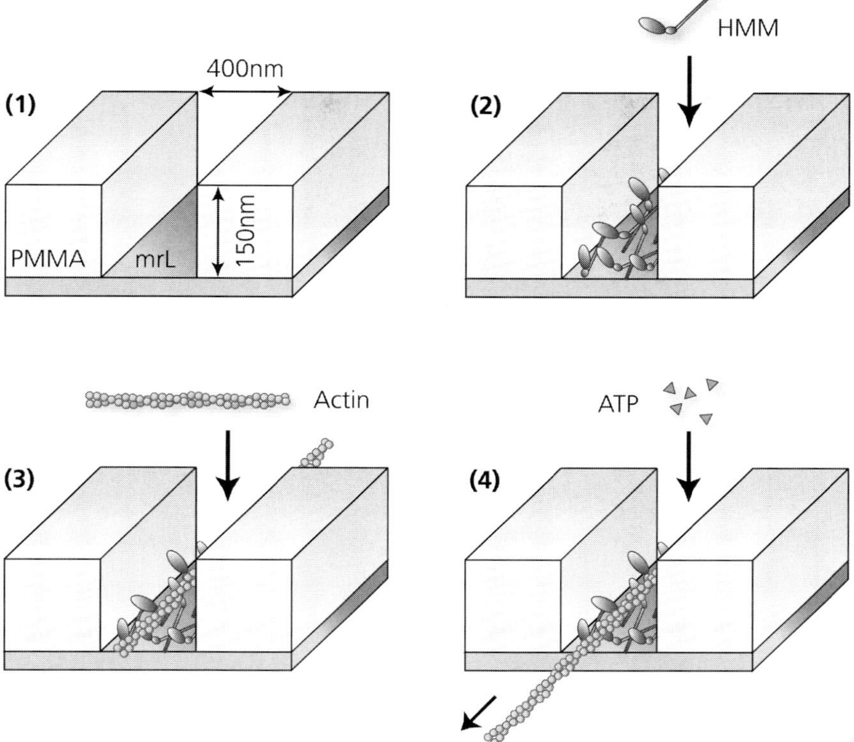

Figure 12.15. Schematic presentation of the nanochannels defined by a lithographic method followed by immobilization of myosin heads to the patterned surface, and addition of fluorescently labeled actin filament and ATP. The actin filaments will, with the proper choice of material for the lines, be effectively transported through the channels. (From Ref. [130].)

a large number of axons on a small area and, indeed, we showed that possibility. Such spatial separation and the possibility to organize axons have important consequences for the possible ability to make electronic connections to the nervous system [146]. State-of-the-art implanted electronic devices found in practical use today

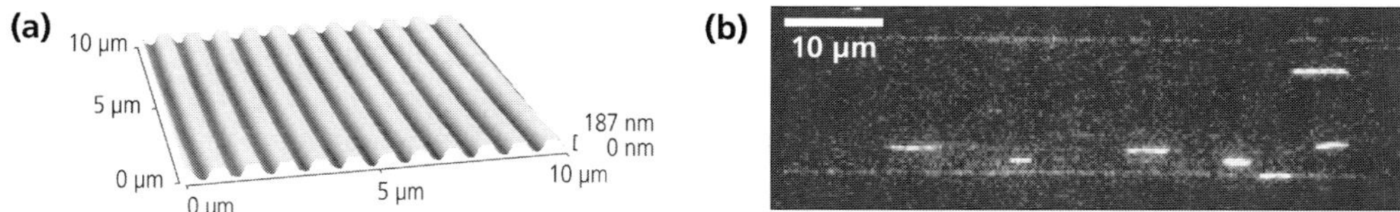

Figure 12.16. Similar nanochannels defined by NIL (a) and a snapshot of fluorescently labeled actin filaments (white colored) being transported in the lines (b). (From Ref. [131].)

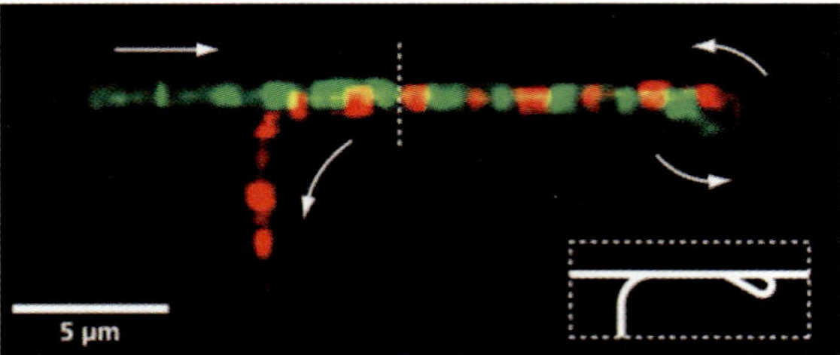

Figure 12.17. Three nanotools, rectifiers, roundabouts and directional couplers, put together to transport actin filaments (fluorescently labeled) first in a rectified way from left to right (green colored), and then in a roundabout structure being forced to make a 360° turn and go back into the same channel as before, moving in the opposite direction (red colored) and then halfway back it will follow a route downwards (directional coupler) and out to a reservoir. Note that the filament cannot enter into this downward branch on its move from left to right. (From Ref. [135].)

are the cortical implant BrainGate [147] and the very well-known cochlear implant. These devices have only a hundred electrodes on a 1-mm^2 area and such a density is not sufficient for making a functional brain–machine interface (BMI). Our results above show that it may be possible, using NIL, to increase the spatial resolution considerably; we found hundreds of axons on an area of 0.4-mm^2 [148]. Our

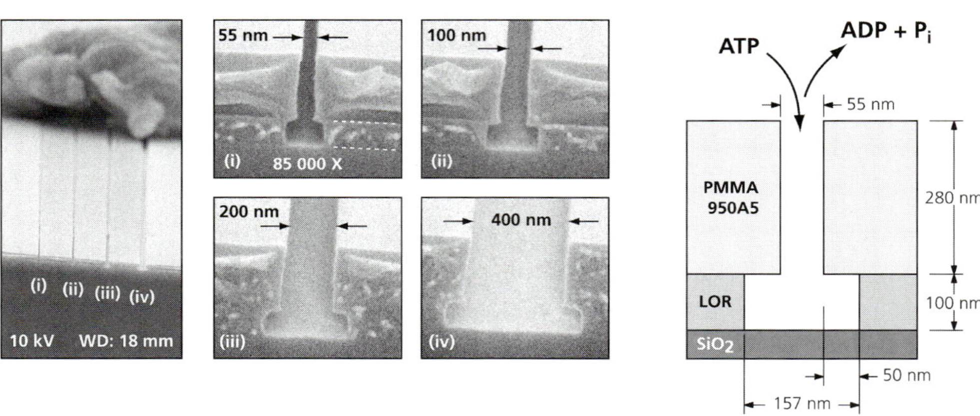

Figure 12.18. Cross-section of the double layer of resists combined with selective silanization of the channels employed to achieve the nanometer control of the movement of actin filaments. It consists of a LOR layer and on top of a PMMA layer. By proper patterning and processing, an undercut as shown in the cross-section picture, can be obtained. As illustrated, the slit along the wire (opening) serves an important role – through this opening ATP can be inferred and products such as ADP can be removed. If we employed sealed nanochannels we would eventually get clogging and we would also be limited by diffusion of ATP, leading to unstable conditions. The bottom layer is functionalized through a vapor-phase silanization procedure to enhance motility. (From Ref. [132].)

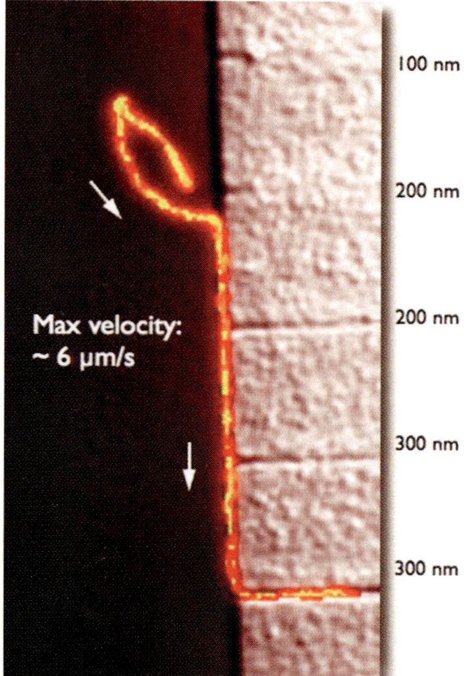

Figure 12.19. A top view of the pattern showing the loading zone to the left and nanochannels going to the right in the picture. An actin filament (fluorescently labeled) is shown to have followed the edge of the loading zone (under the undercut) until a suitable nanochannel made the filament turn right and follow the nanochannel. (From Ref. [132].)

simple, nonoptimized study shows the possibility to concentrate and separate nerve cells on a small area of a surface. Furthermore, the topographical structures are highly stable and NIL, as a method, is scaleable, providing additional support for possible future BMI applications based on nanostructured surfaces. Recently, we have also employed epitaxial nanowires as a support surface showing excellent guidance of neuronal outgrowth (Fig. 12.23). Further details are reported elsewhere [149]. The review by Melechko and coworkers [69] showed various life sciences applications of carbon nanowires. Large sets of interconnected CNTs with cell membranes were recently reported, showing that the biological properties of the membrane were retained and that they could perform electrical measurements [150].

12.3.3
Biomedical Applications utilizing Active Nanostructured Surfaces

In contrast to the previous topics, active nanostructured surfaces can be activated from the outside world in order to investigate the interaction of various forms of

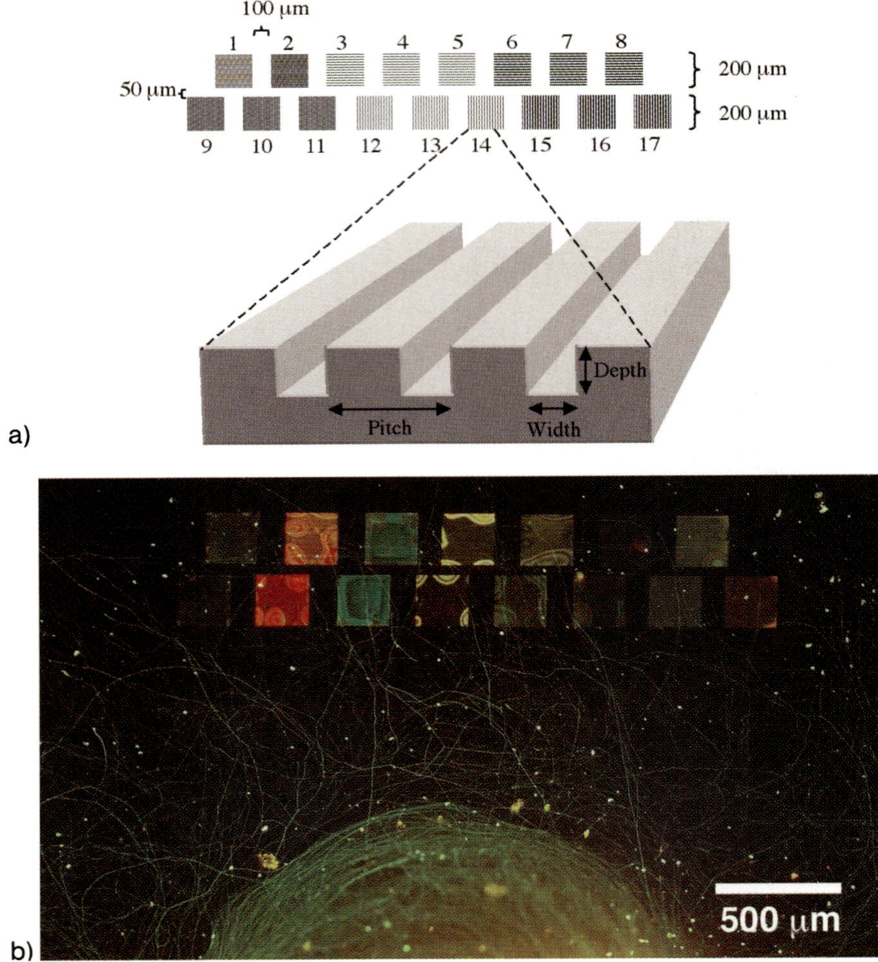

Figure 12.20. A number of NIL-patterned squares (200 µm × 200 µm) containing grooves oriented horizontally (the two at the bottom) and vertically (the two top ones) having different widths and pitches. As can be seen, the axonal outgrowth (mouse cervical ganglion) follows the orientation of the grooves in the squares and changes growth direction when the grooves turn 90°. (From Ref. [145].)

biomolecules that such activation may invoke. A paper by Hamad-Schifferly reported remote inductive control of local DNA hybridization events by the use of a nanocrystal serving as an antenna covalently linked to DNA [151]. Another such class of structures that are made at the nanoscale is the so-called interdigitated array (IDA) of electrodes (Fig. 12.24). Although several groups have been able to fabricate such devices [152–155], few have been able to report any useful information

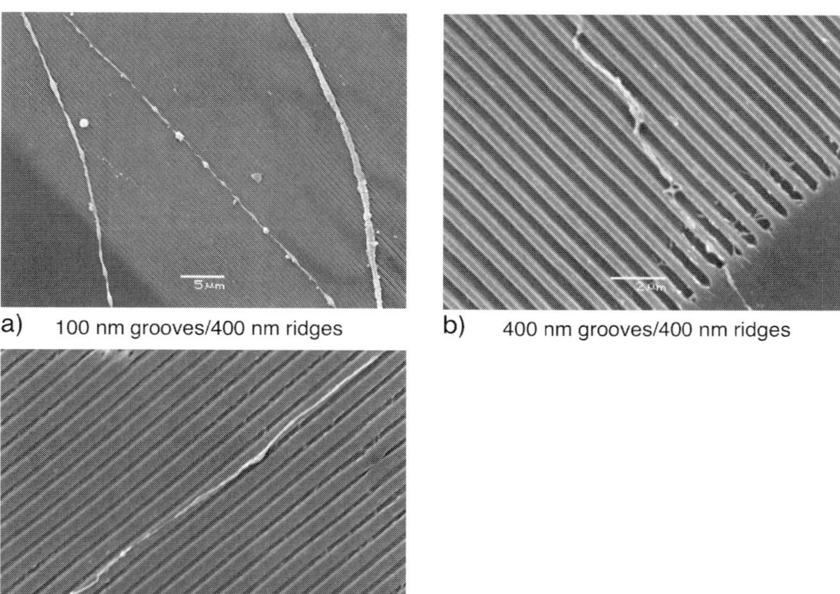

a)   100 nm grooves/400 nm ridges

b)   400 nm grooves/400 nm ridges

c)   400 nm grooves/800 nm ridges

Figure 12.21. SEM images showing axons growing on a positive pattern. The ridges are 400 nm on the (A) and (B), and 800 nm wide on (C)e. The linewidths are 100, 400 and 400 nm, respectively (A–C). The ridges are all 100 nm high. Note especially that all axons seem to grow on top of the ridges and not in the grooves between the ridges. (From Ref. [145].)

having the IDA in the nanometer regime. One of the common obstacles is the noisy environments for all kind of electrical measurements. However, in a set of publications, we have been able to follow some results as a consequence of using nanosized interdigitated electrodes. Choi and coworkers [155] report the possibility to detect DNA binding events by monitoring the impedance (Figs. 12.25 and 12.26). In the paper they report, using NIL in combination with certain nanoprocessing steps, that an array of electrodes having nanogaps of the order of 50 nm provided a way to make label-free detection of DNA hybridization events. Figure 12.25 shows a SEM image of the fabricated structure and the schematics for detection, and Fig. 12.26 displays the frequency-dependent capacitance. Clearly, the hybridized structure of T–A base pairs is detected by the capacitance increase at lower frequencies, in contrast to the nonconjugated base pair reference measurement. The explanation given is that the single- and double-stranded DNAs have different geometrical structures, and hence a different induced counterion profile. In another report [156] using IDA, the authors have investigated the relative increase in signal due to the nanoscale of various scaled IDAs (Fig. 12.27).

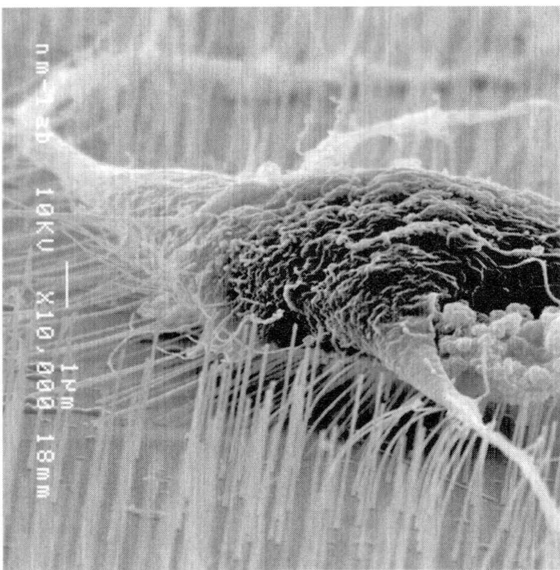

Figure 12.22. Neurons supported on a surface consisting of randomly located epitaxial GaP nanowires on GaP substrate, revealing a healthy appearance. Note that some wires are bent and in some cases wires are inserted into the neuronal cells and processes. (From Ref. [149].)

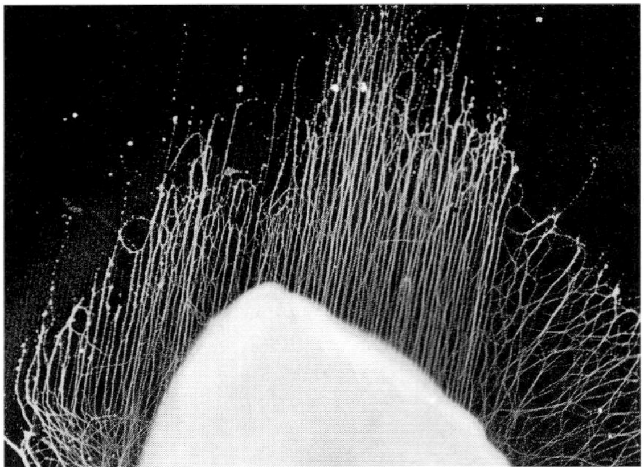

Figure 12.23. Nanowires were placed in rows with varying pitch and wire diameters. As can be seen, the neurons align to the wires and follow them in a very nice way, allowing governance of the growth direction to be engineered by the sample morphology. (From Ref. [149].)

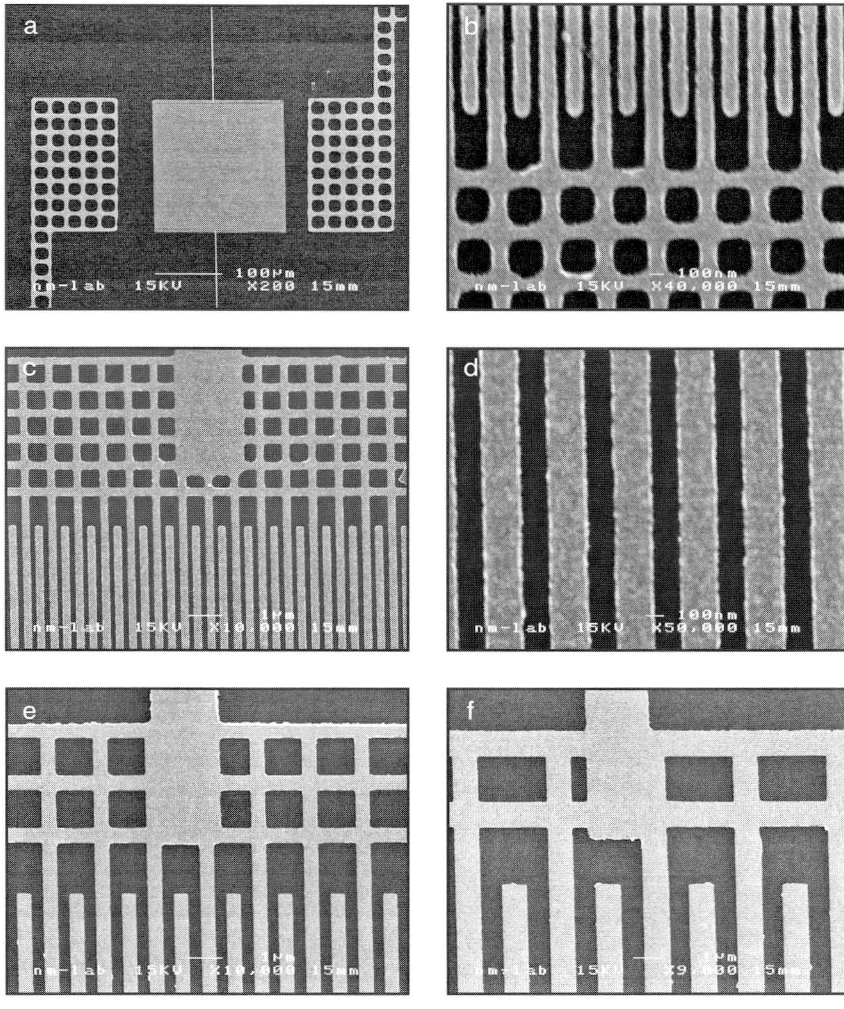

Figure 12.24. An example of an IDA of nanolines as defined by NIL, metallization and lift-off. SEM micrographs from the 2-in. wafer with several IDA electrodes after NIL processing and gold lift-off. (a) One single 3.5×3.5 mm^2 transducer chip, (b) overview of the central part of the transducer chip, (c) area of a 100-nm IDA and (d) part of a 200-nm IDA electrode. Note the contact areas with holes making it possible to print both nanometer lines and micrometer-sized regions in one printing step. (From Ref. [153].)

12.3.4
Protein Chips

Protein-binding detection devices are an important class of devices that may have a large impact on drug screening in conjunction with drug development. They may

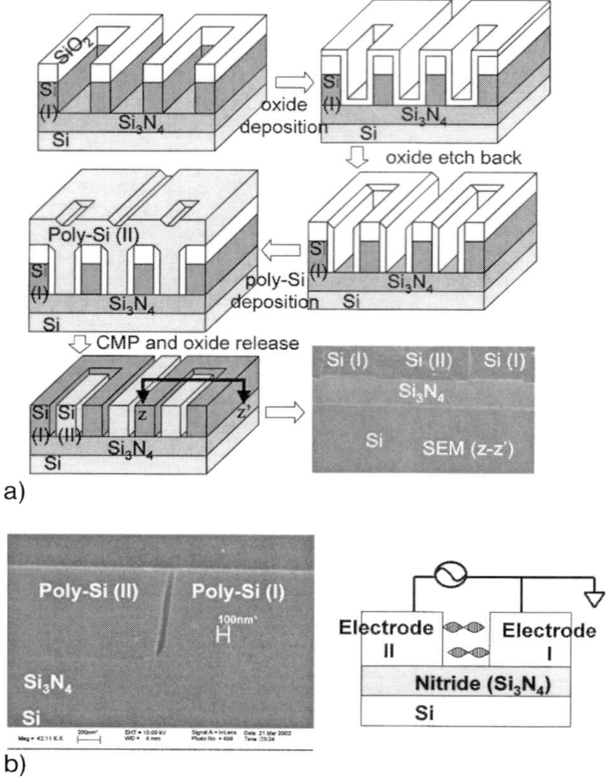

Figure 12.25. (a) The process flow for the creation of the nano-gaps and a SEM image. (b) A schematic cross-section showing the measurement principle. The nanogap can be controlled by the deposited oxide film thickness. (From Ref. [155].)

also have a potential to be developed into small and portable handheld devices for point-of-care analysis or even as home-doctor kits. For this, the main direction is to utilize antibody–antigen detection [157, 158]. Hence, what it boils down to is an ability to detect whether molecular binding between an antibody and an antigen has taken place. Of course, one usually makes measurements of large arrays of antibodies distributed on a surface using some kind of spotting procedure. Then, by spotting a liquid containing an unknown blend of antigens, one will detect which reactions take place and hence a fingerprint will be obtained. By comparing this individual fingerprint with a set of control fingerprints, it is possible by using various decision mechanisms to tell which antigens were present in the liquid and what disease that set of antigens is a representation of. In order to fully understand the various expressions as a consequence of a certain disease, one would like to make the antibody library extensive [159]. Present-day arrays contain hundreds of

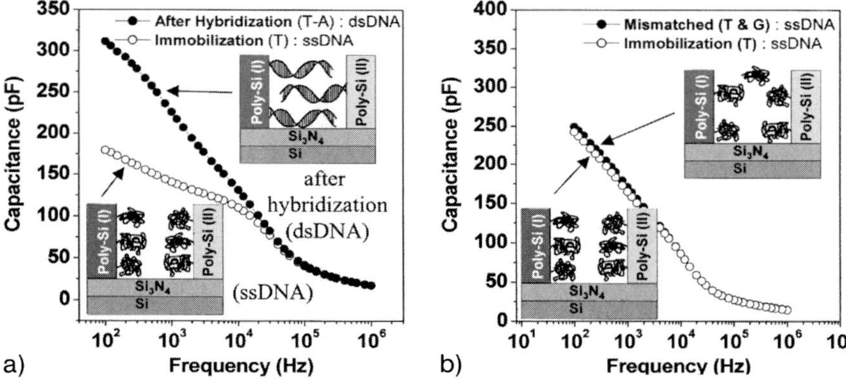

a)   b)

Figure 12.26. Measured capacitance versus input frequency after immobilization and hybridization. Capacitance increases as the frequency decreases for the conjugated pairs, whereas there is no significant effect of the capacitance for the nonconjugated base pairs. (From Ref. [155].)

antibodies, but using nanotechnology it is foreseeable that one could make arrays containing millions of antibodies, if one could confine (pattern) the antibodies onto a small surface spot [159–161]. In a prototype study we have fabricated such structures using EBL [162]. They are fabricated by making nanocontainers in the

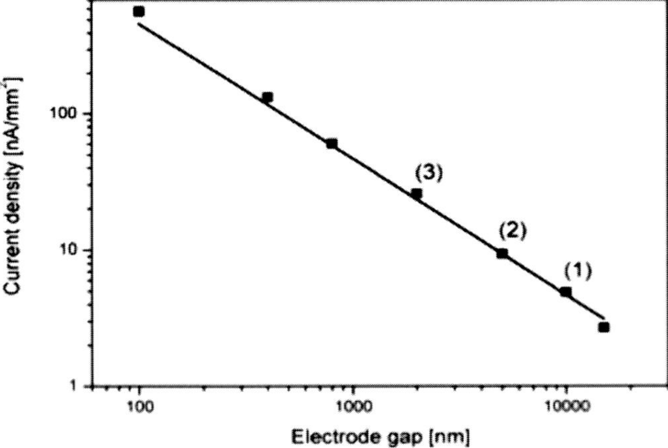

Figure 12.27. Normalized signal strength for redox current as a function of nanogap width showing the increased signal as the gap size is reduced. The increase of current is an effect of more effective redox cycling when the electrodes are closer to each other. (From Ref. [156].)

form of small holes (down to 50 nm in diameter) in a PMMA layer on top of a solid substrate with a waveguide as a surface layer. This means that antibodies can be immobilized inside the nanowells and by excitation through the evanescent field created by establishing a standing wave in the waveguide, one will excite only those molecules situated at the bottom of the well directly on top of the waveguide layer. The other molecules found on top of the polymer around the nanowells will not be excited due to the fast decaying electromagnetic field as a result of the near-field excitation mechanism. This approach has allowed us to study specific antibody–antigen interactions of biological interest [162]. Similar ideas using nanotechnology have been published, but they did not confine the antibodies to the surface effectively, leading to a larger analyzing droplet, or alternatively they employed larger nanowells [160, 161].

12.3.5
Protein Interactions

The field of protein patterning is not only limited to protein chip applications, but also used in other kinds of applications. The reason is that proteins are essential for so many life functions – force production in molecular motors, intra- and extracellular transport, mechanical stability, sensing and signaling, etc. What they all have in common is the procedure to make a surface functionalized in one way or another. Usually, functionalization is based on (covalent) coupling of the primary molecule to a certain surface structure. An example of a popular covalent coupling is thiol coupling to gold surfaces. If the surface is a silicon wafer, then one can utilize the large variety of well-established silanization procedures commonly found in conjunction with separation columns based on silica particles.

An often employed silanization procedure is to covalently bond chlorosilanes to a hydroxylated SiO_2 surface layer, expelling HCl as a result of the binding [163]. The chlorosilanes can then have functional amine, carboxyl or hydroxyl (or fluorine if one likes to have a repelling surface, compare this with NIL stamp fabrication [92–95]) groups at the other end. These functional groups may interact with a second molecular group that one intends to attach to the surface, e.g. enzymes. Hence, many approaches to make patterns in various silane layers on the nanoscale can be found [41, 164–166]. After patterning, one must retain the function of the functional groups and nanoprocessing, in general, is rather harsh to many of these molecular layers. Other approaches found in the literature are results obtained by nanoscale patterning of protein layers (compare also the functional protein layers in the tracks for the motor proteins as described above and in Refs. [130–135]). Several such reports have been made, usually employing some kind of model protein–protein reaction as the probe of success. Most often one employs streptavidin–biotin reactions to show proof of the patterning. In the report by Falconnet and coworkers [166], they combined standard NIL with a lift-off procedure of a biotin functionalized poly(ethylene glycol)–poly(L-lysine) (PEG–PLL) copolymer (Fig. 12.28). After lift-off, a backfilled layer of a non-biotin-

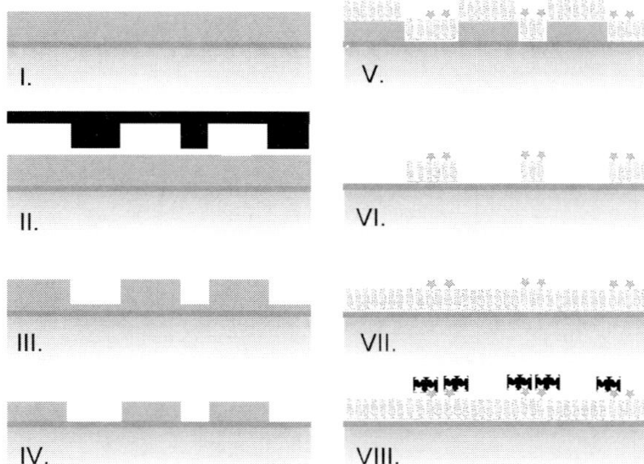

Figure 12.28. The NIL and protein patterning process. (I) Spin coating the PMMA layer and pre-bake. (II) Imprinting of the PMMA layer with the quartz stamp. (III) De-molding of stamp from substrate. (IV) Etching of the residual layer of PMMA opening the niobium oxide surface. (V) Dipping into solution of biotinylated PEG–PLL-mixture. (VI) Lift-off of the remaining PMMA layer leaving a patterned layer of biotinylated PEG–PLL on the surface. (VII) Filling of the background with non-biotinylated PEG–PLL. (VIII) Selective binding of streptavidin to the biotinylated layer. (From Ref. [166].)

Figure 12.29. Scanning near-field micrograph of 100-nm wide biotinylated PEG–PLL lines with Alexa-conjugated streptavidin in a background of nonbiotinylated PEG–PLL showing the successful selective lateral patterning of active protein binding lines. (From Ref. [166].)

functionalized PEG–PLL copolymer was established. In this way, they managed to produce patterned biotin-functionalized lines surrounded by PEG. They then successfully managed to selectively bind streptavidin to the 100-nm wide biotin lines (Fig. 12.29), showing that the biotin properties remained after a common nanoprocessing lift-off procedure. They also discuss the versatility of the process since the PEG copolymer can be functionalized in many different ways. In previous reports by Hoff and coworkers [167] and Gou [41], successful NIL-based protein patterning was displayed. Hoff and coworkers performed NIL and fluorine passivation of the naked silicon surface after reactive ion etching (RIE) exposure, but before removal of the unpatterned resist. A covalent immobilization step of an aminosilane monolayer was performed in combination with a series of subsequent covalent binding events of biotin, streptavidin and, finally, a biotinylated fluorescently labeled protein layer, respectively (Fig. 12.30), creating a sandwich structure. Using epifluorescence, the resulting structures were observed, proving the success of a very uniform sub-100 nm patterning and biofunctional NIL-based process (Figs. 12.31 and 12.32).

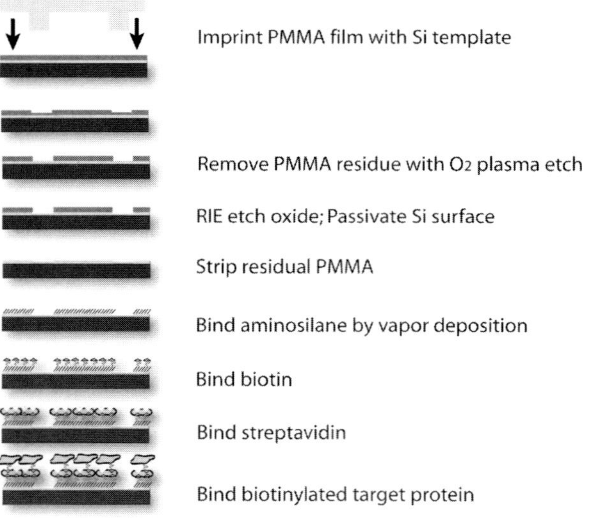

Figure 12.30. Process flow diagram of substrate-selective lateral patterning and subsequent protein immobilization. Spin-coated PMMA polymer is patterned by NIL. Exposed SiO_2 regions are etched and passivated by a RIE-deposited CF-based monolayer. The remaining PMMA is lifted off and an aminosilane is covalently attached to the exposed SiO_2 through a gas-phase deposition procedure. A layer of biotin is covalently linked to the amino layer followed by streptavidin attachment and then, finally, a biotinylated protein is linked to the streptavidin. (From Ref. [167].)

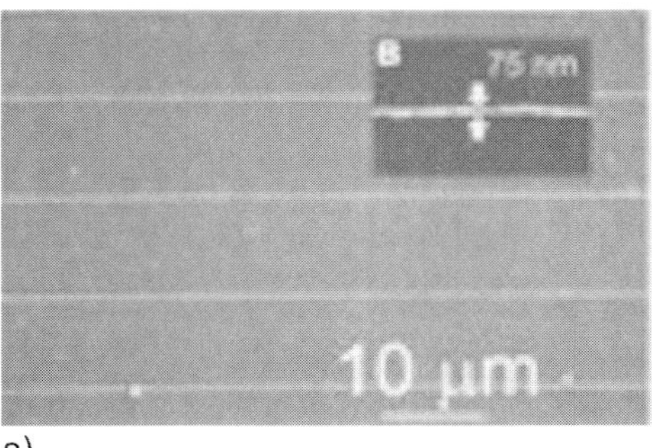
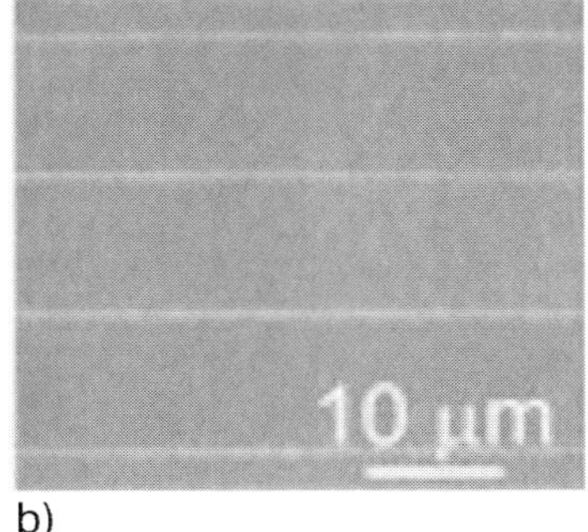

a)   b)

Figure 12.31.   Proteins patterned onto sub-100 nm patterns. (A) SEM image showing the nanolines after NIL, oxidation and passivation. The inset shows that the linewidth is less than 100 nm. (B) Fluorescence micrograph of the lines after linking rhodamine-labeled strepta-vidin to the biotin-modified aminosilane-functionalized lines. (From Ref. [167].)

12.3.6
Biomedical Applications using Nanowires

Nanowires having diameters of the order of tens of nanometers have been employed as extremely sensitive charge sensors. This is based on the fact that if one employs contacts at the two ends of the wire, there will be a possibility (depending on the wire conductivity) for the formation of a narrow 1-D current path along (inside) the wire. If a charge is exposed to the outer surface, this path will be quenched (or opened) (Fig. 12.33) in a way very similar to how a modern CMOS gate functions. By proper functionalization of the wire surface, one may use it as a sensor able to detect specific binding to the wire surface. Wang and coworkers have re-

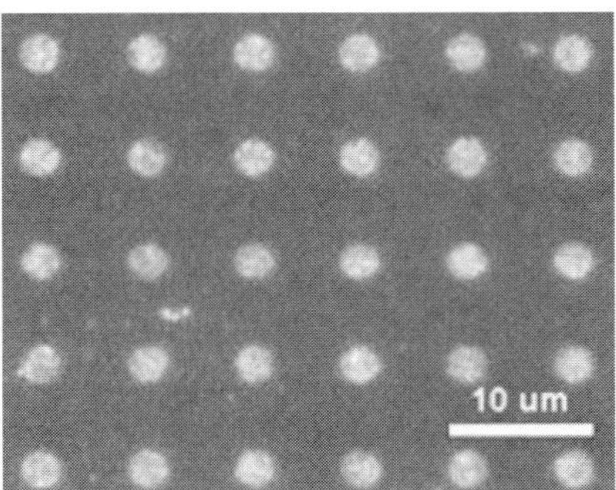

Figure 12.32.   Epifluorescence image showing bioactivity on NIL-patterned surfaces. The contrast arises from rhodamine-labeled catalase bound to the antibody anticatalase patterned in 2-μm dots using the process described in Fig. 12.31. (From Ref. [167].)

ported label-free detection of molecular binding to silicon nanowires [168], and Chen and coworkers [169] discussed non-covalent functionalization of such CNTs and biosensor applications. Other reports have been made dealing with virus detection [170]. Hahm and coworkers [171] showed label-free DNA detection by first attaching streptavidin, then biotin with a linker to peptide nucleic acid, which in turn contained a specific sequence of base pairs that enabled specific binding to DNA. They reported a 2-fold increase of wire conductance attributed to increased negative charge on the nanowires. In the near future, such structures may become very important as direct label-free detectors for clinical use. In a paper by Zhao [172], nanowires consisting of nanotube–metal clusters are discussed, providing proof-of-principle for a novel high-specificity molecular sensor utilizing the possibility to bind various kinds of molecular receptors to metal clusters. Further uses of nanowires are exploited in recent reviews [69, 173].

12.3.7
Biomedical Applications using Nanoparticles

Nanoparticles are excellent transducing components with the ability to report on interactions of biomedical relevance in various ways. The reporting mechanisms are often based on the fact that the magnetic, electronic, mechanical, chemical and optical properties of small objects differ from bulk materials, and small changes in size and shape can often be more easily monitored, since the effects are large. One such class of small objects is the "semiconducting quantum dot" that may change its luminescence (or fluorescence) properties as a function of surface coverage [174–176]. Generally speaking, also for this class of devices, one needs a handshaking procedure in order to utilize them as sensors, for example. This means that different surface functionalization procedures are also highly relevant for these kinds of structures. For instance, a magnetic particle having a certain molecule chemically linked to it can be manipulated using magnetic fields. In a paper by Nam and coworkers [177], such a use was discussed for the detection of proteins; specifically, they showed detection of prostate-specific antigen with a sensitivity 6 orders of magnitude higher than standard clinical assays.

The preparation and use of magnetic particles is a whole scientific field on its own, and interested readers are referred to additional publications [172–181]. Other classes of nanoparticles of great relevance for the biomedical community are soft core–shell nanoparticles and capsules [182–184] with a pharmaceutical content. For example, the core is envisaged to be able to locate a tumor and then by an enzymatic process the core is opened, allowing directed drug delivery to the tumor. Here, readers are also referred to other publications [182–184].

12.3.8
Biomedical Applications using SPM Technology

Here we can distinguish between investigations performed in microscopy mode, chemical force detection and as a sensor structure giving a multitude of possibilities.

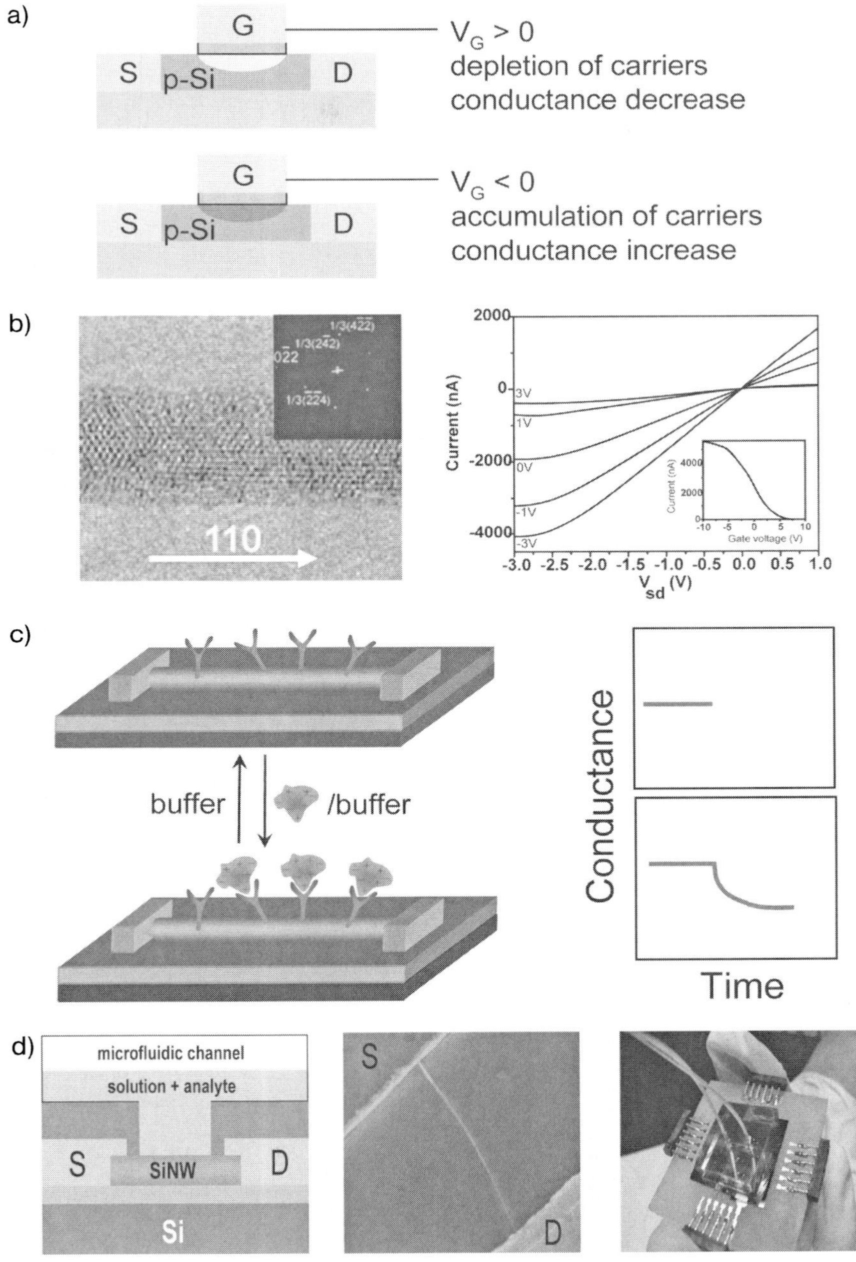
a)
G
S p-Si D
$V_G > 0$
depletion of carriers
conductance decrease
G
S p-Si D
$V_G < 0$
accumulation of carriers
conductance increase
b)
1/3(422)
1/3(242)
022
1/3(224)
110
2000
0
-2000
-4000
Current (nA)
3V
1V
0V
-1V
-3V
-3.0 -2.5 -2.0 -1.5 -1.0 -0.5 0.0 0.5 1.0
V_{sd} (V)
4000
2000
Current (nA)
-10 -5 0 5 10
Gate voltage (V)
c)
buffer /buffer
Conductance
Time
d)
microfluidic channel
solution + analyte
S SiNW D
Si
S
D

12.3.8.1 **Imaging of Biomolecules using SPM**

AFM is often employed for high-resolution microscopy aiming at molecular resolution of soft material, such as tissues, etc. AFM, in contrast to many of the other relatives in the large family of scanning probes, can be used in a liquid environment without losing much of its resolving power [185, 186]. To image soft and delicate materials it is mainly used in one of the noncontact modes or the intermittent-contact mode equipped with super-sharp tips. For this purpose, nanowires have been integrated in various ways with the usually pre-formed tip on the cantilever in order to serve as the sharp tip. A review in this subject area has been published by Hansma [187]. Recently, AFM has been taken to new dimensions, and a publication by Imer and coworkers describes how a minute AFM may be placed inside the body [188] in order to be able to report to an external source about how a knee joint surface develops after a surgical procedure.

12.3.8.2 **Force Detection of Single Molecular Events**

In force detection mode, using a chemically modified tip that will bind to a single molecule on a surface when lowered towards the surface, it is possible to study conformation changes and the mechanics of individual molecules. Such investigations have led to detailed knowledge about DNA recoiling, forces involved in antigen–antibody binding and receptor–ligand interactions, as well as cell–cell adhesion due to discrete adhesion proteins [118–121, 189]. An excellent review of this field is given by Gimzewski and Joachim [190] and many details can be found in the papers by Gaub, who is one of the key drivers in Europe in this field [120, 191]. Gaub's group have, for instance, recently published measurements revealing single base pair mismatches [192].

12.3.8.3 **Cantilever-based Detection of Molecular Events**

The remaining class of scanning probe activities with biomedical relevance utilize the mechanical movement or properties of a cantilever when a chemical interaction has taken place on the cantilever arm [118, 119, 193]. Here, one can distinguish between two classes – a static displacement of the cantilever or a shift of the dynamic motion of the cantilever, e.g. a shift in resonance frequency. For these investigations one can utilize an AFM set-up and a standard AFM cantilever with a tip, but the tip in itself does not play any role since the instrument *per se* is not go-

Figure 12.33.   Nanowire FET sensor. (A) Schematic view of a normal p-type FET device where a positive (negative) voltage applied to the gate (G) led to depletion (or enhancement) of the current path between the source (S) and drain (D) contacts. (B) Transmission electron microscopy image and diffraction pattern of a 4.5-nm diameter silicon single-crystal nanowire and typical electrical characteristics for a p-type nanowire. (C) Schematics of a p-type nanowire employed as a sensor with antibodies and positively charged protein that upon binding quench the conductance of the wire (similar to the positive bias situation in A). (D) Schematics of the nanowire sensor configuration and SEM image showing the nanowire between the two contacts and a photo of the complete biochip with integrated fluidic sample delivery. (From Ref. [172].)

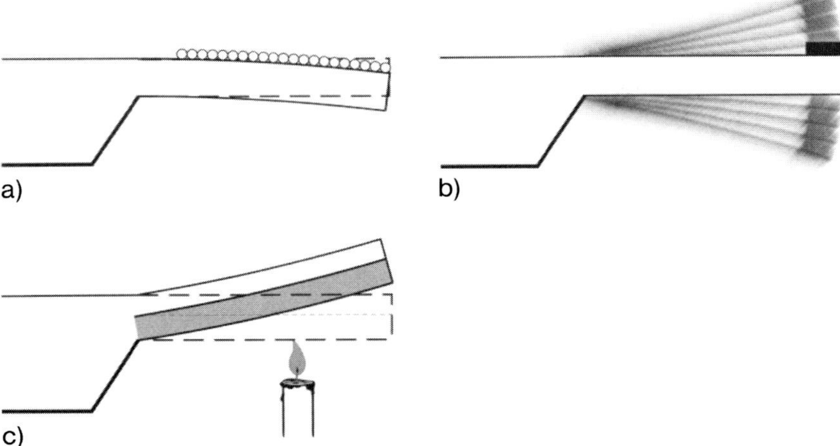

Figure 12.34. Schematic of a cantilever array-based hybridization sensor. (A) Each cantilever is functionalized with different self-assembled monolayers (SAMs) of thiolated oligonucleotides. (B) Upon injection of complimentary sequences to the oligonucleotides on the various cantilevers, recognition and binding occurs to one of the cantilevers, and the cantilever bends down due to the stress. (C) After rinsing, the cantilevers are ready to detect again. (From Ref. [118].)

ing to give any images. There is usually no need to have a scanner able to sweep the cantilever in the lateral directions. In order to achieve a static deflection, one often employs a scheme like a bimetal switch, i.e. when one side of the cantilever is influenced by chemical interactions, such that a strain or compression is experienced, the cantilever will bend. Typical levels of stress are of the order of 10^{-3} N m^{-1}, causing deflections to be around 10 nm or so; the bending will of course depend on the cantilever employed. Therefore, one usually coats one side of the cantilever with a metal that interacts with the molecules of interest, possibly by having a coating on the metal surface permitting only one kind of molecule to bind onto that surface (Fig. 12.34 and 12.35). Using this principle, it has been possible to follow chemical interaction and DNA hybridization, detect explosives, determine pH, detect of bacteria, detect of various vapors, etc. [118]. For dynamic detection, instead of monitoring surface strain or stress, one actually measures the mass being adsorbed on the cantilever surfaces [194]. The basic equation for a cantilever in resonance ($df/dm \sim f/m$) involves the mass, and hence a measurement of the change of the resonance frequency directly measures the added mass (or removed mass if one performs such an experiment). This method resembles quartz crystal mass (QCM) technology, although the QCM crystal is macroscopically large. Nevertheless, the use of QCM, and especially the dissipative mode of QCM, is frequently used for studying various biological processes and reactions to surfaces [195, 196]. The drawback of QCM is the sensitive surface area, which cannot be

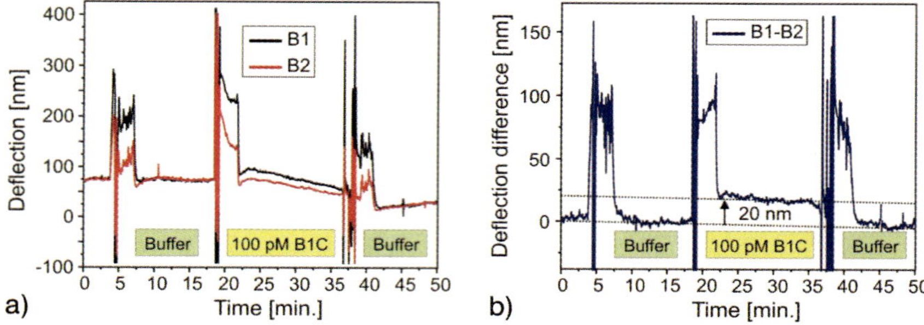

Figure 12.35. (A) Deflection traces of two differently oligonucleotide-functionalized cantilevers upon injection of the complement entity to one of the cantilevers. (B) Differential signal between the two deflection curves. Buffer = rinsing. (From Ref. [118].)

miniaturized in order to study processes happening on a small lateral scale. In order to detect very small masses, interest in the cantilever detection community is, at this stage, directed towards the development of small cantilevers. This is due to the increased mass detection ability that follows if the eigenmass is low and the unloaded resonance frequency is high. However, most of the cantilevers reported in this field have micrometer dimensions. The reason being that there are plenty of opportunities for good research and technology development using these more conventional cantilevers [197–199]. In particular, the coating of functionalized layers with the resulting exclusive binding is still not yet at a fully mature level. There is a necessity to make real systems with arrays of cantilevers (Fig. 12.36) hav-

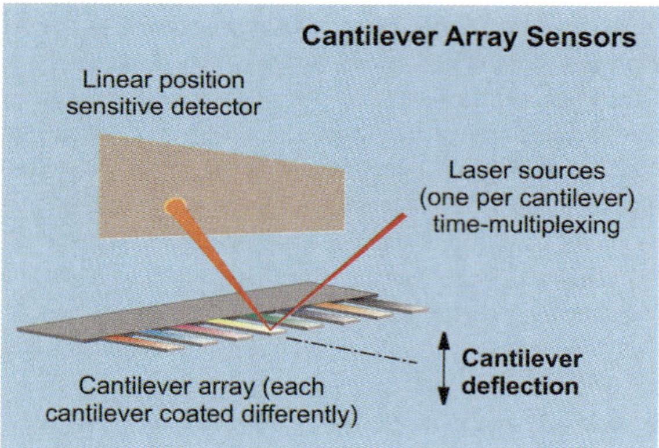

Figure 12.36. Schematic view of the array format used together with cantilever sensors. Each cantilever has a different coating. (From Ref. [118].)

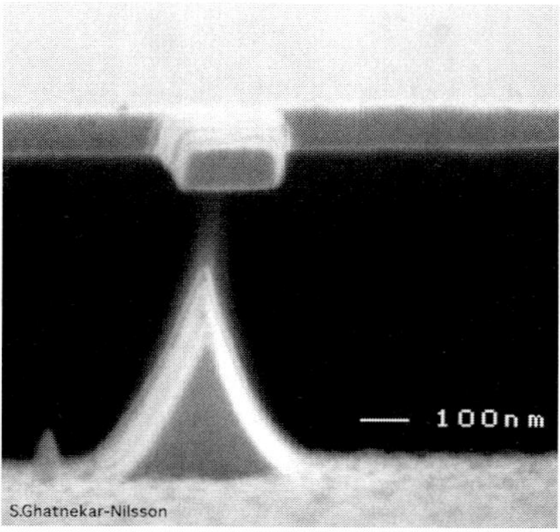

Figure 12.37. Cantilever cross-section view of a true nanosized metal cantilever having a width of 200 nm and a thickness of 50 nm. When the mechanical properties were investi- gated on this and similar metal cantilevers, it was found that they appear to be softer than one could predict from classical mechanics. (From Ref. [201].)

ing different kinds of coatings in order to detect and discriminate various contents in, for example, an unknown liquid [118, 200]. (Compare with Sections 12.3.4 and 12.3.5.) However, a few papers have recently reported [201, 202] cantilevers in the true nanometer regime, with thicknesses well below 100 nm (Fig. 12.37). In these papers, the nanomechanical properties of solid metal cantilevers were investigated as a function of length and thickness. It was found that the properties of thin canti- levers shift as compared to bulk material, making the cantilevers increasingly softer than expected as they get thinner. The cantilever-based nanomechanical con- cept is nice, since it allows investigations to be performed in a liquid, e.g. in body fluid or whole blood (serum). For static measurements, this is rather straightfor- ward [203–205], although care has to be exercised when making a flow system, al- lowing the actual detection to work. For the dynamic case, the situation is more tricky – the main reason being the drastic decrease of the Q-value of such dynamic systems when operated in liquid. During the many years of investigations, very few papers have yet been published. Just recently, two such papers have appeared, one from Professors Gerber's group in Basel and one from the author's group [206, 207]. In the paper by Braun and coworkers they have, in a model experiment, in- vestigated binding of streptavidin-coated latex beads onto biotinylated cantilevers, while in the paper by Nilsson and coworkers [207] they have addressed the chal- lenge to combine detection of adsorption of lipid vesicles onto cantilevers (Fig. 12.38). In that report, it was shown that the lipid vesicles formed a close-packed layer keeping their spherical shape and the mass resolution obtained was of the or- der of 3 fg Hz^{-1}. This low value is promising when compared with similar mea- surements of mass performed in vacuum [208], where a mass sensitivity of 0.2 fg Hz^{-1} was reported. Using an oscillating cantilever of poly-silicon, 1 μm long,

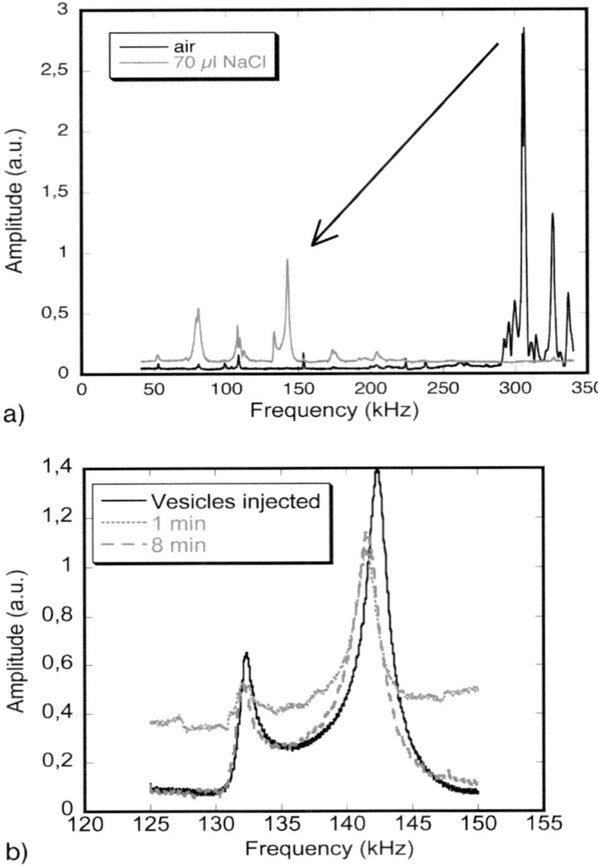

Figure 12.38. (A) The cantilever resonance spectra before and after being immersed in NaCl buffer liquid. (B) The shift of frequency due to immobilization of phospholipid vesicles onto the cantilever. The nonoptimized mass detection sensitivity was found to be 3 fg Hz^{-1}. (From Ref. [207].)

500 nm wide and 150 nm thick with an integrated 1-μm^2 paddle, functionalized with antigen, specific virus detection has been reported with a detection sensitivity of single viruses [209]. The detection of ultra-low masses leads us into another important class of nanotechnologies – nanomechanical systems. Using nanomechanics, it has been possible to make wonderful devices that can elucidate mechanical motion close to the quantum limits. A recent review on this subject can be found in *Physics Today* [210] and in Ref. [211]. Here, the authors, in a fascinating way, describe the development of the nanoelectromechanical systems (NEMS) field and how it interacts with quantum physics. This development leads the way to

be able to construct measurement apparatus that may be able to discriminate and analyze individual macromolecules with atomic-level detection in 3-D. Recently, Roukes and coworkers have shown exceptional weighing measurements down to a few molecules, the sensitivity reported to be 7 zg [212]. The sensitivity of MEMS and NEMS devices, as already mentioned in the Introduction, has also enabled measurements of single electron spins [66] and individual nuclear spins may soon be distinguished.

12.4
Discussion and Outlook

In this chapter, we have seen how modern nanotechnology is increasingly accepted and employed in the biomedical field, and especially in the field of sensors, genomics and proteomics [213–214]. Most of the issues dealt with here and in other similar overviews [215] are of great importance for the development of nanobio research, i.e. controlled surface functionalization, high-resolution imaging, fluidic governance, sensor principles, nanomechanics, etc. In order for nanotechnology to increase its impact on the biomedical field, we have to continue to develop basic technologies, as well as trying to utilize them in the biomedical area in a "play and see" fashion; by doing so one may find uses of the technologies in fields and areas of sciences that one could not otherwise envisage. The Nobel Laureate Herbert Kroemer has formulated a lemma [63] stating that any sufficiently new and innovative technology will find its own principal application created by that technology.

As is obvious from this chapter, most nanotechnologies so far have been employed for sensor purposes or as passive surfaces for molecular interactions. The main reason for this is that nanotechnology offers several orders of magnitude higher sensitivity than present technology. The step for nanoscientists, most often coming from the hard inorganic (top-down) nanotechnology scientific field, is smaller to take, as they can apply well-known concepts in order to study some biological processes, together with the fact that most nanotechnology efforts are presently driven from that field into the medical or biological field, in contrast to the reverse mechanisms. The day when scientists from very different fields start to collaborate in a more mutual way may also witness fundamentally new uses of nanotechnology in the biomedical sector. In this context, it is foreseeable that new nanoscience educational programs that presently are starting all over the world will be instrumental for this interdisciplinary development.

The area of using nanomechanics for biomedical applications and for probing cellular, intracellular as well as intercellular, events and interactions will probably entertain us in the coming years. Surely the class of epitaxial nanowires will be of specific instrumental importance, providing functional tools that, for example, may give full 3-D information when employed as large arrays of point detectors. Another area that will have a large influence on our understanding of biology and biological processes is the development of techniques for achieving fast imaging, using SPM. Such possibilities to gather both nanometer-scale space and millisecond time information will open new avenues for life sciences research. The further development and utilization of printing-like methods, based on nanoimprint and nanoprinting methods, and similar printing nanotechnologies, to create large arrays of nanostructured surfaces of one or another kind will surely be of great future importance, both for high- and low-volume applications. Such printing methods are already able to create small series of nanostructured surfaces economically enough to be used in research applications in the life sciences domain. The development of mega-dense protein chips and DNA sensors may, in combination with hand-held devices, provide us with efficient point-of-care analysis or home-doctor kits that will fundamentally change our way of living. Within the next 10 years the mortality rate in the USA from various cancer forms is predicted to be close to zero due to increased early detection principles with significantly higher sensitivities than at present [215]. We will probably also witness the use of nanotechnology for various forms of prosthetic devices giving freedom back to impaired persons. In this context, however, stem cell research may be of even larger future importance, providing natural means to connect new functional nerve cells to the impaired nervous system, maybe in combination with nanotechnology. Increased abilities in the form of enhanced functions, such as IR vision for use during darkness when driving, may soon be within reach. Research and development activities for increased abilities for learning, enhanced memory and cognitive functions, as well as personalized medical treatments, etc., are presently also being pursued, often under the term "converging technologies" [216].

Most importantly, we should not forget that the tools being developed to handle nanomaterials are still very young, actually being in their caring or nursing stage; typical development time scales are of the order of 10–15 years before a real impact can be judged.

It is my hope that this overview will have inspired the reader to engage in this vigorous research field with so many possibilities open for exploration, where maybe the only limitations are governed by the individuals own ingenuities.

Acknowledgments

This chapter would not have been possible without the collaboration, support, friendship and sharing of knowledge from all my Lund University colleges within the Nanometer Consortium, Division of Solid State Physics, and at the Departments of Physics, ElectroScience, Electrical Measurements Technology, Chemistry, Analytical Chemistry Immunotechnology, Cell & Organism Biology, Neurophysiology, Hand Surgery and Philosophy, as well as all international friends and collaborators in various EU programs (Charge, Nanotech, Chanil, Nanomass, Nano2Life, NaPa, Bionel, Parnass, etc.), and of course also my previous and present graduate students supplying a constant flow of joy and enthusiasm. Last, but not least, the support from my loving daughters Caroline and Cecilia is of great and very special importance for me.

References

1 See for instance: The Internet Technology Roadmap for Semiconductors (ITRS), http://public.itrs.net.

2 See for instance: Moores law, http://www.intel.com/technology/magazine/silicon/moores-law-0405.htm.

3 http://nobelprize.org/physics/laureates/1987/bednorz-lecture.html.

4 http://nobelprize.org/physics/laureates/2000/kroemer-lecture.html.

5 http://nobelprize.org/physics/laureates/2000/alferov-lecture.html.

6 http://nobelprize.org/physics/laureates/1973/esaki-lecture.html.

7 J. Singh, *Physics of Semiconductors and Their Heterostructures.* McGraw-Hill, New York, **1993**.

8 http://nobelprize.org/physics/laureates/1901/rontgen-bio.html.

9 http://nobelprize.org/chemistry/laureates/1991/ernst-lecture.html.

10 http://www.ob-ultrasound.net/ingehertz.html.

11 P. U. Abel, T. von Woedtke, Biosensors for *in vivo* glucose measurement: can we cross the experimental stage, *Biosens. Bioelectron. 17*, 1059, **2002**.

12 http://www7.nationalacademies.org/keck/Keck_Futures_Nano_Conferences_Glucose_Sensor_Description.html.

13 See for instance: http://www.nano.gov.

14 D. C. Ralph, C. T. Black, M. Tinkham, Spectroscopic measurements of discrete electronic states in single metal particles, *Phys. Rev. Lett. 74*, 3241, **1995**.

15 C. Zandonella, The Tiny Toolkit news feature, *Nature 423*, 10, **2003**.

16 A. Gustafsson, M.-E. Pistol, L. Montelius, L. Samuelson, Local probe techniques for luminescence studies of low-dimensional semiconductor structures, *J. Appl. Phys. 84*, 1715, **1998**.

17 G. Binnig, H. Rohrer, Scanning tunneling microscopy, *Helv. Phys. Acta 55*, 6, **1982**.

18 G. Binnig, C. F. Quate, Ch. Gerber, Atomic force microscope, *Phys. Rev. Lett. 56*, 930, **1986**.

19 R. Wiesendanger, Contributions of scanning probe microscopy and spectroscopy to the investigation and fabrication of nanometer-scale structures, *J. Vac. Sci. Technol. B 12*, 515, **1994**.

20 S. Y. Chou, P. R. Krauss, P. J. Renstrom, Imprint of sub-25 nm vias and trenches in polymers, *Appl. Phys. Lett. 67*, 14, **1995**.

21 J. K. Stroscio, D. M. Eigler, Atomic and molecular manipulation with the scanning tunneling microscope, *Science 254*, 1319, **1991**.

22 Ph. Avouris, Manipulation of matter at the atomic and molecular levels, *Acc. Chem. Res. 28*, 95, **1995**.

23 G. Meyer, J. Repp, S. Zöphel, K.-F. Braun, S. W. Hla, S. Fölsch, L. Bartels, F. Moresco, K. H. Rieder, Controlled manipulation of atoms and small molecules with a low temperature scanning tunneling microscope, *Single Molecules 1*, 79, **2000**.

24 S.-B. Carlsson, K. Deppert, T. Junno, M. H. Magnusson, L. Montelius, L. Samuelson, Ångström-level, real time control of the formation of quantum devices, *Semicond. Sci. Technol. 13*, A119, **1998**; T. Junno, S.-B. Carlsson, H. Xu, L. Montelius, L. Samuelson, Fabrication of quantum devices by Ångström-level manipulation of nanoparticles with an atomic force microscope, *Appl. Phys. Lett. 72*, 548, **1998**.

25 T. Junno, M. H. Magnusson, S.-B. Carlsson, K. Deppert, J.-O. Malm, L. Montelius, L. Samuelson, Single-electron devices via controlled assembly of designed nano-particles, *Microelectron. Eng. 47*, 179, **1999**.

26 http://www.samsung.com.

27 http://www.nano-tex.com.

28 S.-W. Lee, A. M. Belcher, Virus-based fabrication of micro- and nanofibers using electrospinning, *Nano Lett. 4*, 387, **2004**.

29 http://www.pilkington.com.

30 A. Torkkeli, Droplet microfluidics on

a planar surface, *PhD Thesis*, VTT, Espoo, **2003**.

31 G. Y. Jung, S. Ganapathiappan, X. Li, D. A. A. Ohlberg, D. L. Llynick, Y. Chen, W. M. Tong, R. S. Williams, Fabrication of molecular-electronic circuits by nanoimprint lithography at low temperatures and pressures, *Appl. Phys. A 78*, 1169, **2004**.

32 T. Tada, T. Kanayama, A. P. G. Robinsin, R. E. Plamer, M. T. Allen, J. A. Preece, K. D. M. Harris, A triphenylene derivative as a novel negative/positive tone resist of 10 nanometer resolution, *Microelectron. Eng. 53*, 425, **2000**; see also: G. O'Sullivan, *EUVL Source Workshop*, Antwerp, **2003**, http://www.sematech.org.

33 A. K. Bates, M. Rothschild, T. M. Bloomstein, T. H. Fedynyshyn, R. R. Kunz, V. Libermanand, M. Switkes, Review of technology for 157-nm lithography, *IBM J Res. Dev. 45*, 605, **2001**.

34 S. M. Sze, *Semiconductor Devices: Physics and Technology*, 2nd edn. Wiley, New York, **2001**.

35 H. I. Smith, M. L. Schattenburg, X-ray lithography from 500 to 30 nm: X-ray nanolithography, *IBM J. Res. Dev. 37*, 319, **1993**.

36 A. N. Broers, Resolution limits for electron-beam lithography, *IBM J Res. Dev. 32*, 502, **1988**.

37 J. Gierak, C. Vieu, M. Schneider, H. Launois, G. B. Assayag, A. Septier, Optimization of experimental operating parameters for very high resolution focused ion beam applications, *J. Vac. Sci. Technol. B 15*, 2373, **1997**.

38 B. Heidari, A. Bogdanov, M. Keil, L. Montelius, Imprint lithography for mass production, in: *Digest of Papers of Microprocesses and Nanotechnology 2001. (2001 International Microprocesses and Nanotechnology Conference)*, Matsue-shi, p. 9, **2001**.

39 T. Bailey, B. J. Choi, M. Colburn, M. Meissl, S. Shaya, J. G. Ekerdt, S. V. Sreenivasan, C. G. Willson, Step and flash imprint lithography: template surface treatment and defect analysis, *J. Vac. Sci. Technol. B 18*, 3572, **2000**.

40 B. Heidari, I. Maximov, E.-L. Sarwe, L. Montelius, Large scale nanolithography using nanoimprint lithography, *J. Vac. Sci. Technol. B 17*, 2961, **1999**.

41 L. J. Gou, Recent progress in nanoimprint technology and its applications, *J. Phys. D 37*, R123, **2004**.

42 A. L. Bogdanov, T. Holmqvist, P. Jedrasik, B. Nilsson, Dual pass electron beam writing of bit arrays with sub-100 nm bits on imprint lithography masters for patterned media production, *Microelectron. Eng. 67/68*, 381, **2003**.

43 W. M. Moreau, *Semiconductor Lithography*, Plenum, New York, **1998**.

44 Khan, L. Mohammad, J. Xiao, L. Ocola, F. Cerrina, Updated system model for x-ray lithography, *J. Vac. Sci. Technol. B 12*, 3930, **1994**.

45 L. Malmqvist, A. Bogdanov, L. Montelius, H. M. Hertz, Nanometer table-top proximity x-ray lithography with liquid-target laser-plasma source, *J. Vac. Sci. Technol. B 15*, 814, **1997**.

46 B. Heidari, I. Maximov, E.-L. Sarwe, L. Montelius, Nanoimprint lithography at the 6-inch wafer scale, *J. Vac. Sci. Technol. B 18*, 3557, **2000**.

47 A. Kumar, G. M. Whitesides, Features of gold having micrometer to centimeter dimensions can be formed through a combination of stamping with an elastomeric stamp and an alkanethiol "ink" followed by chemical etching, *Appl. Phys. Lett. 63*, 2002, **1993**.

48 K. E. Paul, T. L. Breen, J. Aizenberg, G. M. Whitesides, Maskless photolithography: embossed photoresist as its own optical element, *Appl. Phys. Lett. 73*, 2893, **1998**.

49 J. A. Rogers, K.-E. Paul, R. J. Jackman, G. M. Whitesides, Using an elastomeric phase mask for sub-100 nm photolithography in the optical near field, *Appl. Phys. Lett. 70*, 2658, **1997**.

50 M. Coburn, M. Stewart, J. Seawall, T. Michaelson, S. V. Sreenivasan, C. G. Willson, Step and flash imprint lithography: a novel approach to imprint lithography, in: *Proceedings of the SPIE 24th International Symposium on Microlithography: Emerging Lithographic Technologies III*, Santa Clara, CA, p. 379, **1999**, p. 379.

51 T. Bailey, B. J. Choi, M. Colburn, M. Meissl, S. Shaya, J. G. Ekerdt, S. V. Sreenivasan, C. G. Willson, Step and flash imprint lithography: template surface treatment and defect analysis, *J. Vac. Sci. Technol. B 18*, 3572, **2000**.

52 M. D. Stewart, S. C. Johnson, S. V. Sreenivasan, D. J. Resnick, C. G. Willson, Nanofabrication with step and flash imprint lithography, *J. Microlithogr. 4*, 11002, **2005**.

53 K. E. Paul, T. L. Breen, J. Aizenberg, G. M. Whitesides, Maskless photolithography: embossed photoresist as its own optical element, *Appl. Phys. Lett. 73*, 2893, **1998**.

54 M. Komuro, J. Taniguchi, S. Inoue, N. Kimura, Y. Tokano, H. Horishima, S. Matsui, Imprint characteristics by photo-induced solidification of liquid polymer, *Jpn J. Appl. Phys. 39*, 7075, **2000**.

55 B. Vratzov, A. Fuchs, M. Lemme, W. Henschel, H. Kurz, Large scale ultraviolet-based nanoimprint lithography, *J. Vac. Sci. Technol. B21*, 2760, **2003**.

56 M. Otto, M. Bender, F. Richter, B. Hadam, T. Kliem, R. Jede, B. Spangenberg, H. Kurz, Reproducibility and homogeneity in step and repeat UV-nanoimprint lithography, *Microelectron. Eng. 73/74*, 152, **2004**.

57 U. Plchetka, M. Bender, A. Fuchs, B. Vratzov, T. Glinsner, F. Lindner, H. Kurz, Wafer scale patterning by soft UV-nanoimprint lithography, *Microelectron. Eng. 73/74*, 167, **2004**.

58 J. Jeong, Y. Sim, H. Sohn, E. Lee, *Proc. SPIE 5751*, 227, **2005**.

59 B. Heidari, I. Maximov, E.-L. Sarwe, L. Montelius, Nanoimprint lithography at the 6-inch wafer scale, *J. Vac. Sci. Technol. B 17*, 2961, **1999**.

60 L. Montelius, B. Heidari, M. Graczyk, I. Maximov, E.-L. Sarwe, T. G. I. Ling, Nanoimprint- and UV-

lithography: mix and match process for fabrication of interdigitated nano-biosensors, *Microelectron. Eng. 53*, 521, **2000**.

61 Nanoimprint lithography was included in the ITRS roadmap 2002 (see Ref. [1]).

62 MIT: Emerging Technologies, **2003**. MIT news: 10 emerging technologies, March 2003, http://web.mit.edu

63 H. Kroemer, Speculations about future directions, *J Crystal Growth 251*, 17, **2003**.

64 G. Binnig, H. Rohrer, C. H. Gerber, E. Weibel, Surface studies by scanning tunneling microscopy, *Phys. Rev. Lett. 49*, 57, **1982**.

65 G. Binnig, C. F. Quate, C. H. Gerber, Atomic force microscope, *Phys. Rev. Lett. 56*, 930, **1986**.

66 D. Rugar, R. Budakian, H. J. Mamin, B. W. Chui, Single spin detection by magnetic resonance force microscopy, *Nature 430*, 329, **2004**.

67 A. M. Morales, C. M. Lieber, A laser ablation method for the synthesis of crystalline semiconductor nanowires, *Science 279*, 208, **1998**.

68 B. J. Ohlsson, M. T. Björk, A. I. Persson, C. Thelander, L. R. Wallenberg, M. H. Magnusson, K. Deppert, L. Samuelson, Growth and characterization of GaAs and InAs nano-whiskers and InAs/GaAs hetero-structures, *Physica E, 13*, 1126, **2002**; L. Samuelson, C. Thelander, M. T. Björk, M. Borgström, K. Deppert, K. A. Dick, A. E. Hansen, T. Mårtensson, N. Panev, A. I. Persson, W. Seifert, N. Sköld, M. W. Larsson, L. R. Wallenberg, Semiconductor nanowires for 0D and 1D physics and applications, *Physica E, 25*, 313, **2004** and references therein.

69 A. V. Melechko, V. I. Merkulov, T. E. McKnight, M. A. Guillorn, K. L. Klein, D. H. Lowndes, M. L. Simpson, Vertically aligned carbon nanofibers and related structures: controlled synthesis and directed assembly, *J. Appl. Phys. 97*, 041301-1, **2005**.

70 T. Mårtensson, P. Carlberg, M. Borgström, L. Montelius,

W. Seifert, L. Samuelson, Nanowire arrays defined by nanoimprint lithography, *Nano Lett. 4*, 699, **2004**.

71 K. A. Dick, K. Deppert, M. W. Larsson, T. Mårtensson, W. Seifert, L. R. Wallenberg, L. Samuelson, Synthesis of branched "nanotrees" by controlled seeding of multiple branching events, *Nat. Mater. 3*, 380, **2004**.

72 C. E. Flynn, S.-W. Lee, B. R. Pelle, A. M. Belcher, Viruses as vehicles for growth, organization and assembly of materials, *Acta Mater. 51*, 5867, **2003**.

73 C. Mao, D. J. Solis, B. D. Reiss, S. T. Kottmann, R. Y. Sweeney, A. Hayhurst, G. Georgiou, B. Iverson, A. M. Belcher, Virus-based toolkit for the directed synthesis of magnetic and semiconducting nanowires, *Science 303*, 213, **2004**.

74 A. M. Fenninmore, T. D. Yuzvinsky, W.-Q. Han, M. S. Fuhrer, J. Cumings, A. Zetti, Rotational actuators based on carbon nanotubes, *Nature 424*, 408, **2003**.

75 M. L. Roukes, Nanoelectromechanical systems face the future, *Phys. World 14*, 25, **2001**.

76 A. N. Cleveland, M. L. Roukes, Fabrication of high frequency nanometer scale mechanical resonators from bulk Si crystals, *Appl. Phys. Lett. 69*, 2653, **1996**.

77 X. M. H. Huang, C. Zorman, M. Mehregany, M. L. Roukes, Nanodevice motion at microwave frequencies, *Nature 421*, 496, **2003**.

78 Q. A. Pankhurst, J. Connolly, S. K. Jones, J. Dobson, Applications of magnetic nanoparticles in biomedicine, *J. Phys. D 36*, R167, **2003**.

79 L. LaConte, N. Nitin, G. Bao, Magnetic nanoparticle probes, *Mater. Today 8*, 32, **2005**.

80 E. Sackman, Supported membranes: scientific and practical applications, *Science 271*, 43, **1996**.

81 H. M. McConnell, T. H. Watts, R. M. Weiss, A. A. Brian, Supported planar membranes in studies of cell–cell recognition in the immune-system, *Biochim. Biophys. Acta 864*, 95, **1986**.

82 C. A. Keller, B. Kasemo, Surface specific kinetics of lipid vesicle adsorption measured with a quartz crystal microbalance, *Biophys. J. 75*, 1397, **1998**.

83 F. F. Rosetti, I. Reviakine, G. Csúcs, F. Assi, J. Vörös, M. Textor, Interaction of poly(l-lysine)-g-poly(ethylene glycol) with supported phospholipid bilayers, *Biophys. J. 87*, 1711, **2004**.

84 E. Reimhult, F. Höök, B. Kasemo, Intact vesicle adsorption and supported biomembrane formation from vesicles in solution: influence of surface chemistry, vesicle size, temperature, and osmotic pressure, *Langmuir 19*, 1681, **2003**.

85 A. Dahlin, M. Zach, T.Rindzevicius, M. Kall, D. S. Sutherland, F. Hook, Localized surface plasmon resonance sensing of lipid-membrane-mediated biorecognition events, *J. Am. Chem. Soc. 127*, 5043, **2005**.

86 S. Svedhem, D. Dahlborg, J. Ekeroth, J. Kelly, F. Höök, J. Gold, *In situ* peptide-modified supported lipid bilayers for controlled cell attachment, *Langmuir 19*, 6730, **2003**.

87 I. H. El-Sayed, X. Huang, M. El-Sayed, Surface plasmon resonance scattering and absorption of anti-EGFR antibody conjugated gold nanoparticles in cancer diagnostics: applications in oral cancer, *Nano Lett. 5*, 829, **2005** and references therein; E. Reimhult, C. Larsson, B. Kasemo, F. Höök, Simultaneous surface plasmon resonance and quartz crystal microbalance with dissipation monitoring measurements of biomolecular adsorption events involving structural transformations and variations in coupled water, *Anal. Chem. 76*, 7211, **2004**.

88 Chemical and biological point sensors for homeland defense, *Proc. SPIE 5585*, **2004**.

89 I. Maximov, E.-L. Sarwe, M. Beck, K. Deppert, M. Graczyk, M. H. Magnusson, L. Montelius, Fabrication of Si-based nanoimprint stamps with sub-20 nm features, *Microelectron. Eng. 61/62*, 449, **2002**.

90 K. Pfeiffer, F. Reuther, M. Fink, G. Gruetzner, P. Carlberg, I. Maximov, L. Montelius, J. Seekamp, S. Zankovych, C. M. Sotomayor Torres, H. Schulz, H.-C. Scheer, A comparison of thermally and photochemically cross-linked polymers for nanoimprinting, *Microelectron. Eng. 67/68*, 266, **2003**.

91 K. Pfeiffer, M. Fink, G. Ahrens, G. Gruetzner, F. Reuther, J. Seekamp, S. Zankovych, C. M. Sotomayor Torres, I. Maximov, M. Beck, M. Graczyk, L. Montelius, H. Schulz, H.-C. Scheer, F. Steingrueber, Polymer stamps for nanoimprinting, *Microelectron. Eng. 61/62*, 393, **2002**.

92 M. Beck, M. Graczyk, I. Maximov, E.-L. Sarwe, T. G. I. Ling, M. Keil, L. Montelius, Improving stamps for 10 nm level wafer scale nanoimprint lithography, *Microelectron. Eng. 61/62*, 441, **2002**.

93 J.-K. Chen, F.-H. Ko, K.-F. Hsieh, C.-T. Chou, F.-C. Chang, Effect of fluoroalkyl substituents on the reactions of alkylchlorosilanes with mold surfaces for nanoimprint lithography, *J. Vac. Sci. Technol. B 22*, 3233, **2004**.

94 M. Keil, M. Beck, T. G. I. Ling, M. Graczyk, L. Montelius, B. Heidari, Development and characterization of silane antisticking layers on nickel-based stamps designed for nanoimprint lithography, *J. Vac. Sci. Technol. B 23*, 575, **2005**.

95 M. Keil, M. Beck, G. Frennesson, E. Theander, E. Bolmsjö, L. Montelius, B. Heidari, Process development and characterization of antisticking layers on nickel-based stamps designed for nanoimprint lithography, *J. Vac. Sci. Technol. B 22*, 3283, **2004**.

96 F. Reuther, P. Carlberg, M. Fink, G. Gruetzner, L. Montelius, Reactive polymers: a route to nanoimprint lithography at low temperatures, *Proc. SPIE 5037*, 203, **2003**.

97 W. Zhang, S. Y. Chou, Multilevel nanoimprint lithography with submicron alignment over 4 in. Si wafers, *Appl. Phys. Lett.*, 79, 845, **2001**.

98 A. Fuchs, B. Vratzov, T. Wahlbrink, Y. Georgiev, H. Kurz, Interferometric in situ alignment for UV-based nanoimprint, *J. Vac. Sci. Technol. B 22*, 3242, **2004**.

99 R. Huang, X. Fan, Z. Lu, L. Wang, Progress of machining technology, in: *Proceedings of the 7th International Conference on Progress of Machining Technology*, Suzhou, p. 197, **2004**.

100 C. Picciotto, J. Gao, E. Hoarau, W. Wu, Image displacement sensing (NDSE) for achieving overlay alignment, *Appl. Phys. A 80*, 1287, **2005**.

101 L. Montelius, B. Heidari, M. Graczyk, T. G. I. Ling, I. Maximov, E.-L. Sarwe, Large-area nanoimprint fabrication of sub-100-nm interdigitated metal arrays, *Proc. SPIE 3997*, 442, **2000**.

102 C. Perret, C. Gourgon, F. Lazzarino, J. Tallal, S. Landis, R. Pelzer, Characterization of 8-in. wafers printed by nanoimprint lithography, *Microelectron. Eng. 73/74*, 172, **2004**.

103 Q. Xia, C. Keimel, H. Ge, Z. Yu, W. Wu, S. Y. Chou, Ultrafast patterning of nanostructures in polymers using laser assisted nanoimprint lithography, *Appl. Phys. Lett.*, 83, 24, **2003**.

104 B. Michel, A. Bernard, A. Bietsch, E. Delamarche, M. Geissler, D. Juncker, H. Kind, J.-P. Renault, H. Rothuizen, H. Schmid, P. Schmidt-Winkel, R. Stutz, H. Wolf, Printing meets lithography: soft approaches to high-resolution patterning, *IBM J Res. Dev. 45*, 697, **2001**.

105 See the conference series *NNT: Nanoimprint, Nanoprinting and Related Technologies*, http://www.nnt.org.

106 A. A. Yu, T. A. Savas, G. S. Taylor, A. Guiseppe-Elie, H. I. Smith, F. Stellaci, Supramolecular nanostamping: using DNA as movable type, *Nano Lett. 5*, 1061, **2005**.

107 N. A. Melosh, A. Boukai, F. Diana, B. Gerardot, A. Badolato, P. M. Petroff, J. R. Heath,

Ultrahigh-density nanowire lattices and circuits, *Science 300*, 112, **2003**.

108 R. A. Beckman, E. Johnston-Halperin, N. A. Melosh, Y. Luo, J. E. Green, J. R. Heath, Fabrication of conducting Si nanowire arrays, *J. Appl. Phys. Lett. 96*, 5921, **2004**.

109 G. D. Kubiak, D. R. Kania (Eds.), *OSA Trends in Optics and Photonics Vol. 4: Extreme Ultraviolet Lithography*, Optical Society of America, Washington, DC, **1996**.

110 A. A. Tseng, A. Notargiacomo, T. P. Chen, Nanofabrication by scanning probe microscope lithography: a review, *J. Vac. Sci. Technol. B 23*, 877, **2005**.

111 S. Hong, J. Zhu, C. A. Mirkin, Multiple ink nanolithography: toward a multiple-pen nano-plotter, *Science 286*, 523, **1999**.

112 D. S. Ginger, H. Zhang, C. A. Mirkin, The evolution of dip-pen nanolithography, *Angew. Chem. Int. Ed. Engl. 43*, 30, **2004**.

113 A. Meister, M. Liley, J. Brugger, R. Pugin, H. Heinzelmann, Nanodispenser for attoliter volume deposition using atomic force microscopy probes modified by focused-ion-beam milling, *Appl. Phys. Lett. 85*, 6260, **2004**.

114 R. Lüthi, R. R. Schlittler, J. Brugger, P. Vettiger, M. E. Welland, J. K. Gimzewski, Parallel nanodevice fabrication using a combination of shadow mask and scanning probe methods, *Appl. Phys. Lett. 75*, 1314, **1999**.

115 J. Brugger, J. W. Berenschot, S. Kuiper, W. Nijdam, B. Otter, M. Elwenspoek, Resistless patterning of sub-micron structures by evaporation through nanostencils, *Microelectron. Eng. 53*, 403, **2000**.

116 H. W. Li, D.-J. Kang, M. G. Blamire, W. T. S. Huck, Focused ion beam fabrication of silicon print masters, *Nanotechnology 14*, 220, **2003**.

117 P. Vettiger, M. Despont, U. Drechsler, U. Dürig, W. Häberle, M. I. Lutwyche, H. E. Rothuizen, R. Stutz, R. Widmer, G. K. Binnig, The "Millipede" – more than thousand tips for future AFM storage, *IBM J. Res. Dev. 44*, 323, **2000**; P. Vettiger, J. Brugger, M. Despont, U. Drechsler, U. Durig, W. Haberle, M. Lutwyche, H. Rothuizen, R. Stutz, R. Widmer, G. Binnig, Ultrahigh density, high-data-rate NEMS-based AFM data storage system, *Microelectron. Eng. 46*, 11, **1999**.

118 H. P. Lang, M. Hegner, C. Gerber, Cantilever array sensors, *Mater. Today April issue*, 30, **2005**.

119 J. Fritz, M. K. Baller, H. P. Lang, H. Rothuizen, P. Vettiger, E. Meyer, H.-J. Güntherodt, Ch. Gerber, J. K. Gimzewski, Translating biomolecular recognition into nanomechanics, *Science 288*, 316, **2000**.

120 M. Rief, M. Gautel, F. Osterhelt, J. M. Fernandez, H. E. Gaub, Reversible unfolding of individual titin immunoglobulin domains by AFM, *Science 276*, 1109, **1997**.

121 J. M. Fernandez, H. B. Li, Force-clamp spectroscopy monitors the folding trajectory of a single protein, *Science 303*, 1674, **2004**.

122 S. W. P. Turner, M. Cabodi, H. G. Craighead, Confinement-induced entropic recoil of single DNA molecules in a nanofluidic structure, *Phys. Rev. Lett. 88*, 128103, **2002**.

123 J. Han, H. G. Craighead, Entropic trapping and sieving of long DNA molecules in a nanofluidic channel, *J. Vac. Sci. Technol. A 17*, 2142, **1999**.

124 H. Cao, Z. Yu, J. Wang, J. O. Tegenfeldt, R. H. Austin, E. Chen, W. Wu, S. Y. Chou, Fabrication of 10 nm enclosed nanofluidic channels, *Appl. Phys. Lett. 81*, 174, **2002**.

125 R. H. Austin, J. O. Tegenfeldt, H. Cao, S. Y. Chou, E. C. Cox, Scanning the controls: genomics and nanotechnology, *IEEE Trans. Nanotechnol. 1*, 12, **2002**; L. J. Gou, X. Cheng, C.-F. Chou, Fabrication of size-controllable nanofluidic channels by nanoimprinting and its application for DNA stretching, *Nano Lett. 4*, 69, **2004**.

126 J. O. Tegenfeldt, O. Bajakin, C.-F. Chou, S. Chan, R. Austin, W. Pann, L. Liou, E. Chan, T. Duke, T. Cox, Near-field scanner for moving molecules, *Phys. Rev. Lett.* **86**, 1378, **2001**.

127 T. Nielsen, D. Nilsson, F. Bundgaard, P. Shi, P. Szabo, O. Geschke, A. Kristensen, Nanoimprint lithography in the cyclic olefin coloymer, Topas, a highly ultraviolet-transparent and chemically resistant thermoplast, *J. Vac. Sci. Technol. B* **22**, 1770, **2004**; D. Nilsson, S. Balslev, A. Kristensen, A microfluidic dye laser fabricated by nanoimprint lithography in a highly transparent and chemically resistant cyclo-olefin copolymer (COC), *J. Micromech. Microeng.* **15**, 296, **2005**.

128 K.-S. Yun, E. Yoon, Micro/nanofluidic device for single-cell-based assay, *Biomed. Microdevices 7*, 35, **2005**.

129 R. Karnik, R. Fan, M. Yue, D. Li, P. Yang, A. Majumdar, Electrostatic control of ions and molecules in nanofluidic transistors, *Nano Lett. 5*, 943, **2005**.

130 R. Bunk, J. Klinth, L. Montelius, I. A. Nicholls, P. Omling, S. Tågerud, A. Månsson, Actomyosin motility on nanostructured surfaces, *Biochem. Biophys. Res. Commun. 301*, 783, **2003**.

131 R. Bunk, A. Månsson, I. A. Nicholls, P. Omling, M. Sundberg, S. Tågerud, P. Carlberg, L. Montelius, Guiding molecular motors with nano-imprinted structures, *Jpn. J. Appl. Phys. 44*, 3337, **2005**.

132 M. Sundberg, J. P. Rosengren, R. Bunk, J. Lindahl, I. A. Nicholls, S. Tågerud, P. Omling, L. Montelius, A. Månsson, Silanized surfaces for *in vitro* studies of actomyosin function and nanotechnology applications, *Anal. Biochem. 323*, 127, **2003**.

133 R. Bunk, J. Klinth, L. Montelius, I. A. Nicholls, P. Omling, S. Tågerud, A. Månsson, Actomyosin motility on nanostructured surfaces, *Biochem. Biophys. Res. Commun. 301*, 783, **2003**.

134 R. Bunk, M. Sundberg, A. Månsson, I. A. Nicholls, P. Omling, S. Tågerud, L. Montelius, Guiding motor-propelled molecules with nanoscale precision through silanized bi-channel structures, *Nanotechnology 16*, 710, **2005**.

135 R. Bunk, Creation of a nanometer-scale toolbox for molecular transport-circuits, *PhD Thesis*, Lund University, **2005**.

136 H. Suzuki, A. Yamada, K. Oiwa, H. Nakayama, S. Mashiko, Control of actin moving trajectory by patterned poly(methylmethacrylate) tracks, *Biophys. J. 72*, 1997, **1997**.

137 H. Hess, C. M. Matzke, R. K. Doot, J. Clemmens, G. D. Bachand, B. C. Bunker, V. Vogel, Molecular shuttles operating undercover: a new photolithographic approach for the fabrication of structured surfaces supporting directed motility, *Nano Lett. 3*, 1651, **2003**.

138 A. I. Teixera, G. A. Abrams, C. J. Murphy, P. F. Nealey, Cell behavior on lithographically defined nanostructured substrates, *J. Vac. Sci. Technol. B 21*, 683, **2003**; M. J. Dalby, N. Gadegaard, M. O. Riehle, C. D. W. Wilkinson, A. S. G. Curtis, Investigating filopodia sensing using arrays of defined nano-pits down to 35 nm diameter in size, *J. Biochem. Cell Biol. 36*, 2005, **2004**.

139 A. S. G. Curtis, C. D. W. Wilkinson CDW, Topographical control of cells, *Biomaterials 18*, 1573, **1997**.

140 A. S. G. Curtis, Small is beautiful but smaller is the aim: review of a life of research, *Eur. Cells Mater. 8*, 27, **2004**.

141 M. J. Dalby, M. O. Riehle, S. J. Yarwood, C. D. W. Wilkinson, A. S. G. Curtis, Nucleus alignment and cell signaling in fibroblasts: response to a micro-grooved topography, *Exp. Cell Res. 284*, 274, **2003**.

142 B. Wojciak-Stothard, A. S. G. Curtis, W. Monaghan, K. MacDonald, C. D. W. Wilkinson, Guidance and activation of murine macrophages by nanometric scale topography, *Exp. Cell Res. 223*, 426, **1996**.

143 M. J. Dalby, M. O. Riehle, H. Johnstone, S. Affrossman, A. S. G. Curtis, *In vitro* reaction of endothelial cells to polymer demixed nanotopography, *Biomaterials 23*, 2945, **2002.**

144 M. J. Dalby, M. O. Riehle, H. Johnstone, S. Affrossman, A. S. G. Curtis, Investigating the limits of filopodial sensing: a brief report using SEM to image the interaction between 10 nm high nano-topography and fibroblast filopodia, *Cell Biol. Int. 28*, 229, **2004.**

145 F. Johansson, P. Carlberg, N. Danielsen, L. Montelius, M. Kanje, Axonal outgrowth on nano-imprinted patterns, *Biomaterials 27*(8), 1251–1258, **2006.**

146 P. Fromherz, Electrical interfacing of nerve cells and semiconductor chips, *Chem. Phys. Chem. 3*, 276, **2002.**

147 http://www.cyberkineticsinc.com.

148 P. Carlberg, F. Johansson, T. Martensson, R. Bunk, M. Beck, F. Persson, M. Borgstrom, S. Ghatnekar-Nilsson, B. Heidari, M. Grazcyk, I. Maximov, E.-L. Sarwe, T. G. I. Ling, A. Mansson, M. Kanje, W. Seifert, L. Samuelson, L. Montelius, Nanoimprint – a tool for realizing nano-bio research, in: *Proceedings of the 4th IEEE Conference on Nanotechnology*, Munich, p. 199, **2004.**

149 W. Hällström, T. Mårtensson, M. Kanje, L. Montelius, L. Samuelsson, Neurons cultured on gallium phosphide nanowires adherence, survival and interactions, in preparation, **2005**; C. Prinz, W. Hällström, T. Mårtensson, L. Samuelson, M. Kanje, L. Montelius, Aligning neurons, in preparation, **2005.**

150 K. Bradley, A. Davis, J.-C. P. Gabriel, G. Grünner, Integration of cell membranes and nanotube transistors, *Nano Lett. 5*, 841, **2005.**

151 K. Hamad-Schifferli, J. J. Schwartz, A. T. Santos, S. Zhang, J. M. Jacobson, Remote electronic control of DNA hybridization through inductive coupling to an attached metal nanocrystal antenna, *Nature 415*, 152, **2002.**

152 L. Montelius, J. O. Tegenfeldt, T. G. I. Ling, Fabrication and characterization of a nanosensor for admittance spectroscopy of biomolecules, *J. Vac. Sci. Technol. A 13*, 1755, **1995.**

153 M. Beck, F. Persson, P. Carlberg, M. Graczyk, I. Maximov, T. G. I. Ling, L. Montelius, Nanoelectrochemical transducers for (bio-)chemical sensor applications fabricated by nanoimprint lithography, *Microelectron. Eng. 73/74*, 837, **2004.**

154 H. Whitlow, Ng. May Ling, V. Auzelyte, I. Maximov, L. Montelius, J. Van Kan, A. A. Bettiol, F. Watt, Lithography of high spatial density biosensor structures with sub-100 nm spacing by MeV proton beam writing with minimal proximity effect, *Nanotechnology 15*, 223, **2004.**

155 Y.-K. Choi, J. S. Lee, J. Zhu, G. A. Somorjai, L. P. Lee, J. Bokor, Sublithographic nanofabrication technology for nanocatalysts and DNA chips, *J. Vac. Sci. Technol. B 21*, 2951, **2003.**

156 F. Persson, M. Beck, P. Carlberg, T. G. I. Ling, L. Montelius, Redox cycling in nanometer sized interdigitated sensor structures: a case for nanoimprint lithography, *Appl. Nanotechnol. 1*, **2003.**

157 C. A. K. Borrebaeck, Antibodies in diagnostics – from immunoassays to protein chips, *Immunol. Today 21*, 379, **2000.**

158 C. Wingren, J. Ingvarsson, M. Lindstedt, C. A. K. Borrebaeck, Recombinant antibody microarrays – a viable option?, *Nat. Biotechnol. 21*, 223, **2003.**

159 C. Wingren, L. Montelius, C. A. K. Borrebaeck, Mega-dense nano-arrays – the challenge of novel antibody array formats, in: *Protein Microarrays*, M. Schena, S. Weaver (Eds.). Jones & Bartlett, Sudbury, MA, **2004.**

160 M. Lynch, C. Mosher, J. Huff, S. Nettikadan, J. Johnson, E. Henderson, Functional protein nanoarrays for biomarker profiling, *Proteomics 4*, 1695, **2004**.

161 A. Bruckbauer, D. Zhou, D.-J. Kang, Y. E. Korchev, C. Abell, D. Klenerman, An addressable antibody nanoarray produced on a nanostructured surface, *J. Am. Chem. Soc. 126*, 6508, **2004**.

162 S. Nilsson, L. Dexlin, C. Wingren, C. A. K. Borrebaeck, L. Montelius, in preparation.

163 T. G. I. Ling, M. Beck, R. Bunk, E. Forsen, J. O. Tegenfeldt, A. A. Zakharov, L. Montelius, Fabrication and characterization of a molecular adhesive layer for micro- and nanofabricated electrochemical electrodes, *Microelectron. Eng. 67/68*, 887, **2003**.

164 A. Pallandre, K. Glinel, A. M. Jonas, B. Nysten, Binary nanopatterned surfaces prepared from silane monolayers, *Nano Lett. 4*, 365, **2004**.

165 C. K. Harnett, K. M. Satyalakshmi, H. G. Craighead, Low-energy electron-beam patterning of amine-functionalized self-assembled monolayers, *Appl. Phys. Lett. 76*, 2466, **2000**.

166 D. Falconett, D. Pasqui, S. Park, R. Eckert, H. Schift, J. Gobrecht, R. Barbucci, M. Textor, A novel approach to produce protein nanopatterns by combining nanoimprint lithography and molecular self-assembly, *Nano Lett. 4*, 1909, **2004**.

167 J. D. Hoff, L.-J. Cheng, E. Meyhöfer, L. J. Gou, A. J. Hunt, Nanoscale protein patterning by imprint lithography, *Nano Lett. 4*, 853, **2004**.

168 W. U. Wang, C. Chen, K. Lin, Y. Fang, C. M. Lieber, Label-free detection of small-molecule–protein interactions by using nanowire nanosensors, *Proc. Natl Acad. Sci. USA 102*, 3208, **2005**.

169 W. Chen, C. H. Tzang, J. Tang, M. Yang, S. T. Lee, Covalently linked deoxyribonucleic acid with multiwall carbon nanotubes: synthesis and characterization, *Appl. Phys. Lett. 86*, 103114, **2005**.

170 F. Patolsky, G. Zheng, O. Hayden, M. Lakadamyali, X. Zhuang, C. M. Lieber, Electrical detection of single viruses, *Proc. Natl Acad. Sci. USA 101*, 14017, **2004**.

171 J. Hahm, C. M. Lieber, Direct ultrasensitive electrical detection of DNA and DNA sequence variations using nanowire nanosensors, *Nano Lett. 4*, 51, **2004**.

172 F. Patolsky, C. M. Lieber, Nanowire nanosensors, *Mater. Today, April*, 20, **2005**.

173 Q. Zhao, M. Buongiorno Nardelli, W. Lu, J. Bernholc, Carbon nanotube–metal cluster composites: a new road to chemical sensors, *Nano Lett. 5*, 847, **2005**.

174 X. Michalet, F. Pinaud, T. D. Lacoste, M. Dahan, M. P. Bruchez, A. P. Alivisatos, S. Weiss, Properties of fluorescent semiconductor nanocrystals and their application to biological labeling, *Single Molecules 2*, 261, **2001**.

175 A. P. Alivisatos, *Nat. Biotechnol. 22*, 47, **2004**; M. Bruchez, M. Moronne, P. Gin, S. Weiss, A. P. Alivisatos, Semiconductor nanocrystals as fluorescent biological labels, *Science 281*, 2013, **1998**.

176 News Feature, Biologists join the dots, *Nature 413*, 450, **2001**.

177 J.-M. Nam, C. S. Thaxton, C. A. Mirkin, Nanoparticle-based bio-bar codes for the ultrasensitive detection of proteins, *Science 301*, 1884, **2003**.

178 U. Hafeli, W. Schuft, J. Teller, M. Zborowski (Eds.), *Scientific and Clinical use of Magnetic Carriers*, Plenum Press, New York, **1997**.

179 H. Nishibiraki, C. S. Kuroda, M. Maeda, N. Matsushita, M. Abe, Preparation of medical magnetic nanobeads with ferrite particles encapsulated in a polyglycidyl methacrylate (GMA) for bioscreening, *J. Appl. Phys. 97*, 10Q919-1, **2005**.

180 L. Shao, D. Caruntu, J. F. Chen, C. J. O'Connor, W. L. Zhou,

Fabrication of magnetic hollow silica nanospheres for bioapplications, *J. Appl. Phys. 97*, 10Q908-1, **2005**.

181 D. K. Kim, D. Kan, T. Veres, F. Mormadin, J. K. Liao, H. H. Kim, S.-H. Lee, M. Zahn, M. Muhammed, *J. Appl. Phys. 97*, 10Q918-1, **2005**.

182 M. A. N. Coelho, A. Gliozzi, H. Mohwald, E. Perez, U. Sleytr, H. Vogel, M. Winterhalter, nanocapsules with functionalized surfaces and walls, *IEEE Trans. Nanobiosci. 3*, 3, **2004**.

183 C. Loo, A. Lin, L. Hirsch, M. H. Lee, J. Barton, N. Halas, J. West, R. Drezek, Nanoshell-enabled photonics-based imaging and therapy of cancer, *Cancer Res. Treat. 3*, 33, **2004**.

184 H. Moehwald, From Langmuir monolayers to nanocapsules, *Colloids Surf. A 171*, 25, **2000**.

185 T. Ando, N. Kodera, Y. Naito, T. Kinoshita, K. Furuta, Y. Y. Toyoshima, A high-speed atomic force microscope for studying biological macromolecules in action, *Chem. Phys. Chem. 4*, 1196, **2003**.

186 M. B. Viani, T. E. Schaffer, G. T. Paloczi, L. I. Pietrasanta, B. L. Smith, J. B. Thompson, M. Richter, M. Rief, H. E. Gaub, K. W. Plaxco, A. N. Cleland, H. G. Hansma, P. K. Hansma, Fast imaging and fast force spectroscopy of single biopolymers with a new atomic force microscope designed for small cantilevers, *Rev. Sci. Instrum. 70*, 4300, **1999**; T. Fukuma, K. Kobayashi, K. Matsushige, H. Yamada, True atomic resolution in liquid by frequency-modulation atomic force microscopy, *Appl. Phys. Lett. 87*, 0314101, **2005**.

187 P. Hansma, Surface biology of DNA by atomic force microscopy, *Annu. Rev. Phys. Chem. 52*, 71, **2001**.

188 R. Imer, T. Akiyama, N. F. de Rooij, M. Stolz, U. Aebi, N. F. Friederich, D. Wirz, A. U. Daniels, U. Staufer, in: *Proceedings of the 13th International Conference on Scanning Tunneling Microscopy/Spectroscopy and Related Techniques*, Sapporo, see also Stolz, R. Raiteri, A. U. Daniels, M. R. Van Landingham, W. Baschomg, U. Aebi, Dynamic elastic modulus of porcine articular cartilage determined at two different levels of tissue organization by indentation-type atomic force microscopy, *Biophys. J. 86*, 3269, **2004**.

189 M. Benoit, D. Gabriel, G. Gerisch, H. E. Gaub, Discrete interactions in cell adhesion measured by single-molecule force spectroscopy, *Nat. Cell Biol. 2*, 313, **2000**.

190 J. K. Gimzewski, C. Joachim, Nanoscale science of single molecules using local probes, *Science 283*, 1683, **1999**.

191 Gaub homepage: http://www.biophysik.physik.uni-muenchen.de.

192 C. Friedsam, A. K. Wehle, F. Kühner, H. E. Gaub, Dynamic single-molecule force spectroscopy: bond rupture analysis with variable spacer length, *J. Phys. Condens. Matt. 15*, S1709, **2003**.

193 H. P. Lang, M. Hegner, Ch. Gerber, Nanomechanics – the link to biology and chemistry, *Chimia 56*, 515, **2002**.

194 S. Ghatnekar-Nilsson, Nanomechanical studies and applications of cantilever sensors, *PhD Thesis*, Lund University, **2005**.

195 F. Höök, M. Rodahl, C. Keller, K. Glasmästar, C. Fredriksson, P. Dahlqvst, B. Kasemo, The dissipative QCM D-technique: interfacial phenomena and sensor applications for proteins, biomembranes, living cells and polymers, *Proc. IEEE*, 966, **1999**.

196 E. Reimhult, C. Larsson, B. Kasemo, F. Höök, Simultaneous surface plasmon resonance and quartz crystal microbalance with dissipation monitoring measurements of biomolecular adsorption events involving structural transformations and variations in coupled water, *Anal. Chem. 76*, 7211, **2004**.

197 E. Forsén, S. G. Nilsson, P. Carlberg, G. Abadal, F. Pérez-Murano, J. Esteve, J. Montserrat, E. Figueras, F. Campabadal, J. Verd, L. Montelius, N. Barniol, A. Boisen, Fabrication of cantilever based mass sensors integrated with CMOS

using direct write laser lithography on resist, *Nanotechnology 15*, S628, **2004**.

198 S. G. Nilsson, E. Forsén, G. Abadal, J. Verd, F. Campabadal, F. Pérez-Murano, J. Esteve, N. Bariol, A. A. Boisen, L. Montelius, Resonators with integrated CMOS circuitry for mass sensing applications, fabricated by electron beam lithography, *Nanotechnology 16*, 98, **2004**; E. Forsen, G. Abadal, S. Ghatnekar-Nilsson, J. Teva, J. Verd, R. Sandberg, W. Svendsen, F. Perez-Murano, J. Esteve, E. Figueras, F. Campabadal, L. Montelius, N. Barniol, A. Boisen, Ultrasensitive mass sensor fully integrated with complimentary metal-oxide-semiconductor circuitry, *Appl. Phys. Lett. 87*, 034507, **2005**.

199 Gerber homepage: http://www.nccr-nano.org/nccr and http://www.zurich.ibm.com/st/nanoscience/cantilever.html.

200 A. Bietsch, J. Zhang, M. Hegner, H. P. Lang, C. Gerber, Rapid functionalization of cantilever array sensors by inkjet printing, *Nanotechnology 15*, 873, **2004**.

201 S. G. Nilsson, X. Borrisé, L. Montelius, Size effect on Young's modulus of thin chromium cantilevers, *Appl. Phys. Lett. 85*, 3555, **2004**.

202 S. G. Nilsson, E.-L. Sarwe, L. Montelius, Fabrication and mechanical characterization of ultrashort nanocantilevers, *Appl. Phys. Lett. 83*, 990, **2003**.

203 H. P. Lang, M. Hegner, E. Meyer, Ch. Gerber, Tutorial: Nanomechanics from atomic resolution to molecular recognition based on atomic force microscopy technology, *Nanotechnology 13*, R29, **2002**.

204 J. Fritz, M. K. Baller, H. P. Lang, T. Strunz, E. Meyer, H.-J. Güntherodt, E. Delamarche, Ch. Gerber, J. K. Gimzewski, Stress at the solid–liquid interface of self-assembled monolayers on gold investigated with a nanomechanical sensor, *Langmuir 16*, 9694, **2000**.

205 J. Sader, T. Uchihashi, M. J. Higgins, A. Farrell, Y. Nakayama, S. P. Jarvis, Quantitative force measurements using frequency modulation atomic force microscopy – theoretical foundations, *Nanotechnology 16*, S94, **2005**.

206 T. Braun, V. Barwich, M. K. Ghatsekar, A. H. Bredekamp, C. Gerber, M. Hegner, H. P. Lang, Micromechanical mass sensors for biomolecular detection in a physiological environment, *Phys. Rev. E 72*, 03907, **2005**.

207 S. Ghatnekar-Nilsson, J. Lindahl, A. Dahlin, T. Stjernholm, S. Jeppesen, F. Höök, L. Montelius, Phospholipid vesicle adsorption measured *in situ* with resonating cantilevers in a liquid cell, *Nanotechnology 16*, 1512, **2005**.

208 B. Ilic, D. Czaplewski, H. G. Craighead, P. Neuzil, C. Campagnolo, C. Batt, Mechanical resonant immunospecific biological detector, *Appl. Phys. Lett. 77*, 450, **2000**.

209 B. Ilic, Y. Yang, H. G. Craighead, Virus detection using nanoelectromechanical devices, *Appl. Phys. Lett. 85*, 2604, **2004**.

210 K. C. Schwab, M. L. Roukes, Putting mechanics into quantum mechanics, *Phys. Today 58*, 36, **2005**.

211 K. L. Ekinci, M. L. Roukes, Nanoelectromechanical systems, *Rev. Sci. Instrum. 76*, 061101, **2005**.

212 M. L. Roukes, Zeptogram mass detection: weighing molecules, *Phys. News Update 725*, April, **2005**.

213 C.-F. Chou, J. Gu, Q. Wei, J. Liu, R. Gupta, T. Nishio, F. Zernhausern, Nanopatterned structures for biomolecular analysis towards genomic and proteomic applications, *Proc. SPIE 5592*, **2004**.

214 H. G. Craighead, Nanostructure science and technology: impact and prospects for biology, *J. Vac. Sci. Technol. A 21*, S216, **2003**.

215 M. Rocho, Statement by at: *Global Nanotechnology Conference*, Saarbrücken, **2005**.

216 To get an overview of these activities, visit the *Conference Series NanoBioInfoCogno (NBIC)*.

13

Nanodevices in Nature

Alexander G. Volkov and Courtney L. Brown

13.1
Introduction

Living organisms are intricately designed with systems of "checks and balances" which regulate biological processes and minimize malfunctions. Biological nanodevices are largely responsible for the nearly flawless function of various organisms. The use of biological organisms as model systems for the engineering of new technologies is a form of applied case-based reasoning. The wealth of information generated from the study of biological systems creates a database that can be utilized to find solutions to various problems.

In this chapter, we discuss the role of various nanodevices in a wide variety of biological processes. Furthermore, we focus specifically on nanoreactors in multi-electron reactions, the biological function of cytochrome oxidase, nanodevices in photosynthesis and phototropism, membrane transport, molecular motors, and electroreceptors. There are many publications that focus on isolated nanodevices within very specific model systems; however, this approach allows us to analyze the role and significance of nanodevices in a variety of life forms, including plants, animals and bacteria. The study of nanodevices has limitless applications in bioelectronics, biology, chemistry, genetics, biophysics, bioengineering, technology and other fields of scientific study.

Nanodevices are molecules or molecular complexes that have clear and specific functions, and are a few nanometers in size. Millions of nanodevices exist in nature and in this chapter we discuss a few examples. Natural nanodevices include photochemical, electrochemical and synthetic nanoreactors. Photosystem (PS) I and PS II, enzymes, enzymatic systems in the citric acid cycle, and carbon fixation in the reductive carboxylic acid cycle are also common nanodevices in nature. Molecular motors such as ATP synthase, myosin, kinesin, DNA helicases, DNA topoisomerase, DNA helixase, RNA polymerase and bacterial rotary motors are vital nanodevices that serve to regulate biological processes. Molecules in electron transfer chains act as nanorectifiers and nanoswitches. Biological applications for nanodevices include information transfer, molecular computing, mechanosensors, elec-

Nanotechnologies for the Life Sciences Vol. 4
Nanodevices for the Life Sciences. Edited by Challa S. S. R. Kumar
Copyright © 2006 WILEY-VCH Verlag GmbH & Co. KGaA, Weinheim
ISBN: 3-527-31384-2

troreceptors, magnetoreceptors, magnetosomes, neuronal networks, light sensors and ion channels.

13.2
Multielectron Processes in Bioelectrochemical Nanoreactors

Vectorial charge transfer and molecular recognition at the interface between two dielectric media are important stages in many bioelectrochemical processes such as those mediated by energy-transducing membranes [1–5]. Many biochemical redox reactions take place at aqueous medium/membrane interfaces and some of these reactions are multielectron processes. About 90% of the oxygen consumed on Earth is reduced in a four-electron reaction catalyzed by cytochrome c oxidase. Multielectron reactions take place during photosynthesis, which is one of the most important processes on Earth.

Synchronous multielectron reactions may proceed without the formation of intermediate radicals. These radicals are highly reactive, and can readily enter a side-reaction of hydroxylation and destruction of the catalytic complex. Since multi-electron reactions do not pollute the environment with toxic intermediates and are ecologically safe, they are used by nature for biochemical energy conversion during respiration and photosynthesis. In the multielectron reaction that takes place in a series of consecutive single-electron stages, the Gibbs energy necessary for single-electron transfer cannot be completely and uniformly distributed over the stages. The energy demand for various stages is varied and the excess energy in the simpler stages is converted into heat. In a synchronous multielectron reaction, the energy is used very economically [6–9].

An important parameter in the quantum theory of charge transfer in polar media is the medium reorganization energy, E_s, that determines activation energy. The energy of medium reorganization in systems with complicated charge distribution was calculated by Kharkats [10]. Reagents and products can be represented by a set of N spherical centers arbitrarily distributed in a polar medium. The charges of each of the reaction centers in the initial and final state are z^i and z^f respectively. Taking R_k to represent coordinates of the centers and ε_i for dielectric constants of the reagents, it follows that:

$$
E_s = 0.8 \left(\frac{1}{\varepsilon_{opt}} - \frac{1}{\varepsilon_{st}} \right)
$$
$$
\times \left\{ \sum_{p=1}^{N} \left[\frac{(\delta z_k)^2}{2a_p} + \sum_{\substack{k=1 \\ k \neq p}}^{N} \frac{(\delta z_p)(\delta z_k)}{2R_{pk}} + \sum_{\substack{k=1 \\ k \neq p}}^{N} \sum_{\substack{l=1 \\ l \neq p}}^{N} \frac{(\delta z_p)(\delta z_l) a_p^3 (\vec{R}_{pk} \vec{R}_{pl})}{R_{pk}^3 R_{pl}^3} \left(\frac{3\varepsilon_{st}^2}{(2\varepsilon_{st} + \varepsilon_i)^2} - \frac{1}{2} \right) \right] \right\}
$$

$$(1)$$

where $(\delta z_k) = z_k{}^f - z_k{}^i$, $R_{pk} = R_p - R_k$, and $z_k{}^f$ and $z_k{}^i$ are charge numbers of particle k in the initial and final states, respectively. The term a_p is the radius of particle p, R_k is the coordinate of k-particle center and ε_i is the dielectric constant of reactant. Reactions with synchronous transfer of several charges present a particular case of Eq. (1).

It follows from Eq. (1) that E_s is proportional to the square of the number of charges transferred. Homogeneous multielectron processes are unlikely, due to the high activation energy resulting from a distinct rise in the energy of solvent reorganization. For multielectron reactions, the exchange currents of n-electron processes are small compared to those of single-electron multistep processes, which makes the stage-by-stage reaction mechanism more advantageous. Therefore, multielectron processes can proceed only if the formation of an intermediate is energetically disadvantageous. However, conditions can be chosen which reduce E_s during transfer of several charges to the level of the reorganization energy of ordinary single-electron reactions. These conditions require systems with a low dielectric constant and large reagent radii. Furthermore, the substrate must be included in the coordination sphere of the charge acceptor with several charge donors or acceptors bound into a multicenter complex. Recent papers have presented theoretical studies on the kinetics of heterogeneous multielectron reactions at water/oil interfaces, which proved to be capable of catalyzing multielectron reactions and sharply reducing the activation energy.

The most effective coupling of ion and electron transport can be obtained if the activation energy of the coupled process is lower than that of the charge transfer in the electron transport chain. It is obvious from Eq. (1) that this requires a simultaneous transfer of opposite charges, so that the second and the third terms of Eq. (1) are negative. An optimal geometry between the centers of charges of donors and acceptors must also be chosen.

To illustrate this point, we can consider two instances of multielectron reactions: simultaneous transfer of n charges from one donor to an acceptor and simultaneous transfer of several charges (one from each of the centers) to m acceptors ($m \leq n$). In the former case, E_s is proportional to n^2, while in the latter it may be significantly lower (depending on the sign of the charge being transferred and the reciprocal positions of reagents). The concerted multicenter mechanism of multielectron reactions markedly reduces E_s and, hence the activation energy, compared to a two-center multielectron process. With the appropriate arrangement of the reactants, the activation energy associated with electron transfer in a heterogeneous multielectron reaction may be lower than the energy of reorganization of the media.

13.3
Cytochrome Oxidase: A Nanodevice for Respiration

The function of the enzymes of the mitochondrial respiratory chain is to transform the energy of redox reactions into an electrochemical proton gradient across the hydrophobic barrier of a coupling membrane.

Cytochrome oxidase (EC 1.9.3.1, PDB 2OCC) is the terminal electron acceptor of

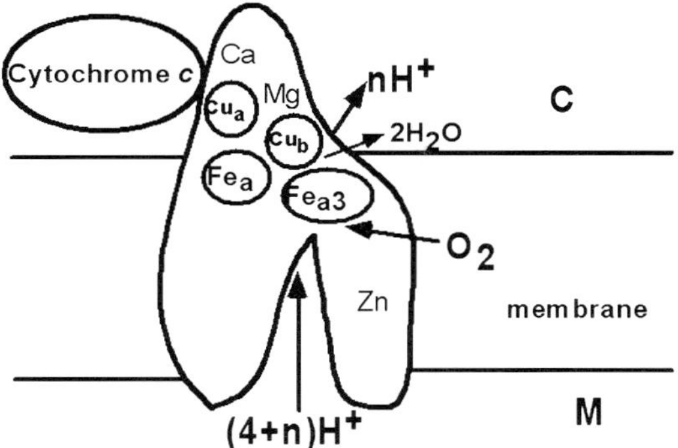

Figure 13.1. Scheme of the structural organization of cytochrome c and cytochrome c oxidase monomer in the inner mitochondrial membrane.

the mitochondrial respiratory chain. Its main function is to catalyze the reaction of oxygen reduction to water using electrons from ferrocytochrome c:

$$4H^+ + O_2 + 4e^- \underset{\text{photosynthesis}}{\overset{\text{respiration}}{\rightleftharpoons}} 2H_2O \tag{2}$$

Reaction (2) is exothermic and this energy can be used to transport protons across the mitochondrial membrane (Fig. 13.1). Mitochondrial cytochrome c oxidase is a dimer; each monomer is composed of 13 subunits. The enzyme contains cytochromes a and a_3, one binuclear copper complex Cu_a, one mononuclear copper site Cu_b, and one bound Mg^{2+} per monomer. It has a molecular weight ranging from 180 000 to 200 000 kDa for the most active form [11–13]. Cytochrome oxidases can transport a maximum of eight protons across the membrane per oxygen molecule reduction [14, 15]. Four of the protons bind to the reaction complex during the reduction of oxygen to water and up to four other protons are transported across the membrane. The resulting chemiosmotic proton gradient is used in ATP synthesis.

There are two types of respiration in photosynthetic organisms – dark respiration and photorespiration. Dark respiration includes O_2 reduction and the oxidation of NADH and $FADH_2$ in mitochondrial membranes, glycolysis, the Krebs cycle, and the oxidative pentose phosphate pathway. Respiration is commonly subdivided into two functional components – growth respiration and maintenance respiration. Growth respiration supplies energy for the production of new biomass; however, maintenance respiration provides the energy needed to maintain the integrity of existing structures and their turnover.

The respiratory chain of mitochondria is an integral part of the inner mitochondrial membrane. It is composed of four electron-transporting protein complexes (NADH dehydrogenase complex I, succinate dehydrogenase complex II, cytochrome reductase complex III and cytochrome c oxidase complex IV), ATP synthase (complex V), and mobile electron carriers ubiquinone and cytochrome c.

Plant mitochondria have additional enzymes not found in mitochondria of animals – the cyanide-insensitive alternative oxidase, an internal rotenone-insensitive NADPH dehydrogenase and an externally located NADPH dehydrogenase, which does not conserve energy. The alternative oxidase catalyzes the oxidation of ubiquinol to ubiquinone and the reduction of oxygen to water. It is inhibited by salicylhydroxamic acid.

Kharkats and Volkov were the first to present proof that cytochrome c oxidase reduces molecular oxygen by synchronous multielectron mechanism without the formation of an O_2^- intermediate [7–9, 14, 15]. The calculations predicted that the first step in oxygen reduction by cytochrome c oxidase should be a concerted multielectron process. As the field progresses, it became clear that the first step of oxygen reduction is a two-electron concerted process. The possible concerted molecular 2:1:1-electron and 2:2 proton pump mechanism of cytochrome c oxidase function is discussed in this chapter.

The 1:1:1:1-electron mechanisms of oxygen reduction by cytochrome oxidase were most frequently discussed in biochemistry. The reaction implies that the Gibbs free energy of the first electron transfer from cytochrome oxidase to O_2 is positive (Fig. 13.2). As a result, this route should be abandoned or the reaction rate should be extremely low. Since the Gibbs free energy of O_2 binding in the catalytic site of cytochrome oxidase is -21 kJ mol^{-1} [16], cytochrome c redox potential is 0.25 V. The Gibbs free energy of the first electron donation to oxygen at pH 7 is $+33$ kJ mol^{-1}. The Gibbs free energy of the reaction

$$O_2 + e^- \rightarrow O_2^- \tag{3}$$

in a cytochrome oxidase catalytic site is equal to $+79$ kJ mol^{-1}. Activation energy for O_2 reduction by fully reduced cytochrome oxidase is equal to 16 kJ mol^{-1} [17].

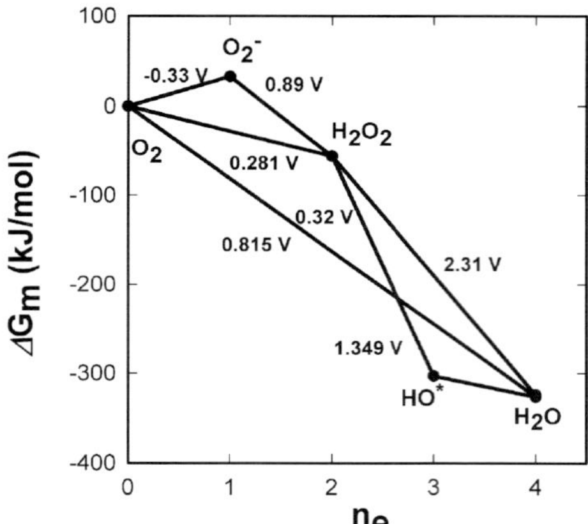

Figure 13.2. Energy diagrams for possible routes of the reaction $O_2 + 4H^+ \leftrightarrow 2H_2O$. G_m is the reaction midpoint Gibbs free energy at pH 7.

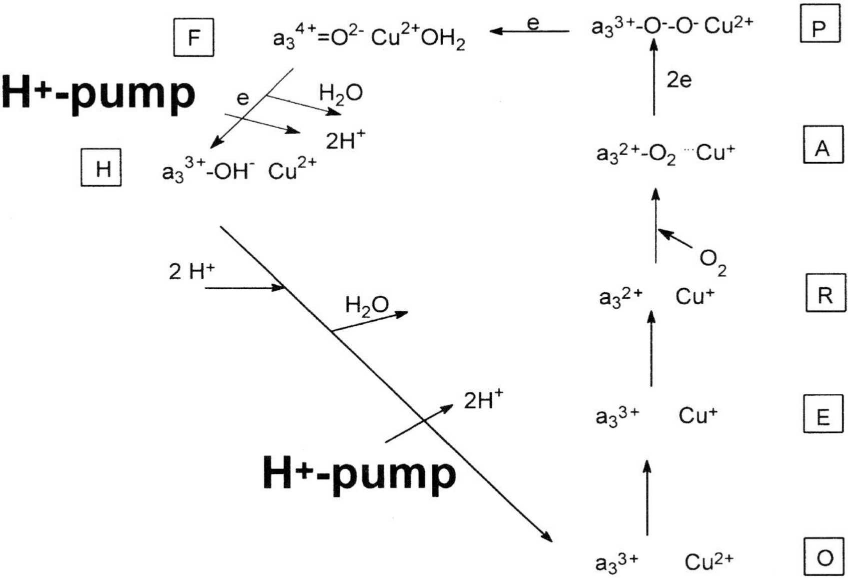

Figure 13.3. Scheme of the 2:1:1-electron reduction mechanism at the cytochrome *c* oxidase active site and its coupling to proton pumping. Starting from the oxidized form (O), the one-electron reduced form (E) and the doubly reduced form (R) are generated. Upon binding, compound (A) is observed. Next the peroxy-intermediate (P) is formed. The oxoferryl state (F) and a hydroxyl state (H) are formed after protonation of the iron-bound oxygen atom. After water formation and release, the O state is regenerated.

Since the Gibbs free energy of the endothermic reaction (3) is 5 times the measured activation energy for O_2 reduction by cytochrome oxidase, the single-electron mechanisms 1:1:1:1, 1:2:1, 1:1:2 and 1:3 at room temperature are unlikely. These reaction mechanisms are favorable when the binding energy of the single-electron intermediate is less than -52 kJ mol^{-1} in magnitude. The significant covalent bonding energy allows this intermediate to be experimentally detected. However, it has not been detected thus far.

The fact that the first electron addition to O_2 is endothermic accounts for the relative chemical inertness of oxygen in nature and it permits the existence of life on Earth.

A possible mechanism of oxygen reduction by cytochrome *c* oxidase is outlined in Fig. 13.3, and is be considered in detail after the discussion of the thermodynamic and kinetic aspects of the problem.

13.3.1
Nanodevice Architectonics

Equation (1) sets the conditions for the structure of cytochrome *c* oxidase catalytic site necessary for oxygen reduction to occur by the concerted *n*-electron mechanism. In order to reduce the reorganization energy of the medium, and thus the activation energy, several conditions must be met.

- The dielectric constant of the medium where oxygen reduction takes place should be low. Simply stated, the catalytic site should be immersed in a hydrophobic phase of the membrane (protein).
- There should be n spatially separated electron donors. For the proposed mechanism, heme and protein–copper complexes satisfy this condition.
- Cation, preferably a proton, transport should accompany electron transport via cytochrome c oxidase. Based on Eq. (1), when opposing charges are simultaneously transferred in close directions, the reorganization energy of the medium may be reduced due to a dependence on the third and fourth terms in Eq. (1). It implies that the coupling of the electron and proton pumps in cytochrome oxidase can be attained if the simultaneous transfer of opposing charges in close directions neutralizes medium reorganization. If electron transfer via cytochrome oxidase is coupled with proton transport across the mitochondrial membrane, then the energy liberated in the second reaction is consumed as opposed to being converted to heat.
- The radii of electron donors should be sufficiently large. This condition is achieved by utilizing the metal ion components of organic complexes (e.g. hemes and cysteines), the systems of conjugated bonds and ligands capable of undergoing redox reactions.

13.3.2
Activation Energy and Mechanism of Oxygen Reduction

The dependence of oxygen reduction rate on temperature reveals that cytochrome oxidase exists in two conformations – "hot" (h) and "cold" (c). The respective activation energies $E_a^{\,h}$ and $E_a^{\,c}$ are 16 kJ mol^{-1} (at 23–35 °C) and 60 kJ mol^{-1} (below 20 °C) [18]. A phase transition accompanied by conformational changes and absorption spectrum takes place between 18 and 23 °C. The temperature T^c depends on the surrounding lipid composition. The low $E_a^{\,h}$ value suggests that the single-electron mechanisms 1:1:1:1, 1:2:1, 1:3 and 1:1:2 are unlikely at temperatures above T^c since the enthalpy for the transfer of first electron from the reduced cytochrome oxidase to oxygen is 5 times more than the measured activation energy.

For the multielectron reaction 2:1:1, according to Eq. (1), E_s for two-electron reactions between O_2, a_3 and Cu_b strongly depend on geometry and distances in a catalytic site. Only the 2:1:1 mechanism of oxygen reduction by cytochrome oxidase can be realized *in vivo* in both "hot" and "cold" conformations.

Consider the molecular mechanism of oxygen reduction outlined in Fig. 13.3 in more detail [9, 14, 15].

The oxidized catalytic site of cytochrome oxidase is composed of cytochrome a_3 and Cu_b. It is reduced via the bridge mechanism by two electrons supplied from the electron reservoir of the respiratory chain. This reduced complex then binds an oxygen molecule. The reaction center is oxidized to the initial state in a double-electron reaction with the formation of a peroxide bridge between a_3 and Cu_b. The partially reduced (to peroxide) oxygen molecule must be bound in the reaction center since cytochrome oxidase is known to reduce oxygen to water without the release of any intermediates from the membrane. Next, the catalytic complex accepts two electrons from the electron reservoir $Fe(c) \rightarrow a_3$. In the next step, the peroxide bridge undergoes 1:1-electron reduction and protonation to water.

13.3.3
Proton Pump

Water molecules released in the course of oxygen reduction are transferred from the hydrophobic catalytic site to the aqueous phase. The continuous movement of the product away from the reaction center causes the equilibrium of the second reaction to shift to the right. Energy liberated in the exothermic reaction (2) is sufficient for transporting 8 H^+ ions across the membrane. Four of the H^+ ions couple with O_2 to form two H_2O molecules. The remaining H^+ ions can be transported across the hydrophobic zone of the membrane and used for ATP synthesis in ATP synthase complex. As follows from thermodynamics (Fig. 13.2), the energy needed for the function of the H^+ pump is liberated only at the last steps of water formation on the addition of third and fourth electrons independently of the reaction route [9, 14]. The functioning of protons pump after formation of ferryl intermediate is possible only if the difference between Gibbs energy of ferryl and peroxy intermediates binding is less than -35 kJ mol^{-1}.

The binding energy of the ferryl intermediate is negative. This energy supports the proton pump function not only during the addition of the fourth electron, but also after the formation of the three-electron oxygen intermediate. The stoichiometry of proton pumping by cytochrome oxidase can be 0:2:2 [14].

As it follows from Eq. (1), media reorganization energy corresponding to simultaneous electron and proton transfer is minimized when the transfer directions are close.

Alternative cytochrome oxidase in green plants can reduce O_2 without concomitant proton transfer. In such a case, the enzymes work like machines converting the energy of electron transfer to heat.

13.4
Photosynthetic Electrochemical Nanoreactors, Nanorectifiers, Nanoswitches and Biologically Closed Electrically Circuits

Life on Earth has been supported by the continuous flow of solar energy over billions of years. The power of this flux is extremely high: 4.14×10^{15} kWh day^{-1} or 1.5×10^{18} kWh year^{-1}. These values are extremely vast and difficult to imagine. According to Einstein's equation $E = mc^2$, the energy equivalent of 1 kg of mass is approximately 2.5×10^{10} kWh. The net daily energy flux incident upon Earth can thus be expressed as 165 tons and the thus net annual flux is 60 000 tons. By comparison, the annual production of electric energy in United States corresponds to an equivalent mass of about 100 kg and the total use of all kinds of energy corresponds to about 800 kg. At present, the annual consumption of energy by mankind is 4×10^{17} kJ, rising rapidly and doubling every 20 years. The known reserves of fossil fuels are limited to an estimated energy equivalent of 5×10^{19} kJ, so new sources of energy are of fundamental importance. One obvious possibility is solar energy. The amount of solar energy incident on the Earth is about 5×10^{21} kJ year^{-1}, of which 3×10^{18} kJ is converted into chemical energy by photosynthesis in plants and microorganisms [5].

The vast majority of the pigments in a photosynthetic organism is not chemically active, but functions primarily as an antenna. The photosynthetic antenna system is a nanodevice that collects and delivers the excited state energy by means of excitation transfer to the reaction center complexes where photochemistry takes place. The antenna system increases the effective cross-section of photon absorption by increasing the number of pigments associated with each photochemical complex. The intensity of sunlight is sufficiently dilute so that any given chlorophyll molecule only absorbs at most a few photons per second.

By incorporating many pigments into a single unit, the reaction centers and electron transport chain can be used to maximum efficiency. A remarkable variety of antenna complexes have been identified from various classes of photosynthetic organisms. Excitation transfer must be fast enough to deliver excitations to the photochemical reaction center and have them trapped in a short amount of time compared to the excited state lifetime in the absence of trapping. Excited state lifetimes of isolated antenna complexes, where the reaction centers have been removed, are typically in the range of 1–5 ns. Observed excited state lifetimes of systems where antennas are connected to reaction centers are generally on the order of a few tens of picoseconds, which is sufficiently fast so that under physiological conditions almost all the energy is trapped by photochemistry.

In water-oxidizing photosynthesis two membrane-integrated protein complexes PS II and PS I are operating in series (Fig. 13.4). The electron transfer starts in both photosystems vectorially across the membrane. Light energy is harvested by photosynthetic pigment systems in which the electronic structure of excited-state chlorophyll donates an electron to a primary acceptor pheophytin, the first component of an electron transport chain. The electron is fortified with it the energy of the original photon of light it absorbed. In the process of electron transport, the energy is captured in two ways. The first involves the coupling a proton pump mechanism to the sequential redox reactions in the electron transport chain, so that a proton gradient is established across the thylakoid membrane. The electrochemical energy of the proton gradient is then used to drive ATP synthesis by the ATP synthase enzymes embedded in the membrane. The second energy capture occurs when an acceptor molecule such as $NADP^+$ is reduced to NADPH, which in turn is used to reduce carbon dioxide in the Calvin cycle. Systems modeling photosynthesis should have the capability of carrying out relatively simple versions of these fundamental reactions.

The redox map of photosynthesis in green plants can be described in terms of

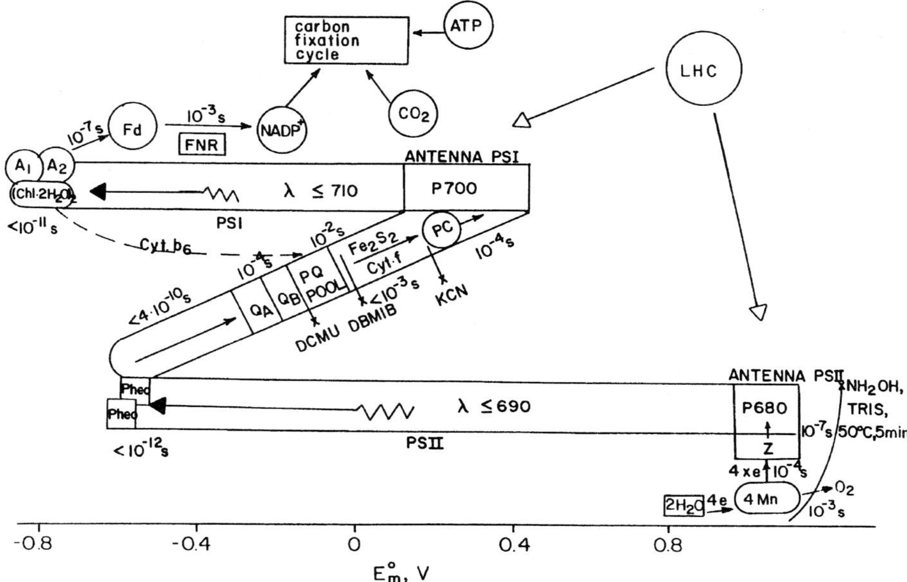

Figure 13.4. Scheme of electron transfer in photosynthesis in higher plants. $E_m{}^o$ on the abscissa stands for midpoint redox potential at pH 7.0. Light quanta ($h\nu$) are absorbed in two sets of antenna chlorophyll molecules, the excitation energy is transferred to the reaction center chlorophyll a molecules of PS II (P680) and PS I (P700) forming (P680)* and (P700)*, and the latter two initiate electron transport. (Reproduced from Ref. [21] with permission from Elsevier Science.)

the well-known Z-scheme proposed by Hill and Bendall [19]. The molecular organization of a thylakoid membrane is shown in Fig. 13.5. The spectral characteristics of PS II indicate that the primary electron donor is the dimer of chlorophyll P680 with absorption maxima near 680 and 430 nm. Water can be oxidized by an oxygen-evolving center (OEC) composed of several chlorophyll molecules, two molecules of pheophytin, plastoquinol, several plastoquinone (PQ) molecules and a manganese–protein complex containing four manganese ions. The OEC is a highly ordered structure in which a number of polypeptides interact to provide the appropriate environment for cofactors such as manganese, chloride and calcium, as well as for electron transfer within the complex. Figure 13.6 shows the electronic equivalent circuit of PS I and PS II.

Manganese-binding centers were first revealed in thylakoid membranes by electron paramagnetic resonance (EPR) methods and it is now understood that four manganese ions are necessary for oxygen evolution during water photooxidation. PQ acts as a transmembrane carrier of electrons and protons between reaction centers of two photosystems in the case of noncyclic electron transfer. It may also serve as a molecular "tumbler" that switches between one- and two-electron reactions. Pheophytin is an intermediate acceptor in PS II. Direct formation of P680

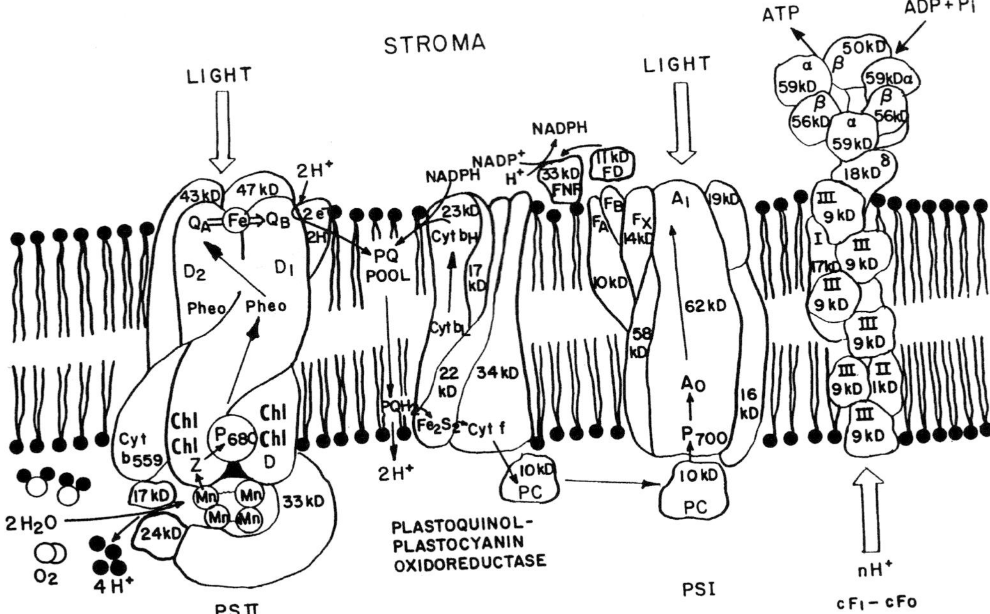

Figure 13.5. A stylized model of the electron transport chain with most of the light-harvesting pigment–protein complexes omitted. (Reproduced from Ref. [21] with permission from Elsevier Science.)

pheophytin ion radical pairs was revealed by experiments on magnetic interactions between pheophytin and PQ as reflected in the EPR spectra.

The photocatalytic oxidation of two molecules of water to oxygen cannot be a single-quantum process since the total energy expenditure of a catalytic cycle cannot be less than 476 kJ mol^{-1}. However, there is no fundamental reason why one quantum process should not induce the transfer of several electrons. For instance, a two-quantum process would require light with a wavelength less then 504 nm, while a four-quantum process would involve a sequential mechanism in which each light quantum is used to transfer one electron from photocatalyst to an electron acceptor. The threshold wavelength for the oxidation of water in this case is 1008 nm. The eight-quantum scheme which is actually used in photosynthesis can be explained by the need to compensate for energy losses in a long electron-transfer chain of redox reactions.

Water oxidation to molecular oxygen is a multielectron process that proceeds with surprisingly high quantum efficiency. The water oxidation reaction can proceed upon illumination at 680 nm – a wavelength of light that excludes one-electron mechanisms using hydroxyl and oxygen radicals. For a three-electron reaction an oxidant stronger than the cation-radical P680$^+$ is needed. A synchronous two-by-two electron pathway of the reaction is thermodynamically possible if the standard

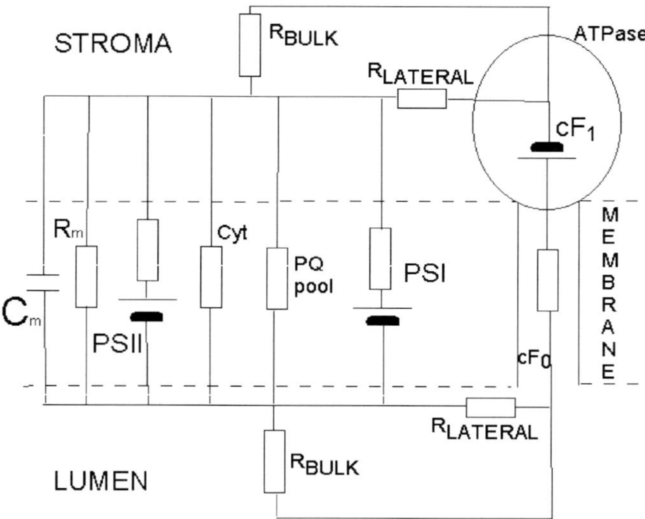

Figure 13.6. The equivalent electrical circuit of thylakoid membrane. C, capacity; R, resistance; cF_1, coupling factor; cyt, cytochrome.

free energy of binding of the two-electron intermediate is about -40 kJ mol^{-1}. This value corresponds to the energy of formation for two hydrogen bonds between H_2O_2 and the catalytic center. For this case a molecular mechanism was proposed [20, 21] and is discussed below (Fig. 13.7).

Membrane-bound P680 enters an excited state upon illumination. In dimers and other aggregated forms of chlorophyll, the quantum efficiency of triplet states is

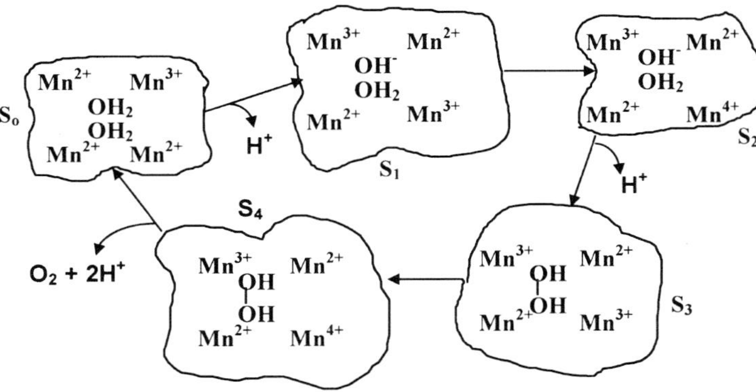

Figure 13.7. Possible 2:2-electron mechanism of water photooxidation by a manganese nanodevice in PS II.

low. It is the singlet excited state that undergoes photochemical transformations. In several picoseconds, an electron is first transferred to pheophytin, then to PQA, and from PQA to polypeptide-bound PQB in thylakoid membranes (Fig. 13.4), resulting in an oxidized pigment and a reduced acceptor. The cation radical P680$^+$ successively oxidizes four manganese ions, which in turn drives the production of molecular oxygen. Formation of a cation radical of chlorophyll or oxidation of manganese ions is accompanied by dissociation of water bound to the reaction center and ejection of protons. A synchronous multielectron process that describes all four oxidizing states of the OEC was proposed earlier. The transfer of electrons in a 1:1:1:1 series from a manganese cluster to the electron transport chain is accompanied by the ejection of 1:0:1:2 protons and the evolution of molecular oxygen [20, 21].

13.5
Phototropic Nanodevices in Green Plants: Sensing the Direction of Light

Plants continually gather information about their environment. Environmental changes elicit various biological responses. The cells, tissues and organs of plants possess the ability to become excited under the influence of environmental factors, referred to as irritants [22–25].

Nerve cells in animals and phloem cells in plants share one fundamental property – they possess excitable membranes through which electrical excitations, in the form of action potentials, can propagate. Plants generate bioelectrochemical signals that resemble nerve impulses and these are present in plants at all evolutionary levels [26].

The conduction of bioelectrochemical excitation is a rapid method of long-distance signal transmission between plant tissues and organs. Plants quickly respond to changes in luminous intensity, osmotic pressure, temperature, cutting, mechanical stimulation, water availability, wounding, and chemical compounds such as herbicides, plant growth stimulants, salts and water [27–30]. Once initiated, electrical impulses can propagate to adjacent excitable cells. The change in transmembrane potential creates a wave of depolarization or action potential, which affects the adjoining resting membrane.

The phloem is a sophisticated tissue in the vascular system of higher plants. Representing a continuum of plasma membranes, the phloem is a potential pathway for transmission of electrical signals. It consists of two types of conducting cells – the characteristic sieve-tube elements and the companion cells. Sieve-tube elements are elongated cells that have end walls perforated by numerous minute pores through which dissolved materials can pass. Sieve-tube elements are connected in a vertical series known as sieve tubes. Sieve-tube elements are alive at maturity; however, before the element begins its conductive function, their nuclei dissipate. The smaller companion cells have nuclei at maturity and are living. They are adjacent to the sieve-tube elements. It is hypothesized that they control the process of conduction in the sieve tubes. Thus, when the phloem is stimulated at any point, the action potential is propagated over the entire length of the cell membrane and along the phloem with a constant voltage.

Electrical potentials have been measured at the tissue and whole-plant level. At the cellular level, electrical potentials exist across membranes, and thus between cellular and specific compartments. Electrolytic species such as K^+, Ca^{2+}, H^+ and Cl^- are actively involved in the establishment and modulation of electrical potentials [31–35]. The highly selective ion channels serve as natural nanodevices [25]. Voltage-gated ion channels, as nanopotentiostats, regulate the flow of electrolytic species and determine the membrane potential [25].

Light is an essential source of energy on which many of the biological functions of plants depend. The sun's radiant energy optimizes germination, photosynthesis, flowering and other processes needed to maintain homeostasis. Plants contain specific photoreceptors that perceive light ranging from UV to far-red light. Natural radiation concurrently excites multiple photoreceptors in higher plants. Specific receptors initiate distinct signaling pathways leading to wavelength-specific light responses. Photoreceptors, phototropins, cryptochromes and phytochromes have been identified at the molecular level [36–42].

Phototropins, such as PHOT1 and PHOT2, are the flavoprotein photoreceptor that responds to light with a wavelength of 360–500 nm (blue light). It regulates phototropism and intracellular chloroplast movements. PHOT1 contains two 12-kDa flavin mononucleotide (FMN)-binding domains. LOV1 (light, oxygen and voltage) and LOV2 are located within its N-terminal region and a C-terminal serine/threonine protein kinase domain. Phototropin, when activated by light, undergoes a conformational change. PHOT1 and PHOT2 bind FMN, and undergo light-dependent autophosphorylation. PHOT2 is localized in the plasma membrane. Cryptochromes and phototropin have different transduction pathways, but similar traits.

Phototropism is one of the best-known plant tropic responses. A positive phototropic response is characterized by a bending or turning toward the source of light. When plants bend or turn away from the source of light, the phototropic response is considered negative. A phototropic response is a sequence of the four following processes: reception of the directional light signal, signal transduction, transformation of the signal to a physiological response and the production of directional growth response.

After 1–2 min of irradiation, a change in the direction of irradiation generates action potentials in soybean (Fig. 13.8) depending on the wavelength of light irradiation. Irradiation at wavelengths 400–500 nm induces fast action potentials in soybean with duration time of about 0.5 ms; conversely, the irradiation of soybean at wavelengths between 500 and 600 nm fails to generate action potentials. Irradiation between 500 and 600 nm does not induce phototropism. Irradiation of soybean by blue light induces positive phototropism.

The sensitive membranes in phloem cells facilitate the passage of electrical excitations in the form of action potentials. The action potential has a stereotyped form and an essentially fixed amplitude – an "all or none" response to a stimulus. Each impulse is followed by the absolute refractory period [43, 44]. The fiber cannot

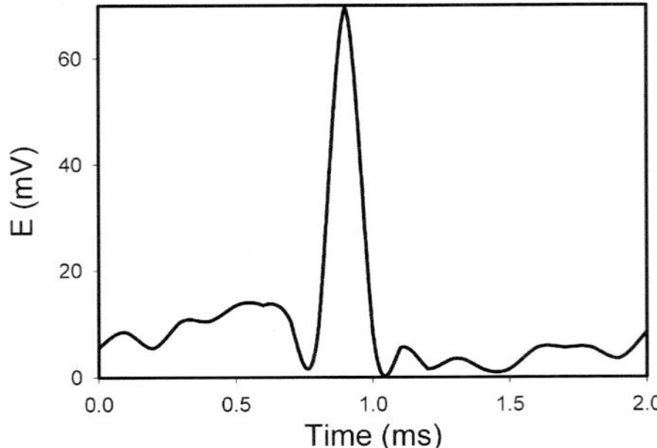

Figure 13.8. Action potentials in soybean induced by irradiation at 450 nm 2 min after changing the direction of irradiation. Irradiance was 10 μE $(m^2s)^{-1}$. Distance between electrodes was 5 cm. The soil was preliminary treated by water every day. Volume of soil was 0.5 L. Frequency of scanning was 50 000 samples s^{-1}.

transmit a second impulse during the refractory period. The integral organism of a plant can be maintained and developed in a continuously varying environment only if all cells, tissues and organs function in concordance.

These propagating excitations are theoretically modeled as traveling wave solutions of certain parameter-dependant nonlinear reaction–diffusion equations coupled with some nonlinear ordinary differential equations. These traveling wave solutions can be classified as single- and multiple-loop pulses, fronts and backs waves or periodic waves of different wave speed. This classification is matched by the classification of the electrochemical responses observed in plants. The experimental observations also show that under the influence of various pathogens, the shapes and speeds of the electrochemical responses undergo changes. From the theoretical perspective, the changes in the shapes and wave speeds of the traveling waves can be accounted by appropriate changes in parameters in the corresponding nonlinear differential equations.

Hodgkin and Huxley's membrane model [45] accounts for K^+, Na^+ and ion leakage channels in squid giant axons (Fig. 13.9A). The membrane resting potential for each ion species is treated like a battery and a variable resistor models the degree to which the channel is open. In an axon there is the K^+ and Na^+ transmembrane transport; conversely, in phloem cells the K^+, Ca^+ and, more than likely, H^+ channels are involved in this process (Fig. 13.9B).

Some voltage-gated ion channels work as plasma membrane nanopotentiostats. Blockers of ion channels such as tetraethylammonium chloride and $ZnCl_2$ stopped the propagation of action potentials in soybean plants induced by blue light and inhibited phototropism. Voltage-gated ionic channels control the plasma membrane potential and the movement of ions across membranes, thereby regulating various biological functions. These biological nanodevices play vital roles in signal transduction in higher plants.

All processes of life have been found to generate electric fields in every organism that has been examined with suitable and sufficiently sensitive measuring tech-

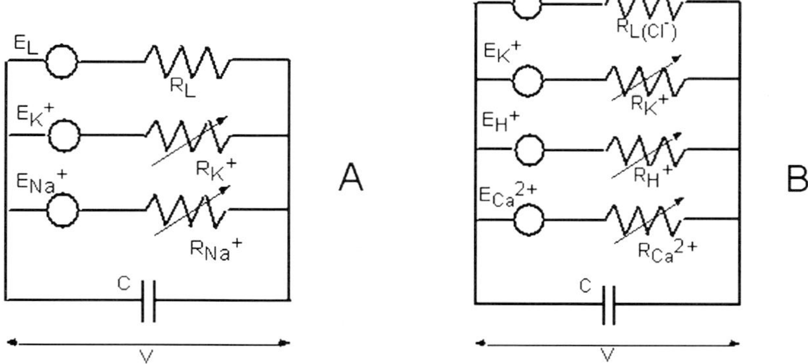

Figure 13.9. The Hodgkin–Huxley (HH) equivalent circuit for an axon (A) and the modified HH circuit for sieve tubes in phloem (B).

niques. The conduction of electrochemical excitation is regarded as one of the most universal properties of living organisms. It arose in connection with a need for the transmission of a signal in response to an external influence from one part of a biological system to another. The nature of regulatory relations of the plant organism with the environment is a basic bioelectrochemical problem, one that has a direct bearing on tasks of controlling the growth and development of plants.

13.6
Membrane Transport and Ion Channels

Membrane transport is vital to cell survival. Two major mechanisms used to transport ions and solutes across biological membranes are ion pumps and ion channels [1, 5, 46–48]. Ion transport is essential to the generation of membrane potentials, signal transduction and other biological processes. A membrane potential is a difference in electrical potential between intercellular and extracellular aqueous solutions. The membrane potential is influenced by the unequal distribution of electrolytic species inside and outside of the cell. Many intercellular proteins are negatively charged and remain inside the cell. The leakage of K^+ and H^+ ions is largely responsible for the generation of membrane potentials. The open K^+ ion channel facilitates the outward diffusion of K^+ ions without hydrolyzing ATP.

Ion channels are integral proteins that quickly facilitate the movement of specific ions across a biological membrane down their electrochemical gradient. Ion channels can facilitate the movement of approximately 10^6–10^8 ions s^{-1} [49]. These channels are classified as mechanically gated, non-gated, voltage gated or ligand gated. Non-gated channels remain permeable to specific ions. Voltage-gated chan-

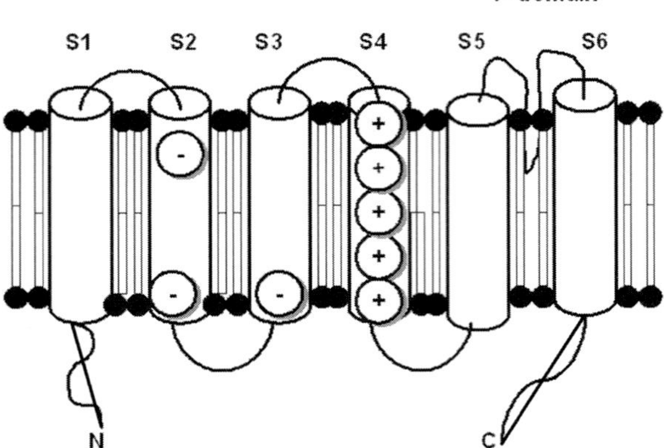

Figure 13.10.   General architecture of the voltage-gated inward rectifying K$^+$ channel AKT1.

nels become permeable when the membrane voltage is modulated above its threshold. Ligand-gated channels become permeable when the bound ligand is removed. Ion pumps employ a different mechanism. Ion pumps undergo conformational changes and they require energy to move specific ions against the electrochemical gradient. Ion pumps can facilitate the movement of approximately 10–100 ions s^{-1}. Ion channels are highly specific filters, allowing only desired ions through the cell membrane. Ion channels are devices in the engineering sense – they have signal inputs, power supplies and signal outputs. They use their complex structure to convert input signals to output signals [50, 51].

Voltage-gated channels open or close depending on the transmembrane potential. Examples include the sodium and potassium voltage-gated channels of nerve and muscle cells that are involved in the propagation of action potentials, and voltage-gated calcium channels that control neurotransmitter release in presynaptic endings. Voltage-gated channels are found in neurons, muscle cells and plant cells. Voltage-gated ion channels are membrane proteins that conduct ions at high rates regulated by the membrane potential [52–54]. Voltage-gated channels consist of three major parts – the gate, the voltage sensor and the ion-selective conducting channel (Fig. 13.10). The voltage sensor is a region of protein-bearing charged amino acids that relocate upon changes in the membrane potential. The movement of the sensor initiates a conformational change in the gate of the conductive pathway thus controlling the flow of ions. The voltage-gated K$^+$, Na$^+$ and Ca^{2+} channels have a common domain of six helical transmembrane segments S1–S6. The fourth segment, S4, is the voltage sensor of the channel and has a symmetrical arrangement of charged residues, with each third residue being arginine or lysine. A voltage-sensing domain consists of membrane segments S1–S4 and controls the conformation of gates located in the pore domain S5–S6.

Ligand-gated channels open in response to a specific ligand molecule on the external face of the membrane in which the channel resides. Examples include the "nicotinic" acetylcholine receptor, AMPA receptor and other neurotransmitter-gated channels. Cyclic nucleotide-gated channels, Calcium-activated channels and others open in response to internal solutes, and they mediate cellular responses to second messengers. Stretch-activated channels open or close in response to mechanical forces that arise from local stretching or compression of the membrane around them. Such channels are believed to underlie touch sensation and the transduction of acoustic vibrations into the sensation of sound. G-protein-gated channels open in response to G-protein activation via its receptor. Inward-rectifier K channels allow potassium to flow into the cell in an inwardly rectifying manner, e.g. potassium flows into the cell, but not out of the cell. They are involved in important physiological processes such as the pacemaker activity in the heart, insulin release, and potassium uptake in glial cells. Light-gated channels like channelrhodopsin 1 and channelrhodopsin 2 are directly opened by the action of light [55]. Resting channels remain open at all times.

13.7
Molecular Motors

A molecular motor is a protein that uses energy from ATP hydrolysis or the gradient of electrochemical potentials of protons or cations to generate directed movement along filamentous track, or rotation [56–59]. There are three different classes of motor proteins that move along either actin or microtubule tracks – myosin moves along actin filaments; the kinesins and dyneins move along microtubules. Protein motors are used in nature for force generation and motion. Motor proteins convert chemical energy into mechanical force via conformational changes. One important difference between molecular motors and macroscopic motors is that molecular motors operate in an environment where thermal noise is significant relative to the motor's energy consumption.

Myosins contain common motor domains that are responsible for muscle contraction. Myosin, like other molecular motors, uses energy obtained from ATP to travel along the action filament. Normally, myosin is bound to ADP. In the process of muscle contraction, the ADP molecule is freed when the myosin head binds to actin. An ATP molecule replaces ADP and induces a conformational change. Once changed into a ready state, the ATP is hydrolyzed. This process influences the protein to migrate from the negative end to the positive end. This migration, in addition to other processes, compels the muscle fibers to contract.

Dynein is another type of motor protein that also transforms energy from ATP hydrolysis into a form of energy that may be used to do mechanical work. This motor protein complex is composed of multiple heavy chains, intermediate chains and light chains. The heavy chain weighs approximately 530 kDa, and has four ATP-binding sites and a microtubule-binding site. The intermediate chains range from 53 to 79 kDa. The larger of the intermediate chains binds the protein to the cargo site. This binding equips the dynein with the ability to move the cargo to the negative end of a microtubule. Additional motor proteins known as kinesins move the cargo in the opposite direction to the positive end.

DNA helicases are classified as nanodevices, because are molecular motors. One of the functions is to detach conjoined strands of DNA during genetic processes. Another function of DNA helicases is to transform the chemical energy produced from ATP into a form that may be used to perform mechanical work. A hexamer is a common type of helicase found in many organisms. Traditionally, this motor protein utilizes its multimeric structure to provide numerous DNA-binding sites.

Rotary motor proteins present in flagella are another class of motor proteins and, thus, nanodevices. The electrochemical energy is provided from the H^+ or Na^+ transmembrane gradient (Fig. 13.11). *Escherichia coli* and other bacterium employ rotary proteins in conjunction with the flagella to propel them in aqueous solutions. This small, yet powerful, protein complex has the capacity to rotate at a speed of approximately 20 000 r.p.m. and it is extremely energy efficient. These

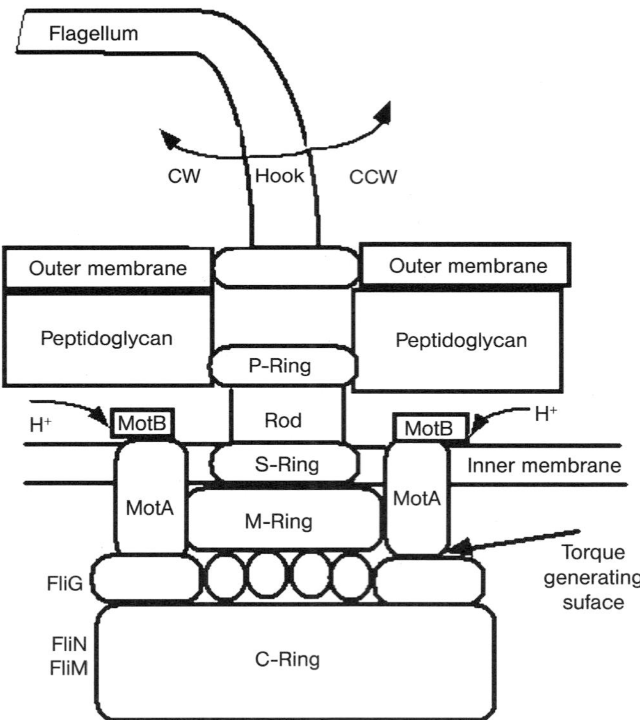

Figure 13.11. The bacterium flagellar motor is a rotary motor that sits in the cell envelope of bacteria. It is driven by the flow of ions (H^+ or Na^+) across the cytoplasmic membrane, and its purpose is to rotate long helical filaments that protrude from the cell and propel swimming bacteria. The diagram depicts a Gram-negative envelope. Torque is generated by the flow of ions across the inner membrane via ion channels MotA–MotB and by interactions between MotA (stator) and FliG (rotor).

proteins reverse their rotation patterns by inducing conformational changes in the filament and the uncoupling of the associated flagellar complex, to influence the traveling direction of the bacteria. *E. coli* flagellar motors have the ability to oscillate between counterclockwise and clockwise motions, while bacteria like *Rhodobacter sphaeroides* employ clockwise motions and then pause.

13.8
Nanodevices for Electroreception and Electric Organ Discharges

Living organisms have the ability to gather, translate and respond to information regarding their environment. Electroreceptors are also classified as nanodevices; they play a crucial role in the sensory systems of various categories of animals. Sensory systems that rely on electroreception mechanisms are sophisticated in certain families of fish [60–63]. However, this sensory system is significantly more primitive in small families of amphibians and mammals. In its passive form, the electroreceptors are used to filter and map electric fields present in their surroundings. This mechanism aids in an organism's awareness of other organisms and objects within close proximity. The more active form of electroreceptor-based sensory involves the production of currents that work in harmony with sensory organs to distinguish the organism's electric field from that of any objects in the surrounding area. Once the perceived electric fields are distinguished, they are analyzed with respect to spatial and temporal structure.

Species such as the *Torpedo*, commonly referred to as the electric ray, and *Electrophorus*, also known as the electric eel, have well-developed electric organs which aid in the visualization of their present venue. The electric sensory organs dedicated to the production of high and low currents arise from altered muscle cells and nerve endings; yet the primary mechanism for the sustained electromotive force is the ion pump – a well-known nanodevice.

An electric discharge is generated when one side of the electromotor cell is stimulated, causing a potential difference to develop across the faces of the cell. The continuum of charged membranes within the electric organ is able to discharge an electric current into it the environment.

The arrangements of the nerve endings in the electric organ determine the discharge patterns. In the case of the electric eel, the stimulated face discharges while the other face is at rest. Shortly following the discharge, the resting face is stimulated and then discharges. This alternating sequence allows for a series of currents to be released. Other aquatic organisms send low and irregular currents. Species employing this tactic are referred to as "pulse fish". However, some fish expel currents at regular time intervals and are known as "wave fish." This sensory mechanism allows for a continuous exchange of information between the organism and the environment in real-time. Lateral line nerves innervate the electric organ. Ampullary receptors are unable to perceive stimuli above 20 Hz. Conversely, tuberous receptors are unable to perceive stimuli below 30 Hz. Various species increase the rate of impulses or shorten their response times.

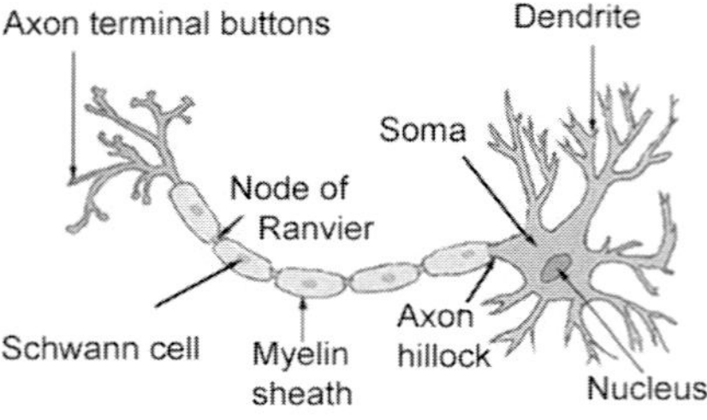

Figure 13.12. Morphology of a neuron.

13.9
Neurons

The human body is made up of approximately 10^{13} cells and roughly 10^{11} of them are neurons [64]. The brain is the major component of the central nervous system; it is a highly sophisticated network of neurons. There are three standard classes of neurons – afferent, efferent and interneurons. Afferent networks channel information from the surroundings to the central nervous system and efferent networks channel information away from the central nervous system to the peripheral nervous system.

Neurons are classified as microdevices, which include many nanodevices such as ion channels, enzymatic systems and different proteins. The average neuronal cell has a body, axon and dendrites (Fig. 13.12). Often referred to as the soma, the body houses the nucleus and is not extremely active in the conduction of impulses. The axons are slender projections of the soma that allow signals to travel away from the soma. In most neurons, the axon is protected by a myelin sheath. Glial cells are the main components of the sheath. Not only does the sheath serve as a protective covering, it also facilitates the rapid propagation of action potentials. The dendrites are small branches at the ends of the neuron. They are intricately connected to other dendrites forming a plexus or dendritic tree. Primarily, dendrites are responsible for receiving information. Gaps between dendrites are known as synaptic gaps and they serve as avenues for neurotransmitters to travel.

Neurons have the capability to become excited in response to various internal and external stimuli. The excitations induced are characterized as action potentials. These characteristic responses can be induced by stimuli such as applied pressure, chemical substances, thermal stimuli, electrical stimuli and mechanical stimuli.

References

1 VOLKOV, A. G., DEAMER, D., TANELIAN, D., MARKIN, V. S. *Liquid Interfaces in Chemistry and Biology.* Wiley, New York, **1998**.

2 VOLKOV, A. G., DEAMER, D. (Eds.). *Liquid–Liquid Interfaces: Theory and Methods.* CRC Press, Boca Raton, FL, **1996**.

3 VOLKOV, A. G. (Ed.). *Liquid Interfaces in Chemical, Biological, and Pharmaceutical Applications*. Dekker, New York, **2001**.

4 VOLKOV, A. G. (Ed.). *Interfacial Catalysis*. Dekker, New York, **2003**.

5 KSENZHEK, O. S., VOLKOV, A. G. *Plant Energetics*. Academic Press, New York, **1998**.

6 VOLKOV, A. G. Molecular mechanism of the photooxidation of water during photosynthesis: cluster catalysis of synchronous multielectron reactions, *Mol. Biol.* **1986**, *20*, 728–736.

7 KHARKATS, Yu. I., VOLKOV, A. G. Interfacial catalysis: multielectron reactions at liquid/liquid interface, *J. Electroanal. Chem.* **1985**, *184*, 435–439.

8 KHARKATS, Yu. I., VOLKOV, A. G. Membrane catalysis: synchronous multielectron reactions at the liquid–liquid interface. Bioenergetical mechanisms, *Biochim. Biophys. Acta* **1987**, *891*, 56–67.

9 KHARKATS, Yu. I., VOLKOV, A. G. Cytochrome oxidase: the molecular mechanism of functioning, *Bioelectrochem. Bioenerg.* **1989**, *22*, 91–103.

10 KHARKATS, Yu. I. The calculation of the solvent energy reorganization of reactions with complex distribution of charges in reactants, *Sov. Electrochem.* **1978**, *14*, 1721–1724.

11 EINARSDOTTIR, O. Fast reactions of cytochrome oxidase, *Biochim. Biophys. Acta* **1995**, *1229*, 129–147.

12 TSUKIHARA, T., AOYAMA, H., YAMASHITA, E., TOMIZAKI, T., YAMAGUCHI, H., SHINZAWA-LTOH, K., NAKASHIMA, R., YAONO, R., YOSHIKAWA, S. The whole structure of the 13-subunit oxidized cytochrome *c* oxidase at 2.5 A, *Science* **1996**, *272*, 1136–1144.

13 SUCHETA, A., GEORGIADIS, K. E., EINARSDOTTIR, O. Mechanism of cytochrome *c* oxidase-catalysed reduction of dioxygen to water: evidence for peroxy and ferryl intermediates at room temperature, *Biochemistry* **1997**, *36*, 554–565.

14 KHARKATS, Yu. I., VOLKOV, A. G. 2:1:1 Molecular mechanism of cytochrome oxidase functioning, in: *Charge and Field Effects in Biosystems 4*, ALLEN, M. J., CLEARY, S. F., SOWERS, A. E. (Eds.). World Scientific, Hackensack, NJ, **1994**, pp. 70–77.

15 KHARKATS, Yu. I., VOLKOV, A. G. Cytochrome oxidase at the membrane/water interface: mechanism of functioning and molecular recognition, *Anal. Sci.* **1998**, *14*, 27–30.

16 CHANCE, B., SARONIO, C., LEIGH, I. S. Functional intermediates in the reaction of membrane-bound cytochrome oxidase with oxygen, *J. Biol. Chem.* **1975**, *250*, 9226–9237.

17 ERECINSKA, M., CHANCE, B. Studies on the electron transport chain at subzero temperatures: electron transport at site III, *Arch. Biochem. Biophys.* **1972**, *151*, 304–315.

18 ORII, Y. M., MANABE, M., YONEDA, M. Molecular architecture of cytochrome oxidase and its transition on treatment with alkali or sodium dodecyl sulfate, *J. Biochem.* **1977**, *81*, 505–517.

19 HILL, R., BENDAL, P. Function of the two cytochrome components in chloroplasts: a working hypothesis, *Nature* **1960**, *186*, 136–137.

20 VOLKOV, A. G. Thylakoid membrane: electrochemical mechanisms of photosynthesis. The mechanism of oxygen evolution in the reaction center of photosystem II of green plants, *Biol. Membr.* **1987**, *4*, 984–993.

21 VOLKOV, A. G. Oxygen evolution in the course of photosynthesis, *Bioelectrochem. Bioenerg.* **1989**, *21*, 3–24.

22 VOLKOV, A. G. Green plants: electrochemical interfaces, *J. Electroanal. Chem.* **2000**, *483*, 150–156.

23 VOLKOV, A. G., MWESIGWA, J. Interfacial electrical phenomena in green plants: action potentials, in: *Liquid Interfaces in Chemical, Biological, and Pharmaceutical Applications*, VOLKOV, A. G. (Ed.). Dekker, New York, **2001**, pp. 649–681.

24 VOLKOV, A. G. Electrophysiology and phototropism, in: *Communication in Plants. Neuronal Aspects of Plant Life*, BALUSHKA, F., MANUSCO, S., VOLKMAN, D. (Eds.). Springer, Berlin, **2006**, pp. 351–367,

25 VOLKOV, A. G., DUNKLEY, T., LABADY, A., BROWN, C. Phototropism and electrified interfaces in green plants, *Electrochim. Acta* **2005**, *50*, 4241–4247.

26 GOLDSWORTHY, A. The evolution of plant action potentials, *J. Theor. Biol.* **1983**, *103*, 645–648.

27 MWESIGWA, J., COLLINS, D. J., VOLKOV, A. G. Electrochemical signaling in green plants: Effects of 2,4-dinitrophenol on resting and action potentials in soybean, *Bioelectrochemistry* **2000**, *51*, 201–205.

28 VOLKOV, A. G., DUNKLEY, T. C., MORGAN, S. A., RUFF II, D., BOYCE, Y., LABADY, A. J. Bioelectrochemical signaling in green plants induced by photosensory systems, *Bioelectrochemistry* **2004**, *63*, 91–94.

29 VOLKOV, A. G., HAACK, R. A. Insect induces bioelectrochemical signals in potato plants, *Bioelectrochem. Bioenerg.* **1995**, *35*, 55–60.

30 VOLKOV, A. G. (Ed.). *Plant Electrophysiology*. Springer, Berlin, **2006**.

31 VOLKOV, A. G., DUNKLEY, T. C., LABADY, A. J., RUFF, D., MORGAN, S. A. Electrochemical signaling in green plants induced by photosensory systems: molecular recognition of the direction of light, in: *Chemical Sensors VI: Chemical and Biological Sensors and Analytical Methods*, BRUCKNER-LEA, C., HUNTER, G., MIURA, K., VANYSEK, P., EGASHIRA, M., MIZUTANI, F. (Eds.). The Electrochemical Society, Pennington, NJ, **2004**, pp. 344–353.

32 SHVETSOVA, T., MWESIGWA, J., VOLKOV, A. G. Plant electrophysiology: FCCP induces fast electrical signaling in soybean, *Plant Sci.* **2001**, *161*, 901–909.

33 VOLKOV, A. G., MWESIGWA, J., LABADY, A., KELLY, S., THOMAS, D'J., LEWIS, K., SHVETSOVA, T. Soybean electrophysiology: effects of acid rain, *Plant Sci.* **2002**, *162*, 723–731.

34 LABADY, A., THOMAS, D'J., SHVETSOVA, T., VOLKOV, A. G. Plant electrophysiology: excitation waves and effects of CCCP on electrical signaling in soybean, *Bioelectrochemistry* **2002**, *57*, 47–53.

35 SINUKHIN, A. M., BRITIKOV, E. A. Action potentials in the reproductive system of plant, *Nature* **1967**, *215*, 1278–1280.

36 CASAL, J. J. Phytochromes, cryptochromes, phototropin: photoreceptor interactions in plants, *Photochem. Photobiol.* **2000**, *71*, 1–11.

37 QUAIL, P. H. An emerging molecular map of the phytochromes, *Plant Cell Environ.* **1997**, *20*, 657–665.

38 SHORT, T. W., BRIGGS, W. R. The transduction of blue light signals in higher plants, *Annu. Rev. Plant Physiol. Plant Mol. Biol.* **1994**, *45*, 143–171.

39 AHMAD, M., JARILLO, J. A., SMIRNOVA, O., CASHMORE, A. R. Cryptochrome blue-light photoreceptors implicated in phototropism, *Nature* **1998**, *392*, 720–723.

40 CASHMORE, A. R., JARILLO, J. A., WU, Y. J., LIU, D. Cryptochromes: blue light receptors for plants and animals, *Science* **1999**, *284*, 760–765.

41 FRECHILLA, S., TALBOTT, L. D., BOGOMOLNI, R. A., ZEIGER, E. Reversal of blue light-stimulated stomatal opening by green light, *Plant Physiol.* **2000**, *122*, 99–106.

42 SWARTZ, T. E., CORCHNOY, S. B., CHRISTIE, J. M., LEWIS, J. W., SZUNDI, I., BRIGGS, W. R., BOGOMOLNI, R. A. The photocycle of a flavin-binding domain of the blue light photoreceptor phototropin, *J. Biol. Chem.* **2001**, *276*, 36493–36500.

43 FROMM, J., BAUER, T. Action potentials in maize sieve tubes change phloem translocation, *J. Exp. Botany* **1994**, *45*, 463–469.

44 VOLKOV, A. G., SHVETSOVA, T., MARKIN, V. S. Waves of excitation and action potentials in green plants, *Biophys. J.* **2002**, *82*, 218a–218a.

45 HODGKIN, A. L., HUXLEY, A. F. A quantitative description of membrane current and its application to conduction and excitation in nerve, *J. Physiol. London* **1952**, *117*, 500–544.

46 HILLE, B. *Ion Channels of Excitable Membranes*. Sinauer, Sunderland, MA, **2001**.

47 DeCOURSEY, T. E. Voltage-gated proton channels and other proton transfer pathways, *Physiol. Rev.* **2003**, *83*, 475–550.

48 MacKinnon, R. Potassium channels, *FEBS Lett.* **2003**, *555*, 62–65.

49 Jordan, P. C., Miloshevsky, G. V., Partenskii, M. B. Energetics and gating of narrow ionic channels: the influence of channel architecture and lipid–channel interactions, in: *Interfacial Catalysis*, Volkov, A. G. (Ed.). Dekker, New York, **2003**, pp. 493–534.

50 Ottova, A. L., Tien, H. Ti supported planar BLM (lipid bilayers): formation, methods of study, and applications, in: *Interfacial Catalysis*, Volkov, A. G. (Ed.). Dekker, New York, **2003**, pp. 421–459.

51 Eisenberg, B. Ionic channels as natural nanodevices, *J. Comput. Electron.* **2002**, *1*, 331–333.

52 Arhem, P. Voltage sensing in ion channels: a 50-year-old mystery resolved? *Lancet* **2004**, *363*, 1221–1223.

53 Gandhi, C. S., Clark, E., Loots, E., Pralle, A., Isacoff, E. Y. The orientation and molecular movement of a K^+ channel voltage-sensing domain, *Neuron* **2003**, *40*, 515–525.

54 Bezanilla, F. Voltage gated ion channels, *IEEE Trans. Nanobiosci.* **2005**, *4*, 34–48.

55 Nagel, G., Szellas, T., Huhn, W., Kateriya, S., Adeishvili, N., Berthold, P., Ollig, D., Hegemann, P., Bamberg, E. Channelrhodopsin-2, a directly light-gated cation-selective membrane channel, *Proc. Natl Acad. Sci. USA* **2003**, *100*, 13940–13945.

56 Titus, M. A., Gilbert, S. P. The diversity of molecular motors: an overview, *Cell. Mol. Life Sci.* **1999**, *56*, 181–183.

57 Schmitt, R. Helix rotation model of the flagellar rotary motor, *Biophys. J.* **2003**, *85*, 843–852.

58 DeRosier, D. J. The turn of the screw: the bacterial flagellar motor, *Cell* **1998**, *93*, 17–20.

59 Berry, R. M., Berg, H. C. Torque generated by the flagellar motor of *Escherichia coli* while driven backward, *Biophys. J.* **1999**, *76*, 580–587.

60 Bullock, T. H., Hopkins, C. D., Popper, A. N., Fay R. R. (Eds.). *Electroreception*. Springer, New York, **2005**.

61 Kalmijn, A. J. Electric and magnetic field detection in elasmobranch fishes, *Science* **1982**, *218*, 916–918.

62 Hopkins, C. D. Lightning as background noise for communication among electric fish, *Nature* **1973**, *242*, 268–270.

63 Lissmann, H. W., Machin, K. E. Electric receptors in a non-electric fish (*Clarias*), *Nature* **1963**, *199*, 88–89.

64 Shepherd, G. M. *Neurobiology*. Oxford University Press, New York, **1994**.

Index

a

activated hopping conductor, DNA conductivity 163–169
adsorption
– of motor proteins 253–255
– protein immobilization 71–74
AFM *see* atomic force microscopy
aldehyde-terminated SAMs, nanopatterns 95
alkanethiols
– electrostatic immobilization 70
– microcantilever 110
– SAM chemistry 74–75
alkylsilane
– electrostatic immobilization 70
– SAM chemistry 76
aluminum mask, diamond probes 124
aminopropyltriethoxysilane (APTS), monolayer 286–292
analyte capture, kinetics 19–24
analyte detection, fluorescence microscopy 269
analyte–receptor binding 19–20
Anderson and Chaplain's model 40–45
anisotropic wet etching 118
antiangiogenic drugs, chemotherapy model 57
antibody capture domains, virus patterning 143
apertured pyramidal tips, nanofluidic systems 126–128
APTS *see* aminopropyltriethoxysilane
arrays
– BSA 93
– cantilever 8–11, 97–101
– DPN probe 118
– fluid-coupled 16
– microcantilever sensor 112–115, 118–121
– microneedle 136–137
– microtubule 261–263

– millipede 100
– protein dot 81
assays
– development 94
– immuno- 142
– kinesin 248–249
– surface 69–76
associative memories, holographic 180
atomic force microscopy (AFM) 2, 67
– cantilever arrays 98–99
– diamond probes 123
Au(111)-surfaces, SAM chemistry 75
avidin–fluorescein isothiocyanate (FITC) 81
azurin, protein devices 182

b

B–Z device 197–198
background noise, cantilever 4
bacteriorhodopsin (BR), protein devices 179
bandgap behavior, DNA conductivity 169–170
base pair, DNA electron transfer 162
bead assay, kinesin 248
Bell's deterministic model, receptor–ligand-mediated binding 49–52
bias-induced electrochemistry, DPN 91
bias-induced lithography 78, 81
biased-induced SPL 81
bifurcation, flow simulations 45
binding
– analyte–receptor 19–21
– electrostatic-mediated 70
– receptor–ligand 47–54
binding affinity, analyte capture 21
bio-building blocks, fundamentals 153–155
bio-self-assembly 150
bioanalysis, kinesin–microtubule-driven systems 245–269
bioassays development 94

Nanotechnologies for the Life Sciences Vol. 4
Nanodevices for the Life Sciences. Edited by Challa S. S. R. Kumar
Copyright © 2006 WILEY-VCH Verlag GmbH & Co. KGaA, Weinheim
ISBN: 3-527-31384-2